DYNAMIC
ASTRONOMY

Fourth Edition

DYNAMIC ASTRONOMY

Robert T. Dixon

Associate Professor of Astronomy
Riverside City College

Prentice-Hall, Inc., Englewood Cliffs, New Jersey 07632

Library of Congress Cataloging in Publication Data

Dixon, Robert T.
  Dynamic astronomy.

  Includes index.
  1. Astronomy. I. Title.
QB43.2.D58 1984   520   83-19197
ISBN 0-13-221333-8

Editorial/production supervision: Paula Martinac
Interior and cover designs: Maureen Eide
Manufacturing buyer: John Hall
Cover illustration: A record of what the infrared satellite, IRAS, saw in the region of the Orion constellation. Huge dust and molecular clouds are revealed even though their temperature is far below the freezing point of water. The brightest region toward the lower right of center represents the much photographed Orion nebula. (JPL–NASA)

Printed in the United States of America

10 9 8 7 6 5 4 3 2 1

ISBN 0-13-221333-8

Prentice-Hall International, Inc., *London*
Prentice-Hall of Australia Pty. Limited, *Sydney*
Editora Prentice-Hall do Brasil, Ltda., *Rio de Janeiro*
Prentice-Hall Canada Inc., *Toronto*
Prentice-Hall of India Private Limited, *New Delhi*
Prentice-Hall of Japan, Inc., *Tokyo*
Prentice-Hall of Southeast Asia Pte. Ltd., *Singapore*
Whitehall Books Limited, *Wellington, New Zealand*

Contents

Preface   *xi*

History of Astronomy   *2*

Babylonian "Astronomy" 6   Egypt, Land of the Nile 10
A Megalithic Legacy 11   Stonehenge 14   Early Concepts of the
Universe 16   Planets in Motion 20   How Good was Ptolemy's
Model? 22   Arabic Influence 22   Nicolaus Copernicus, Founder
of Modern Astronomy 23   Tycho Brahe, the Diligent Observer 27
Johannes Kepler, the Inventive Mathematician 29   Galileo Galilei,
Father of Experimental Science 35   Isaac Newton, the Young
Genius 38   The Force of Gravity Between Two Objects 42
Our Place in the Universe 46   Questions 49

Methods of Astronomy   *50*

The Wave Nature of Light 56   Polarization of Light 59   Does Light
Travel in a Straight Line? 61   Reflection of Light 61   Refraction
of Light 62   Dispersion by Refraction 64   Diffraction of Light 65
Interference of Light 66   The Dual Nature of Light 68
Telescope Design 70   Limitations in Refractors 72
The Reflecting Telescope 73   Multiple-Mirror Telescopes 78

The Schmidt Camera 80   Schmidt Cassegrain Telescope 81
Telescope Performance 81   Atmospheric Turbulence: A Resolution
  Spoiler 85   Seeing 85   Magnification 85   The Binocular 86
The Spectrograph 87   The Atom Signs its Name 89   Doppler
  Shift 91   A Very Important Generalization 92   Radio Astronomy 93
Radio Interferometry 96   Radar Astronomy 99
Infrared Astronomy 100   Ultraviolet Astronomy 103
Space Telescope 104   X-ray Astronomy 105
Gamma-ray Astronomy 106   Gravitational Wave Astronomy 107
Questions 108

Three

The Earth *110*

Volume, Mass, and Density 112   Volume of the Earth 114   Mass of
  the Earth 114   Density of the Earth 116   Crust of the Earth 119
Age of the Earth 121   Atmosphere of the Earth 122   Why is the
  Sky Blue? 124   The Earth's Magnetic Field 125   Rotation of the
  Earth 127   Revolution of the Earth 130   The Seasons 132
Time: How Long is a Day? 133   Latitude and Longitude 135
Right Ascension and Declination 138   Apparent Motions of the
  Sky 140   Altitude-Azimuth System 141   Precession of the Earth 142
The Calendar 143   Other Motions of the Earth 144
Questions 144

Four

The Moon *146*

Bouncing a Laser Beam off the Moon 148   Surface Features 151
Moon Rocks 157   History of the Moon 160   Magnetic Field 161
Origin of the Moon 162   Revolution of the Moon 162
Tides Produced by the Moon 166   Eclipses 169
Solar Eclipse 170   Lunar Eclipse 175   Eclipse Seasons 176
Questions 177

Five

The Planets *180*

The Planets in General 182   Rotation of the Planets 183   Spacing
  of the Planets 183   Earthlike versus Jupiterlike Planets 185
Configurations of the Inferior Planets 187   Configurations of the
  Superior Planets 188   Mercury 189   Venus 193   Earth 203
Mars 204   Questions 220

Six

The Jovian Planets 222

Jupiter 224   Saturn 233   Uranus 242   Neptune 245
Pluto 246   Planet X 250   Origin and Evolution of the Planets 251
Questions 253

Seven

Asteroids, Comets, and Meteors 254

Origin of Asteroids 259   Comets 260   Origin of Comets 263
Physical Nature of Comets 264   Lifetime of a Comet 267
Comet Spin and Perturbation 268   How are Comets Discovered and
Recorded? 269   Meteors 271   Observing Meteors 272   Meteor
Showers 273   Physical Properties of Meteorites 276   Orbit and
Origin of Meteorites 278   Meteorite Craters 278   The Siberian
Event 280   Tektites 281   Micrometeorites 283   Questions 284

Eight

The Sun 286

Physical Characteristics of the Sun 289   The Sun: A Hydrogen Fusion
Reactor 291   Neutrinos from the Sun 293   Layers of the Sun 295
The Solar Spectrum 301   Spectroheliogram 304
Magnetic Fields of the Sun 306   Sunspots 308   Maunder
Minimum 310   The Cycle of Sunspots 310   Rotation of the
Sun 311   Source of the Sun's Magnetism 311   Plages 312
Spicules 313   Prominences 313   The Solar Flare 314
Our Nearest Star 317   Questions 318

Nine

Stars in General 320

Motion of Stars 325   Proper Motion 325   Space Velocity 327
Peculiar Velocity 328   Distance to Moving Clusters 328
Brightness of Stars 331   The Scale of Brightness 332   Absolute
Magnitude 334   Temperature of Stars 336   Hertzsprung-Russell
Diagram 339   Spectroscopic Parallax 343   Size of Stars 344
The Great Variety of Stars 346   The Infrared Sky 347
X-ray Astronomy 348   Questions 348

Ten

Multiple Star Systems *350*

Optical and Visual Binaries 351   Spectroscopic Binaries 352
Spectrum Binaries 353   Astrometric Binaries 353   Eclipsing
Binaries 354   Close (Contact) Binaries 355   Determination of the
Mass of Binary Systems 356   Mass-Luminosity Relationship 358
Origin of Binaries 358   Clusters 359   Origin of Clusters 360
Cluster Types 360   X-ray Bursters 363   Associations 365
Questions 365

Eleven

Variable Stars *366*

Cepheid Variables 368   Cepheids: Indicators of Distance 370
RR Lyrae Stars 372   Long-Period Variables 374
Irregular Variables 375   Recurrent Novae 375   Supernovae 377
R Coronae Borealis Stars 378   Flare Stars 378   Questions 379

Twelve

The Interstellar Medium *380*

Interstellar Gases 381   21-cm Radio Emission 383
Interstellar Dust 385   Interstellar Molecules 391
Giant Molecular Clouds 393   Cosmic Rays 394   Questions 396

Thirteen

Stellar Evolution *398*

Gas Laws 400   Initial Energy Source 401   Model Building 401
Formation of a Star 402   A Star is Born 406
Main-Sequence Stage 409   The Red Giant 412
The Supernova 416   Questions 418

Fourteen

Death of Stars *420*

The White Dwarf 422   Stellar Winds 422   The Neutron Star 425
Pulsars 426   The Black Hole 432   Rotating Black Holes 435
Exploding Black Holes 436   Questions 437

Fifteen

The Milky Way *438*

Rotation of the Milky Way Galaxy 447   21-cm Radio Observations
Show Spiral Structure 448   Mass of the Galaxy 450
Distribution of Stars in the Milky Way Galaxy 450
The Galactic Nucleus 452   Questions 453

Sixteen

The Realm of Galaxies *454*

Classification of Galaxies 455   Hubble's Classification Scheme 460
The Local Group 461   Clustering of Galaxies 463
Intergalactic Media 464   Measuring the Distance to Galaxies 464
Hubble's Red-shift Law 467   Radio Galaxies 468
Seyfert Galaxies 469   Quasars 470   The Double Quasars 476
BL Lacertae Objects 478   A Modern Classification of Galaxies 478
Cosmology 479   Cosmic Background Radiation 482
The Primordial Atom: Infinite or Finite? 483
Curvature of Space-Time 485   Tests for Cosmological Models 489
Mass-Energy Density of the Universe 490
Deuterium Abundance 491   The Universe in Infrared 491
Questions 493

Seventeen

Relativity *494*

Relative Motion 495   Michelson-Morley Experiment 497
The Lorentz Contraction 499   Albert Einstein 500
A Thought Experiment 501   Questions 505

Eighteen

Extraterrestrial (ET) Life *506*

Is it Worth the Effort? What Are the Chances that ETI Exists? 509
What Language Would We Use? 513   What Shall We Say to ETIs? 513
How Shall it Be Done? 514   Questions 515

Glossary *517*

Appendixes *527*

Appendix 1 Temperature Conversion Charts 527   Appendix 2
International System of Units (SI) with English Equivalents 527

Appendix 3 Powers-of-ten Notations 528   Appendix 4 Constants with Useful Approximations 528   Appendix 5 Orbital Data of the Planets 529   Appendix 6 Physical and Rotational Data for the Planets 529   Appendix 7 Satellites of Planets 530   Appendix 8 The Twenty Brightest Stars 531   Appendix 9 The Messier Catalogue of Nebulae and Star Clusters 532   Appendix 10 The Greek Alphabet 534   Appendix 11 The Constellations 535 Appendix 12 The Periodic Tables 536 Appendix 13 Sky Maps 538

Index   543

Preface

Like the first three editions of this text, which have been used in over two hundred colleges and universities in the United States and elsewhere, the fourth edition is designed to help the liberal arts student and the general reader to gain an understanding of how the astronomer studies his or her subject and to appreciate the grandeur of the universe in which we live. No prior background in science is presumed. Concepts are developed in the light of common experiences, and what little mathematics is needed is developed as the need arises. Relationships that are usually presented only as mathematical expressions are also stated verbally in order to dispel any anxiety on the part of the reader as to the mysteries of mathematics.

The real beauty of astronomy lies not only in the facts that surround the subject but also in the development of an understanding of relationships within the universe. The background of thought that has brought us to our present view of the cosmos is presented together with the methods we use to make pertinent observations.

This revision represents not only an updating of information relative to a fast-moving science, but also a move in the direction of a more comprehensive treatment of astronomy. Significant portions of the text have been completely rewritten to reflect the current state of the art.

While the text is designed primarily for a one-semester course, it also lends itself naturally to division for a two-quarter or two-semester program. Chapters 1 through 7 emphasize the methods of astronomy and deal with the solar system in particular, while the remaining chapters treat the stars and galaxies.

To a large extent, the astronomy student's understanding is dependent upon his or her ability to visualize the *motion* of whatever is being discussed in terms of both space and time. To assist the student in visualizing certain rather intricate motions of objects in the universe, a unique kind of illustration is presented in the margins of this text. By flipping the successive pages, the illustrations appear to move.

For their constructive criticisms and suggestions, I would like to thank the following reviewers: Bart Bok, University of Arizona; Gladwin Comes, Broward Community College; Joel Mann Martin, Broward Community College; and, Michael M. Shurman, University of Wisconsin, Milwaukee. I would also like to thank the staff at Prentice-Hall, especially Paula Martinac, production editor; Maureen Eide, designer; and Doug Humphrey, editor, for all their help and encouragement in the making of this book.

This book is dedicated to my wife, Marian. Without her constant encouragement and patience, this book would not have been written.

Robert T. Dixon

KEY TO FLIP PAGES

The eight flip-page sequences that occupy the upper margins of the pages of this text are a unique instructional tool dynamically illustrating the basic motions of astronomy. To use the flip pages on the right-hand margins, grasp the desired section with the right hand, bending these pages as illustrated here and allowing the pages to flip, one by one, from beneath your thumb. Left-hand flip pages may be handled in a similar fashion, using the left hand. Remember, flip pages starting on an odd-numbered page (right-hand) are flipped toward higher page numbers; those starting on an even-numbered page (left-hand) are flipped toward lower page numbers. (See the illustrations here in the left margin, which make this clear.)

Odd-numbered pages

Right-hand flip pages

Even-numbered pages

Left-hand flip pages

A brief introduction to each flip sequence follows, together with an indication of its location in the text.

The Ptolemaic system (Right-hand flip pages beginning on page 25) Ptolemy envisioned the earth to be the center of the universe. In order to explain the apparent retrograde (westward) motion of the planets among the stars, he utilized the concept of the epicycle. Each planet was thought to revolve on its epicycle as the epicycle revolved on its primary orbit, called the deferent. By assigning the proper speed to each of these motions, he was able to create a retrograde motion. This motion also explains the fact that at certain times a given planet is closer to the earth. Since it had been observed that Mercury and Venus always remained close to the sun, the centers of their epicycles must always align with the earth and the sun. (This sequence of flip pages will be useful in your study of Chapter 1.)

Retrograde motion of Jupiter (Left-hand flip pages beginning on page 114) This sequence shows Jupiter moving first in a direct (eastward) motion among the stars, then apparently stopping and moving in a retrograde (westward) motion among the stars. Later the planet appears to stop again and resume its direct motion. By careful observation over an extended period of time, this apparent retrograde motion of Jupiter or of any other planet may be observed in the real sky. (This sequence will be helpful in your study of Chapters 1, 5, and 6.)

The Copernican system (Right-hand flip pages beginning on page 115) In this sequence, the planets will be seen to move at their proper speeds in relation to the earth's motion. A period of approximately one year is depicted. The period of revolution of Mercury is 88 days, hence it can make four complete revolutions in one year. Within the period of 116 days, it returns to inferior conjunction with the earth. This is called its synodic period. Venus, on the other hand, makes a complete revolution in 225 days but does not return to inferior conjunction in the period shown. This planet requires 584 days for one synodic revolution. Jupiter needs approximately 12 years for one revolution, hence it is seen to move only about 30° during the period shown. Other configurations such as maximum elongation, superior conjunction, conjunction, quadrature, and opposition may be seen on certain pages individually. See if you can find all possible configurations. (This sequence will be helpful in your study of Chapters 1, 5, and 6.)

A binary system (Left-hand flip pages beginning on page 182) This sequence demonstrates the fact that the earth and the moon form a binary (two-body) system and that it is the barycenter of the system which follows a smooth elliptical orbit. The barycenter of the system is located approximately 3000 miles from the center of the earth. As the moon revolves about this barycenter, the earth also deviates up to 3000 miles on either side of the system's orbit. In a very similar way, two binary stars, stars which lie in each other's gravitational field, orbit around a barycenter. The position of the barycenter is determined by the way in which the material (mass) is distributed in the two stars. (This sequence of flip pages should be helpful in your study of Chapters 4 and 10.)

A comet in motion (Right-hand flip pages beginning on page 235) This sequence depicts Halley's Comet, moving in its highly elongated elliptical orbit about the sun. This comet returns to the region of the sun every 76 years, its next expected return being in 1986. Halley's Comet travels in a retrograde direction that carries it beyond the orbit of Neptune. When a comet is at such a great distance from the sun, it possesses no coma (head) nor tail but exists only as a swarm of frozen gas bodies. As the comet approaches the sun, however, the warmth of the sun vaporizes a portion of the gas, thus producing the coma and tail. The outflow of particles from the sun (the solar wind) continually pushes the gases of the tail in a direction away from the sun. The speed with which the comet travels increases as it approaches the sun, hence only a relatively short time is spent in the vicinity of the sun. (This section of flip pages will be helpful in your study of Chapter 7.)

The proper motion of stars (Left-hand flip pages beginning on page 284)   Over a 100,000 year-period, the familiar constellation of the Big Dipper will change in appearance until it no longer resembles a dipper. This change results from the fact that stars are in constant motion. The stars that make up this constellation are moving in different directions. The apparent change in the position of a star in 1 year is very small, perhaps in the order of 1 second of arc per year. This is called the proper motion of the star. (This sequence of flip pages will be helpful in your study of Chapter 9.)

Motion of globular clusters (Right-hand flip pages beginning on page 345)   The upper view in this series shows the motion of globular clusters that form the halo of the Milky Way galaxy. Each globular moves along an elliptical path which causes it to periodically "dip" into the nucleus of the Galaxy; however, it spends a relatively short time there. (In this sequence, it is best to choose a particular globular and follow its motion, say, the one marked by the double circle.) *Rotation of the Milky Way galaxy* The lower half of the pages shows the revolution of the Galaxy. The sun participates in this revolution and makes a circuit around the center of the Galaxy in 200 million years. From the sun's position, we see one spiral arm beyond and two arms toward the center of the Galaxy. (This sequence of flip pages will be helpful in your study of Chapter 15.)

An eclipsing binary system (Left-hand flip pages beginning on page 410)   The top view in this section shows the motion of a binary system, seen from above. The brighter star (light in color) is about five times as massive as its cooler (darker) component.

The middle view shows the same system, seen from our position on earth. One star is seen periodically to eclipse the other, for the earth lies very nearly in plane of their orbit.

The lower view shows the light curve which is generated as these stars move in their orbit. When the cooler star almost completely eclipses the hot star, the lowest light output is apparent. When the hotter (brighter) star is in front, only a slight dip occurs. While you are seeing the motion of these stars and the light curve generated simultaneously, the astronomer usually observes only the light curve and must infer the actual motion from that curve. (This sequence will be helpful in your study of Chapter 10.)

CHAPTER-OPENING ILLUSTRATIONS

Chapter 1   Stonehenge, in England. (British Tourist Authority)

Chapter 2   The 64-m Goldstone radio antenna, near Barstow, California. (NASA-JPL)

Chapter 3   The planet earth as seen from Apollo 11 at a distance of 178,000 km. (NASA)

Chapter 4   Astronaut James Irwin of Apollo 15 walking on the moon, August 1971. (NASA)

Chapter 5   Surface of the planet Mercury, photographed by Mariner 10 from a range of 86,800 km, March 29, 1974. (NASA-JPL)

Chapter 6 Saturn and three of its satellites, as seen by Voyager 2. (NASA)

Chapter 7 Comet West. (Lick Observatory)

Chapter 8 Sunspot group. (Big Bear Solar Observatory)

Chapter 9 An open star cluster in Cancer (M67). (Palomar Observatory Photograph)

Chapter 10 Pleiades, an open star cluster in Taurus. (Mount Wilson and Las Campanas Observatories, Carnegie Institute of Washington)

Chapter 11 The Small Magellanic Cloud. (Mount Stromlo and Siding Spring Observatories, The Australian National University)

Chapter 12 The Horsehead Nebula in Orion. (Palomar Observatory Photograph)

Chapter 13 The Lagoon Nebula in Sagittarius (M8). (Lick Observatory)

Chapter 14 The Crab Nebula (M1). (Palomar Observatory Photograph)

Chapter 15 The Milky Way galaxy, Sagittarius region. (Palomar Observatory Photograph)

Chapter 16 Spiral galaxy (M74). (Palomar Observatory Photograph)

Chapter 17 Relativity.

Chapter 18 The Cyclops proposal for a radio telescope array. (NASA-Ames)

DYNAMIC
ASTRONOMY

History
of Astronomy

One

The history of astronomy is characterized by the ever-expanding human awareness of the cosmos. Your study of astronomy will no doubt expand your own awareness. Through observations of celestial objects, humans have been able to relate the rhythms of their lives to those of the visible universe. In order that you may understand the development of this harmonious relationship, place yourself backward in time about 6000 years and try to imagine what was taking place then. Early humans were nomadic; they were constantly searching for new supplies of food and shelter from the weather and unfriendly animals. Even as nomads, humans surely recognized several obvious cycles of nature: the cycle of day and night, the monthly phasing of the moon (see Figure 1.1), and the seasonal changes which made hunting more difficult at certain times. These cycles became yet more important, however, when humans turned from nomadic patterns to become farmers. The domestication of animals and planting of crops called for a greater awareness of seasonal patterns; crops had to be planted at the right time during the spring for harvest in the fall.

The most natural locations for agriculture lie within river valleys; they are generally fertile and provide a ready supply of irrigation water. At least four early civilizations arose in such fertile valleys: that of China, along the Hwang

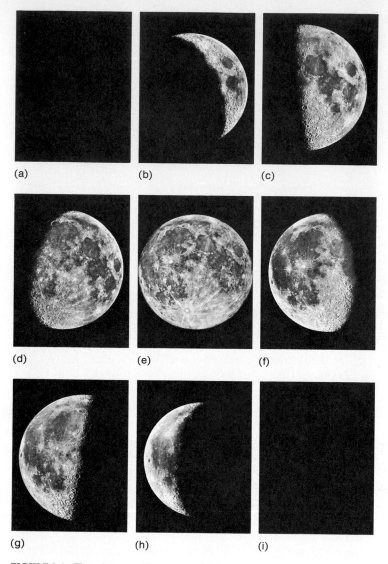

FIGURE 1.1 The phases of the moon: (a) "new moon"; (b) waxing crescent, 4 days old; (c) first quarter, 7 days old; (d) waxing gibbous, 10 days old; (e) full moon; (f) waning gibbous; (g) last quarter; (h) waning cresent; (i) "new moon" again. (Lick Observatory)

Ho river; that of India, along the Indus river; that of Babylonia and Sumeria, along the Tigris and Euphrates rivers, and that of Egypt along the Nile. We have clear evidence dating from these early cultural beginnings that early humans watched the skies and charted the motion of the sun, moon, and planets among the stars. In some cases, these objects were thought to possess godlike qualities and became objects of worship; the watchers were thus motivated to record their rhythmic motions.

The people of China produced one of the more accurate determinations of the length of the year. Their findings were recorded by means of a number of circular objects divided into 365.25 parts—corresponding to the number of days in a year. What kind of experiment could they have performed in order to determine the number of days in a year so accurately? Visualize yourself standing under the imaginary dome of the sky, facing south. Let an imaginary curved line run on that dome north and south directly over your head. This line is called your *local meridian*. The point directly over your head is called your *zenith*. Choose the time of year when the sun passes closest to your zenith and place a stick in the ground so that it points directly toward the sun as it crosses your meridian. At this time it will have no shadow (see Figure 1.2). Note that if the sun is east or west of the meridian, the stick would cast a shadow, and the time would not be noon. You can then easily tell when a full day has passed, because at that moment the stick will cast no shadow again. As the days go by, however, it will become evident that the sun is crossing your meridian farther from your zenith and as a consequence the stick casts a shadow even at noon. If you count the number of days until the sun again passes closest to your zenith and the stick casts no shadow at noon, you will have determined the number of days in a year, which is approximately 365 days. After several years of counting, your record might look like this:

First year	365 days
Second year	365 days
Third year	366 days
Fourth year	365 days

FIGURE 1.2 A stick is placed in the ground so as to cast no shadow when the sun appears highest in the sky.

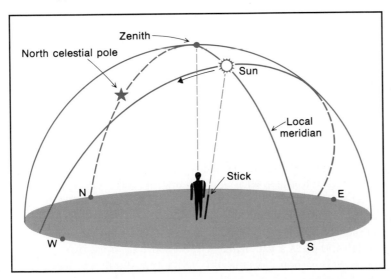

Fifth year	365 days
Sixth year	365 days
Seventh year	366 days
Eighth year	365 days
Average	2,922 divided by 8 = 365.25 days per year

By averaging the counts of eight consecutive years, we see that a year contains 365.25 days. Even this is not exact and must be corrected periodically, but we will consider this further adjustment later.

As with so many early civilizations, Chinese observers were motivated to follow the motions of the sun, moon, and planets in order to predict the future of human and governmental affairs. After many years of such observations, it became possible to predict eclipses—times when the earth, sun, and moon would align so perfectly that the shadow of one object would fall on the other. The Chinese were particularly superstitious about solar eclipses. A story relates that two observers named Hi and Ho became lax in their priestly duties and failed to predict a given eclipse which was unexpectedly viewed by the citizens of their province. Because an unheralded eclipse was thought to be a bad omen, the two priests were beheaded.

The great temple to the sun at Peking is oriented to point toward the rising point of the midwinter sun (at the time of the winter solstice). Sun alignments are quite common among the temples and monuments of many cultures. St. Peter's in Rome, for example, is oriented toward the rising sun at the times of the equinoxes (approximately March 21 and September 21). At these times sunlight floods the entire length of the basilica and illuminates the high altar. At the setting of the sun, the window of the Holy Ghost is illuminated, producing a spectacular, almost mystical appearance.

Early Indian culture is recorded in a work called the Vedas. Included are hymns to the sun that are at least 3500 years old—hymns that recognize human dependence upon the sun for life and light.

BABYLONIAN "ASTRONOMY"

The Sumerians were very early inhabitants of the region we refer to as the "fertile crescent," a region dominated by the Tigris and Euphrates rivers in present-day Iraq. These people produced a written record using cuneiform script. In these records one finds a very early recognition that the cycle of seasons, as it relates to the sun's apparent position in the sky, is also directly related to the sowing and reaping of crops. This relationship is expressed in a calendar which has for its fourth month, the time of sowing, a name that is a contraction of the words for hand and seed. The name of its eleventh month indicates a time of reaping, and that of its twelfth month indicates a gathering of the corn into storage.

In constructing their calendar, the Babylonians faced the same problem we still face today: that of having months that closely reflect the cycle of the moon's phases and also years that accurately reflect seasonal changes. To grasp this problem, suppose one were to count the number of days required for the moon to go through six complete cycles of phasing, starting with its appearance as a new crescent. This would require 177 days. The average lunar cycle then is 177 divided by 6 or 29.5 days. We have already charted the seasonal cycle, as illustrated in Figure 1.2, and have found it to contain 365.25 days. Is it possible to construct an ideal calendar based on moon phases, something like the one that follows, and still have 365.25 days in a year?

Month	Number of Days
1st	30
2nd	29
3rd	30
4th	29
5th	30
6th	29
7th	30
8th	29
9th	30
10th	29
11th	30
12th	29
TOTAL	354

Placing first 30 days, then 29 days, in each successive month would make a given phase of the moon fall on the same day each month because the average of 29.5 days per month would be preserved. But note that the 12-month total of 354 days is 11.25 days short of the 365.25 day cycle of seasons, so that the planting season would not fall on the fourth month in succesive years. Even if an approximate lunar month of 30 days were used, 12 such months would only total 360 days, and the calendar year would be 5.25 days short of a seasonal year.

The Babylonians solved this problem by simply inserting an extra month in certain years, whenever it was necessary to bring the calendar back in tune with the seasons. This they did, not according to a regular plan but sporadically as they realized that the eleventh month did not coincide with the ripening of the corn.

The Assyrians, who became dominant in the land of Babylon about 800 B.C., employed a more accurate method for keeping their calendar in tune with the seasons. They noted that as the year passed, certain stars were visible in the winter evening, others in the spring evening, and still others in the summer and fall (see Figure 1.3). By noting the return of the stars that were characteristic of a given season, they were able to keep their calendar more closely synchronized with the year of seasons.

The Assyrians also believed that the events of nations and individuals were influenced by events in the sky. From such beliefs came the religion

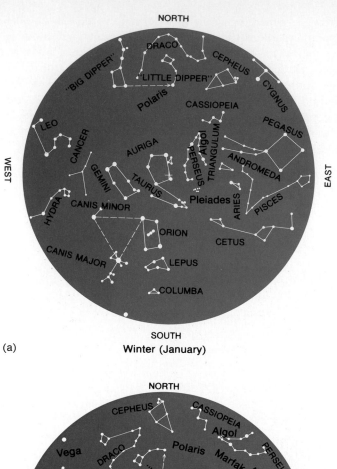

(a) **Winter (January)**

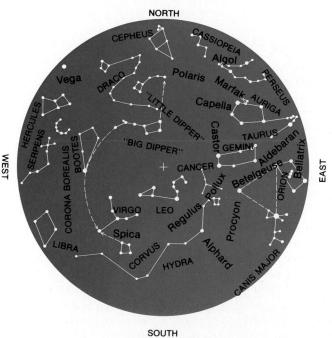

(b) **Spring (April)**

FIGURE 1.3 Evening constellations of winter, spring, summer, and fall. (Griffith Observatory)

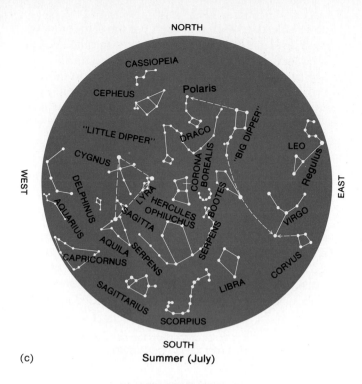

(c)   **Summer (July)**

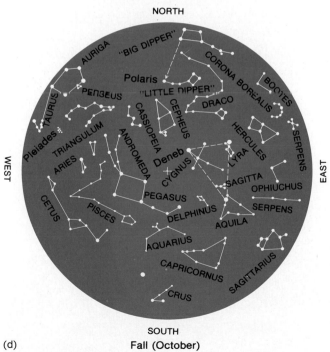

(d)   **Fall (October)**

called *astrology*, with its priests of Zoroaster. Because the location of the sun, moon, and planets against the background of stars became such an important factor in this new religion, observations became much more precise. The 12 constellations through which the sun appears to move in one year are called the signs of the zodiac; besides these, many more constellations were named by ancient observers. The religion of astrology and the science of astronomy had a long marriage, parting ways only in recent times. Certainly without the strong motivation of astrology, the modern astronomer would not have the wealth of ancient observations available today.

EGYPT, LAND OF THE NILE

The Nile river played a dominant role in the life of the Egyptians, largely because it rose each year and overflowed a large delta region. This action of the river deposited fertile silt in which crops could be grown, and the flood waters were retained behind a series of dams, to be released as needed. Because the food supply of the entire nation depended upon the

FIGURE 1.4 The zodiacal constellations, as depicted on the ceiling of the Temple of Dendera, now in the Louvre. (This drawing, first published in 1822, is by an unknown author.)

proper utilization of this natural occurance, it was vital that the people be forewarned as to when this flooding would begin.

The Egyptians had a very accurate calendar. They were well aware of the signs of the zodiac through which the sun appears to move in a year's time. See how many of the zodiacal signs you can find in Figure 1.4, a reproduction of the ceiling painting from the temple of Dendera. When one cannot see the stars in the daytime, how can one observe the position of the sun in relation to the background of stars? The Egyptians observed the star or constellation which appeared about one hour ahead of the rising sun, when dawn was just "breaking." Each morning they noticed a slight change in star positions, relative to the sun. Only on one day per year would a given star rise exactly one hour before sunrise. For example, the Egyptians noticed that when the bright star Sirius rose exactly one hour before sunrise, the Nile would soon begin to rise; hence the celestial event became a useful predictor. Certainly, after the Egyptians had observed this cycle for a number of years, it would have been clear that the seasons recur in a cycle of 365.25 days, yet their calendar contained only 365 days. Perhaps we can understand this inaccuracy when we realize that it was the priests who observed such cycles. By purposely leaving the calendar in error, they retained for themselves the secret of the proper correction to make—a secret that enhanced their power.

A MEGALITHIC LEGACY

No other culture has left so many huge and enduring reminders of its astronomical observations as did the Egyptians. Perhaps the most obvious are the pyramids. As shown in Figure 1.5, the pyramids of Giza are so

FIGURE 1.5 The Great Pyramids of Giza are so accurately aligned with the cardinal points (north, south, east, west) that we can be certain that their builders must have observed the stars. (Robin Rector Krupp)

accurately oriented with the cardinal points (north, south, east, and west) that one must assume that the alignment could have been made only by observing the stars; the greatest deviation is less than 1/12 of one degree.

The temples of the Egyptians reveal the importance of the sun in their religious beliefs. The temple builders took advantage of the fact that the sun follows a predictable pattern throughout the year, in terms of its rising points and setting points. As illustrated in Figure 1.6, the sun rises directly out of the east on the dates we call the *equinoxes* (usually March 21 and September 21); it rises considerably to the north on June 21, known as the *summer solstice*, and to the south on December 21, known as the *winter solstice*. These dates may vary from year to year by plus or minus one day, but within this degree of accuracy they return year after year. The reason for this variation lies in the fact that we do not correct our calendar by one-fourth day each year, but rather we keep our modern calendar in tune with the seasons by adding a day every four years, making our average year 365.25 days. We call those years in which we add a day a *leap year*. The Egyptian calendar of 365 days allowed the date of the summer solstice to move through the calendar, occurring 25 days earlier every century. Only the priests knew the correction to be made. If the temples of Egypt had merely been oriented along the cardinal directions, as were the great pyramids of Giza, this would not necessarily indicate a fascination for the sun; however, we find many temples oriented toward one of the solstice points. Note the plan of Karnak, illustrated in Figure 1.7. A long series of doorways was definitely oriented toward the setting sun at the time of the summer solstice. On the one day of the year when the sun appeared to travel across the sky in an arc farthest to the north, the sunset event produced a most spectacular flash of light upon the holy of holies. You can imagine the mystical quality of such an event, especially if the king stood in that darkened location just before the flash occurred. The time interval between two successive flashes marked the passage of a year very accurately.

FIGURE 1.6 During the course of one year the sun appears to rise over different points along the eastern horizon.

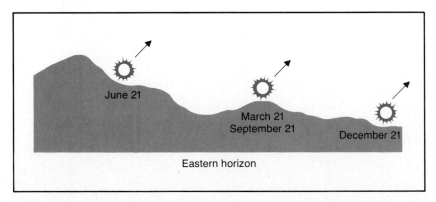

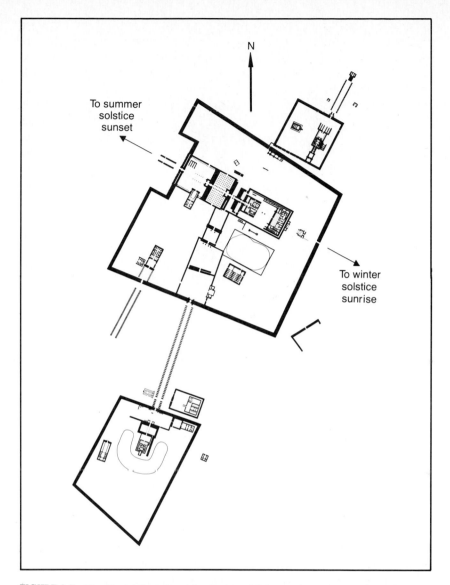

FIGURE 1.7 The Great Temple of Amen-Ra at Karnak, Egypt, shows an alignment of many corridors and doorways with the setting sun at the time of the summer solstice. On the longest day of the year, the sun's last rays would seem to flash upon the holy of holies, creating a rather mystical phenomena. (Griffith Observatory, Joseph Bieniasz)

Further evidence of Egyptian sun worship is the written hieroglyphic record left as inscriptions on temples and other monuments. The name of their sun god was Ra, who brought life to all things. Each day Ra would appear and travel in his barge up over the back of the sky goddess, Nut (pronounced "Noot"). We might call this representation, as illustrated in Figure 1.8, an Egyptian cosmology; it was their concept of the universe.

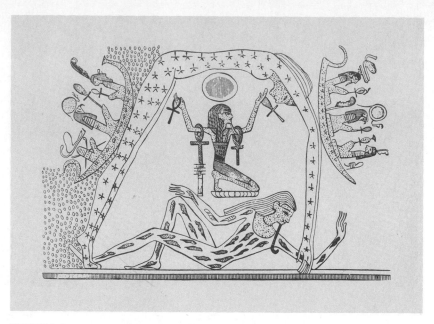

FIGURE 1.8 An Egyptian concept of the sun god traveling in his barge over the back of a starry goddess and down into the underworld. (Yerkes Observatory)

STONEHENGE

Far away from all the "cradles of civilization" we have mentioned, yet dating from a similar period of time (some 4000 to 6000 years ago), is Stonehenge. This megalithic monument to human intelligence is set out on the Salisbury Plain of southern England (Figure 1.9). Here massive stones were aligned to point to the extreme rising or setting positions of the sun at the summer and winter *solstices*; that is, they were the means whereby the shortest and longest days of the year could be noted and the length of the year accurately determined. If the megaliths of Stonehenge marked only the extremes of the sun's apparent motions, then we might assume that they served primarily a religious function. Stonehenge, however, also marks the extremes in the rising and setting points of the moon—hence it could also be considered a lunar observatory.

Whenever we think of the sun and the moon simultaneously, we think of eclipses. Eclipses occur on any occasion in which the sun, moon, and earth are aligned. The subject of eclipses is treated more fully in Chapter 4; suffice it to note here that eclipses occur in cycles and only when the moon is new or full. Every 18.61 years, a given phase of the moon occurs on the same day of the year. This is known as the *Metonic* cycle. Three such cycles (3 × 18.61) yield a number very close to 56 (55.83). In that length of time a given eclipse would occur again at the same season of the year. Some observers feel that this is the significance of the 56 Aubrey holes which mark the perimeter of Stonehenge [see Figure 1.9(c)]. By moving a marker one hole per year, such a seasonal cycle of

(a)

(b)

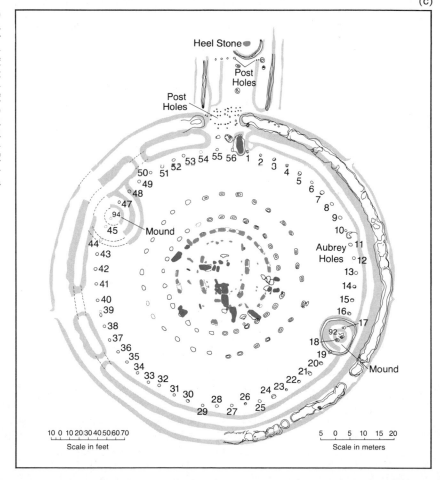

(c)

FIGURE 1.9 (a) and (b) Megaliths at Stonehenge, in England, possibly used by early man for celestial observations. (c) Plan view of Stonehenge. The alignment of certain positions (for example, Heel Stone, center of circle, mounds 91, 92, 93, 94) pointed to key rising and setting points of the sun and the moon. (a, b, British Tourist Authority; c, Controller of Her Britannic Majesty's Stationery Office. British Crown Copyright)

Heel Stone

Post Holes

Post Holes

Mound

Aubrey Holes

Mound

Scale in feet

Scale in meters

eclipses may be charted. At a number of other sites in England and in Brittany (in northwest France), similar meglithic structures were erected, although of somewhat less massive stones. It is interesting to contemplate the degree of intelligence and sophistication that the builders of such monuments possessed.

EARLY CONCEPTS OF THE UNIVERSE

The way in which humans conceive the earth appears to influence greatly the way in which they conceive the universe, and logically so. Can you imagine your own interpretation of the earth if you had never traveled farther than you were able to walk in a day? Undoubtedly you would be convinced that the earth was flat and that the stars simply formed a tent or canopy overhead. It might appear as though the sky were a huge brass shield stretched over your head. If hundreds of small holes were punched in that shield and light from a fire on the outside shone through the holes, it would give the appearance of the stars in the sky. This was an explanation given by the ancient Greeks.

To explain the apparent motion of the sun, the Greeks held that Helios, the sun god, drove his chariot across the sky each day from east to west and sailed around the northerly stream of ocean each night to arise in the east again at dawn. We have already seen that the Egyptians favored a scheme in which the sun god (Ra) sailed in his barge up over the back of his mother (Nut), the sky goddess, who arched her body over the earth (Figure 1.8). As you can see, the flat-earth concept caused humans to create explanations for the motions of the sun which might appear absurd to us today but which reflected their personal feelings for objects in the sky.

Figure 1.10 depicts the early Greek concept of the universe in general terms. Various ideas were set forth by major schools of Greek philosophers. Thales of Miletus (636–546 B.C.) said that water was the first principle of all things, and he visualized the earth as a flat disk floating on water; Anaximander (611–547 B.C.) thought of the world as infinite and viewed the earth as a cylinder floating free in space; and Anaximenes (585–526 B.C.) pictured the earth as a disk supported by air.

It is interesting to note that from the same school of training which produced these imaginative schemes came a man whose ideas were so advanced for his day that they were branded as sacrilegious and he was exiled for his teachings. This was Anaxagoras (499–428 B.C.), who more accurately pictured the moon and planets as earthlike in nature—having a solid, crusty surface and shining by reflected sunlight. Furthermore, he rightly explained lunar eclipses as the result of the moon moving into the earth's shadow.

Another school was established about the same time by the philosopher-mathematician Pythagoras of Samos (about 582–507 B.C.). It is thought that Pythagoras was the first to recognize the shape of the earth as spherical (ball-like). A ship appears to "sink" into the water as it sails away from an observer and to "rise" from the water as it sails toward an ob-

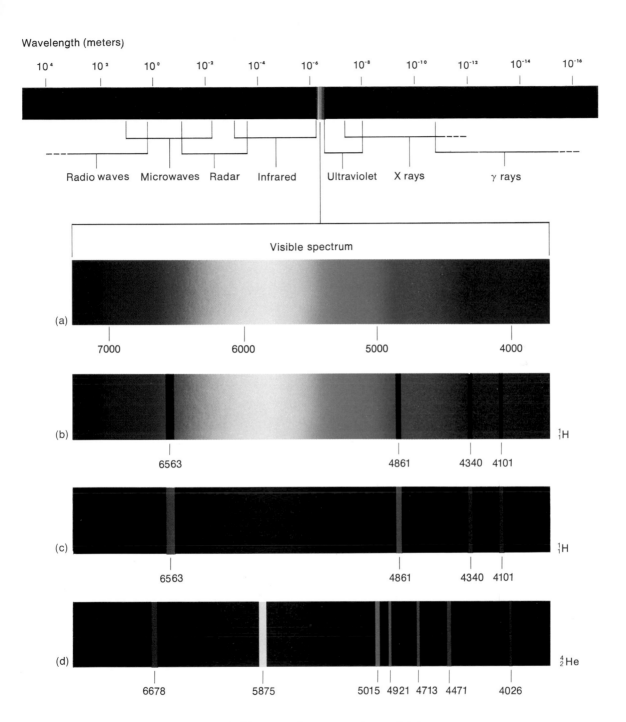

PLATE 1 Spectrograms of (a) white light—a continuous spectrum; (b) hydrogen—an absorption spectrum; (c) hydrogen—an emission spectrum. Numbers on the scale under the spectrograms indicate wavelengths corresponding to each color and are given in angstroms (Å); Å equals 10^{-10} m. See pages 87 to 91 of the text.

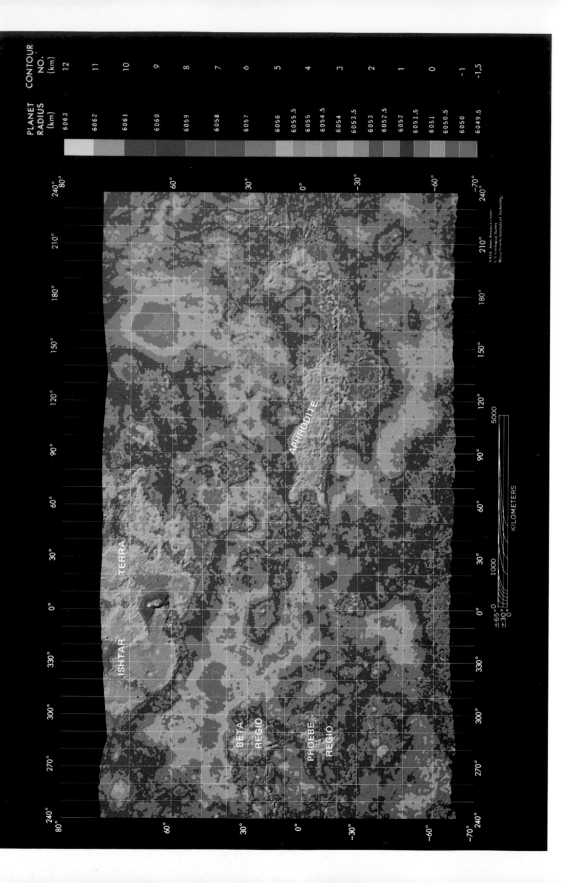

PLANET RADIUS (km)	CONTOUR NO.
6063	12
6062	11
6061	10
6060	9
6059	8
6058	7
6057	6
6056	5
6055.5	
6055	4
6054.5	
6054	3
6053.5	
6053	2
6052.5	
6052	1
6051.5	
6051	0
6050.5	
6050	-1
6049.5	-1.5

ISHTAR TERRA

BETA REGIO

PHOEBE REGIO

APHRODITE

NASA Ames Research Center
U.S. Geological Survey
Massachusetts Institute of Technology

KILOMETERS

±65°
±30°

1000   5000

PLATE 3 This first color photograph, taken by Viking 2, shows the rocky surface of Mars. The color is more accurately described as yellowish-brown, with the sky a lighter hue of the same color. The horizon is actually level although it appears tilted due to the 8° tilt of the lander.

PLATE 4 Jupiter, as seen by Voyager 1 while 33 million kilometers from the planet. Vigorous swirling activity surrounds the Great Red Spot. Objects as small as 600 km across can be identified—the best resolution of Jupiter ever achieved. (JPL–NASA)

<PLATE 2 A radar view of the surface of Venus. (NASA, Ames Research Center, U.S. Geological Survey, Massachusetts Institute of Technology)

PLATE 5 (a) Io, as seen by Voyager 1 at a distance of 900,000 km. Surface activities identified on this moon of Jupiter include cratering and erupting volcanoes. This moon is bathed in the intense radiation of its mother planet. (JPL–NASA)

PLATE 5 (b) Europa, smallest of Jupiter's four Galilean moons, appears bright due to a mantle of ice perhaps 100 km thick. Cracks that developed appear to be filled with dark material from below. (JPL–NASA)

PLATE 5 (c) Ganymede, Jupiter's largest satellite, as seen by Voyager 1 from a distance of 2.6 million kilometers. Ganymede is probably composed of ice and rock. Some of its surface markings remind one of the maria and the rayed impact craters of the moon. The small colored dots (blue, green, and orange) are not present on the satellite but are simply camera markings. (JPL–NASA)

PLATE 5 (d) Callisto, a little smaller than Ganymede, is thought to have a similar composition; however, its darker appearance may be due to "dirty ice" on the surface. Visible near the upper left is a large basinlike structure, the center of which is much brighter than its surroundings. The brighter spots may represent clean ice. (JPL–NASA)

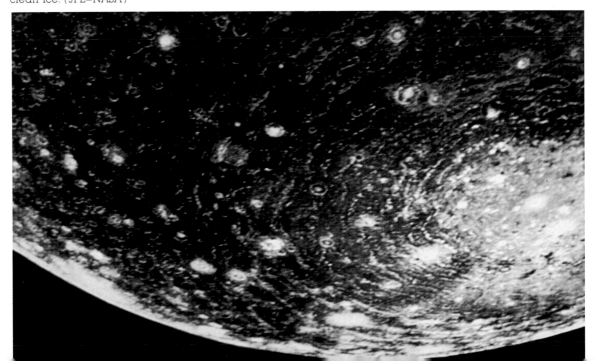

PLATE 6 (a) Voyager 1 returned this color-enhanced view of Saturn. It combines views in green, violet, and ultraviolet light, to emphasize the brighter features in the rings and on the cloudy surface. (JPL–NASA)

PLATE 6 (b) This Voyager 2 view shows the hundreds of ringlets that compose Saturn's rings.

PLATE 7 Comet West (1975n) showing a great amount of detail in its dust and gas trails. (Clifford Holmes, Riverside Astronomical Society)   >

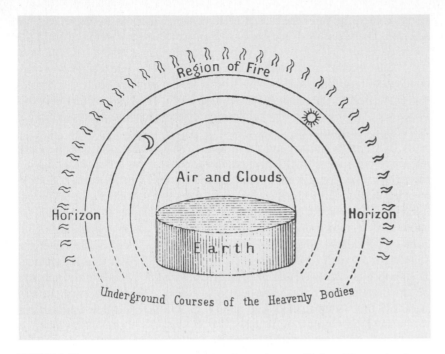

FIGURE 1.10 An early Greek concept of the universe. (Yerkes Observatory)

server, indicating the curvature of the earth's surface. Pythagoras may also have noticed that during a lunar eclipse, as the moon passes into the earth's shadow, the curvature of the edge of that shadow is suggestive of the curved nature of the earth itself (Figure 1.11).

The concept of a spherical earth was a great step forward, for it changed people's entire concept of the universe. In place of a canopy of stars, they could now view the stars as surrounding the earth in all direc-

FIGURE 1.11 (a) The moon passing into the shadow of the earth. (b) A partially eclipsed moon.

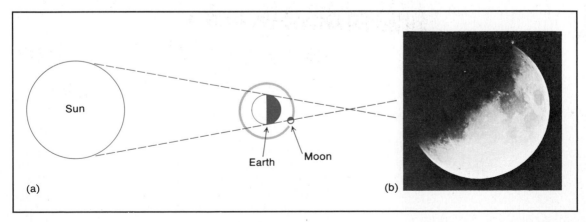

tions; some thought of them as fixed to a vast crystal sphere. Now, too, the sun could be visualized as moving in a smooth orbital motion around the earth.

This is an example of how a single discovery may topple an entire conceptual structure that is based on a false premise. Such discoveries serve to produce a scientific revolution. As you progress through this book, see if you can identify other such discoveries that produced major revolutions of thought.

In the sixth century B.C. there was still no recognition that the earth itself was in motion (rotating on its axis or revolving around the sun). If the earth were not moving in space, then all apparent movements of objects in the sky had to be explained in terms of their own motion. It was not until about 450 B.C. that another Greek scholar, Philolaus, made an attempt to loosen the earth from its fixed position in the minds of his contemporaries by suggesting that there existed a "central fire" (not the sun) around which revolved the earth, the moon, the sun, and the planets. The earth was supposed to make one trip around this central fire each day, producing the apparent motion of all objects around the earth in that period of time. When asked if he had ever seen the central fire, he replied in the negative, indicating that Greece was always turned away from it. Although the system of Philolaus did not find many adherents, it served the significant purpose of suggesting that the earth was in motion. Today we realize that the apparent daily motion of all objects in the sky is due to the rotation of the earth on its axis.

Also of the Pythagorean school, Democritus (about 450 B.C.) recognized that the fuzzy band of light we call the Milky Way was in reality numerous distant stars which appeared close together. It is interesting to compare the wildly imaginative scheme of Philolaus on the one hand and the very keen insight of Democritus on the other. Perhaps it is only our advantage of hindsight that allows us to judge between these ideas.

Plato's Academy was founded in Athens in the fourth century B.C., and one of its most famous pupils was Aristotle (384–322 B.C.). Aristotle, who became the predominant philosopher of his age and tutor to the young Alexander of Macedon (Alexander the Great), was to influence astronomical thinking for almost 2000 years. He firmly established the idea that the spherical earth was the center of the universe and stood stationary in that position, with the sun, moon, and planets moving around it in circular orbits. Aristotle's contemporary, Heraclides (388–315 B.C.), logically suggested that the apparent daily motion of these objects was a result of the rotation of the earth; but because his idea went counter to the thinking of the dominant academic group, it was cast aside.

In 332 B.C. Alexander the Great founded the city of Alexandria on the Mediterranean coast of Egypt. This city was established as a cultural center, a place where theoreticians and practical observers alike could find a level of support unheard of in other lands. Here Aristarchus (about 270 B.C.) challenged the teachings of Aristotle by asserting that the sun was the center of the solar system and that the earth and other planets revolved around the sun. He explained the apparent daily motions of all objects to be the result of the earth's rotation, and he believed the stars to be very distant. These ideas were far too radical for his contemporaries. Because the erroneous theory of Aristotle had become so deeply ingrained by that time, the challenge failed. It might have been argued by the Aristotelians that if the earth traveled around the sun, then nearby stars should appear to shift in their alignment with more distant stars—a phenomenon called *stellar parallax*. No such shift could be seen with the naked eye. The distances are too great. But had observers of that day been able to measure (as we can with our modern instruments) the very slight shift in the apparent position of nearby stars against the background of distant stars, as the earth revolves around the sun, the validity of Aristarchus's *heliocentric* (sun-centered) model would have been apparent. One can visualize this shift in the alignment of a star (A) with respect to more distant stars (B, C, or D) as the earth moves to various positions as shown in Figure 1.12. Even in the case of the nearest star, Alpha Centauri, the apparent shift is only 1/4800 of one degree to either side of its apparent central position.

Hipparchus, (c. 190–120 B.C.) also of the Alexandrian school, made several very significant contributions to astronomy toward the end of the second century B.C. His methods of observation were very advanced for his time. He is often called the father of positional astronomy, for he constructed one of the first systematic catalogues of stars. He accurately stated the positions of more than 1000 stars and ranked them according to their apparent brightness. He placed them into six categories: the brightest were called *first magnitude* stars; the dimmest (to the naked eye) were called *sixth magnitude* stars. This cataloguing project appears most remarkable when we realize that Hipparchus had no telescope or other optical aids with which to work. We still use his scale of apparent magnitudes today, with only slight modification (see Appendix 13, Star Maps). Hipparchus observed that the sun appeared to move faster through the stars during one part of the year and slower during the other part. He reasoned that when it moved more slowly the sun must be farther from the earth and when it moved more quickly the sun must be nearer the earth. He concluded that its orbit could not be circular and devised the model depicted in Figure 1.13. The sun moves on a large orbit called

FIGURE 1.12 The apparent shift of a nearby star (A) against the background of more distant stars, due to the revolution of the earth, as suggested by Aristarchus.

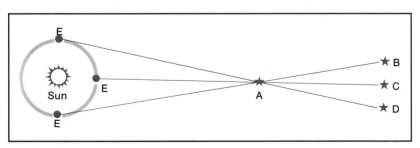

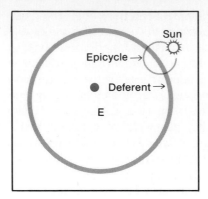

FIGURE 1.13 The epicycle of Hipparchus.

the *deferent* and on a second, smaller orbit called the *epicycle*. His model, you might say, is a wheel turning on a wheel. By choosing the proper speed of turning for the deferent and the epicycle, the apparent variations in the sun's rate of motion and distance from earth could be explained.

PLANETS IN MOTION

The model conceived by Hipparchus for the motion of the sun could also be applied to the moon with only minor adjustments, but the apparent motions of the sun and moon were really quite simple compared to those of the planets. Not only do planets vary in their rate of travel against the background of stars, but they even appear to stop and reverse their direction among the stars at certain intervals. The casual observer would not recognize this reversal, for one only notices the daily westward motion of the sky. However, if one were to plot the position of a given planet on a map of the sky night after night, it would be obvious that planets usually move eastward among the stars. Sometimes they appear to stop, move westward for a few weeks, then stop again and proceed in an eastward direction (Figure 1.14). The astronomer speaks of the westerly motion of planets among the stars as being their *retrograde* (backward) motion. The flip pages beginning on page 114 show the retrograde motion of Jupiter, seen among the stars in the constellation of Leo, in a very dynamic way. Hipparchus not only recognized this motion of a planet but made detailed observations and kept accurate records of planetary positions—records that were invaluable in later years.

FIGURE 1.14 The apparent retrograde motion of a planet among the stars (as charted over several months).

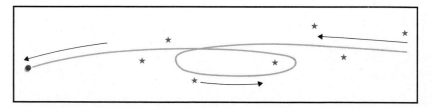

In A.D. 140 Ptolemy (Claudius Ptolemeus), the last great astronomer of the Alexandrian school, gathered together all the accumulated astronomical knowledge of the ancient world and published a series of 13 volumes which reflected the thinking of such men as Aristotle, Pythagoras, and Hipparchus, in combination with a few of Ptolemy's own ideas, and this combined picture of the universe is called the *Ptolemaic system*. Remember that Aristotle viewed the earth as standing still in space at the center of the cosmos. Pythagoras had proposed a spherical earth, and Hipparchus had introduced the concept of the epicycle, as applied to the sun. The observation which demanded Ptolemy's main attention, and that of his immediate predecessors, was the fact that planets sometimes appear to retrograde (back up) against the background of stars. In an attempt to preserve the concept of uniform circular motion, thought by the Greeks to be an absolute necessity, Ptolemy assigned each planet its own epicycle, the center of which moved along the larger orbit called the deferent (Figure 1.15). As the sun moved about the earth, so the outer planets (Mars, Jupiter, and Saturn) moved on their respective epicycles. This model predicted both the times the planets appeared to retrograde, and the times they sometimes appeared brighter and thus closer to the earth. The righthand flip pages beginning on page 25 show Ptolemy's model of planetary motion.

It should be particularly noted that Mercury and Venus move so the centers of their respective epicycles always lie on an imaginary line joining the earth and the sun. This is consistent with the fact that Mercury is always seen within 28 degrees of the sun. From our point of view on earth, Mercury would appear to flit back and forth on either side of the sun. Likewise Venus, while it has a much slower apparent motion among the stars than Mercury, also moves from one side of the sun to the other and never appears more than 47 degrees from it. The epicycle of the sun has been omitted in Figure 1.15 for the sake of simplicity. Of course, the sun never appears to retrograde among the stars, and the turning of its epicycle had

FIGURE 1.15 The Ptolemaic system.

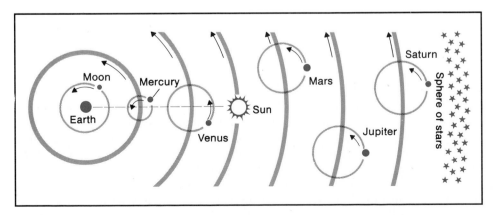

to be synchronized with its motion on the deferent to explain its lack of retrograde motion.

HOW GOOD WAS PTOLEMY'S MODEL?

We should now ask ourselves some interesting questions: Just how good was the Ptolemaic system? How well did it work as a model of the universe? Did it seem to agree with the observations of Ptolemy's time? Could one use this model to predict the location of any planet at some prescribed time in the future? To the latter two questions we can answer, "Yes, within fairly good accuracy," and to that degree the Ptolemaic system was a good model. There were no observations at the time that seemed to invalidate the model. As time passed, however, it became evident that the model did not predict the position of the planets accurately enough. Small corrections were frequently necessary, and these multiplied until the model became quite cumbersome.

Despite these difficulties, Ptolemy's model was destined to remain the principal model of the universe for more than 1300 years. How did a theory based on so many erroneous ideas stand unchallenged for so long? Ptolemy's model was one of the last to emerge from the classic age of philosophy and natural science. Successor to the Greek power, the once mighty Roman Empire had already begun its gradual decline in Ptolemy's time. By the middle of the fifth century, Rome had fallen prey to barbarian hordes from the north—the Vandals, Visigoths, and other tribes. The plundering that ensued destroyed much of classical European culture. This is sometimes called the "Dark Ages" of Europe. However, when the city of Alexandria fell to Arabs in A.D. 642, the culture and "science" that characterized that city were preserved. The Arabic translations of Ptolemy's work, the *Almagest*, found numerous readers among the scholars in Bagdad, a new center of learning.

ARABIC INFLUENCE

One of the important roles of the Arabs involved the establishment of a basis for the exchange of ideas from many nations. They unified many nations under the Islamic religion and established Arabic as the common language. For several centuries, that common language provided the means whereby works from China, India, Egypt, Greece, Syria, and Mesopotamia could be shared. Scholars from many countries went to Bagdad to share their ideas. One might characterize the work of the Arabs as encyclopedic in nature—that of compiling facts. The Arabs, however, also made a number of original contributions in the field of mathematics and science. It was most likely their motivation to spread their religion that led Moslems to travel across North Africa and the Straits of Gibraltar into Spain. We still see the Moorish (Arabic) influence in the architecture of cities like Toledo and Cordova. In Spain, a new awakening may be traced to Alphonso X, king of Castile (northern Spain), who in 1222 commissioned the preparation of new tables describing the motion of the sun, moon, and planets. Within these Alphonsine Tables we see the Greek influence transmitted through Ptolemy's *Almagest*; however certain improvements had been made by both Arab and Jewish astronomers working in the observatory of Toledo. Spain was the site for yet another translation of the multicultural works brought there by the Moors—a translation by Jewish and Christian scholars into Latin. This made available to all of Europe the intellectual achievements of many cultures and may have been a crucial factor in the rebirth of learning that soon took place; the translation represented an important bridge between eastern and western cultures.

In the thirteenth century St. Thomas Aquinas, a prominent Dominican educator, philosopher, and theologian, succeeded in elevating the Aristotelian model to the level of religious dogma. This model placed the stationary earth at the center of the universe. From then on, any person who taught a contrary view might be tried before the Inquisition as a heretic and, if found guilty, might be put to death. Within this atmosphere Nicolaus Copernicus, Tycho Brahe, Johannes Kepler, and Galileo Galilei carried on their work. As late as 1600 Giordano Bruno, an opponent of Aristotelianism and champion of the new Copernican cosmology, was found guilty of heresy by the Inquisition and burned at the stake.

Renewed interest in science was evident in Europe as early as the fourteenth century, but that interest blossomed fully in the fifteenth and sixteenth centuries, especially in the universities of Italy and Germany. With the successful voyages of explorers around Africa, India, and East Asia, the discovery of the Americas, and the circumnavigation of the globe in 1522, there was a revived interest in astronomy and mathematics as aids to ocean navigation. The astronomical–philosophical classics of ancient Greece were examined in their original form, and study of these works produced a flood of reactions. Many of the concepts of Aristotle, Philolaus, Aristarchus, and Ptolemy, together with the writings of the Arab world, were now scrutinized very critically. The time was ripe for the emergence of a great Polish astronomer, Nicolaus Copernicus (1473–1543).

NICOLAUS COPERNICUS, FOUNDER OF MODERN ASTRONOMY

Under the influence of a learned uncle who adopted him at the age of 10, Nicolaus Copernicus attended the University of Cracow to study the classics, philosophy, theology, law, medicine, and mathematics (Figure 1.16). Continuing his education in Italy, he acquired a doctoral degree in canon law and also became competent in the medical practices of the time, but his contact with a noted astronomer at the University of Bologna sparked a lasting interest in astronomy. In 1500 he lectured in that subject at the Vatican in Rome. His training in the Greek language allowed him to pursue a firsthand study of the classics. These interests, together with the errors he had found in the Alphonsine Tables, spurred him to spend some 30 years in the development of a model of the solar system (Figure 1.17), published

FIGURE 1.16 Nicolaus Copernicus. (Yerkes Observatory)

in a six-volume work called *De Revolutionibus Orbium Coelestium*. Although the manuscript of this great work was virtually complete in 1530, Copernicus delayed publication until the final years of his life because of political and religious considerations. The final volume only reached him at his deathbed.

The essense of his scheme is expressed by his own words:

At rest in the middle of everything is the sun. For in this most beautiful temple, who would put this lamp in another or better position than from which it can illuminate the whole thing at the same time? Thus indeed, as though seated on a royal throne, the sun governs the family of planets revolving around it.

Copernicus had thus displaced the earth from its favored position at the center of the universe and had relegated it to the position of one of

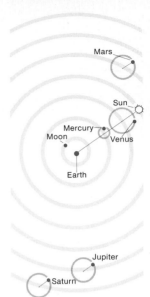

several planets that orbit the sun. He might well have incurred the wrath of the Inquisition for such a heretical idea, if he had not died within a few days of the publication of his theory.

Copernicus believed that the earth and all the other planets both rotated on their axes and revolved around the sun. He viewed the stars as being very distant and fixed in place. It was their great distance, he argued, that explained the lack of any stellar parallax (apparent shifting in position of nearby stars as the earth moves in its orbit around the sun). Copernicus explained that the daily "rising" and "setting" of all objects in the sky could be explained by the daily rotation of the earth on its axis. The apparent yearly motion of the sun among the stars could be explained by the revolution of the earth around the sun. But what about the retrograde motion of the planets? Could Copernicus explain why planets sometimes appear to move backwards among the stars?

The right-hand flip pages starting on page 115 illustrate the motions of the planets in the Copernican model. As you flip these pages, you will see that Mercury and Venus periodically pass the earth and the earth likewise passes the outer planets of Mars, Jupiter, and Saturn periodically. This phenomenon of passing produces the apparent retrograde motion of the planets, even as one train passing another makes the slower train appear to be backing up in relation to the faster train. Consider the earth and Mars as

FIGURE 1.17 The Copernican system, as shown in De Revolutionibus Orbium Coelestium, 1566. (Yerkes Observatory)

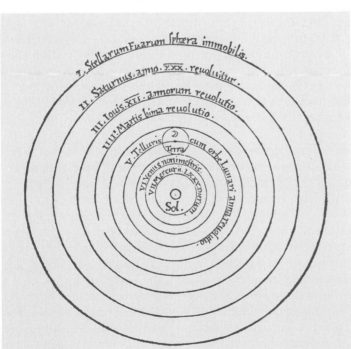

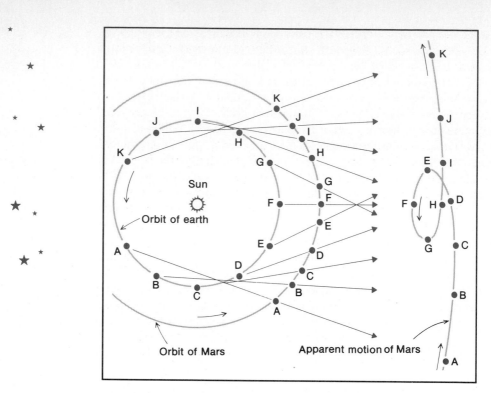

FIGURE 1.18 The apparent retrograde motion of Mars (shown at the right).

seen in Figure 1.18. Note that as the earth moves from positions A through K, Mars also moves through the corresponding positions on its orbit; however, the earth moves faster. At any one position, the stars with which Mars appears to align vary. Moving from A to E, Mars seems to move eastward among the stars, but while moving from E to G it seems to move westward. Finally, moving from G to H, it appears to move eastward again. A similar explanation applies to the apparent retrograde motion of the other planets.

It is also easy to see from the flip pages that planets are closer to the earth on some occasions and farther from it at others. This explains their apparent changes in brightness. While Copernicus no longer needed the epicycle to explain the retrograde motion of planets, he retained the idea for other reasons. He believed that all objects move in a uniform way (at the same rate) on circular orbits or in combinations of circular orbits. Yet he saw planets sometimes moving rapidly against the background of stars and sometimes slowly; he thought that through a set of epicycles, which were assigned the proper uniform motion, the complex motion of the planets could be explained. Some of his earlier work contained epicycles upon epicycles. The model of Copernicus appears to explain the observations of the time, but the model of Ptolemy did so as well. Although there was not yet any substantial evidence as to which model most closely represented reality, the publication of the Copernican theory created a furor which was to last more than 200 years. The daring of Copernicus encouraged others to devise better means for observation so that his theory might be put to the test. Typical of this new spirit of observation was the work of Tycho Brahe.

TYCHO BRAHE, THE DILIGENT OBSERVER

Tycho Brahe (1546–1601), a Danish nobleman (Figure 1.19), was trained as an astronomer and, under the sponsorship of the king of Denmark, he devoted himself to a systematic observational approach to the subject. He was granted the tiny Baltic island of Hven (Figure 1.20), where he established an observatory and constructed instruments capable of measuring angles between stars quite accurately (Figure 1.21). Brahe made numerous observations of the position of individual stars and averaged his readings to produce a catalogue of approximately 800 stars. The positions of many stars were accurate to within 1/100 of a degree. Brahe's catalogue was far more accurate than that of Hipparchus or Ptolemy, and it stood as a reliable reference source for more than a century. Brahe also kept precise records of the changing positions of planets, believing that only through an understanding of the orderliness and unity of the universe might man better order his own affairs. In a lecture at the University of Copenhagen in 1574, Brahe said:

> To deny the forces and influence of the stars is to undervalue firstly the divine wisdom and providence and moreover to contradict evident experience. For what could be thought more unjust and foolish about God than that He should have made this large and admirable scenery of the skies and so many brilliant stars to no use or purpose—whereas no man makes even his least work without a certain aim.

FIGURE 1.19   Tycho Brahe, the observer. (Yerkes Observatory)

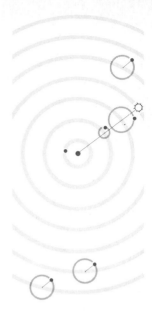

FIGURE 1.20 Uraniborg, Tycho Brahe's observatory. (Yerkes Observatory)

FIGURE 1.21 Tycho Brahe at Uraniborg: (a) Tycho observing with his great mural quadrant; (b) sextant used to measure angles between celestial objects. (Rare Book Division, The New York Public Library, Astor, Lexox and Tilden Foundations)

(a)   (b)

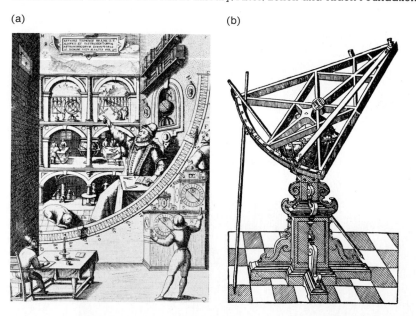

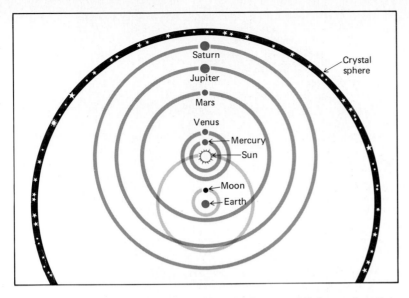

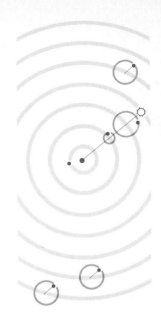

FIGURE 1.22 Tycho's Brahe's system. Note that the sun orbits the earth as the center of the celestial sphere (an idea from the Greeks), but the planets orbit the sun. Would the planets appear to back up (retrograde) periodically in this system?

Brahe's labors epitomized an important aspect of the scientist's work that is often overlooked: that of keeping absolutely honest and careful records, which are often valuable regardless of the correctness of conclusions drawn at the time. Brahe's observational records were his greatest contribution to astronomy, for they assisted those who followed him to present a clearer picture of the solar system. He could not accept Copernicus' premise that the earth was in motion, so he devised a very interesting scheme that pictured the planets as circling the sun but had the sun orbiting around the earth. An associate of Brahe's is now credited with the variation on that theme (Figure 1.22). See if you can visualize how such a scheme could produce the observed motions of the planets.

In his later years Brahe fell out of favor with the new king of Denmark and lost his island sanctuary. He sought and found the support of Rudolf II, the scholarly Holy Roman emperor, and built a new observatory near Prague in Bohemia. Here the master astronomer took as his aide a young German philosopher and mathematician, Johannes Kepler (1571–1630), who after Brahe's death fell heir to Brahe's voluminous observational records.

JOHANNES KEPLER, THE INVENTIVE MATHEMATICIAN

Kepler was educated at Tubingen University and taught at the University of Graz in Austria (Figure 1.23). When he was relieved of his professorship for religious reasons, he received an invitation from Tycho Brahe to be-

FIGURE 1.23 Johannes Kepler. (Yerkes Observatory)

come his assistant. Kepler accepted and soon acquired a keen respect for the high degree of accuracy with which Brahe observed the positions of the stars and planets.

Brahe was overly protective of his records, fearing that Kepler would steal them. While this frustrated Kepler, he soon recognized that Mars did not move against the background of stars at a constant rate but moved faster during one part of its orbit and then slowed down during another part. This fact alone must have suggested to him that the planets did not move around the sun in a circular orbit as Copernicus had proposed, for if the orbit were circular, a constant rate of motion would have been expected. Kepler attempted to fit various orbital shapes to the observed motion of Mars, shapes which would provide for the variation in speeds he had witnessed. Only after much effort did Kepler conclude that the best fit was that of an *ellipse* (Fig. 1.24).

FIGURE 1.24 Drawing ellipses of different eccentricities.

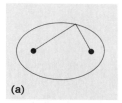

(a)

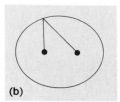

(b)

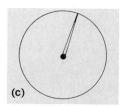

(c)

A device for drawing an ellipse can be fashioned using two thumb-tacks and a string of fixed length. By attaching each end of the string to a thumbtack and by stretching the string taut with a pencil, it is possible to draw an ellipse. By varying the spacing of the thumbtacks, one may draw ellipses of varying eccentricity. Figure 1.25 shows a number of terms associated with an ellipse. The two points (thumbtacks) are called the *foci* (plural for *focus*), and a line drawn through these points and extending to the curve itself is called the *major axis* of the ellipse. One-half that distance is called the *semimajor axis*, and this is equivalent to the average distance (radius) from one focus to any point on the ellipse. In Table 1.1 the semi-major axis for each planet is given, thus expressing its average distance from the sun. The *eccentricity* of an ellipse is found by dividing the distance between the foci by the length of the major axis. A circle has an eccentricity of zero, and the eccentricities of the planetary orbits range from 0.007 (Venus) to 0.25 (Pluto), with that of the earth being 0.017, not very much different from a circle.

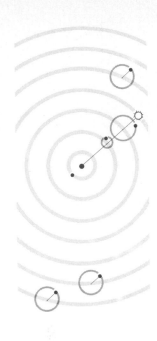

Kepler found that he could best fit the observed motions of Mars to an elliptical orbit with the sun at one focus, the other focus being only an imaginary point in space. He generalized this idea to include all the planets as follows:

> Each planet moves around the sun in an orbit whose shape is that of an ellipse, with the sun at one focal point.

Although this law is called Kepler's first, this numbering may not represent the order of discovery. That which is called his second law was

FIGURE 1.25 Axis and average radius of an ellipse.

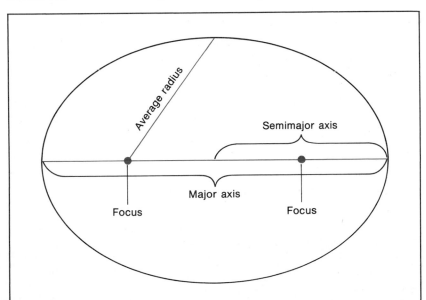

TABLE 1.1

			PLANETARY DATA		
Planet	Semimajor Axis (r), A.U.	r^3	Period (p), Years	p^2
Mercury	0.39	0.058	0.24	0.058
Venus	0.72	0.378	0.62	0.378
Earth	1.00	1.000	1.00	1.000
Mars	1.52	3.54	1.88	3.54
Jupiter	5.20	140.7	11.86	140.8
Saturn	9.54	867.7	29.46	867.9

actually more obvious—the fact that a planet moves more slowly when far from the sun and more rapidly when nearer the sun. This lack of uniformity in motion suggests something other than a circular orbit. Kepler's second law is stated:

> A straight line joining the planet and the sun sweeps out equal areas in space in equal intervals of time.

As an illustration of Kepler's second law, let us suppose that Mars moves from points A to B in a month (Figure 1.26). Some time later it also moves from C to D in a month, but now it is closer to the sun, and so the distance from C to D must be greater in order for Mars to sweep out an equal area in the same time. To cover this greater distance in the same time, Mars must move faster. Likewise, when the planet travels from E to F, still closer to the sun, it must move still faster to sweep out an equal area in a month; however, as it moves from G to H it is moving more slowly again, and as it passes from A to B it has returned to the speed that it had at the start. Thus Kepler's second law provides a mathematical model whereby the speed of a planet may be computed for any given position on its orbit.

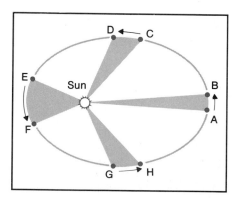

FIGURE 1.26 Illustration of Kepler's law of equal areas in equal time.

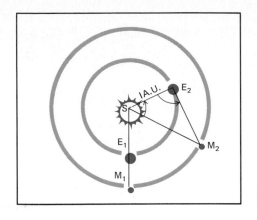

FIGURE 1.27 Calculating the distance to Mars.

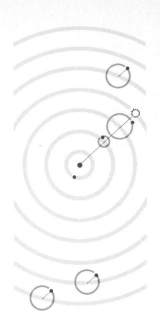

To recognize a third basic relationship required several years of effort on Kepler's part. Of course, it was obvious to him that the farther a planet was from the sun, the longer it took to complete one revolution around the sun. But Kepler sought to find the exact mathematical relationship between a planet's period of revolution and its average distance from the sun (the semimajor axis of its orbit), if such a relationship did indeed exist. Although the actual distances between the sun and the planets were not known then, it was possible to calculate the ratio between a planet's distance from the sun and the earth's distance from the sun.

We can see how this was possible by making several observations of Mars (Figure 1.27). First note the day when Mars is in opposition (directly opposite the sun from our point of view, as shown by the line SE_1M_1). The earth requires 365.25 days to orbit the sun and will move approximately $1°$ per day along its orbit; because Mars requires 687 earth-days to orbit the sun, it will move approximately $0.524°$ per earth-day. Choose a later time when the earth has moved significantly ahead of Mars and compute the angle between the two planets (angle E_2SM_2) from their rates of movement above. Then measure the angle between the sun and Mars, as viewed from earth (angle SE_2M_2). Knowing these two angles and defining the distance from the sun to the earth (line SE_2) as 1 A.U. (astronomical unit), we can use trigonometry to find the length of the side of the triangle that represents the distance from the sun to Mars (side SM_2). This distance is approximately 1.52 A.U.

Table 1.1 lists the semimajor axis for each planet, as measured in astronomical units. This table also shows the period of revolution for each planet, measured in years. Note that for any planet, if the semimajor axis (r) is cubed and the period of revolution (p) is squared, the two numbers produced are the same, with only minor discrepancies for Jupiter and Saturn. This fact may be expressed by the simple equation $p^2 = r^3$, where p is measured in years and r is measured in astronomical units. Should we choose to measure the period of the planet in days or the distance in miles, then we would not expect $p^2 = r^3$; however, we can allow for this

possibility by writing $p^2 = kr^3$, where k is a constant whose value is determined by the units of measure used. Still another equation may be written to specify Kepler's third law:

$$\frac{(p_1)^2}{(p_2)^2} = \frac{(r_1)^3}{(r_2)^3}$$

which may be stated as:

> The square of the periods of any two planets have the same ratio as the cubes of their semimajor axes.

Assuming that this relationship would hold true for any planet in our solar system, let us suppose an imaginary planet X has a period of 8 years. Can we find its average distance from the sun in astronomical units? Let the period and semimajor axis of planet X replace p_1 and r_1 in the equation, and let the period and semimajor axis of the earth replace p_2 and r_2. For the earth, we know that $p_2 = 1$ year and $r_2 = 1$ A.U. (see Table 1.1). The period of planet X has been given as 8 years, so $p_1 = 8$. If we make these replacements in the equation and solve for r_1, we have:

$$\frac{(p_1)^2}{(p_2)^2} = \frac{(r_1)^3}{(r_2)^3}$$

$$\frac{(8)^2}{(1)^2} = \frac{(r_1)^3}{(1)^3}$$

$$\frac{64}{1} = \frac{(r_1)^3}{1}$$

$$4 = r_1$$

Therefore the imaginary planet X would have a semimajor axis of 4 A.U.

A WARNING.   All of Kepler's laws are highly idealized, for they neglect the fact that each planet influences every other, so that the orbits are never smooth ellipses nor are the areas swept out in equal time exactly the same. Furthermore, Kepler's third law is limited to a restricted set of conditions, namely, to very low-mass objects going around very high-mass objects, such as planets around the sun or moons around a planet. Nonetheless, Kepler's discoveries were remarkable approximations that were later confirmed and generalized by Sir Isaac Newton, whose astronomical knowledge and mathematical techniques were more advanced.

Although Kepler was a mystic, his work served as the foundation for the removal of astronomy from the realm of mysticism and for the establishment of its cause-and-effect nature. He sensed that some unknown force, emanating from the sun, kept the planets moving in their orbits. He was on the verge of discovering universal gravitation. We know that Kepler communicated with Galileo, a man in whom these thoughts were to find fruition.

GALILEO GALILEI, FATHER OF EXPERIMENTAL SCIENCE

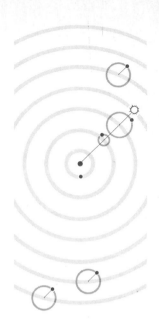

The seventeenth century was one of discovery, experimentation, and invention. The invention that had the largest impact on astronomy was the telescope, of which the exact history is unclear. A Dutch spectacle maker, Lippershey, is usually credited with having combined several lenses to produce an enlarged image of a distant object. Although he produced the instrument in 1608, he apparently did not fully grasp its enormous potential. The following year in Padua, Italy, Galileo Galilei (1564–1642) heard of this invention and, without any detailed knowledge of how the task had been accomplished, set out to produce a telescope of his own (Figure 1.28). Galileo's first success resulted in an instrument that enlarged objects three times, a *three-power* telescope. Later refinements resulted in a 30-power instrument. These instruments he put into immediate use. By directing them skyward, Galileo soon discovered four moons circling the planet Jupiter, each moving around the planet in its own particular period (Figure 1.29). This sight reminded Galileo of the solar system itself, as Copernicus had described it, and it surely demonstrated that the earth is not the only center of revolution in the universe.

Using the telescope, Galileo also found that what had appeared as a fuzzy patch in the sky, a nebula, could now be seen as a group of separate stars. In fact, in all regions of the sky the instrument revealed many stars that were too faint to be seen with the naked eye. Concerning a familiar constellation, Galileo wrote:

FIGURE 1.28   Galileo Galilei. (Yerkes Observatory)

FIGURE 1.29 Galileo's drawings showing the motions of the moons of Jupiter. (Yerkes Observatory)

I had determined to depict the entire constellation of Orion, but I was overwhelmed by the vast quantity of stars and by want of time, and so I have deferred attempting this to another occasion, for there are adjacent to, or scattered among, the old stars more than five hundred new stars.

Galileo also observed that the planet Venus goes through phases (see Figure 1.30), sometimes appearing as a thin crescent and at other times as a disk almost fully lighted. This was a startling observation; it provided evidence for the first time that the Ptolemaic system could not represent

FIGURE 1.30 The phases of Venus. (Lowell Observatory)

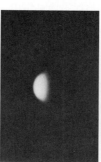

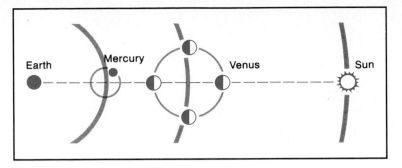

FIGURE 1.31 The phases of Venus-Ptolemaic system.

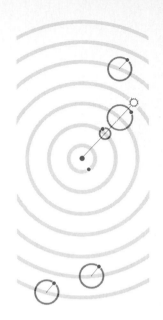

reality. Remember that in the Ptolemaic system the center of Venus' epicycle must always remain on an imaginary line joining the earth and the sun. Under this restriction, the planet Venus could never appear from the earth any fuller than a mere crescent (see Figure 1.31). Galileo observed that Venus sometimes appears almost full. Could this fact be better explained by the Copernican theory? Yes, as you can see in Figure 1.32. When viewed from the earth, Venus may be seen in all possible phases, except when it is closely aligned with the sun.

The revelation that Venus goes through phases dealt the death blow to the Ptolemaic theory. Confident in his new-found proof, Galileo tried to convince the leadership of the Roman Catholic Church that its interpretation of the Holy Scriptures was inconsistent with observed facts. He failed in this attempt. In fact, the church hierarchy pronounced his findings to be false and heretical and forbade Galileo or anyone else to teach them. How blind are the eyes of humans when they do not want to see!

Galileo also observed sunspots and identified them as being on the surface of the sun itself. By charting their movement, he determined the rotation period of the sun (Figure 1.33). To the philosophers of that time,

FIGURE 1.32 The phases of Venus-Copernican system.

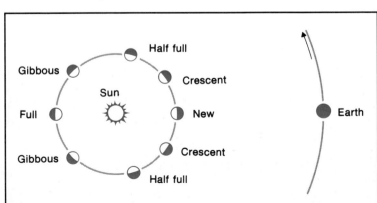

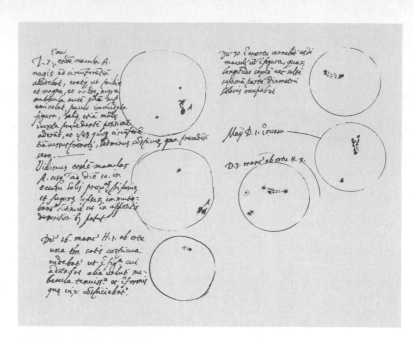

FIGURE 1.33 Galileo's drawings of sunspots. (Yerkes Observatory)

however, any "imperfection" on a celestial body was unthinkable; so his report was again ridiculed.

Despite all these obstructions, Galileo opened up a new era of scientific experimentation. For example, he wondered about Aristotle's claim that when two objects of different weights are dropped from a high place, the heavier object hits the ground first. According to legend, Galileo climbed the Leaning Tower of Pisa (in the city of his birth) and from the top dropped two round stones, one of which was much heavier than the other. To the amazement of the crowd that gathered, the stones fell side by side, hitting the ground at the same time. In order to measure the effect of gravity more accurately, Galileo allowed a ball to roll down an inclined ramp and timed its acceleration. He caused an object to slide along different surfaces and observed that when friction was least, the object would travel farther. He may have come to the realization that an object that is moving through space, where no friction exists, will travel onward at its given speed forever. This concept, however, is usually ascribed to Sir Isaac Newton. That great British scientist was surely crediting the work of Galileo, Kepler, Brahe, and Copernicus, when he said, "If I have been able to see farther than other men, it is because I was standing on the shoulders of giants."

ISAAC NEWTON, THE YOUNG GENIUS

Newton (1642–1727) entered Cambridge at 19. Four years later, in 1665, the university closed during the great London plague, and Newton returned home (Figure 1.34). During this time he conceived an explanation for the fact that a raindrop or a prism creates a rainbow—a phenomena we call dispersion (we will return to that idea in Chapter 2). Newton also began to formalize his thoughts about the way moving objects behave; he produced three laws of motion. Newton stated his first law, the law of inertia, as follows:

> Every body continues in its state of rest, or of uniform motion in a right (straight) line, unless it is compelled to change that state by force impressed thereon.

This says that if a body is at rest, it will remain at rest unless acted upon by an external force. That statement probably does not surprise you or anyone else. The law, however, does contain an idea that was really new for Galileo's time. The law of inertia also says that if a body is moving, it will continue to move at the same speed and in a straight line forever unless acted upon by an external force. Even Kepler had thought that some force was needed to keep a planet moving; but if Newton's first law is true, then one may assume that because a planet is moving, it will keep on moving unless a force acts to stop it. Furthermore, it follows that the natural motion of an object moving in space is not circular (nor elliptical, as Kepler had said) but straight, providing that no other force is acting upon that object. Now, in place of a force to keep the planets "going," Newton needed a force to make them "fall away" from their natural straight-line motion. We will see the solution to that problem shortly.

In Newton's own words, his second law reads:

> The alteration of motion is ever proportional to the motive force impressed, and is made in the direction of the right (straight) line in which it is impressed.

FIGURE 1.34 Sir Isaac Newton. (Yerkes Observatory)

He explained this law further by saying, "If any force generates a motion, a double force will generate a double motion, a triple force will generate a triple motion." But what did Newton mean by the word "motion"? There are two words that are often associated with the concept of motion. One is called *speed*. If someone says, "I am traveling at 80 km/hr," this would tell you the person's speed. That is a rather incomplete statement of motion, however, for it gives only the magnitude of the motion. By contrast, the statement, "I am traveling 80 km/hr due north," would tell you the person's *velocity*, which includes both magnitude and direction. Any quantity that specifies both magnitude and direction is called a *vector quantity* and can be represented graphically as an arrow whose direction agrees with the direction of motion and whose length is proportional to its magnitude.

The term *acceleration* is defined as any change in velocity, whether it be a change in magnitude or a change in direction or a change in both magnitude and direction, and this is what Newton meant by the phrase "alteration of motion." We often refer to the gas pedal of our automobile as an "accelerator." When we press on it the car typically speeds up; that is one form of acceleration. When we press on the brake pedal, the car typically slows down; that is another form of acceleration. The brake pedal then could also be called an accelerator, but in the negative sense. Suppose we allowed the speed of our car to remain the same and merely caused it to change direction; that is also a form of acceleration. The principal relationship Newton was trying to convey in his second law is that a *force* is required to produce an *acceleration*. He said that whenever a force is applied to an object, its velocity will be changed: it will be speeded up, slowed down, or its direction will be changed (or a combination of these). Furthermore, if that force is doubled then the acceleration that it produces will also be doubled, and if that force is tripled then the acceleration that it produces will also be tripled. He expressed this relationship very simply: $F = ma$, where F = force; m = mass (a measure of the amount of material in an object); and a = acceleration.

Suppose that you were able to drop an object from a high cliff and mark how far it fell at the end of each half-second. It would then be possible to compute its acceleration due to gravity, and near the surface of the earth you would find that acceleration to be 9.8 m/sec^2 (meters per second per second). This means that the velocity of a free-falling object will be 9.8 m/sec at the end of the first second, 19.6 m/sec at the end of the second second, 29.4 m/sec at the end of the third second, and so on. We are neglecting the fact that the velocity of the object is usually less than these numbers indicate because of the retarding effect of friction as the object literally has to push air molecules out of the way as it falls. If we define the *weight* of an object as the force with which the earth attracts that object due to gravity, then $F = ma$ becomes $w = mg$, where w = weight and g = acceleration due to gravity. Your weight, when standing on the surface of the earth, is given by $w = 9.8 \times m$. We will develop this relationship and the units of force and mass in Chapter 3.

Legend has it that, while taking tea in the garden one afternoon, Newton saw an apple fall to the ground and he asked himself, "Could the force that accelerates the apple as it falls to the ground be the same force that keeps the moon in her orb (orbit)?" Galileo had repeatedly demonstrated the attractive force that exists between the earth and other objects near its surface, and Robert Hooke, a contemporary of Newton, suggested that planets move under the influence of a force that "falls off with the square of the distance." It was up to Newton, however, to demonstrate mathematically that the force of gravity responsible for the acceleration (the speeding up) of the apple as it fell to the ground was the same force responsible for the acceleration (the change in direction) of the the moon as it orbits the earth—and likewise the acceleration of each planet as it orbits the sun. Following the statement of Newton's third law of motion, we will return to his efforts to express the force of gravity in a mathematical relationship.

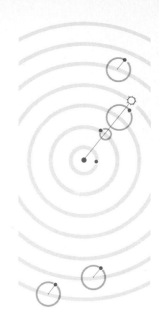

Newton stated in his third law:

> To every action there is always opposed an equal reaction; or the mutual actions of two bodies upon each other are always equal, and directed to contrary parts.

To illustrate this law, imagine shooting a gun. A force acts to accelerate the bullet from the gun (call that the action force), but an equal force creates an acceleration on the gunstock, usually referred to as the "kick" of the gun (call that the reaction force). This law also explains how a rocketship is propelled forward, even in empty space (see Figure 1.35). As particles are expelled from the rear of the rocket, the equal but opposite reactive force propels the ship forward. The action and reaction forces always act on different bodies (the *action* force acts on the particles being expelled the *reaction* force acts on the rocketship to move it forward). We live with action-reaction forces every day; for example, the act of walking can only be accomplished by the fact that when we exert a force backward on the sidewalk, the sidewalk exerts a force forward on us. The fact that we are not thrown off a spinning earth results from action-reaction forces that are due to gravity—the earth pulls on us and we pull on the earth with an equal force. But what determines the magnitude of that force?

FIGURE 1.35 Newton's third law in action-reaction.

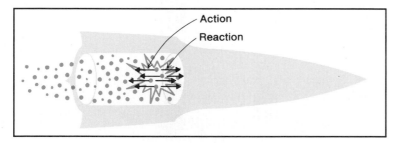

THE FORCE OF GRAVITY BETWEEN TWO OBJECTS

Galileo certainly knew that the force of gravity between two objects is directly proportional to the product of their masses. This fact can be illustrated as follows: Support a number of different masses on spring balances like those illustrated in Figure 1.36. By observing the amounts by which each spring is stretched, one can conclude that the force with which the earth pulls on each is proportional to its mass (the amount of material it contains); that is, the mass of object 2 is twice that of object 1, and you can see that the earth pulls on object 2 with twice the force that it pulls on object 1 (the spring is stretched twice as far). Likewise, the mass of object 3 is three times that of object 1, and the earth pulls on object 3 with three times the force it pulls on object 1 (its spring is stretched three times as far as is spring 1).

Neither Galileo nor Hooke had gone so far as to determine the effect of distance on the gravitational forces between two objects, and this is where the story of the apple and the moon seems most pertinent. Newton ascertained that the moon was about 60 times as far away from the center of the earth as the apple (see Figure 1.37). If gravitational force between two objects weakens in proportion to its distance from the center of that object, then the acceleration of the moon in its orbit should be 1/60th of the acceleration of the apple. His calculations showed that the acceleration of the moon is only 1/3600th the acceleration of the apple, thus demonstrating that the force of gravity does "fall off with the square of the distance" [$(1/60)^2 = 1/3600$]. Newton expressed his findings as follows:

$$F = G \frac{m_1 \times m_2}{r^2}$$

FIGURE 1.36 The force due to gravity is proportional to the mass of the object.

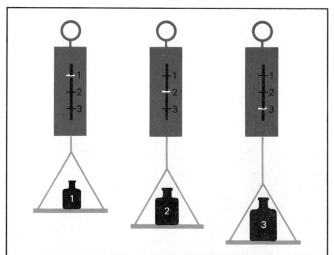

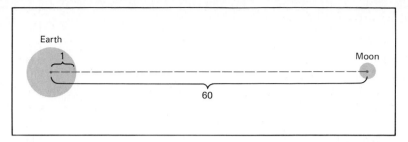

FIGURE 1.37 The moon orbits the earth at an average distance which is 60 times the earth's radius.

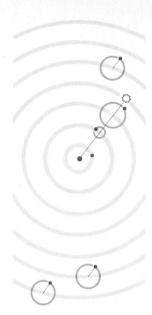

The force (F) due to gravity is proportional to the product of the masses (m_1 and m_2) of the two bodies being considered and is inversely proportional to the square of the distance (r) between their centers (see Figure 1.38). The letter G stands for the gravitational constant, the value of which has been subsequently determined and can be found in Appendix 1. Newton's expression for the force of gravity tells us that if either mass is increased, the mutual attraction of the two bodies also gets stronger, but if the distance between the two objects is increased, their mutual attraction gets weaker—it "falls off as the square of their distance between centers."

This expression, "falls off as the square of their distance," will appear in a number of concepts in this text, and for good reason—it is the nature of the three-dimensional space in which we live. Suppose we allow a portion of the light from a 200-watt lamp to fall on a square screen which is 1 m by 1 m (see Figure 1.39). The screen will be illuminated with a certain intensity. Now if the screen is moved twice as far away, the same portion of light will be spread over four such screens (with a surface area of $4\,m^2$) and one would expect the intensity of illumination (the illumination falling on each screen) to be one-fourth that of the first example. When the distance is tripled, the same portion of light illuminates nine such screens (with a surface area of $9\,m^2$) and one would expect the intensity of illumi-

FIGURE 1.38 The factors which influence the gravitation force (g) between two objects are: the mass of object 1 (m_1), the mass of object 2 (m_2), and the separation of centers (r).

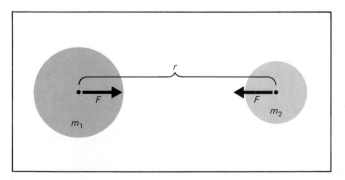

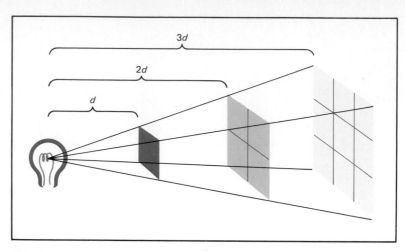

FIGURE 1.39 The same light illuminates four squares located at distance (2*d*) as compared to one square at distance (*d*), therefore the intensity at 2*d* will be ¼th that at *d*.

nation to be one-ninth that of the first example. The meaning of the expression "falls off with the square of the distance" becomes clear as we summarize these findings:

Distance	(Distance)^2	Fraction of light per 1-m^2 screen
1	$(1)^2 = 1$	1
2	$(2)^2 = 4$	1/4
3	$(3)^2 = 9$	1/9

Now replace the 1-m^2 screens with entire spheres surrounding the lamp, the first sphere having a radius of 1 m, the second a radius of 2 m, and the third a radius of 3 m (see Figure 1.40). The first sphere would be illuminated with a certain intensity, but if it were removed, the second sphere would be illuminated with an intensity one-fourth that of the first—the surface area of the second sphere is four times that of the first (surface area of a sphere = $4\pi r^2$). If the first two spheres were removed, the total light would fall on the third sphere, which has a surface area nine times that of the first, and the intensity of illumination will be one-ninth as great as the first. Thus the "falling off with the square of the distance" is as natural as space itself. Now replace the lamp with an object of a given mass. That object can be thought of as modifying the space in which it exists, to the point of creating a gravitational *field*. Any other object will experience a force of attraction between the first object and itself, because it is in the field of the first object. The intensity of that field "falls off with the square of the distance," just like the intensity of the lamp in our earlier illustration, and for the same reason, because it is the nature of space itself. We can then conclude that the reason for the term r^2 in the denominator of the gravitational formula is that an r^2 appeared in the formula for the

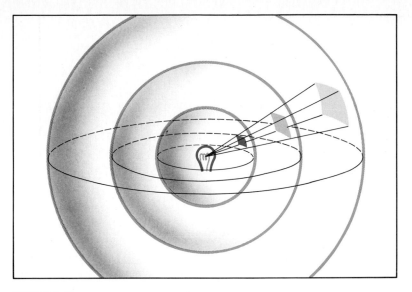

FIGURE 1.40 The intensity of light "falls off with the square of the distance"—it is a characteristic of three-dimensional space.

surface area of a sphere. When you have occasion to study electricity and magnetism, you will see that our line of reasoning carries into those areas, because of the nature of space, the formulas for electrical and magnetic forces will also have an r^2 in the denominator.

Let us apply Newton's laws to the planets and their orbits around the sun. His first law says that the natural tendency is for the planet to move in a straight line, thus "flying off" along the tangent (AB) to the circle (see Figure 1.41). However, the force of gravitation between the planet and the sun acts to accelerate the planet in a direction (AC) toward the sun. The effect of this acceleration due to gravity is such that the planet continually falls away from its natural path by an amount (BC) that just maintains it in an elliptical orbit.

Kepler had already described the motions of the planets based on what he had observed, but Sir Isaac Newton was able to deduce these same motions from his universal laws of gravitation (the derivation is beyond the level of this text). He realized that the total mass of the objects in question also entered into the relationship, as shown in the following equation:

$$(m_{sun} + m_{planet})p^2 = kr^3$$

Since the mass of the sun (m_{sun}) is so much greater than the mass of any one of its planets, we may say that ($m_{sun} + m_{planet}$) is approximately the same as the mass of the sun alone, therefore

$$(m_{sun})p^2 = kr^3$$
$$(m_{sun}) = \frac{kr^3}{p^2}$$

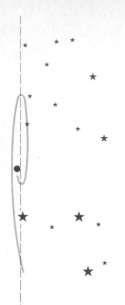

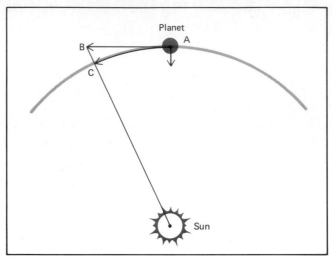

FIGURE 1.41 The planet continually "falls" toward the sun, in the sense that after a given period of time it is closer to the sun, by the distance BC, than it would be if free to move along a straight (inertial) line AB.

This shows that the mass of the sun can be found with only the knowledge of the period of a single planet and its average distance from the sun. Furthermore, Newton's law of planetary motion reduces to that of Kepler when we solve for p^2 and then let k/m_{sun} be set equal to a new constant K:

$$(m_{sun})p^2 = kr^3$$
$$p^2 = \frac{kr^3}{m_{sun}}$$
$$p^2 = Kr^3$$

Thus Newton was able to confirm Kepler's laws by first deriving a more general relationship from the basic laws he had discovered. Using the refined statements, it is possible not only to describe the motions of natural bodies accurately but also to predict with a high degree of accuracy the orbits of today's artificial satellites.

OUR PLACE IN THE UNIVERSE

The sixteenth and seventeenth centuries were years of tremendous change in human understanding of the "universe," but they can hardly be compared to the twentieth century. Until about sixty years ago, our view of the universe was not much greater than that of Galileo. Humans were convinced that they lived on one of eight planets that circled the sun—Uranus having been discovered in 1781 and Neptune in 1846. We seemed satisfied to occupy this off-center position in the solar system, and we recognized that the sun was merely one of millions of stars that could be seen with the creation of ever larger telescopes. These same instruments revealed fuzzy patches of light that were dubbed nebulae, some of which could be seen to contain multiple stars and others that seemed to blend into a fuzzy mass. Not until the early twentieth century did astronomers begin to suspect that some of those fuzzy objects were actually collections of stars separated from our own collection by vast expanses of empty space. Thus the concept of galaxies was born and the realization that our own Milky Way galaxy is only one of many galaxies. The American astronomer Edwin Hubble referred to these separate collections as "island universes." This revelation was a giant leap forward in our understanding of the overall size of the universe.

Some fuzzy objects seen by telescopic observation turned out to be clouds of glowing gas—today we call these *nebulae*. Some turned out to be stars that are relatively crowded together—we call these *clusters*. As we have mentioned, some turned out to be *galaxies*—collections of stars and gas and dust (and perhaps solar systems) separated from our own Milky Way galaxy by vast distances of empty space. The milky-looking band of light that surrounds the entire sky, called the Milky Way, was revealed to be millions of stars crowded together in a flattened disklike shape. It is not simply a band of glowing gas as it appears to the naked eye. In reality, the Milky Way band reveals the shape of our Galaxy. Were the stars of our Galaxy distributed in a spherical (ball-like) shape, we would undoubtedly see about the same numbers of stars in all directions, but this is not the case (see Figure 1.42). Rather, we believe that our Galaxy resembles one called M 31, the galaxy in the constellation of Andromeda (see Figure 1.43). It has a flattened disklike shape with a bulge of brightness in the center and spiral arms almost like a spinning "fourth of July" pinwheel. We will see that this description is not far from being true for the Milky Way galaxy. Because you are actually sitting inside the Milky Way galaxy, it is more difficult to tell its shape and to ascertain your own location.

To help us with this last question, astronomers have noticed yet another class of fuzzy objects, called *globular clusters*, most of which lie to one side of us. We could think of the objects as being off-center in relation to the galaxy as a whole, or it could be that we are off-center. We now assume that these globular clusters form a "halo" around the nucleus (center) of our Galaxy and they then become an indicator of just how far off-center we lie. Again, we have little cause to be egotistical about our

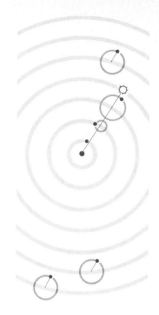

FIGURE 1.42 The Milky Way from Sagittarius to Cassiopeia. (Palomar Observatory Photograph)

FIGURE 1.43 The Andromeda galaxy, a spiral galaxy considered to be very much like the Milky Way galaxy. (Palomar Observatory Photograph)

position in the universe. We believe the disk of our Galaxy to be slightly over 100,000 light-years in diameter and we (the solar system) are located about 30,000 light-years from the center (see Figure 1.44). One light-year is equal to approximately 9 trillion km (6 trillion miles). The modern estimate of the number of stars in the Milky Way galaxy is in excess of 100 billion, with the number of galaxies in the universe of the order of 1 billion. This means that the universe is exceedingly vast; furthermore, the universe is also known to be expanding, with the most distant objects yet discovered going away from us at almost the speed of light (300,000 km

FIGURE 1.44 The sun is located approximately 30,000 light years from the center of the Milky Way galaxy.

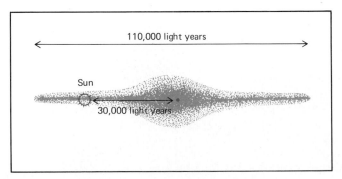

110,000 light years

Sun

30,000 light years

per second). In the second half of this book, we will explore these ideas in detail and develop methods by which they are known. In fact, we will begin our study of the tools and methods of astronomy in the next chapter.

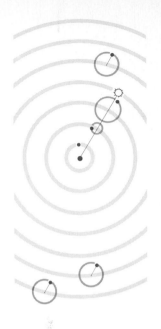

QUESTIONS

1. List several facts indicating that humans made astronomical observations before 1000 B.C.
2. Why would early humans have had an interest in observing cycles of nature?
3. What object in the sky appears to determine the length of the month?
4. Of what practical value were astronomical observations in Egypt?
5. Why did the Mesopotamian calendar of twelve 30-day months fail to record the length of a year accurately?
6. What are several similarities and differences between astronomy and astrology?
7. What evidence led Pythagoras to conclude that the earth was shaped like a ball?
8. While Philolaus' model of the planetary system did not represent reality, it did contribute an important true concept. What was that concept?
9. In what sense was Aristarchus' model many years ahead of its time?
10. List the principal points of Aristotle's picture of the sun, earth, planets, and the universe.
11. By what scheme did Hipparchus explain that the sun was not always the same distance from the earth?
12. How did Ptolemy explain the apparent retrograde (backward) motion of the planets?
13. How can the apparent retrograde motion of the planets be explained in terms of the Copernican system?
14. Why did the incorporation of the Aristotelian model into church dogma tend to delay the gaining of a more accurate picture of the solar system?
15. How was Tycho Brahe able to make a catalogue of accurate star positions when the telescope had not yet been invented?
16. What do you think was Brahe's motivation for studying the planets and stars?
17. What is the shape of the orbits of the planets?
18. Express Kepler's second law in common language.
19. Explain the significance of Galileo's discovery that Venus goes through phases like the moon.
20. What did Newton view as the most natural tendency for a body moving in empty space?
21. Design a calendar that would make the new moon fall on the first day of each month. Would your calendar make the shortest day of the year fall at the same date (give or take one day) each year? Why?
22. If at noon of the shortest day of the year you placed a stick in the ground so that it had no shadow, how could you use that stick to tell when a year had passed?
23. Explain in your own words why both the intensity of light and the intensity of a gravitational field "fall off with the square of the distance."

Methods
of Astronomy

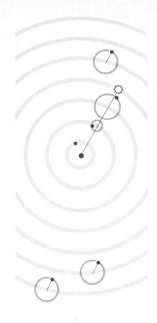

Two

The study of astronomy is different from the study of physics or chemistry or geology in that astronomers cannot get their hands on much of what they study. There are a few exceptions: meteorites come to us from outside the earth, perhaps from the outer reaches of the solar system and moonrocks have been brought back with space missions. In a sense we have touched Mars by using a mechanical probe to analyze soil samples there. For the most part, however, astronomers have to be content to study the universe through one class of phenomena called electromagnetic radiation. We immediately think of light as illustrating this kind of radiation— we see stars because they radiate light in all directions and because our eyes are sensitive to light. However, light is only one of several forms of electromagnetic radiation that reaches the earth from objects such as stars and galaxies and nebulae. Consider the sun for instance—it is a star and it sends us infrared radiation (we feel it as radiant heat that warms our bodies) and ultraviolet radiation (we receive a suntan or sunburn from this form of radiation). Our bodies are not prepared to detect the radio energy of the sun but special radio telescopes reveal this form of radiation. At times the sun also radiates X rays and gamma rays that would reach the surface of the earth in lethal doses were it not for their absorption by the earth's upper atmosphere.

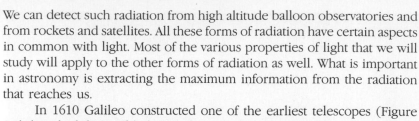

We can detect such radiation from high altitude balloon observatories and from rockets and satellites. All these forms of radiation have certain aspects in common with light. Most of the various properties of light that we will study will apply to the other forms of radiation as well. What is important in astronomy is extracting the maximum information from the radiation that reaches us.

In 1610 Galileo constructed one of the earliest telescopes (Figure 2.1), by which he could observe four moons of Jupiter. He recorded their periods of revolution as follows: Io, 1.769 days; Europa, 3.551 days; Ganymede, 7.155 days; and Callisto, 16.689 days. By 1675 the Danish astronomer Ole Roemer (Figure 2.2) became interested in the possibility of using Jupiter's moons as a natural clock. Jupiter's moons could be visualized as the hands of a clock (a four-handed clock) and from their position at any time, one might learn to tell time. No clock was yet available that would keep time accurately on a pitching, rolling ship. Time is an essential factor in navigation and if a seafaring navigator could observe the moons of Jupiter for a significant part of the year using only a modest telescope, he could tell the time. From his land-based observatory (Figure 2.3), Roemer began to refine Galileo's observations, starting with the period of Io. He recorded successive times when Io was eclipsed as it revolved behind Jupiter (Figure 2.4). From his position (on the earth) near E_0, Roemer observed several revolutions of Io and averaged his results. Now if Io's revolution was regular (always in the same period), as Roemer hoped, he

FIGURE 2.1 Galileo's telescopes. (Yerkes Observatory)

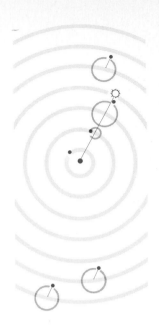

FIGURE 2.2 Ole Roemer, discoverer of the speed of light. (Yerkes Observatory)

could predict eclipses of Io into the future, say when the earth had moved to E_5. But he was disappointed to find that, compared to his predicted time, Io eclipsed almost 1000 seconds late. Then he realized that his position at E_5 was much farther from Jupiter than when at E_0; he concluded that the accumulated delay of 1000 seconds resulted because the light that signaled the eclipse took longer to travel the extra distance. If Roemer had known that the major axis of the earth's orbit was 186,000,000 miles, he could have computed the speed of light as follows:

$$\frac{186,000,000 \text{ miles}}{1000 \text{ sec}} = 186,000 \text{ miles/sec}$$

Although Roemer had apparently failed to find a regular clock in the moons of Jupiter, he had provided the basis for finding the speed of light when the distance across the earth's orbit became known.

We usually think in terms of miles per hour, and therefore we may not at first be impressed with how fast light travels—it is measured in miles per second. Perhaps an imaginary trip will help. Suppose that you could climb into a jet which had unlimited speed capabilities. Flying over the equator of the earth, you attained a speed of 186,000 miles/sec. At that speed it would be possible to make more than seven complete trips around the earth in one second. If you can comprehend such a speed, it will help you relate to the size of the universe in later sections of the book, because the unit of length used to measure the distances to stars and galaxies depends directly on the speed of light. If light travels 186,000

FIGURE 2.3 Roemer's telescope. (Yerkes Observatory)

FIGURE 2.4 Roemer's determination of the velocity of light.

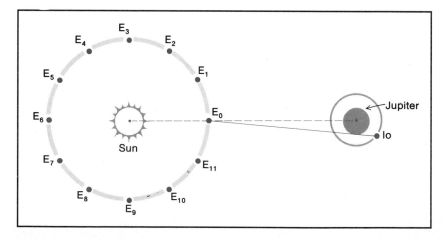

miles in a second, and there are 60 seconds in a minute, then light travels 186,000 × 60 miles in a minute. Since there are 60 minutes in an hour, light travels 186,000 × 60 × 60 miles in an hour, 186,000 × 60 × 60 × 24 miles in a day, and 186,000 × 60 × 60 × 24 × 365.25 miles in a year. By multiplying this latter set of numbers, we have the distance light travels in a year—a *light-year*. Perform this arithmetic yourself and your answer will be just under 6 trillion (6,000,000,000,000) miles/year. The closest star, Alpha Centauri, is 4.3 light-years away; thus its distance from the earth is 4.3 × 6,000,000,000,000 miles, or roughly 26 trillion miles.

In 1924 Albert A. Michelson, an American physicist, developed an experiment by which he could more accurately determine the velocity of light. In order to determine velocity, one must know two things: the distance traveled and the time required. He set up part of his apparatus on Mt. Wilson in California and part on Mt. San Antonio, approximately 22 miles away. It is not an easy task to measure the distance from one mountain peak to another; however, by carefully using standard surveying techniques, a US Geodetic Survey team was able to calculate the distance between the two parts of the experiment with an error of less than one inch. The device whereby Michelson measured the time for light to travel from one mountain top to the other and return consisted of a rotating mirror set on a motor which would turn it at any desired speed. A bright source of light was directed onto mirror face A, one of eight mirrors on the eight-sided wheel (Figure 2.5). Mirror face A reflected the light to a flat mirror on Mt. San Antonio. The flat mirror returned the light to mirror face C, when the wheel was standing still, and into the viewing tube. Once the wheel was rotating, the speed was adjusted so that the flash of light sent by face A would be caught by face B, which had turned to the position that face C had occupied by the time the flash traveled to Mt. San Antonio and back—a distance of 43.978 miles. If the wheel turned at the rate of 529.37 rps (revolutions per second), the computation would look like this:

$$529.37 \ \frac{\text{revolutions}}{\text{sec}} \times 8 \frac{\text{flashes}}{\text{revolution}} \times 43.978 \ \frac{\text{miles}}{\text{flash}} = 186,250 \ \frac{\text{miles}}{\text{sec}}$$

FIGURE 2.5 Michelson's rotating mirror experiment.

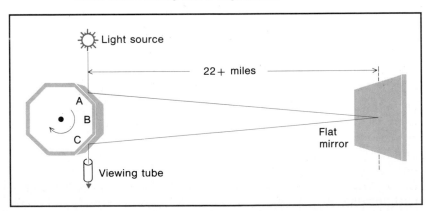

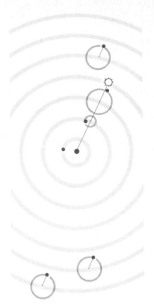

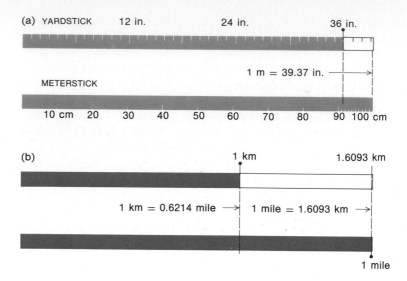

FIGURE 2.6 Relationships between units of the International System (Système International-SI) and the English system are shown in graphic form.

Expressed in metric units, (Figure 2.6) this velocity is 299,729 km/sec, or almost 300,000 km/sec. Of course this experiment was not performed in a vacuum, and so the result represents the velocity of light in air, near the surface of the earth. Modern determination of the speed of light in a vacuum produces a velocity only slightly faster: 299,793 km/sec (186,288 mi/sec). This is the accepted value for the velocity of light in empty space. Light travels considerably slower than that in media such as water or glass.

THE WAVE NATURE OF LIGHT

We are all familiar with the expanding wave pattern that is set up when a rock is dropped into a quiet pond. Energy from the falling rock is transferred to the water to deform its surface and this energy radiates outward in all directions as a wave. Molecules of the water rise and fall periodically in a very definite and predictable manner; thus some energy that the rock possessed could now reach a toy boat near the edge of the pond, causing it to rise and fall. This illustrates how energy may be transferred from one point to another by means of waves. Although light may be created by oscillation (up-and-down motion) of a charged particle, it is not transferred from one place to another by causing particles to move up and down. We know this is true because light travels through empty space, where no particles exist. When we speak of the wave nature of light, we are speaking of a sequence of electrical and magnetic changes, the graphical representation of which appears as a wave form.

Perhaps you have experienced the phenomenon of becoming electrically charged by scuffing your feet on a nylon rug. Picture a small object that is charged and standing still. The space around the charged particle (its "region of influence") can be thought of as a "field." If another charged (test) particle is brought into the field, the test particle will experience a constant force. If the test charge is like that of the fixed charge, the test particle will experience a force that repels. If the test particle is charged in a manner opposite to the fixed charge, it will experience a force of attraction. If the first particle is set into oscillation, then the test particle will experience a changing field and will respond by oscillating also.

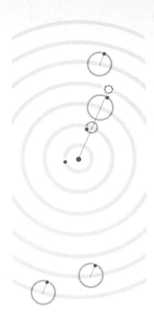

Thus some of the energy of one oscillating particle can be transferred to another particle without there being any material between the two. This is a partial model of how light travels through empty space. To complete the model we should recognize that whenever a changing electrical field occurs, a changing magnetic field accompanies it. The two always go hand in hand. To demonstrate a changing magnetic field, hold a bar magnet in one hand and set a compass nearby. Now rotate the magnet back and forth. You are creating a changing magnetic field, which is evidenced by the corresponding motion of the compass needle (Figure 2.7). Again, some of the energy of the oscillating magnet will be transferred to the compass even if no material exists between the two. Now we have a more complete picture of light; we envision it as an electromagnetic disturbance, produced by an oscillating charge that creates a changing electrical and a changing magnetic field simultaneously. This idea is represented graphically in Figure 2.8. Because the electric component of the wave is responsible for all optical effects, and because the magnetic component will always accompany a changing electrical field, we will henceforth speak only of the electrical component.

How does the boat know that a rock was dropped into the pond? By the energy it received to produce its rising and falling motion. How do we know a star exists? By the energy we receive from it to interact with our eye. This energy is transferred from the star to us by electromagnetic waves.

Light is only a small portion of the electromagnetic spectrum, which

FIGURE 2.7 A changing magnetic field.

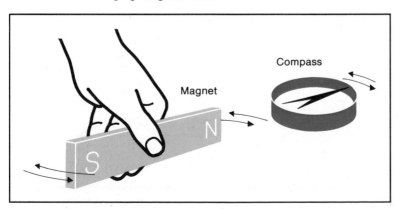

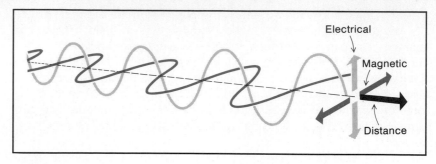

FIGURE 2.8 A graphic representation of electrical and magnetic disturbances.

includes radio, radiant heat (infrared), ultraviolet radiation, X rays, and gamma rays. What makes these forms of radiation different from one another? It is their wavelength. We may picture one *wavelength* in a changing electrical field as the distance the disturbance travels as it goes through one complete oscillation—say, from the crest of one wave to the crest of the next (Figure 2.9). Remember that all forms of electromagnetic disturbances travel through empty space at approximately 300,000 km/sec; accordingly, if a radio wave travels 1000 meters while going through one cycle, we speak of its wavelength as 1000 m. It should be apparent at this point, that if a wave travels 1000 m in one cycle and it travels 300,000,000 m in one second, then it must go through 300,000,000/1000 = 300,000 cycles/sec. This is the *frequency* of the wave disturbance. While we will continue to emphasize wavelength, it will always be possible for you to obtain the frequency of the wave by dividing the velocity of light by the wavelength (using the same units, of course):

$$\text{frequency} = \frac{\text{velocity}}{\text{wavelength}}$$

The wavelengths of radio signals which we receive on AM radio fall within the range of 200 to 500 m, whereas the amateur radio operator typically uses wavelengths in the range of 2 to 160 m. Radio astronomers are particularly interested in radio waves between 0.001 m (1 millimeter) and 1 m. The wavelengths of these and other forms of electromagnetic radiation are shown in Figure 2.10. Note that the expression 10^{-2} m

FIGURE 2.9 The waveform.

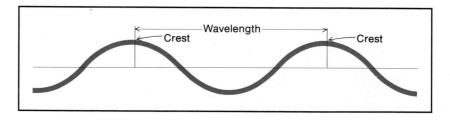

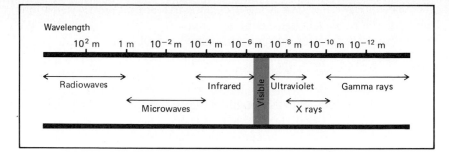

FIGURE 2.10   The electromagnetic spectrum.

means 0.01 m, 10^{-3} m means 0.001 m, and so on. (Appendix 3 provides a full explanation of powers of 10.) As you can see, the wavelength of visible light is very short, between 0.0000004 and 0.0000007 m, with ultraviolet, X rays, and gamma rays being characterized by even shorter wavelengths. Rather than measuring such small lengths in meters, we use a much smaller unit called the *angstrom* denoted by the symbol Å:

$$1 \text{ Å} = 10^{-10} \text{ m} = 0.0000000001 \text{ m}$$

The range of visible light, measured in angstroms, is thus 4000 to 7000 Å. For slightly longer infrared wavelengths, a unit called a micron is used:

$$1 \text{ micron} = 10^{-6} \text{ m} = 0.000001 \text{ m}$$

POLARIZATION OF LIGHT

In Figure 2.8, we showed the changing electrical component of an electromagnetic wave as oriented vertically—varying up and down. However,

FIGURE 2.11   An analogy for the polarization of light: (a) both vertical gates; (b) vertical and horizontal gates.

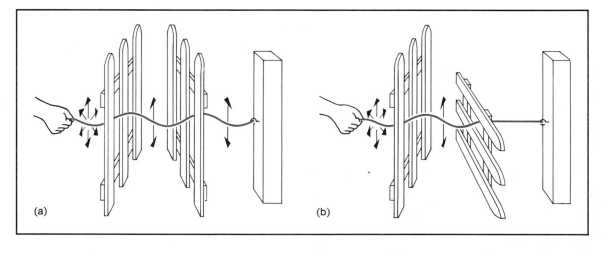

(a)   (b)

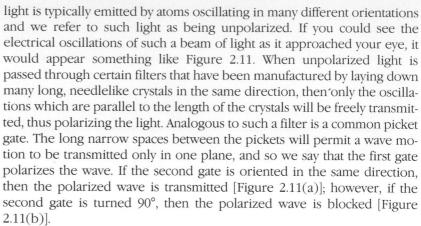

light is typically emitted by atoms oscillating in many different orientations and we refer to such light as being unpolarized. If you could see the electrical oscillations of such a beam of light as it approached your eye, it would appear something like Figure 2.11. When unpolarized light is passed through certain filters that have been manufactured by laying down many long, needlelike crystals in the same direction, then only the oscillations which are parallel to the length of the crystals will be freely transmitted, thus polarizing the light. Analogous to such a filter is a common picket gate. The long narrow spaces between the pickets will permit a wave motion to be transmitted only in one plane, and so we say that the first gate polarizes the wave. If the second gate is oriented in the same direction, then the polarized wave is transmitted [Figure 2.11(a)]; however, if the second gate is turned 90°, then the polarized wave is blocked [Figure 2.11(b)].

When sunlight strikes a flat surface, such as a lake or the hood of a car, the reflected rays tend to be horizontally polarized because the surface absorbs oscillations that are perpendicular to it but reflects oscillations parallel to it. Such reflected light is called *glare*, and beacuse it is horizontally polarized, it may be prevented from reaching the eye by use of polarized sunglasses which have vertical "gates" (Figure 2.12).

The astronomer is particularly interested in the polarization of starlight because it provides clues to the material through which the light has passed in reaching the observer. For instance, dust is thought to polarize starlight and the quantity of dust through which starlight has passed may be indicated by the degree to which the light reaching the earth has been polarized. The subject of polarization will be more fully discussed in Chapter 12.

FIGURE 2.12 Sunlight which is reflected from the surface of the water is partially polarized in a horizontal orientation, and the vertical "gates" of the polarized sunglasses prevent its entry into the eyes.

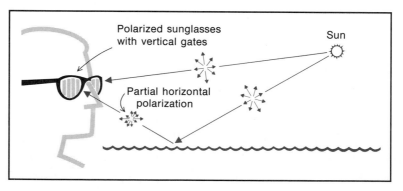

DOES LIGHT TRAVEL IN A STRAIGHT LINE?

Newton articulated the idea that a moving object tends to move in a straight line unless influenced by an outside force and so we might guess that electromagnetic waves also tend to move in a straight line, at least in empty space far from any object. Later we will demonstrate that electromagnetic waves also have a particlelike nature; this would seem to strengthen the above line of reasoning. But we do not live in empty space far from other objects, so light may be expected to bend for several reasons. It may be reflected from a mirror, refracted as it moves from one media into another (from air into water, for example), or bent by gravitational forces, say when it passes near a very massive object. Let us first consider reflection.

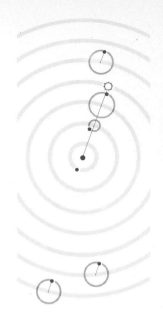

REFLECTION OF LIGHT

In order to describe the reflection phenomenon more accurately, we must first understand what is meant by the expression "a normal to the surface." The *normal* is a line perpendicular to a surface at a given point. A vertical flagpole erected on level ground is an example of a normal to a surface. The angle between the incident (incoming) ray of light and the normal is called the *angle of incidence*, and the angle between the reflected ray of light and the normal is called the *angle of reflection*. The law of reflection may be stated as follows:

> The angle of reflection is equal to the angle of incidence.

This law allows us to predict where a reflected ray will go, and thus it forms the basis for the design of a reflecting telescope. This law is true for every point on the surface of a mirror, even if the mirror is curved (Figure 2.13).

FIGURE 2.13 The angle of reflection always equals the angle of incidence (a), even for a curved mirror (b).

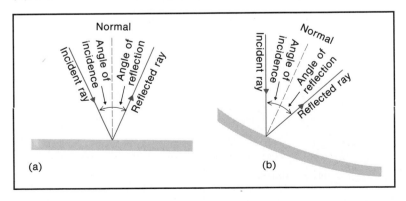

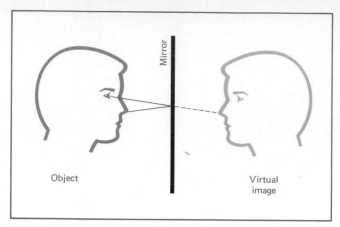

FIGURE 2.14 As you look at yourself in a mirror, you perceive the tip of your nose as being located behind the mirror, as illustrated, because the ray of light from your nose seems to originate there, however, this is merely a virtual image and actual rays do not travel to that point. A photograph could not be taken at the point indicated behind the mirror.

Let us apply the law of reflection to the simple process of looking at yourself in a mirror. In Figure 2.14, we have drawn four rays from the tip of your nose and we have shown their reflected rays diverging (spreading outward) as if emanating from the image of your nose behind the mirror. Your image seems to be as far on the other side of the mirror as you are on the front side. No light actually passes through the mirror, hence no photograph could be taken at the location of the image—we call such an image a *virtual* image. If you could trace the light paths of every point on the object (you), a composite of those points would "build up" the total image.

REFRACTION OF LIGHT

Another way in which light is influenced by its environment is by *refraction*. We have shown that light travels at 299,793 km/sec (186,288 miles/sec) in a vacuum. It travels at about 224,000 km/sec (140,000 miles/sec) in water, and at still different velocities in glass or other transparent materi-

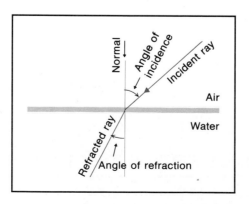

FIGURE 2.15 Refraction of light.

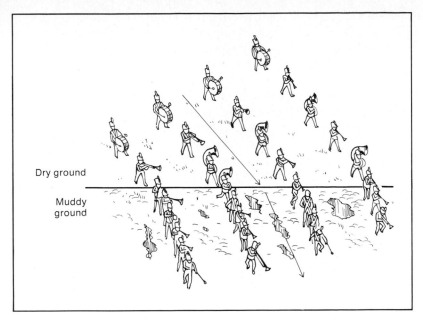

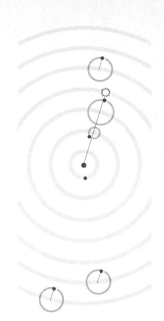

FIGURE 2.16 As band members march from dry ground into muddy ground, the mud sticks to their feet and slows their pace. Because the right-hand members enter the muddy ground first, their row is gradually diverted to move in a new direction. Light behaves in a similar manner.

als. The refraction (bending) of light results from this fact—that light travels at different velocities in different materials. For light of a given wavelength, the ratio of its velocity in a vacuum to its velocity in a given medium is called the *index of refraction* for that medium. The angle the refracted ray makes with the normal is called the *angle of refraction* (see Figure 2.15). As the light passes from air into water, the angle of refraction is

FIGURE 2.17 Refraction by the earth's atmosphere.

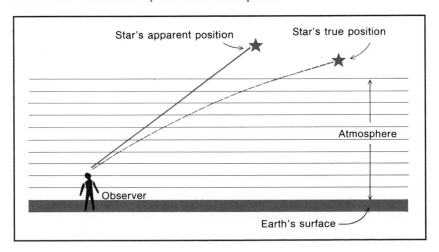

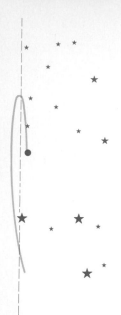

smaller than the angle of incidence; the reason is that the velocity of light is less in water than in air. The principle of refraction is clearly illustrated by a marching band practicing on dry ground but then marching into muddy ground and experiencing the retarding influence of the mud on the movement of their feet (see Figure 2.16). You can see how the direction of the band was changed because the band members on the right-hand side of the page enter the mud before those on the left. Perhaps you have noticed that a spoon, when placed in a glass of water and viewed from an angle, appears to be bent. This illusion is caused by refraction. A less obvious result of refraction occurs in the atmosphere of the earth. As light from a star enters the atmosphere, it passes through "layers" of increasing density and is slowed by increasing amounts. Thus the light is refracted. Because the observer judges the star's position by the direction in which the light enters his eye, he sees the star as if it were higher in the sky than its true position (Figure 2.17).

The effects of atmospheric refraction are quite obvious when the sun is setting over the ocean. As it approaches the horizon, its bottom part is refracted more than its top part and so the sun appears to be flattened (Figure 2.18).

FIGURE 2.18 Photographs of the setting sun, showing effects of atmospheric refraction. (Lick Observatory)

DISPERSION BY REFRACTION

A phenomenon that occurs simultaneously with refraction is *dispersion*. White light is composed of many wavelengths and each wavelength represents a different color. Since the angle of refraction depends upon the wavelength of the light as well as the medium, each color is bent at a slightly different angle when refracted by a transparent medium. Among visible colors, violet light has the shortest wavelength and is bent more than red light. By use of the prism, white light may be separated into the full range of color, an array called the *spectrum* [Figure 2.19(a); see also Color Plate 1]. Reflection, refraction, and dispersion take place in raindrops, producing a rainbow [Figure 2.19(b)].

As we shall see, the phenomenon of dispersion creates a problem in the simple refracting telescope or camera, but it also serves as the basis for one of the most useful tools in astronomy, the spectrograph.

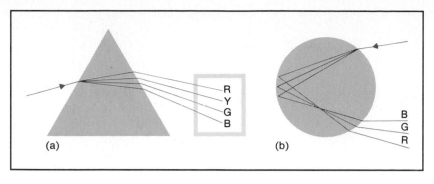

FIGURE 2.19 Dispersion of white light by (a) a prism and (b) a raindrop.

DIFFRACTION OF LIGHT

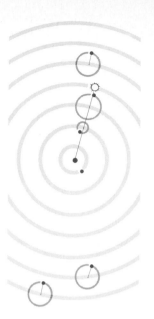

As waves pass the edge of an object or through a small opening, they spread out in all directions, as though the edge were a new source of waves. This phenomenon, called *diffraction*, is illustrated by the water waves shown in Figure 2.20 (a). Diffraction may be useful in certain applications but is somewhat detrimental in astronomical instruments, for as light enters the telescope it must pass numerous edges. A telescope generally has a round opening at the front and braces to support the secondary

FIGURE 2.20 (a) Diffraction in water waves. (b) Interference in water waves. Destructive interference occurs along line AB; constructive interference, along line CD. (Educational Development Center)

(a)   (b)

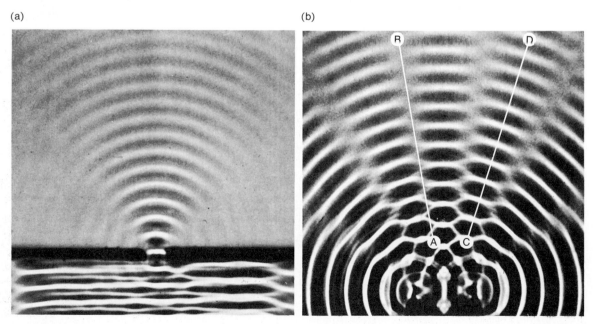

FIGURE 2.21 Diffraction spikes are produced by the structure which supports the secondary mirror of the telescope. (Mount Wilson and Las Campanas Observatories, Carnegie Institute of Washington)

mirror (see Figure 2.37, page 75). Each of these parts acts as a new source of light. Figure 2.21 reveals the pattern of diffraction spikes (cross shape) seen on bright star images. Diffraction prevents any telescope from forming perfect images of stars, since only about one-half of the light falls in the central spot.

INTERFERENCE OF LIGHT

When two or more small openings or obstacles act as new sources of waves, a pattern of interference is seen, as in Figure 2.20(b). Along the line AB no wave motion is evident, which indicates that the waves from the two sources have cancelled each other. This is called *destructive interference*. We visualize this as occurring when the *crest* (high point) of one wave meets the *trough* (low point) of the other, as in Figure 2.22.

On the other hand, along line CD in Figure 2.20(b) the wave action is distinct. This is a consequence of *constructive interference*; the two waves

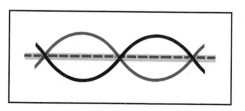

FIGURE 2.22 Crest of one wave meets trough of another wave, resulting in destructive interference.

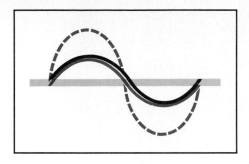

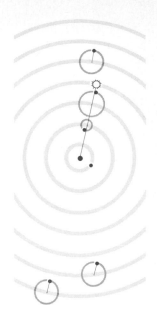

FIGURE 2.23 Crest of one wave meets crest of another wave, resulting in constructive interference.

interact with crest meeting crest and trough meeting trough, as shown in Figure 2.23.

Both diffraction and interference are easily seen in water waves. What evidence do we have that similar phenomena occur in light? Suppose that many fine lines are ruled (scratched) on a sheet of glass, in effect providing many edges. Such a device is called a *diffraction grating*. When the light from a single source passes these edges, they act as many new sources. Consider the light that passes two adjacent edges, lines A and B, in Figure 2.24. The wave nature of light tells us that at certain angles the light waves from adjacent lines will interfere with each other destructively, thereby producing dark bands, and that at other angles the light waves will interfere constructively, producing bright bands on the screen. The spacing between the bright and dark regions will depend on the spacing of the lines and the wavelength of the light (Figure 2.25). Furthermore, could these bright bands be seen in color, the full spectrum would be evident in each band, because different colors (different wavelengths) interfere constructively at slightly different places on the screen. Thus the diffraction grating becomes an important device whereby the light of a given source may be dispersed into its spectrum. Typically, the diffraction grating disperses light more than a prism.

FIGURE 2.24 The diffraction grating.

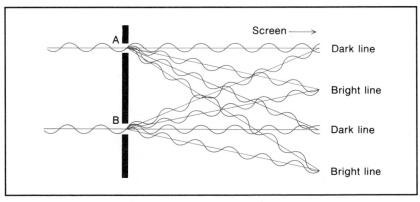

FIGURE 2.25 The interference of light. The light bands represent constructive interference; the dark bands destructive interference. (Don Jenkins)

THE DUAL NATURE OF LIGHT

We have noted several experiments demonstrating that light behaves like waves. Some experiments, however, demand another kind of explanation. Such an experiment may be performed using an electroscope with a clean zinc plate attached to its top. An electroscope is a device that is capable of detecting the presence of charged particles. When there is no charge on the zinc plate, the foil leaves hang downward together; but if the zinc plate becomes charged, that charge is conducted to the foil leaves and they experience a like charge, causing them to be repelled from each other.

Suppose that a negative charge (a surplus of electrons) is placed on the plate, causing the foil leaves to separate, and then ultraviolet light is allowed to fall onto the plate. The leaves will be seen to fall back together, indicating that electrons have been knocked free of the plate, leaving it neutralized. To explain this phenomenon, called the *photoelectric effect*

FIGURE 2.26 The photoelectric effect.

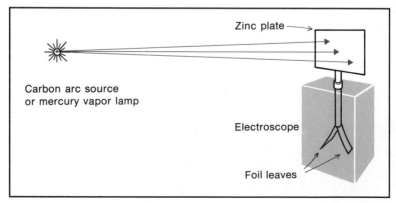

Zinc plate

Carbon arc source
or mercury vapor lamp

Electroscope

Foil leaves

(Figure 2.26), imagine that light behaves like particles, called *photons*, which are capable of knocking electrons free. Unlike ordinary particles, however, photons cannot be placed on a scale and weighed—that is to say, they have zero rest mass. Their masslike quality derives from the fact that photons are in motion. Their energy is inversely proportional to their wavelength: thus ultraviolet light, characterized by short wavelengths, carries more energy per photon than blue light. Blue light of any intensity can be directed onto the plate of the electroscope and no electrons will be freed. When ultraviolet light falls onto the plate, however, electrons are ejected instantaneously. Evidently, the ultraviolet photon carries enough energy to free such electrons, whereas no collection of blue light photons would accomplish the task. If light behaved only like a wave (with energy distributed uniformly over a wave front), and not as particles, then the photoelectric effect could not be explained.

Could it be possible that all matter exhibits a dual nature, that of a wave and that of a particle? The following experiment suggests that it does. When electrons (particles) are shot through a very small opening of a crystal lattice, a diffraction pattern is formed [Figure 2.27(a)] that closely resembles the diffraction pattern created when X rays (usually thought of as a wave phenomenon) are passed through the small openings within the structure of aluminum foil [Figure 2.27(b)].

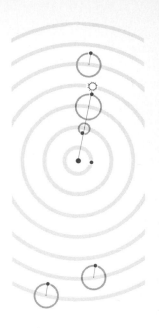

FIGURE 2.27 (a) An electron diffraction pattern created by passing a beam of electrons through a crystal lattice, in this case beryllium. (b) The diffraction pattern created by X rays directed through polycrystalline aluminum. (a, RCA Laboratories, Princeton, New Jersey: b, courtesy of Mrs. M. H. Read, Bell Telephone Laboratories, Murray Hill, New Jersey)

(a)   (b)

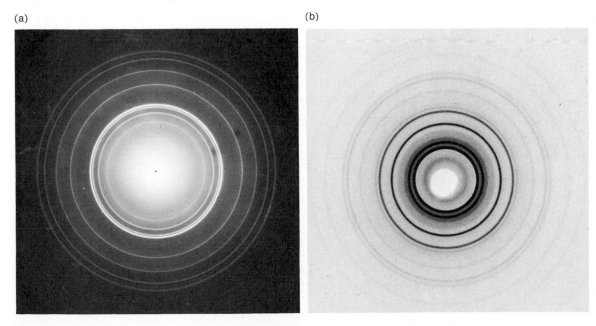

TELESCOPE DESIGN

When Galileo viewed the sky with his telescope, he realized that he could see stars which were too faint to be seen with the naked eye. This fact illustrates the real purpose for assembling either lenses or mirrors to make a telescope—that is, to gather light over a large surface and concentrate its energy into a small area to produce an image that is brighter than the object appears to the naked eye. The effects of refraction were known before the time of Galileo, but mere refraction, say into water or a flat piece of glass, would not concentrate the light energy at one point. Galileo demonstrated that when he gave a piece of glass the proper curvature, light rays from a distant star that passed through the glass near its edge would be bent (refracted) more than those which passed near its center, causing the rays of light from a given source to bend toward a focal point (Figure 2.28). The distance between the lens and the focal point is called the *focal length* of the lens.

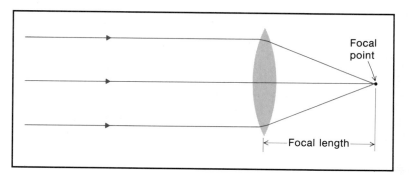

FIGURE 2.28   A simple convex lens.

FIGURE 2.29   Formation of the extended image.

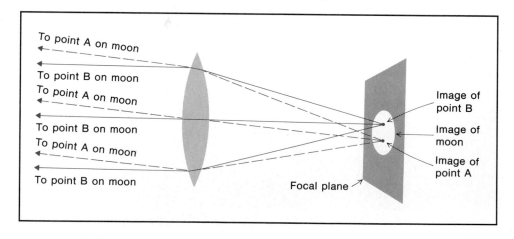

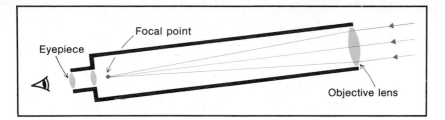

FIGURE 2.30 The refracting telescope.

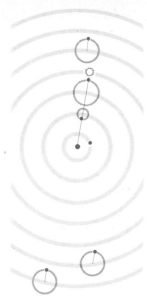

Were we to view an object like the moon, it would present many points from which light travels to our telescope. Each point on the object is brought to a focus at a different point, and the collection of all such focal points forms the image of the moon on what is called the *focal plane* of the telescope (Figure 2.29).

FIGURE 2.31 The 1-m (40-in.) refractor at Yerkes Observatory in Wisconsin. (Yerkes Observatory)

Today a photographic plate can be placed at the focal plane and a picture of the moon may be made in which every point is said to be in focus. Using a photographic plate, images can be recorded of objects too dim to be seen with the eye, even through a telescope. Light from a dim object is allowed to fall on the photographic plate over a period of several hours and an image is "built up" on that plate in a cumulative way. By contrast, the light energy that enters your eye is dissipated as fast as it is received and has no cumulative effect. Thus the astronomer can see objects on the photographic plate that could not be seen otherwise and the plate represents a permanent record of nighttime observations.

If one desires to use a telescope for visual observation, it is necessary to place an eyepiece just behind the focal plane (Figure 2.30). The eyepiece serves as a magnifying glass, making the image appear larger, say in the case of the moon. For a given focal length, the larger the objective (first) lens is made, the more light will be collected and the brighter the image will be. In an attempt to see fainter objects, opticians have constructed several large objective lenses, the largest of which is approximately 1 m (40 in.) in diameter located at Yerkes Observatory in Wisconsin (Figure 2.31).

(a) Chromatic aberration in a lens, (b) corrected by the addition of a second lens.

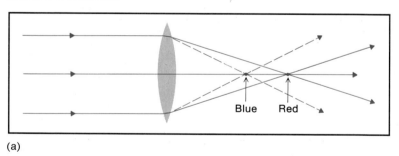

(a)

(b)

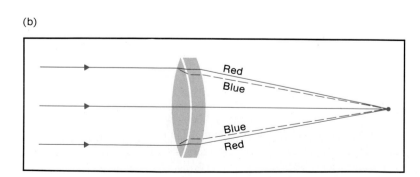

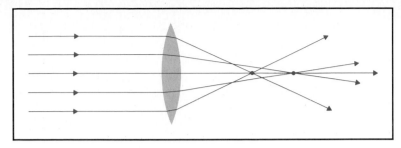

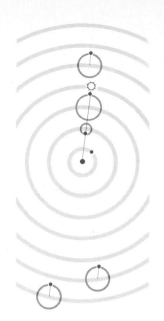

FIGURE 2.33 Spherical aberration in a lens.

LIMITATIONS IN REFRACTORS

Refractor telescopes are limited in size because large objective lenses are difficult to support without sagging, which changes their shape. The glass from which the objective is ground must be of high quality, free of bubbles and other imperfections. Still further limitations are imposed by the design of the lens itself.

CHROMATIC ABERRATION.   Since white light is composed of all colors and blue light is refracted more than red light, blue light is brought to a focus ahead of red light. Similarly, all colors between red and blue are brought to slightly different focal points by a single lens, resulting in the formation of a blurred image. This produces a color defect called *chromatic aberration* [Figure 2.32(a)]. This defect may be partially corrected by the addition of a second lens, which because of its shape and a slightly different index of refraction, will diverge the rays and bring all the colors back to the same focal point, producing a sharp image. A compound lens of this type is called an *achromatic lens* [Figure 2.32(b)]. An inexpensive camera, for example, generally has only a single element lens, and so will suffer from chromatic aberration. A more expensive, multiple-element lens will produce a sharper image; its achromatic nature is usually stated in writing on the frame of the lens itself.

SPHERICAL ABERRATION.   While it is much easier to grind a lens of spherical shape, a simple lens of this design suffers a defect called *spherical aberration* [(Figure 2.33)]. The rays of light from a distant object are not brought to the same focal point. Those rays passing through the lens near its outer edge are brought to a closer focus than those passing nearer the center. This defect may also be corrected by additional elements in the lens system.

THE REFLECTING TELESCOPE

Sir Isaac Newton was the first to recognize fully that whenever light is refracted, it is also dispersed, causing the phenomenon of chromatic aberration in refractors. He also recognized that light is not dispersed when it

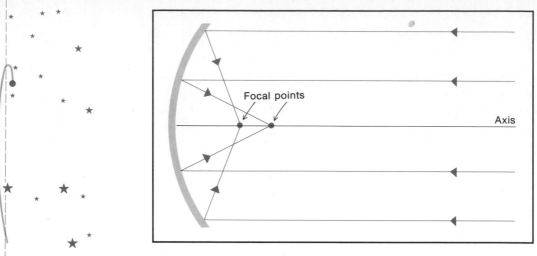

FIGURE 2.34 Spherical aberration in a mirror.

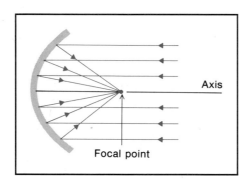

FIGURE 2.35 The parabolic reflector.

FIGURE 2.36 Sections of a cone: (a) circle; (b) ellipse; (c) parabola; (d) hyperbola.

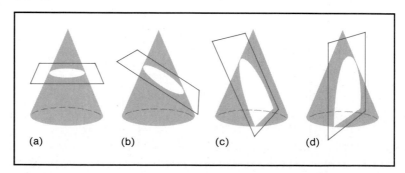

is reflected, hence that a telescope depending only on reflection will not be plagued by chromatic aberration. Obviously a flat mirror would not concentrate the light. Newton may have tried a spherical mirror but would have found that it possessed spherical aberration, because the rays of light (from a star) which struck the outside portions of the mirror did not reflect to the same focal point as those that struck the mirror toward its center [Figure 2.34]. Newton deduced that if he were to give the mirror a parabolic shape by hollowing out its center (Figure 2.35), he could cause all rays from a given star to converge on a point, the focal point. As you can see in Figure 2.36(c), the parabola is one of the distinctive curves which may be obtained by cutting a cone parallel to one edge. Other curves which are useful in our discussion of astronomy may also be obtained by cutting the cone in various ways, as illustrated in Figure 2.36(a,b,c,d). Planetary orbits follow ellipses, for example.

There are a few basic designs used in reflecting telescopes, differing primarily in the placement of the eyepiece (Figure 2.37). The *prime focus* is used only on larger instruments like the 5-m (200-in.) Hale telescope at Mt. Palomar Observatory, since the placement of the observer's head at that position on a small instrument would obscure all incoming rays of light. The *Newtonian focus* was the original design used by Sir Isaac Newton and is still one of the most popular among amateur astronomers who build their own. The *Cassegrain focus* allows the light to make one additional trip through the tube. This fact, together with the curvature of the secondary mirror, allows the design of a long effective focal length in a short tube, thus increasing the portability of the instrument. The *Coudé focus* utilizes a third mirror which directs the light down the polar axis of

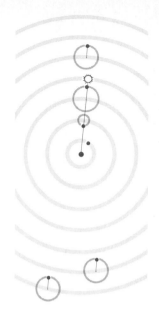

FIGURE 2.37 Four reflecting telescopes, utilizing different focal points: (a) prime focus; (b) Newtonian focus; (c) Cassegrain focus; (d) Coudé focus.

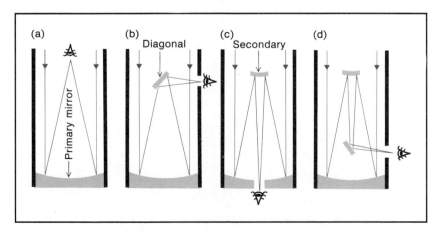

the telescope into a fixed eyepiece or spectrograph. Because this focal point does not move, even when the telescope is driven to compensate for the earth's rotation, the heavy equipment that characterizes spectroscopic studies can be place in the Coudé focus room. Most modern telescopes, like the 4-m Mayall at Kitt Peak (Figure 2.38), utilize the prime focus, the Cassegrain focus, or the Coudé focus in the same instrument by moving the proper secondary mirror or mirrors into place. The focal length of the Palomar 5-meter is 16.8 m (55 ft); thus the prime focus position is 16.8 meters from the mirror. At this position the astronomer moves with the telescope while making long exposures on the photographic plate. An astronomer may also ride at the Cassegrain focus while the telescope is in motion (Figure 2.39). Figure 2.40 shows the largest single-mirror telescope in the world, the 6-m reflector operated by the Soviet Union.

FIGURE 2.38 The 4-m Nicholas U. Mayall telescope at Kitt Peak National Observatory in Tucson, Arizona. (Kitt Peak National Observatory)

(a)

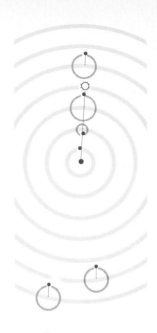

(b)

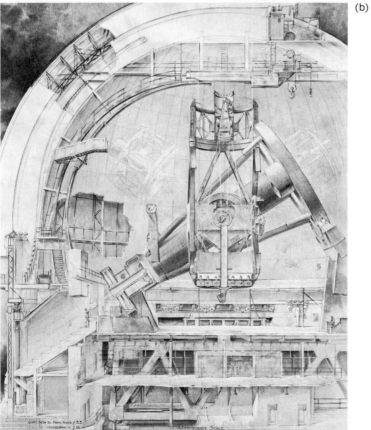

FIGURE 2.39 (a) Looking down the tube of the 200-in. (5.1-m) Palomar telescope, past the observer at the prime focus position, to the mirror at the far end. (b) A drawing of the Palomar telescope. (Palomar Observatory Photograph)

FIGURE 2.40 The largest telescope in the world, a 6-meter reflector, operated by the USSR Academy of Sciences near Zelenchukskaya in the southern part of Russia. (TASS Fotokhronika)

MULTIPLE-MIRROR TELESCOPES

Astronomers have been motivated to build larger and larger reflecting telescopes for two reasons. First, when more light is collected, dimmer (and hence more distant) objects may be seen. Second, larger mirrors produce sharper images, as you will see in forthcoming sections. On the other hand, the cost of a finished mirror is proportional to its diameter cubed. That is, if a 25.4-cm (10-in.) mirror roughly costs $200, then a 76.2-cm (30-in.) mirror would cost 3^3 or 27 times as much ($5400). This relationship will demonstrate the fiscal motivation to assemble six 1.8-m (72-in.) mirrors in a single structure, as was done at Mt. Hopkins near Tucson, Arizona (see Figure 2.41). The effective light-gathering power of this instrument is equal to that of a 4.5-m (176-in.) telescope, and it was achieved at far less cost than a single 4.5-m mirror. Obviously, problems were faced that were not encountered in a conventional design, such as making seven mirrors of identical focal length (the seventh is a guide telescope—see the center opening in Figure 2.41) and directing each mirror precisely to the same focus point. These difficulties have been overcome and the telescope is performing well.

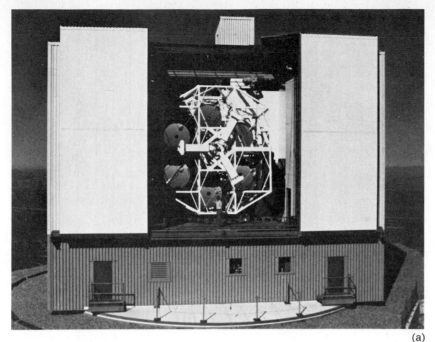

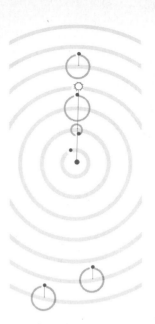

(a)

(b)

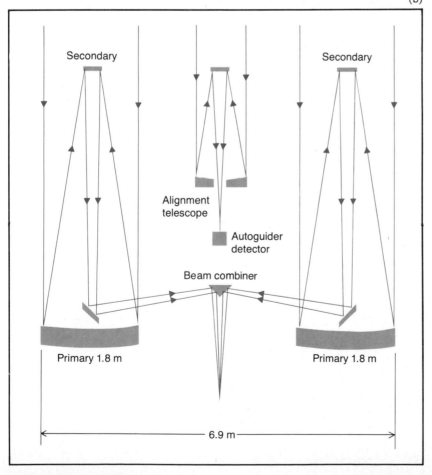

Secondary

Secondary

Alignment
telescope

Autoguider
detector

Beam combiner

Primary 1.8 m

Primary 1.8 m

6.9 m

FIGURE 2.41 (a) The Multiple-Mirror telescope on Mt. Hopkins in Arizona. The object of this design is to provide the light-gathering power and the resolution of a very large telescope in a smaller package and at less cost. (b) This drawing shows how the light of two of the six telescopes is combined to form a single image. The central scope is used to guide the entire array. (Smithsonian Astrophysical Observatory and Steward Observatory, University of Arizona)

THE SCHMIDT CAMERA

We speak of the Schmidt design as a camera because it is used exclusively for photographic work and has no provision for visual observation. All guiding is done through an auxiliary scope mounted on the exterior. The principles of refraction and reflection are both employed in the Schmidt camera. Light enters the system through a weak lens called a corrector plate, where it is refracted slightly, and then the rays are reflected by the spherical primary mirror to the focal point. A photographic plate that measures 35.6 by 35.6 cm (14 × 14 in.) is exposed at this point to record a wide-angle view of the sky. The 1.22-m (48-in.) Schmidt camera at Mt. Palomar records a square portion of the sky 6° on a side, covering 36 square degrees in a single exposure. The 5-m (200-in.) telescope covers only about one square degree per exposure. The Schmidt camera is very

FIGURE 2.42   The 48-in. (1.2-m) Schmidt telescope of the Palomar Observatory. (Palomar Observatory Photograph)

well suited to survey the entire visible sky, a program that requires seven years to complete, with over 1800 photographs (Figure 2.42).

SCHMIDT CASSEGRAIN TELESCOPE

One of the most popular amateur telescope designs is a variation on the Schmidt camera (see Figure 2.43). Its popularity stems partially from its compact design. For instance, a telescope with an effective focal length of 2 m [79 in.] can be built into a tube only 0.5 m [20 in.] long. The angle at which the rays of light converge from the secondary determines the effective focal length of this design. Note that four surfaces (either lens or mirrors) must be ground to a very close tolerance if this design is to perform properly. This design is not one to choose for your first "do-it-yourself" project in telescope making.

TELESCOPE PERFORMANCE

The factors ultimately affecting the size, brightness, and quality of an image formed in a telescope include the aperture (diameter of lens or mirror), the focal length, and the quality of the lens or mirror and the eyepiece. Of first importance is the quality of the lens or mirror, determined basically by its shape. We have already stressed that there is a correct shape for a lens or a mirror; the degree to which this shape is achieved determines the quality of the telescope. If the mirror is ground so that its surface is like that of a parabola with no variations larger than one-tenth of the wavelength of blue light [no more than 0.00000004 m, or 40 nanometers (nm)], then its quality is specified as being "tenth-wave." This is considered to be very adequate for an amateur telescope. Some mirrors have a superior "twentieth-wave" quality, with no deviations greater than one-twentieth the wavelength of blue light.

Before the factors which determine size, brightness, and sharpness of the image are discussed, the distinction must be made between two classes of objects viewed by astronomers: those which have an apparent size, such as the sun, moon, planets, clusters, nebulae, and galaxies, called *extended objects*, and those which have no apparent size, such as stars, called *point sources*. While the earth's atmosphere sometimes blurs images to produce an apparent size in stars, no amount of magnification will actually produce a real size in star images; stars are simply too far away to be seen as disks. Exceptions to this general rule, however, do exist; a few stars are both large enough and close enough to be resolved into a disk by speckle photometry. Such a star is Betelgeuse in the constellation of Orion (see Color Plate 8b).

SIZE. For extended objects at the same distance, the size of the image formed at the prime focus depends upon the focal length of the lens or mirror. The longer the focal length, the larger the image formed. If f is the focal length of the lens or mirror expressed in centimeters, then the size of

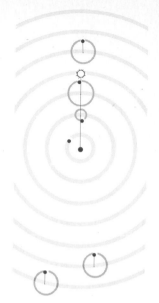

FIGURE 2.43 (a) The Schmidt Cassegrain design, providing a relatively long focal length in a very portable (short tube) telescope. (b) Optical diagram of the Celestron Schmidt Cassegrain lens system: The light enters the system through a thin aspheric corrector lens (1); it is then reflected by a large spherical primary mirror (2) toward the prime focus (3). The light from the primary is intercepted by the convex secondary mirror (4) and reflected back throgh a hole in the primary mirror to the Cassegrain focus (5). The effect of folding the optical path back with a convex secondary mirror also increases the effective focal length by a factor of three to seven times that of the primary alone. (Celestron International)

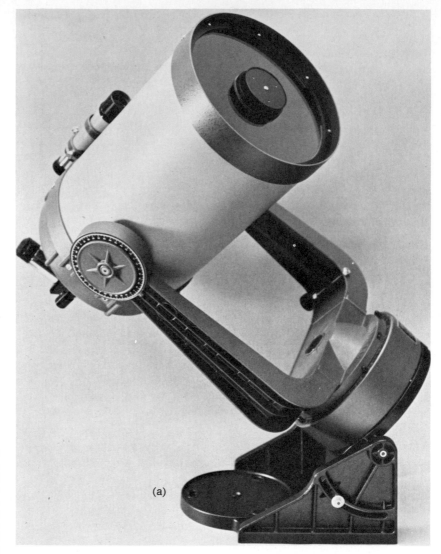

(a)

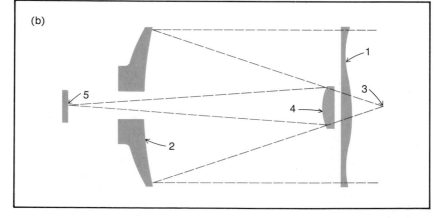

(b)

the image (s) expressed in centimeters that will be produced by an object 1° in diameter is given by the formula $s = 0.01744\,f$. The scale of a typical amateur telescope—for example, one of 152-cm (60-in.) focal length—would be:

$$s = 0.01744 \times 152\ \text{cm} = 2.67\ \text{cm/degree}$$

Using the full moon as an object that has a diameter of 0.5°, we would expect its image to be one-half the figure computed above—that is, 1.33 cm (0.5 in.) in diameter. This would be the size of the image that could be recorded on film without the aid of an eyepiece. A telescope of twice the focal length would produce an image twice as large on a piece of film, without the aid of an eyepiece. Thus the longer focal lengths are particularly useful in observing the planets. Planets are typically bright because of their nearness to us, so our emphasis is on making them look as large as possible in order to see detail.

BRIGHTNESS. In the case of a point source (a star), the *brightness* of the image depends only upon the aperture (diameter) of the mirror or lens. The larger the aperture, the brighter the image. For an extended source, the brightness of the image produced by various telescopes of equal focal length depends upon the aperture of the mirror or lens. The brightness of the image increases with the square of the aperture, meaning that if the diameter of the mirror or lens is doubled, the total light concentrated in the image will be multiplied by four. On the other hand, if the aperture of several telescopes is the same and they differ only in focal length, the one with the longer focal length will produce the dimmer image. This is an inverse relationship. The combined effect of aperture and focal length upon the brightness of the extended image is expressed by the formula

$$B = C\left(\frac{a}{f}\right)^2$$

where B represents brightness, a represents aperture, f represents the focal length of the telescope, and C is a constant. The ratio of focal length to aperture (f/a) is called the *focal ratio* or *f-stop*, as when used in connection with a camera. A focal ratio of $f/10$ means that the focal length of the lens or mirror is ten times the aperture. A telescope in which the focal ratio is $f/5$ will provide images approximately four times as bright as with the $f/10$ ratio.

We have been referring to the brightness of the image as viewed by the human eye. The effective brightness may be greatly increased by photographic and/or electronic devices. Imagine an object so dim that it cannot be seen through the telescope. If the light from that object is allowed to fall on a sensitive photographic plate for a prolonged period, an image will be "built up" on the plate itself. When developed, the plate will reveal the presence of that object. Dim objects that normally require hours to expose on a photographic plate may now be intensified by means of an electronic device that senses their presence and builds the photographic image at a much faster rate. Thus the effective aperture of the telescope has

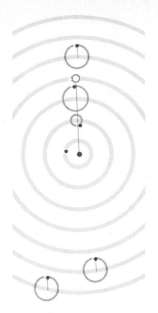

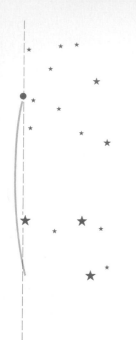

been increased (Figure 2.44). Representing what might be called a quantum leap forward in terms of sensitivity to light is the *charge-coupled device* (c.c.d.) that incorporates a silicon chip to record an array of dots (of varying intensity) that make up the image of a distant object. We will discuss this particular device later in this chapter under infrared astronomy where its usefulness is particularly dramatic.

RESOLUTION. The *resolution* of a telescope measures its ability to show detail in an image—that is, to separate objects that appear to be close together. The resolving power of a telescope is expressed in terms of the smallest angle between two stars that can be distinguished as separate objects. The resolving power of an optical telescope depends primarily upon its aperture and is expressed by the formula $a = 11.58/d$ measured in seconds of arc, where d is the diameter of the lens or mirror measured in centimeters. A telescope with a 15.24-cm (6-in.) aperture can theoretically separate star images that are at least 0.76 seconds of arc apart. (In angular measures, $1° = 60$ minutes and 1 minute $= 60$ seconds.) A 25.4-cm (10-in.) telescope can theoretically separate stars that are at least 0.5 sec-

FIGURE 2.44 Region of the Orion Nebula, showing the effect of aperture and time exposure on the formation of star images: (a) small aperture and/or short exposure time; (b) large aperture and/or long exposure time. (Lick Observatory)

(a) (b)

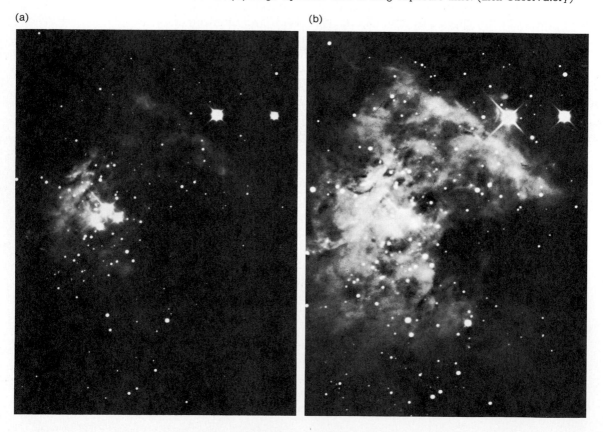

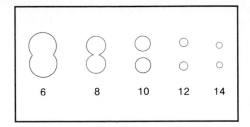

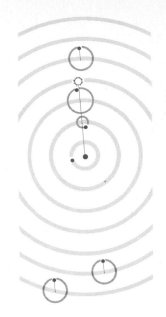

FIGURE 2.45 Resolution of star images using different apertures (apertures shown in inches).

onds of arc apart. Figure 2.45 indicates how two images might appear in telescopes of the indicated aperture under ideal conditions, neglecting the effect of the earth's atmosphere.

ATMOSPHERIC TURBULENCE: A RESOLUTION SPOILER

Ground-based telescopes have always been plagued by the fact that the light from a star (or other source) must pass through the earth's atmosphere, which is in constant motion. In moving from the vacuum of outer space to the increased density of the atmosphere, individual rays of light experience a random effect in relation to their reduction in velocity, producing a random effect in relation to their refraction. When all the rays of light from a single point source are brought to a focus by the lens or mirror of a telescope, the individual rays arrive at the focal point with their waves out of step with one another and they tend to destroy each other partially. Recall the discussion on interference on page 66. The random nature of the wave interference causes the star to appear to twinkle and it also spoils the resolution, spreading the light which would normally fall into a very small dot into a larger disk of light (see the brighter stars in Figure 2.44). Under such atmospheric conditions, a 25.4-cm (10-in.) telescope can separate stars only if they are at least 20 arc sec apart. The Palomar 5 m (200 m) telescope has an effective resolution of 1 arc sec.

SEEING

The term *seeing* applies directly to the concept we have been discussing. If astronomers say the seeing is good tonight, they mean that the atmosphere is unusually calm (not as turbulent as usual) and star images are small—binaries are easier to separate. *Poor seeing* would, of course, mean the opposite. You can judge the seeing conditions yourself by noticing the degree to which stars appear to twinkle—where less twinkling indicates better seeing.

MAGNIFICATION

While the size of an extended image, formed at the prime focus, is determined by the focal lenght of the objective lens or mirror, the *power* (magnification) of the telescope can only be determined after choosing an eye-

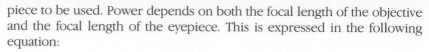

piece to be used. Power depends on both the focal length of the objective and the focal length of the eyepiece. This is expressed in the following equation:

$$\text{power} = \frac{\text{focal length of the objective}}{\text{focal length of the eyepiece}} = \frac{f_o}{f_e}$$

Suppose that for a given telescope, $f_o = 2m$, and we choose an eyepiece of 10 mm (millimeters) focal length. We must first express the focal length of the objective in millimeters:

$$2m = 2000 \text{ mm}$$

The effective power of this combination is then found:

$$\text{power} = \frac{f_o}{f_e} = \frac{2000 \text{ mm}}{10 \text{ mm}} = 200$$

When we use this telescope with the 10-mm eyepiece, an extended object will appear 200 times as large as with the naked eye—stars will not appear enlarged but will appear 200 times as far apart. Frequently, the useful power of a telescope is limited by the object being viewed and/or by the atmospheric conditions. If an eyepiece with a focal length of 20 mm is chosen, the power will be 100:

$$\text{power} = \frac{f_o}{f_e} = \frac{2000}{20} = 100$$

As a general rule, the useful limit of power for any telescope is 20 times its aperture in centimeters.

THE BINOCULAR

Binoculars are simply a pair of refracting telescopes that have the added advantage of presenting images erect (right side up). Figure 2.46 shows

FIGURE 2.46 The binocular may be considered to be a small telescope. Because of its unique optics (its multiple reflection and refraction), it produces an erect (right side up) image and correct image (right to left). Obviously it would not be very suitable for watching a football game if this were not true. (Bushnell Optical Co.)

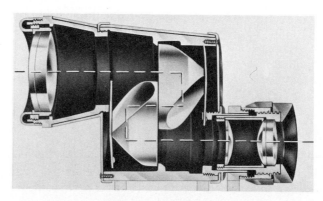

how this is accomplished through multiple reflection in the corner prisms. Binoculars are available in various sizes and powers, for example

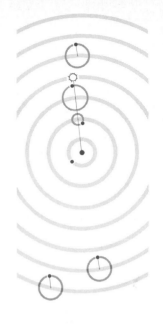

7×35, meaning 7 power with 35 mm objective lenses
7×50, meaning 7 power with 50 mm objective lenses

The "7 power" means that objects appear seven times larger than to the unaided eye. The 50 mm objective will gather approximately twice as much light as the 35 mm lenses; however, the human eye can take advantage of this additional light only when dark-adapted—that is, only during nighttime viewing. It should be noted that a large number of celestial objects—the major features of the moon, the disklike appearance of Jupiter and Saturn, the beauty of open clusters like the Pleiades, and certain nebulae (gas clouds)—are easily viewed in binoculars.

THE SPECTROGRAPH

Spectroscopy is one of the most important areas in astrophysics. Its operations depend upon the same phenomenon that was shown to be the cause of chromatic aberration in the simple refractor: the fact that light of different colors is bent by differing amounts when refracted by glass or some other transparent medium. Thus white light, which is composed of all colors, may be separated into these various colors by refraction, producing a full spectrum of colors (Figure 2.47; see also Color Plate 1).

Light from a star is passed through a narrow slit, then through a collimating lens, which bends the rays so that they will be parallel as they pass through the prism. The light, which has been broken into its various colors, is then focused onto a photographic plate.

The picture of the spectrum, called a *spectrogram*, consists of a series of adjacent images of the slit, each representing a slightly different wavelength. When sunlight is viewed through such an instrument, dark lines will appear at certain places in the spectrum. At first these were not recognized as being significant, but in 1814 the German optician Joseph Fraunhofer recorded the position of several hundred of these dark lines, which are still referred to as *Fraunhofer lines*. In 1859 Gustav Kirchhoff, a German physicist, discovered that this same phenomenon could be produced

FIGURE 2.47 The spectrograph.

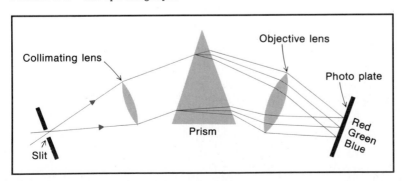

in the laboratory by passing white light through various gases and then through the spectrograph. He particularly noticed two close lines in the yellow portion of the spectrum and found that he could produce these same dark lines by passing white light through sodium vapors. To date, more than 30,000 *absorption (dark) lines* have been found in the visible portion of the solar spectrum, and more than half of these have been identified with elements known on earth (Figure 2.48).

Kirchhoff also observed that a thin glowing (excited) gas, under reduced pressure, produced a spectrum consisting of a series of bright lines on a dark background. As a result of these observations, he stated the following basic laws:

> Continuous spectrum. A heated liquid or solid, or a gas under high pressure, emits light of all wavelengths, producing a continuous spectrum consisting of all colors.

> Bright-line spectrum. A low-pressure gas that has been excited (say, by an electrical current or heat) produces a bright-line spectrum consisting of only certain colors.

FIGURE 2.48 The solar spectrum, with certain lines of known elements identified. (Mount Wilson and Las Campanas Observatories, Carnegie Institute of Washington)

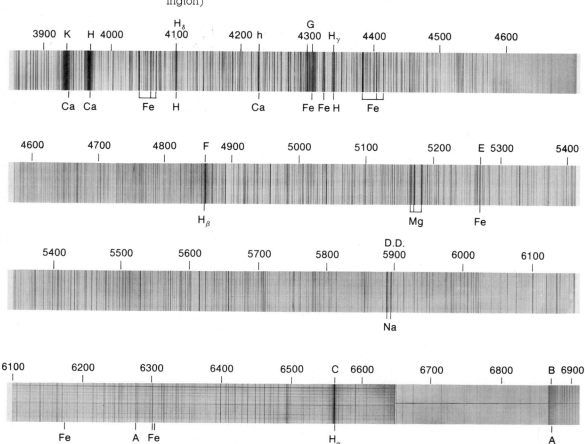

Absorption spectrum. If light of all wavelengths is passed through a gas, the gas may absorb certain wavelengths, producing an absorption spectrum consisting of dark lines on a continuous background.

One might suspect that the wavelengths absorbed by the gas would be reemitted as bright lines and simply fill in the dark lines immediately. It is true that they are reemitted, but they are reemitted in all directions. Only a small fraction of their energy continues in the original direction of the light ray, and so the lines appear less bright than the rest of the spectrum.

THE ATOM SIGNS ITS NAME

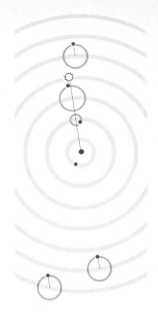

Suppose we place some hydrogen gas in a tube from which the air has been removed. Then we pass an electric current through this low-pressure gas by means of electrodes sealed at either end of the tube. Light will be radiated by the gas, and if that light is passed through a spectroscope, only four dominant bright lines in its spectrum are visible (Figure 2.48 and Color Plate 1(C)). Each line represents one wavelength: the red line has a wavelength of 6563 Å; the blue line, a wavelength of 4861 Å; one violet line, of 4341 Å; and the other violet line, of 4101 Å. But why are these four lines produced in the hydrogen atom?

We visualize the atom, the simplest unit of a given element, as being composed of a central nucleus around which electrons move—not in a completely predictable manner, but at least so that we can predict their most probable levels of energy. These energy levels are represented in Figure 2.49 in rather oversimplified terms as circles surrounding the nucleus. Circle 1 represents the lowest energy the electron may possess, circle 2 represents the second energy level, and so on. In this model of the hydrogen atom, proposed by the prominent Danish physicist Niels Bohr (1885–1962), no intermediate levels are possible. The single electron of a hydrogen atom would normally be at the lowest energy level. If an electric current were passed through the tube of gas, however, the electrons of some of the atoms would receive energy from the electric current, permitting them to move to a higher energy level. An electron would not remain at that higher level but would return to the lowest energy level. In the process of making such a downward transition, the electron may stop temporarily at any level, releasing energy as it makes the downward transition. In other words, the atom has absorbed energy initially to make the upward transition, and now it emits energy as it makes a downward transition. If the electron stops at energy level 2, even temporarily, it will produce visible light, but only those colors of visible light that correspond to the amount of energy the electron lost in making a specific downward transition. In other words, each possible change in energy level has a corresponding definite amount of energy and emits a photon of a definite wavelength.

In the hydrogen atom, when the electron makes a transition from the third to the second energy level (a "3 to 2" transition), it produces the red line (6563 Å), called the *hydrogen-alpha line*. A "4 to 2" transition pro-

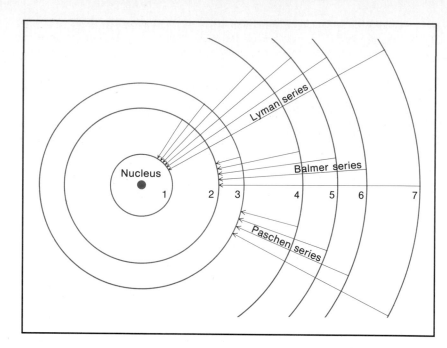

FIGURE 2.49 Electron transitions in the hydrogen atom.

duces the blue line (4861 Å), called the *hydrogen-beta line*; a "5 to 2" transition produces the first violet line (4341 Å), called the *hydrogen-gamma line*; and a "6 to 2" transition produces the second violet line (4101 Å), called the *hydrogen-delta line*. Other transitions from still higher energy levels that stop temporarily at level 2 produce numerous faint violet lines near the end of the visible spectrum. Only these specific lines are possible for the neutral hydrogen atom at low pressure; no partial transitions are possible between energy levels. For example, the electron cannot move from level 4 to $2\frac{1}{2}$ or from level 5 to $3\frac{1}{4}$, for only whole number transitions are possible. One series of visible lines produced by the hydrogen atom is called the Balmer series. Of course, some downward transitions stop temporarily at energy level 3; these produce a series of lines in the infrared portion of the spectrum called the Paschen series. Transitions to the lowest level produce lines in the ultraviolet portion; these lines are called the Lyman series.

Whereas downward transitions *emit* the light of definite wavelengths, upward transitions *absorb* light of the same wavelengths. Suppose we cause white light, which is a mixture of all visible wavelenghts, to pass through hydrogen gas at low pressure. The photons having wavelengths corresponding to transitions from level 2 to 3 (or from 2 to 4, or from 2 to 5, and so on), would be absorbed, causing such upward transitions in the atoms. We would see the same dominant four lines of hydrogen, but now they would be dark lines on the full rainbow background of white light. The spectrum of white light is called a continuous spectrum because it contains all colors; but when certain colors (wavelengths) are removed by absorption, it is called an *absorption spectrum* (also called a *dark-line spectrum*). Since the lines occur in an absorption spectrum at the same position as the bright lines of an emission spectrum, either type is equally useful to the astronomer. Both types occur in stars, nebulae, galaxies, and the like.

Now suppose we look at the spectrum of helium, which consists of an entirely different set of lines from that of hydrogen. Although the model of the helium atom or of other heavier elements is more complex, it is evident from experimentation that each kind of atom has its own characteristic set of spectral lines. Thus the atom "signs its name" in its spectrum. One of the great values of spectroscopy to the astronomer now becomes clear—the atoms which compose a distant object can be identified by recording its spectrum and comparing that to the spectra of different elements produced in the laboratory. For example, some of the lines of calcium (Ca), iron (Fe), hydrogen (H), magnesium (Mg), and argon (A) have been identified in the solar spectrum (see Figure 2.48, page 88).

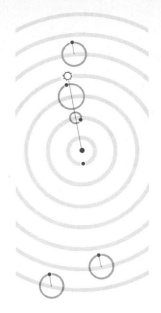

DOPPLER SHIFT

The spectrum can also be used to tell something about the motion of an object that is emitting light, owing to a phenomenon called the *Doppler effect*. The Doppler effect is a change in wavelength caused by the motion of the source or of the observer or both (Figure 2.50). Imagine that a fire truck is approaching you with its siren operating at a certain note (emitting a certain wavelength of sound). The firefighters riding in the truck hear that wavelength, but you hear a higher pitched note (shorter wavelength) than they do as the truck approaches you. Why? Because each time the siren emits a new wave crest it is closer to you, and therefore the wave crests occur closer together from your point of view, producing a shortened wavelength. Then as the truck passes and recedes from you, the note you hear is lower (longer in wavelength). This occurs because each time

FIGURE 2.50 The Doppler effect: (a) stationary source; (b) moving source.

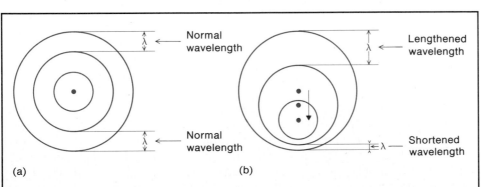

the siren emits a new wave crest of sound, it is farther from you and the wave crests appear to be farther apart from your point of view. The faster the truck is moving, the greater the change in pitch.

The Doppler effect also occurs in electromagnetic phenomena. The wavelengths of a given line in the spectrum of a star will be lengthened if the star is moving away from the observer, causing the line to appear shifted in position toward the red end of the spectrum. If the star is moving toward the observer, a given line will be shifted in position toward the blue end of the spectrum. The amount by which the wavelength is changed ($\Delta\lambda$) depends only on the relative velocity of the star with respect to the earth, along the observer's line of sight. If c represents the velocity of light and λ represents the laboratory wavelength of the given line, then the change in wavelength due to the motion is given by:

$$\Delta\lambda = \frac{v \times \lambda}{c}$$

This formula, when solved for v, indicates the method by which the *radial velocity* of a star is determined:

$$v = c \times \frac{\Delta\lambda}{\lambda}$$

Here v, a radial velocity, is that part of the relative velocity that is in the direction of the observer's line of sight.

The first line in the hydrogen spectrum normally has a wavelength of 6563 Å. If in the spectrum of a star this line is observed to have a wavelength of 6565 Å, we may calculate the radial velocity of the star as follows:

$$\Delta\lambda = 6565 \text{ Å} - 6563 \text{ Å} = 2 \text{ Å}$$
$$v = c \times \frac{\Delta\lambda}{\lambda} = 300{,}000 \text{ km/sec} \frac{2 \text{ Å}}{6563 \text{ Å}}$$
$$v = 91.4 \text{ km/sec} \ (56.7 \text{ mi/sec})$$

Since the wavelength of the line from the stellar source appears longer than that of the laboratory source, we know that the star is moving away from us. Thus a shift toward the red end of the spectrum—a "red shift"—indicates that the star is receding from us, whereas a shift toward the blue end of the spectrum—a "blue shift"—would indicate its approach toward the earth. It should be observed that the Doppler shift tells us nothing about the distance to the star, only its radial velocity.

A VERY IMPORTANT GENERALIZATION

We have discussed several concepts related to the phenomenon called light, but light is only a very small portion of the electromagnetic spectrum; in addition to visible light, the spectrum includes radio, infrared, ultraviolet, X rays, and gamma rays. All forms travel through empty space in a straight line at the same speed but may be reflected or refracted by the proper materials. They all can be collected or concentrated in some fash-

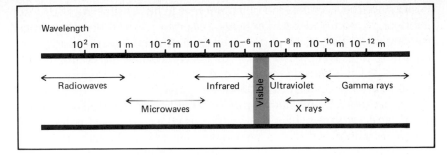

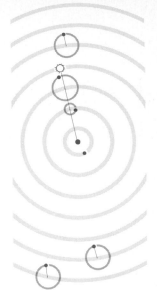

FIGURE 2.51 The electromagnetic spectrum.

ion, even as light is concentrated by an optical telescope. Objects in space emit certain wavelengths better than others, creating a spectrum in radio, infrared, visible light, ultraviolet, X rays, and gamma rays (Figure 2.51), and that spectrum reveals Doppler shifts, even as in the visible spectrum.

Why then has the astronomer concentrated so much on the visual observations and utilized only a small portion of the spectrum? There are two very logical reasons. Until the advent of radio receivers, the only natural "receiver" through which the universe could be perceived was the eye, which only "tunes in" the visible spectrum. With the perfection of the radio telescope, astronomers gained another portion of the spectrum: radio wavelengths between 10 and 0.001 m. This gave them two narrow portions of the spectrum in which to work. But what of the remaining portions? With the exception of a small portion in the infrared part of the spectrum, the earth's atmosphere reflects or absorbs most of the other wavelengths. Thus only these three small "windows" existed through which astronomers could observe the universe. They began to conquer the atmospheric problem by sending balloons high into the atmosphere or rockets above it. The real breakthrough came when scientists were able to place satellites in orbit around the earth. On those satellites they can now install telescopes, some with people aboard (as in the case of Skylab or the Apollo series) but most without (as in the case of OAO–C, HEAO, and others which will be mentioned later). This capability has opened up the entire spectrum for study, and an explosion of data has taken place in what might be called the new astronomy: infrared astronomy (already begun from earth-based observatories), ultraviolet astronomy, X-ray astronomy, gamma-ray astronomy, and cosmic-ray astronomy.

In order to take advantage of the broadened window, astronomers are developing new tools and techniques at a rapid pace. With the information gathered, they are beginning to unravel some of the mysteries of the universe. We will consider each of these new techniques briefly.

RADIO ASTRONOMY

The discovery that radio energies are emitted by objects in space came quite unexpectedly in 1931 when Karl Jansky of the Bell Telephone Laboratories noticed a certain kind of interference in his sensitive receiver. This interference occurred a few minutes earlier each day. When he learned that any given star or galaxy appears to rise four minutes earlier each night according to our clocks, he correctly concluded that the source of this radio energy was outside the earth—in fact the source of interference is located near the center of the Milky Way galaxy.

Scientists reasoned that, if a device could be constructed to "collect" a very weak radio signal and concentrate its energy at a focal point, and if a sensitive amplifier could strengthen that signal without mixing in its own noise, the astronomer would have a fine tool with which to expand his knowledge of the universe. In view of the similarity in properties of all electromagnetic disturbances, it seemed logical to follow the parabolic design of the optical mirror in planning construction of a *radio telescope*. Owing to the longer wavelengths of radio energy, the radio telescope can be built of steel and simply covered with a wire mesh or perforated metal sheets. It is thus possible to construct very large models, such as the 64-m (210-ft) movable antenna at Goldstone, California (Figure 2.52), or the 305-m (1000-ft) fixed antenna at Arecibo, Puerto Rico. At Arecibo in a

FIGURE 2.52 The 210-ft. (64-m) Goldstone antenna, located near Barstow, California. This instrument is used primarily to tract and communicate with deep space probes. (JPL-NASA)

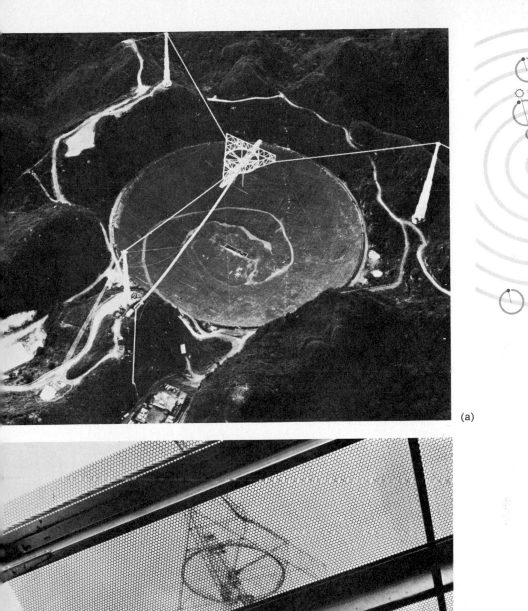

(a)

(b)

FIGURE 2.53 (a) The 1000-ft. (304.8-m) Arecibo antenna, in Puerto Rico. The focal point for the telescope is suspended by cable from the surrounding elevations and is movable for purposes of scanning the sky. (b) A recent updating of the Arecibo antenna included a new mesh covering for the dish itself. The resulting accuracy with which the surface is shaped provides added resolution. In this view you are looking upward through the mesh to the collector point. (The Arecibo Observatory is part of the National Astronomy & Ionosphere Center which is operated by Cornell University under a contract with the National Science Foundation.)

mountainous area, a natural depression was bulldozed into the shape of a spherical basin. This basin allowed the suspension of a parabolic "dish" composed of perforated metal sheets with a linear collector overhead (see Figure 2.53). Although this radio telescope is relatively fixed with relation to the ground, it is not fixed as to what it "sees." As the earth rotates, the antenna scans a path through the sky every day, and the focal collector overhead can be moved north or south. This has the effect of pointing the telescope south or north (respectively) about 20° from the zenith. Thus a very significant portion of the sky may be covered.

Certain objects may be detected by a radio telescope even though they cannot be observed by an optical telescope. The more distant members of a class of objects called *quasars* (quasi-stellar radio sources) are a good example. Their name suggests the fact that they are starlike in appearance and that they radiate a significant portion of their energy in the radio portion of the spectrum. Some of these objects have such large red shifts in their spectra as to suggest that they are receding from us at speeds approaching that of light. These may be the most distant objects in the universe. A radio telescope can be used day and night, on cloudy or clear days, and it can penetrate many regions of our galaxy which were previously hidden from us by clouds of dust through which optical telescopes could not see. Also, whereas the optical astronomer is plagued by a background of light radiation that obscures the fainter stars, the radio astronomer has virtually no background radiation to worry about, although interference is sometimes caused by automobile ignition or radio stations. The radio astronomer's biggest problem is how to amplify the weak radio signal coming from a distant source without introducing artificial noise that will obscure the signal. To solve this problem, the scientist uses a device (called a *maser*) submerged in liquid helium, which cools it to approximately −270°C (almost absolute zero). In the maser, the weak radio signal triggers a series of downward electron transitions in atoms that are kept in a state of continuous excitation, thus amplifying the original radio signal (Figure 2.54).

Radio telescopes in general have poor resolution because the wavelengths with which they must deal are much longer than the wavelengths of light. The length of a typical radio wave is 3 cm, compared to 0.00006 cm for a typical light wave. The formula that expresses the resolution of any telescope is given by

$$\alpha = \frac{2.1 \times 10^5 \times \lambda}{d} \text{ seconds of arc}$$

The wavelength of the wave received is denoted by the Greek letter λ (lambda), the telescope diameter by d. For a telescope 30,480 cm (1000 ft) in diameter, receiving a 3-cm signal, the resolution is

$$\alpha = 2.1 \times 10^5 \times \frac{3 \text{ cm}}{30,480 \text{ cm}} \text{ seconds of arc}$$

$$\alpha = 21 \text{ seconds of arc}$$

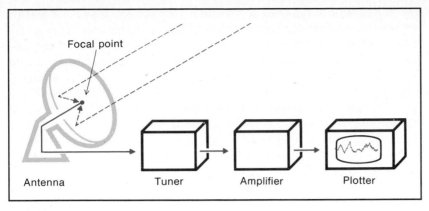

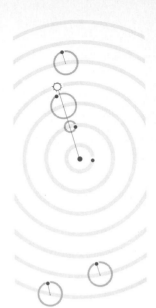

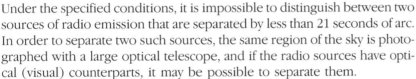

FIGURE 2.54 Diagram showing the essential components of a radio telescope: the antenna collects the radio energy and concentrates it at the focal point; the tuner selects the wavelength desired; and the amplifier builds up the signal, which is then plotted as a permanent record.

Under the specified conditions, it is impossible to distinguish between two sources of radio emission that are separated by less than 21 seconds of arc. In order to separate two such sources, the same region of the sky is photographed with a large optical telescope, and if the radio sources have optical (visual) counterparts, it may be possible to separate them.

The advent of the radio telescope has led to the discovery of many radio sources, some of which are too faint or too distant to be seen.

RADIO INTERFEROMETRY

The poor resolution characteristic of radio telescopes may be improved considerably by interconnecting two or more radio telescopes that are separated by some distance (Figure 2.55). Since the radio signal from a given source will arrive at the two antennae at slightly different times, the received waves will interfere with each other. From these interference patterns it is possible to locate a source much more accurately (Figure 2.56). The Very Large Array (VLA) radio telescope system (interferometer) in central New Mexico (Figure 2.55), consists of 27 movable antennae, each 85 feet in diameter and each weighing 214 tons. The radio signal from each antennae is fed by cable or wave guide into a computer, and the combined effect of these antennae produces a resolution rivaling the largest optical telescopes. Specifically, the VLA is able to separate objects that lie within one second of arc, but this is only the beginning of greater accomplishments.

Imagine radio telescopes hundreds or thousands of miles apart; if used in a similar relationship to those in the VLA, their effective resolution could be a thousand times greater. The primary problem is in detecting the difference in the arrival times of a given signal at each location. This is called the *phase difference*. Modern atomic clocks are capable of keeping time to the nearest billionth of a second. They can be used as a standard whereby the signals arriving at several telescopes can later be synchronized, without the necessity for interconnecting cables. Thus telescopes scattered over the entire earth can serve as a single huge instrument, and using this technique, resolutions of 0.001 arc-sec or better are regularly achieved.

The method we have just described is called Very Long-Baseline Interferometry (VLBI). To illustrate its usefulness, let's consider the early discovery of quasars (quasi-stellar radio sources), now thought to be the cores of very distant galaxies. With only the limited resolution of a single dish, it was impossible for radio astronomers to pinpoint a quasar within the field of view, but using VLBI, astronomers are now mapping the very structure of the quasar—differentiating the core from its outer portions. We shall discuss the quasar in more detail in Chapter 16.

RADAR ASTRONOMY

Many radio telescopes, including those mentioned above, can be used for radar studies as well as for radio astronomy. Radar astronomy involves a two-way communication with relatively nearby objects such as the moon

FIGURE 2.55 Artist's conception of the Very Large Array (VLA) radio telescope system which was completed in 1980. The Array is located in the Plains of Saint Augustin, in central New Mexico. This array of instruments permits a detailed mapping of radio sources with a resolution equal to or better than any optical telescope. (National Radio Astronomy Observatory)

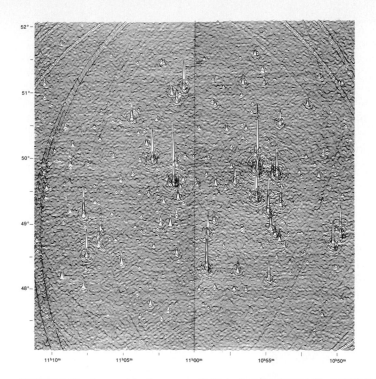

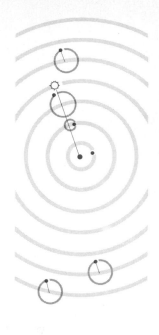

FIGURE 2.56 A graphical representation of radio sources in a region of the sky 4° in diameter, as recorded by the Cambridge One-Mile interferometer. (From Pooley and Kenderdine, Monthly Not. R.A.S. 139, 529 (1968); Mullard Radio Observatory University of Cambridge)

and the planets. A radio signal is sent by the antenna toward an object, then the same antenna is used to listen for the echo or reflection of that signal. Since the velocity of a radio signal is known, the length of time required by the signal to travel to a given planet and back reveals its distance. Other types of information such as the planet's velocity, its rotation, and its surface contour may also be ascertained by analyzing the Doppler shifts in the returning signals. Radar astronomy was used to penetrate the clouds of Venus, for instance, to give us our first indications of its rate and direction of rotation and its surface features.

INFRARED ASTRONOMY

The development of infrared astronomy lagged behind that of radio astronomy by approximately 20 years. The reason was that no one had perfected a detector that would record very small changes in the amount of infrared energy falling on it and that would also discriminate as to the direction from which the radiation came. Infrared radiation has wavelengths that fall between radio and red light. Neither the photographic plate used by the optical astronomer nor the antennae used by the radio astronomer will record infrared radiation. The standard optical telescope may be employed, however, if a suitable detector is placed at its focal point.

One of the latest detector designs utilizes the charge- coupled device (c.c.d.), a marvel of the electronic age (Figure 2.57). Picture a silicon "chip," perhaps $1.25 \, cm \times 1.25 \, cm$ ($\frac{1}{2}$ in. $\times \frac{1}{2}$ in.) in size, with up to 800 separate photon counters arranged along a side (like a checker board) thus providing 640,000 photon counters (called *pixels*) on this one small "chip." The output of each pixel is fed individually into a computer where it is stored and may be added over any given time interval. The computer can then produce a composite "picture" of what the telescope is seeing in infrared.

Among the very exciting objects which can be detected almost exclusively in infrared are stars in the making and stars that are dying. These objects are often surrounded by gas and dust clouds to the point of being totally obscured in visible light. The dust grains typically absorb the shorter wavelengths of light especially of ultraviolet, reemitting that energy in the longer wavelengths of infrared. In the shells of gas and dust expelled by dying stars, astronomers find the heavier elements thought to have been manufactured in their cores, some of which take the form of organic molecules that make life in the universe possible.

A researcher, sitting at a console like that shown in Figure 2.58, is not the usual picture one has of the astronomer who spends the night at the prime focus of a large telescope. This is typical of more and more ground-based instruments utilizing electronic equipment and permitting more comfortable observing conditions in adjoining offices. The water vapor in the earth's atmosphere tends to absorb a large amount of infrared radiation; hence it would be advantageous to get above at least a portion of that atmosphere. This can be accomplished by means of a high-flying aircraft

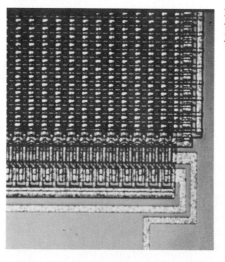

FIGURE 2.57 Change-coupled device (C.C.D.) (© AURA, Inc. Kitt Peak National Observatory, 1982)

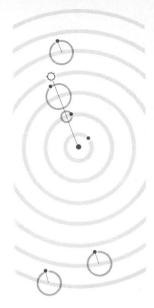

FIGURE 2.58 Larry Ott at the console of the 4-meter telescope which employs many electronic detectors. (Kitt Peak National Observatory)

called the Kuiper Airbourne Observatory (KAO) that flies at elevations in excess of 12 km (40,000 ft)—above 99 percent of the atmosphere's water vapor. The aircraft is a modified Lockheed C-141 with a 91-cm (36-in.) telescope and is sponsored by NASA. This facility is made available for research in a number of wavelengths spanning from infrared to ultraviolet. Great effort was expended to isolate the telescope from the vibrations inherent in the aircraft. Both the pointing of the telescope and the flight of the aircraft during observations are controlled by a computer, providing a stability factor equal to a ground-based observatory.

The most recent advance in infrared astronomy came with the launch of the satellite called IRAS (Infrared Astronomical Satellite) (Figure 2.59). It consists of a 60-cm, supercooled telescope, operating at a temperature of −271°C, making it the most sensitive infrared telescope ever built. It represents a joint effort of the United States, the Netherlands, and England, with more than 150 scientists and engineers directly involved. Early observations transcended the expectations of its developers, revealing star birth in the interior of a dark nebula and providing preliminary indications of another solar system. As can be said of almost any orbiting telescope, it is impossible to overestimate its value as a research tool. In the next few years, observations by IRAS and the Space Telescope may demand explanations totally different from those that characterize astronomy today.

FIGURE 2.59 Artist's conception of the Infrared Astronomical Satellite (IRAS) in earth orbit. (IPL; NASA)

ULTRAVIOLET ASTRONOMY

The greatest impediment to detection of radiation in the ultraviolet portion of the spectrum has been the earth's atmosphere. Success in this area has come with the use of rockets and orbiting satellites such as the Orbiting Astronomical Observatory named *Copernicus* (OAO-C, Figure 2.60). This observatory carries the largest reflecting telescope ever sent into space. Its 0.8-m mirror, spectrograph, and allied sensors were designed to detect the ultraviolet spectra of stars and of interstellar molecules with higher resolution than previously achieved. The satellite does not actually photograph the spectrum but reads out the photon count as it scans the source at a variety of wavelengths. A plot of such a readout is seen in Figure 2.61. The dips in this graph correspond to absorption features in the visible spectrum. Ultraviolet radiation is typically associated with very

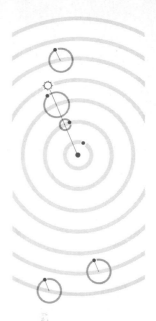

FIGURE 2.60 The Copernicus satellite (OAO-C) carries a 1-m telescope designed to sense ultraviolet wavelengths and two smaller X-ray telescopes. Also illustrated here are the two solar energy panels and the knoblike star trackers near the base of the satellite. (NASA)

hot objects, such as stars with surface temperatures in the range of 50,000 to 100,000°K or very hot gas clouds found between the stars.

SPACE TELESCOPE

Planned for launching in the mid-1980's is a large orbiting telescope, dubbed simply Space Telescope (ST) (Figure 2.62). Operating above the earth's turbulent atmosphere, this instrument with its 2.4-m (94.4-in.) pri-

FIGURE 2.61 A readout of a portion of the ultraviolet spectrum by OAO-C, the Copernicus satellite, showing absorption lines of ionized argon and hydrogen gas (H_2 molecules). (From Rogerson et al., The Astrophysical Journal (Letters) 181 (3), L97-L102 (1973); courtesy of the American Astronomical Society and University of Chicago Press)

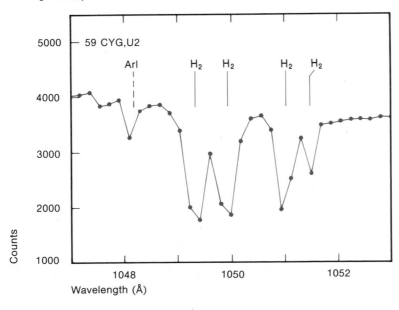

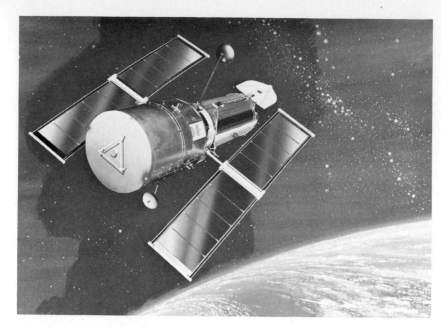

FIGURE 2.62 The 96-inch Space Telescope, to be launched by Space Shuttle, will provide the clearest photos ever taken of the universe. It will respond to a wide range of wavelengths (from 1000 Å to 1 mm). (NASA)

mary mirror will represent a major leap forward in optical astronomy. It will provide a resolution several hundred times better than that of any ground-based telescope—namely 0.05 arc-sec. Its CCD detectors (described on page 100) will "paint" an enhanced picture of the universe in the infrared, visual, and ultraviolet portion of the spectrum, yielding new information regarding faint objects, quasars, nuclei of galaxies, chemical evolution of galaxies, and tests for the models of cosmology (these topics will be treated later in this text). The ST is to be launched by the space shuttle, and if necessary it may be revisited or recovered by the same means, for refitting or repair.

X-RAY ASTRONOMY

The earth's atmosphere absorbs almost all X-ray radiation directed toward the earth. While this is fortunate for our existence, it limits any effective examination of the universe in wavelengths between 0.1 and 100 Å. The first succesful attempts to sense sources of X rays came with the firing of rockets high above the obscuring layers of our atmosphere, and 30 to 40 strong discrete sources of X rays were immediately found within our Galaxy. The launching of a satellite exclusively designed for X-ray observation created a massive amount of additional data contributing to the emerging X-ray "picture" of the universe. The satellite was dubbed UHURU, meaning "freedom" in Swahili, in honor of Kenya's independence (Figure 2.63).

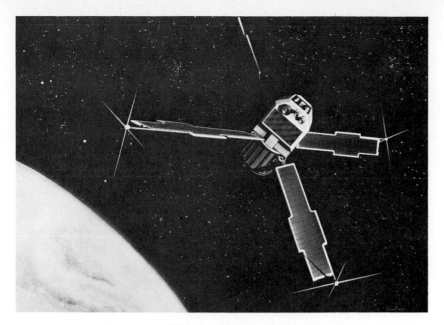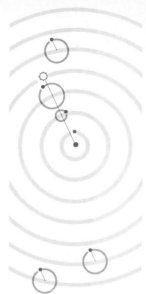

FIGURE 2.63 Artist's view of UHURU, an X-ray satellite. (From R. Fiacconi, X-ray astronomy. The Physics Teacher 11 (3), 135-143 (1973))

FIGURE 2.64 A plot of X-ray sources, based on observations by satellite. The larger dots indicate the stronger sources; several are identified. (Frederick D. Seward, Lawrence Livermore Laboratory, University of California)

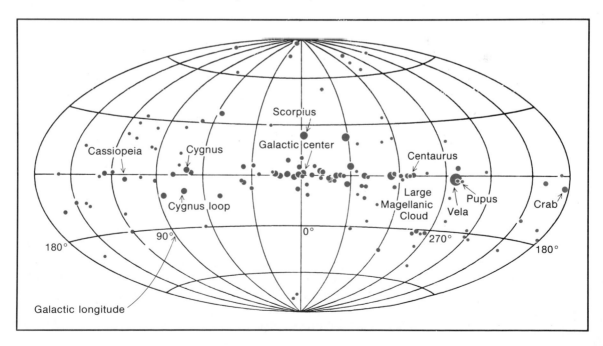

The craft spins slowly on its axis, scanning the sky continually for X-ray sources, and its orientation (pointing direction) can be controlled from ground-based control centers. The ability to detect X-ray sources was also greatly enhanced by the launching of a series of orbiting observatories called HEAO (High Energy Astronomical Observatory). Significantly more sensitive than UHURU, HEAO has now detected X-ray pulsars, X-ray emitting quasars, Seyfert galaxies, and BL Lacertae objects, in addition to the celebrated black holes. Each of these topics will be treated in later chapters. A plot of the principal X-ray sources in the Milky Way is shown in Figure 2.64.

GAMMA-RAY ASTRONOMY

The latest chapter in the opening up of the entire electromagnetic spectrum to the astronomer's view involves the detection of gamma rays and particularly gamma-ray bursts. X-ray and gamma-ray photons carry more energy than any other portions of the spectrum, hence we may assume that the events with which they are associated represent some of the most dramatic events in the universe. Bursts of gamma rays were first detected quite unintentionally by a number of Vela satellites launched in the 1960s and early 1970s. These satellites were designed to detect atomic blasts violating the nuclear test ban treaty—tests which would have produced gamma-ray radiation. Instead, two or three Vela satellites, orbiting at various points around the earth, sensed gamma-ray bursts almost simultaneously, indicating a source outside the earth. This and subsequent observations of gamma-ray bursts have opened a wide range of speculation as to the source of this radiation. One such source appears to coincide with the site of a black hole (Cygnus X-1), hence we may feel quite certain that gamma rays are produced whenever material falls onto a very dense object. Other possible sources will be discussed in their appropriate sections later in the text. Of one thing we may be sure: the energy poured out in one second by these events exceeds the total energy emitted by the sun in many years.

GRAVITATIONAL WAVE ASTRONOMY

We have seen that surrounding every charged particle is an electrical field. If that charged particle is set into *oscillation* (vibration), a changing electrical field results. If another charged particle exists in that field, it will react by oscillating, and this oscillation permits us to detect the existence of that changing electrical field. In 1916 Albert Einstein predicted that a similar thing would happen in the gravitational field of an accelerating object. First, picture an object—say, an uncharged steel ball—placed in the center of an empty room. We could say that a certain gravitational field exists in the room because of the presence of the steel ball. We could demonstrate that fact by showing that another steel ball, when brought into the room, would experience a force of attraction toward the original ball. This force is so small in comparison to the electrical and magnetic forces with which we have been dealing that it would be difficult to measure. Suppose that the second ball were hung from a high ceiling by a steel wire so that it was very close to the first ball. The second ball would be deflected slightly toward the first, so that it did not hang vertically from the ceiling. That deflection proves there is gravitational attraction between the two steel balls; there is an unseen gravitational field which is static (unchanging). Now suppose that the first ball were set into a rapid vibration; then the gravitational field around it would become a changing field, and the second ball would reveal that changing field by vibrating itself.

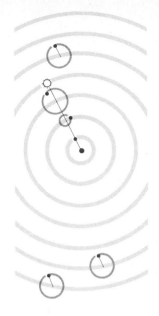

Einstein predicted that an accelerating mass emits gravitational waves; astronomers reasoned that if this was correct, they might detect the collapse or explosion of gigantic galaxies, the collision of two black holes, or other cataclysmic events by designing a detector for the resulting waves. Such a detector would have to be free from any other kind of influence, such as earthquakes, changes in electrical or magnetic fields, and air disturbances.

One of the first gravitational wave detectors consisted of a 3000-kg aluminum cylinder suspended from an arched support system. A very sensitive recording device had been placed on the cylinder, and the very small vibrations that gravity waves were thought to produce appeared to be within the detection capability of this instrument. The cylinder could not be isolated completely from its physical surroundings, however, and consequently it recorded vibrations from the surrounding environment. In order to separate these vibrations from those that might represent gravitational waves, experimenter Joseph Weber installed a second detector 1600 km (1000 miles) away, reasoning that a disturbance originating outside the earth would be recorded almost simultaneously by both detectors, whereas local disturbances would show up only on one detector. In 1969 Weber detected what he believed to be the first gravitational waves ever recorded.

As is required in science, confirmation of Weber's observations is being sought by other scientists. One such effort centers around the work of William M. Fairbank and William Hamilton at Stanford University. These two men bring their study of *cryogenics* (the branch of physics dealing with the behavior of materials at very low temperatures) to bear upon the problem of isolating the detector from its surroundings in order to reduce the effect of background vibrations. The new detector (Figure 2.65) consists of a six-ton cylinder of aluminum encased in several containers of coolants to reduce the temperature in the innermost container to within 0.5° of absolute zero (-272.5°C). At this temperature vibrations due to heat virtually cease. What is more important, the cylinder can be made to "float" in the middle of the innermost cylinder without touching anything. This is possible because when certain metals are supercooled, an electrical current, once started, will flow through them forever. The cylinder can thus be made to float on swirls of electrical current and yet can be isolated

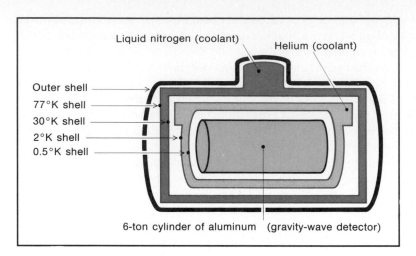

FIGURE 2.65 Gravitational wave detector.

from any electrical or magnetic disturbances from without. Two such detectors, located at Stanford University and Louisiana State University, approximately 3200 km apart, are to operate simultaneously in an attempt to confirm vibrations (gravity waves) originating in our galaxy or beyond.

Using the tools we have briefly surveyed in this chapter, astronomers have increased their ability to receive data over the full spectrum of electromagnetic and particle wavelengths, and they have used these data to project a richer and clearer picture of the universe than has ever been conceived before.

QUESTIONS

1. Perform the necessary arithmetic to determine the length of a light year in kilometers, based on the fact that light travels at the speed of approximately 300,000 km/sec, then convert this distance into miles.

2. In any experiment to determine the speed of light, what two factors must be known?

3. Albert A. Michelson used a rotating mirror to measure one of the two essential factors mentioned in Question 2. Which factor did he measure with the turning mirror?

4. Other than visible light, in what ways do electromagnetic disturbances exhibit themselves?

5. Why is sound not included among electromagnetic disturbances?

6. The telescope mirror is curved so as to bring light rays from a given source to the same focal point. What basic law or principle is utilized in the design of this curvature?

7. Why do stars near the horizon appear higher in the sky than their true position?

8. The earliest telescopes utilized the principle of_____(refraction, reflection).

9. List four examples of extended sources.

10. List two defects that may exist in the refracting telescope if it is not correctly designed.

11. Which telescope design is best suited for a sky survey?

12. Discuss the advantages of using film to photograph dim objects over merely observing them with the naked eye.

13. What is the primary reason that radio telescopes have relatively poor resolution even though they are larger than optical telescopes?

14. The functioning of a spectrograph depends upon the fact that whenever light is refracted, it is also_____.

15. The primary spectrum of the sun, as observed from the earth, is a_____(dark-line, bright-line, continuous) spectrum.

16. A bright-line spectrum is produced by_____(upward, downward) transitions of the electron.

17. What is indicated by a very large red shift in the spectral lines of a distant galaxy?

18. Find the size of the image of a full moon (0.5° in diameter) photographed at the prime focus of the 5-m (200-in.) telescope; focal length is 16.8 m (667 in.) (Answer: 14.6 cm or 5.8 in.)

19. What advantage does a radio telescope have over its optical counterpart?

20. What advantage does a telescope aboard an orbiting observatory have over a ground-based telescope?

21. Why should the infrared sky look different from the visible sky?

22. What kinds of events are expected to produce gravitational waves?

23. If a given star is moving through space at the same speed and direction as the sun, we do not expect to see a Doppler shift in its spectrum; but because the earth is moving around the sun at a velocity of 30 km/sec, the earth's motion creates a periodic blue shift when it approaches the star and a red shift when it recedes from the star. Find the maximum Doppler shift produced by the earth's motion.

24. Given two telescopes having the same focal length, the one with an aperture of 25.4 cm (10 in.) collects_____times as much light as the one with an aperture of 12.7 cm (5 in.).

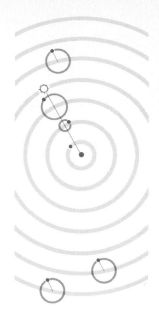

The Earth

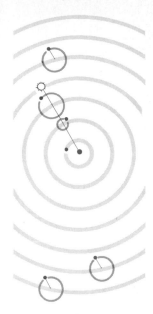

Three

Although the study of the earth is often considered to be the exclusive domain of the geologist, the astronomer is also very much concerned about our planet's origin and its properties such as size, mass, density, layering, atmosphere, magnetic field, and motion. Through knowledge of the earth, the astronomer is better prepared to ask significant questions concerning the other planets and their moons.

Pythagoras in the sixth century B.C. recognized the shape of the earth as being spherical. Another Greek astronomer, Eratosthenes (c. 275–195 B.C.) went a step further— he estimated the earth's circumference. He observed that on the longest day of the year the rays of the noon sun illuminated the bottom of a vertical well in Syene (now Aswan), Egypt, which indicated that the sun was directly overhead. On the same day in Alexandria, 800 km to the north of Syene, the rays of the sun at noon made an angle of 7.2° with a vertical post (Figure 3.1).

Assuming the earth to be spherical, Eratosthenes ventured to compute its circumference by the following line of reasoning. The angle formed between the sun's rays and a vertical post at Alexandria had to be the same as the angle formed by these lines extended to the center of the earth. This is an application of the geometric theorem which states that if two parallel lines are cut by a transversal (a sloping

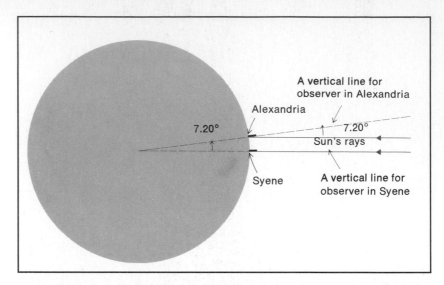

FIGURE 3.1   The method by which Eratosthenes measured the size of the earth.

line), the corresponding angles are equal. Eratosthenes reasoned that if an angle of 7.2° at the center of the earth spanned a distance of 800 km on the surface of the earth and 50 such angles fit in the circle (360°/7.2° = 50), then the total distance around the earth would be equal to 50 × 800 km—that is, 40,000 km. This is very close to the value for the circumference of the earth accepted today. From this value, it is possible to find the diameter, remembering that $C = \pi d$ and $d = C/\pi$:

$$d = \frac{40{,}000 \text{ km}}{3.1416} = 12{,}732 \text{ km } (7{,}912 \text{ mi})$$

In 1687 Sir Isaac Newton suggested that the earth was not a perfect sphere since its rotation tends to flatten it. Today we are able to measure that flattening and find the radius of the earth to the north or south pole to be approximately 22.5 km (14 mi) shorter than the radius of the earth to the equator. This flattening amounts to only about one-third of one percent.

VOLUME, MASS, AND DENSITY

The concepts of volume, mass, and density are fundamental to all of science. A short exercise will help you grasp these concepts before they are applied to the earth, planets, and other objects. Picture a brick of dimensions 12 cm × 6 cm × 4 cm. How could you find its volume, mass, and density? To answer, you must know the precise definition of each term.

Volume is a measure of the space an object occupies. What an object is made of has no bearing on its volume. The volume of a rectangular solid (box) is simply: $V =$ length × width × height. Therefore the volume of the brick is 12 cm × 6 cm × 4 cm = 228 cm³ (cubic centimeters). Our unit of volume is the cubic centimeter, a cube which is 1 cm × 1 cm × 1 cm.

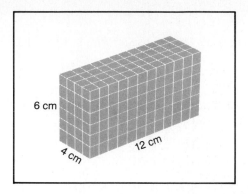

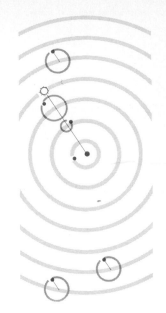

FIGURE 3.2 Measuring the volume of a rectangular solid.

The formula, $V = l \times w \times h$, merely counts the number of cubic centimeters that would occupy the same space as the brick (Figure 3.2).

The *rest mass* of an object is a measure of how much material it contains, a factor independent of its volume. We define the unit of mass to be the *gram*. This is the amount of material in 1 cm³ of pure water at 4°C. How can the mass of an object be determined? A simple balance will be sufficient for an object of reasonable size. Place the object on one side and sufficient standardized masses on the other side until a balance is achieved (Figure 3.3).

Density combines the concepts of volume and mass: it is a measure of the amount of material packed into a given space:

$$\text{density} = \frac{\text{mass}}{\text{volume}}$$

The density of the brick is

$$D = \frac{m}{v} = \frac{552 \text{ g}}{168 \text{ cm}^3} = \frac{3.29 \text{ g}}{\text{cm}^3}$$

$$= 3.29 \text{ grams per cubic centimeter}$$

FIGURE 3.3 Determining the mass of the brick.

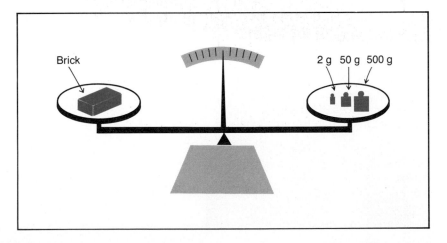

Ecliptic

LEO

Regulus

Planet

Path
of
planet

VOLUME OF THE EARTH

Because the earth is so nearly spherical, we may use a formula which has been derived especially for a sphere:

$$V = \tfrac{4}{3}\pi r^3 \ (r = \text{radius})$$

Using a modern value for the radius of the earth, 6375 km, we find the earth's volume to be

$$V = \tfrac{4}{3}(3.1416) \ (6375 \text{ km})^3$$
$$= 1.085 \times 10^{12} \text{ km}^3 \text{ (cubic kilometers)}$$

The cubic kilometer is not a practical unit for use in the laboratory or for everyday situations, hence the volume might better be expressed in cubic centimeters (6375 km = 637,500,000 cm):

$$V = \tfrac{4}{3}(3.1416)(637,500,000 \text{ cm})^3$$
$$= 1.08 \times 10^{27} \text{ cm}^3 \text{ (cubic centimeters)}$$

MASS OF THE EARTH

We cannot place the earth on one side of a balance to determine its mass, rather we must devise an indirect method. We will use the law of gravity as stated by Sir Isaac Newton (Figure 3.4):

$$F = \frac{Gm_1m_2}{r^2}$$

where .

$F =$ force due to gravity

$G =$ gravitational constant

$m_1 =$ mass of object 1

$m_2 =$ mass of object 2

$r =$ distance between centers of mass

FIGURE 3.4 Sir Isaac Newton. (Yerkes Observatory)

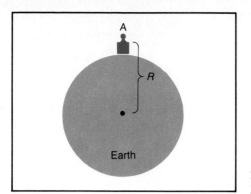

FIGURE 3.5 The weight of an object (A) is the force due to gravity.

Jupiter

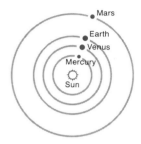

Mars

Earth

Venus

Mercury

Sun

Newton would have expressed the law of gravity as follows:

> The mutually attractive force between two objects is directly proportional to the product of their masses and inversely proportional to the square of the distance between their centers.

As shown in Figure 3.5, when an object A is located on the surface of the earth, its weight is actually the force (F) due to gravity:

$$F = \frac{Gm_A m_E}{r^2}$$

where
$$G = \text{gravitational constant}$$
$$m_A = \text{mass of object } A$$
$$m_E = \text{mass of earth}$$
$$r = \text{radius of earth}$$

Suppose that we place equal masses A and A' on opposite pans of a long-beam balance [Figure 3.6(a)]. The force that the earth exerts on A is stated above. The force on A' would have a similar form:

$$F = \frac{Gm_{A'} m_E}{r^2}$$

These forces are obviously equal, because the mass of A is equal to the mass of A', and so the system is in balance.

Now we roll a large mass B under one pan but not touching it. Mass B exerts an additional force on mass A due to gravitational attraction in the amount of $Gm_A m_B/d^2$, where d is the separation of centers of A and B [see Figure 3.6(b)]. In order to balance this force, a small mass C is added to the other pan, thus adding the force $Gm_E m_C/r^2$. If a balanced condition is achieved in this way, then the force of gravity between objects A and B must just equal the force of gravity between object C and the earth (E).

$$G\frac{m_A m_B}{d^2} = G\frac{m_E m_C}{r^2}$$

Solving this equation for m_E we get

$$m_E = \frac{m_A m_B r^2}{m_C d^2}$$

The Earth   **115**

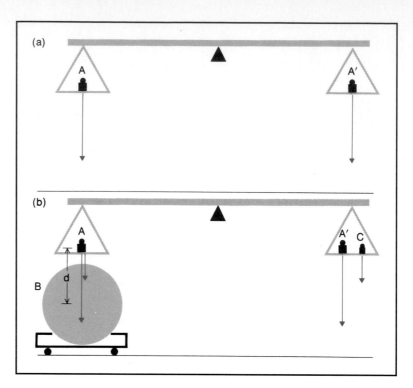

FIGURE 3.6 Measuring the mass of the earth.

Since each member on the right is a known quantity in the experiment, we could determine the mass of the earth (m_E) if such an experiment could be performed. The answer is: $m_E = 5.98 \times 10^{27}$ g. (This value was actually derived from a more complicated method.)

DENSITY OF THE EARTH

We have already defined density as a measure of how much material is packed into a given space:

$$\text{density} = \frac{\text{mass}}{\text{volume}} = \frac{\text{the amount of material}}{\text{the amount of space}}$$

Using the values derived from the mass (m_E) and volume (V_E) of the earth, we may find its average density

$$D_E = \frac{m_E}{V_E} = \frac{5.98 \times 10^{27}}{1.08 \times 10^{27}} = 5.5 \text{ g/cm}^3$$

How this density compares to the density of various common materials is shown in Table 3.1.

Material such as ice, the density of which is less than 1 g/cm³, will float in water. The average density of Saturn is 0.687 g/cm³, less than that of

TABLE 3.1

DENSITIES OF SOME COMMON MATERIALS	
Material	Density (g/cm³)
Air	0.00129
Oxygen	0.00143
Cork (average)	0.25
Gasoline	0.69
Ice	0.91
Water	1.00
Sea water	1.02
Blood	1.04
Aluminum	2.70
Granite (average)	2.70
Iron	7.86
Copper	8.89
Lead	11.35
Mercury	13.59
Gold	19.27
Platinum	21.37

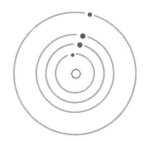

water, whereas Mars has an average density of 3.82 g/cm³, more like that of the earth. Thus the density of an object tells us something of its physical nature. Saturn is more like a ball of gas, having a thick atmosphere, with perhaps a liquid layer beneath and a relatively small dense core in the center. Mars, on the other hand, is characterized by a hard crusty nature, primarily a solid, with only a very thin atmosphere.

The average surface material of the earth has a density of only 2.7 g/cm³, whereas the average density for the entire earth is 5.5 g/cm³. What must the interior of the earth be like in order to make up for this difference? The only logical answer is that the interior must be more dense. The density of the central portion of the earth is thought to range as high as 15 g/cm³ (Figure 3.7). We generally expect increasing density in any body toward its center because of the force of gravity. All parts of a body experience a mutual attraction that has the effect of pulling these parts toward the center. Material near the center experiences increased pressure that tends to force it into a smaller space, thus increasing its density. Furthermore, elements such as iron and nickel, which are naturally more dense, experience a greater force toward the center. When the earth was in a molten state and material was free to distribute itself according to the forces upon it, the more dense material tended to move toward the center of the earth.

How can we verify this trend of increasing density with increasing depth? We have studied the fact that light, a wavelike disturbance, changes

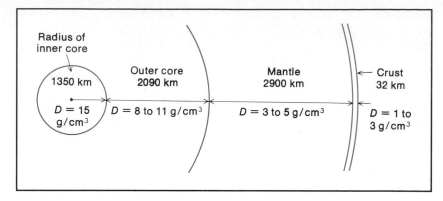

FIGURE 3.7 Layers of the earth.

FIGURE 3.8 Earthquake waves travel at different velocities in various layers of the earth. The refraction which results is an indicator of the density of these layers.

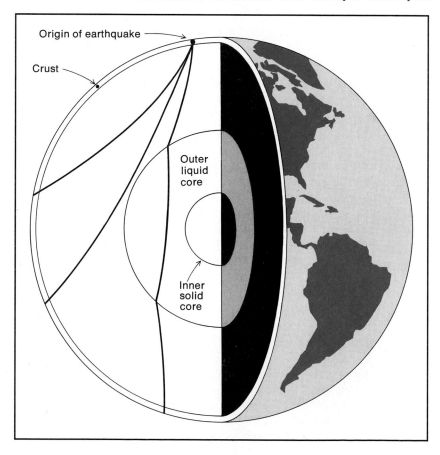

velocity as it passes from air into water and is thereby refracted. A similar phenomenon happens to earthquake waves as they travel through layers of increasing density within the earth. By timing the arrival of earthquake waves at various points over the surface of the earth, researchers can detect distinct layers within the earth to a fair degree of accuracy. The density appears to change quite abruptly at the boundaries of these layers, as indicated by distinct refraction effects at these points (see Figure 3.8). Earthquakes send out two kinds of waves: those with wavelike motion perpendicular to the line of travel, called S waves (shear waves), and those with wavelike motion parallel to the line of travel, called P waves (pressure waves); S waves will not pass through a liquid. Because S waves do not pass through the outer core of the earth, that layer is thought to be liquid in nature. This fact is very significant in terms of the earth's magnetic field, which will be discussed later). .

CRUST OF THE EARTH

As indicated in Figure 3.8, the crust of the earth is relatively thin, and because it is less dense than the mantle, the crust floats on the mantle much as ice floats in water. On this floating crust the arrangement of

FIGURE 3.9 The six major plates of the earth's crust with the convergence or separation boundaries indicated.

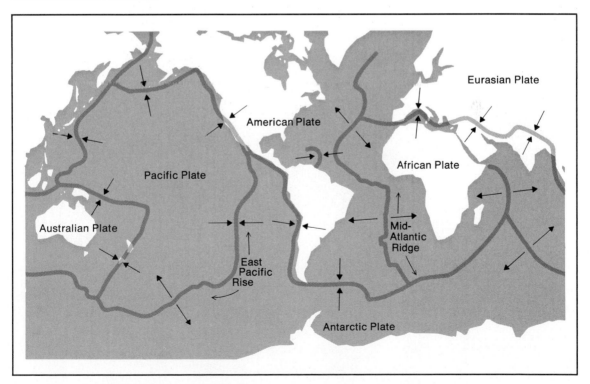

continents is not static. Geologists have traced the relative positions of the continents for at least 200 million years. Six major plates have been identified as shown in Figure 3.9. Along certain plate boundaries, compressional forces are producing mountain-building, volcanic, and earthquake activity; almost the entire Pacific coast of North and South America lies along such a boundary. Centers of earthquake activity virtually outline the crustal plates.

Other boundaries mark plates that are separating. Such a boundary is the Mid-Atlantic Ridge (see Figures 3.9 and 3.10). As plates separate new material wells up to fill in the gap that is created. Thus samples taken from the Mid-Atlantic ridge are very "young." Another indication that the sea floor is spreading is scars that have been left in the bottom of the Atlantic Ocean (Figure 3.10). Studies of the similarity in geologic structure and fossil remains between the land masses of North and South America and those of Africa and Europe confirm that these regions adjoined one another 200 million years ago (Figure 3.11).

FIGURE 3.10 The mid-Atlantic ridge, showing the effects of seafloor spreading as the plates drifted apart. Numbers indicate feet below sea level. (Photograph courtesy of Alcoa)

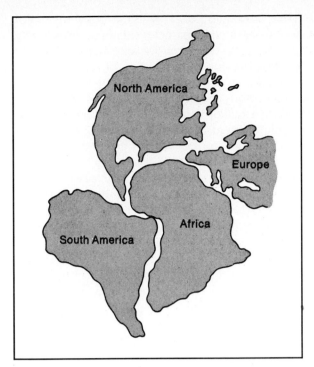

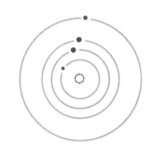

FIGURE 3.11 Approximately 200 million years ago North and South America may have been in very close proximity to the African and European continents.

One of the latest techniques of measuring the motions of crustal plates involves the process of radio interferometry described in Chapter 2, page 96. A radio signal is received by two different radio antenna from a radio source outside our own galaxy. Because of the different path lengths, the signal arrives out of phase and the waves interfere in a partially destructive manner. If the two antennas are located on opposite sides of a plate boundary (a fault), then even a slight change in the relative positions of the two plates (or two sides of a fault) can be detected by a change in the interference pattern. This technique may lead to earthquake prediction based on the theory that slight dilations in the fault take place before a quake.

AGE OF THE EARTH

The rocks of the earth have a story to tell. Within them lie clues to the age and process of formation of the earth. The surface material of the earth is by weight approximately 47 percent oxygen, 28 percent silicon, 8 percent aluminum, and 5 percent iron, with lesser amounts of magnesium, calcium, potassium and other elements.

Many rocks contain traces of radioactive elements—elements that tend to break down into lighter elements at a set rate. If we could observe a certain quantity of uranium, say 8 grams, over a period of 4.5 billion years, we would find that half of that quantity, 4 grams, is transformed through several intermediate steps into lead. How is this possible? Radio-active elements spontaneously eject alpha particles (helium nucleii), beta particles (electrons), and gamma rays from their nucleus. An alpha particle contains two protons and two neutrons. The ejection of such a particle from a uranium atom would therefore reduce the number of protons by two and the total number of protons and neutrons by four, resulting in an atom of thorium. After ejecting a total of 32 neutrons and protons, the atom is changed into lead, a stable element, and no further decay occurs. Another 4.5 billion years would reduce the 4 grams to 2 grams, and so forth; 4.5 billion years is thus spoken of as the *half-life* of uranium. When a rock crystallizes, the participants in this process are locked into place, and the scientist may estimate the age of the rock by determining the proportion of uranium to lead-206. Since lead-206 derives entirely from the radioactive decay of uranium-238, we may assume that the lead-206 present has been produced since the time of crystallization. The age of the rock may also be confirmed by the presence of elements such as radioactive potassium-40, which turns into argon-40, a gas stored in the crystalline structure of the rock. We may assume that any argon gas formed before crystallization of the rock would have been lost into the air.

The oldest rocks found on the surface of the earth date back approximately 3.7 billion years to their time of crystallization. Most meteorites and some of the oldest moon rocks date back 4.6 billion years. A very tempting generalization follows: if the oldest solid material of our solar system dates back 4.6 billion years, then the entire solar system is at least that old. This idea stems from the theory that all the members of the solar system—the sun, nine planets, many moons, numerous asteroids, comets and meteoroids—had a common origin. Details of this theory will be presented in Chapter 6.

ATMOSPHERE OF THE EARTH

Although the atmosphere of the earth is not divided into distinct layers, it will be helpful to think of it in this way. The first layer, the one that contacts the earth's surface, is called the *troposhere*. This layer, varying from 8 to 16 km (5 to 10 mi) in thickness, is the location of all weather disturbances. Three-fourths of the total atmosphere is "packed" into this layer due to gravity, creating a normal sea-level pressure of 1 atmosphere (1 atmosphere (atm) is equivalent to the pressure exerted by a column of mercury 76 cm high at 0°C, or about 15 lb/in^2). Perhaps you have heard atmospheric pressure expressed in terms of millibars on weather forecasts. One thousand millibars is equivalent to 1 atm, hence a pressure of 1024 millibars is slightly higher than normal. Atmospheric pressure is exerted in all directions equally at any given point and so tends to be equalized over a body. We do not normally sense this pressure, but if we dive into a swimming pool to a depth of 3 m (10 ft), we sense the additional pressure on our ear drums, due to the overlying layer of water in addition to the normal atmospheric pressure. The troposphere is composed of 78 percent

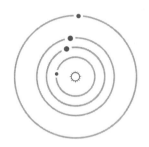

nitrogen and 21 percent oxygen by volume, with water vapor, carbon dioxide, neon, and argon making up most of the remaining 1 percent. Traces of other elements are also found in the atmosphere. The troposphere is characterized by an average temperature that decreases from about +60°F (Fahrenheit) at sea level to −60°F near the top of the layer.

The next layer, the *stratosphere*, extends from about 16 km (10 mi) to a height of 50 km (31 mi) and is characterized by a temperature increasing from −60°F to +32°F. Within this layer there exists a special form of oxygen called *ozone*. Three atoms of oxygen combine to form a molecule of ozone. It has a rather pungent odor that is sometimes noticeable during a lightning storm. The ozone layer absorbs ultraviolet light from the sun, thus protecting us from these dangerous rays. Although it performs this essential function, the ozone layer is nevertheless a handicap to the study of the ultraviolet portion of the spectrum. Observations in this part of the spectrum must be carried out from above the ozone layer with the aid of a balloon or rocket, or from an observatory orbiting in space. All of these methods have been utilized.

Above the stratosphere is the *mesosphere*, which extends from about 50 km (31 mi) to a height of 90 km (56 mi) and is characterized by decreasing temperatures with increasing height (+32°F to −130°F).

Above the mesosphere is the *ionosphere*, a series of layers of *ionized* gases. An *ion* is an atom that has lost or gained one or more electrons, hence is a charged atom (Figure 3.12). Ionization is caused by solar radiation, therefore the degree to which the layers are formed varies from daytime to nightime. These layers range in elevation between 80 km (50 mi) and 320 km (200 mi). They reflect radio waves that have wavelengths longer than 15 m (9.3 ft) (Figure 3.13). These long waves travel around the curvature of the earth by reflection from the ionospheric layers; they are the basis for long-distance radio communication. Wavelengths shorter than 15 m generally penetrate the ionosphere from the earth or from space. This layer, therefore, does not prevent the study of very short wavelength

FIGURE 3.12 Ionized atoms (neutrons not shown).

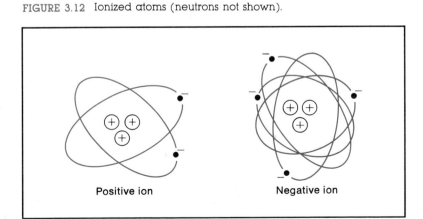

Positive ion   Negative ion

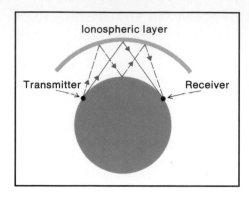

FIGURE 3.13 The ionosphere reflects certain radio waves.

radio energies in the universe. This is often referred to as the radio "window" into space. (TV and FM wavelengths fall in the 10 m to 11 m band; thus they freely pass from earth to outer space.)

The fact that the earth's atmosphere contains ozone, water vapor, and carbon dioxide causes it to act like a greenhouse. The earth receives many forms of energy from the sun including ultraviolet light, visible light, and infrared light. The crust of the earth absorbs this energy and reemits it, primarily in the form of infrared radiation. The atmosphere tends to absorb this form of radiation quite efficiently, and so the earth retains a large amount of heat: in fact, were it not for this "greenhouse" phenomenon, the earth's surface temperature would probably be 45°C cooler.

WHY IS THE SKY BLUE?

Scattering refers to the process whereby the molecule absorbs certain colors of light and then reemits that light in all directions. Because of their very small size, the gas molecules that compose the earth's atmosphere

FIGURE 3.14 The scattering of sunlight produces the blue sky.

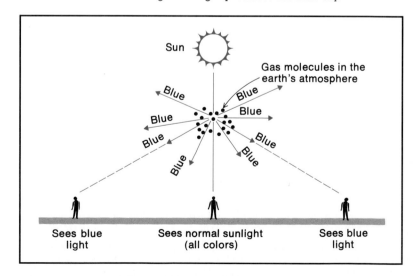

tend to vibrate with a wavelength corresponding to blue light more readily than with that of any other color. If we viewed a collection of molecules directly in line with the sun, we would see the usual color of the sun. However, if that same collection of molecules was viewed from the side, we would see the predominantly blue light which that collection had scattered (Figure 3.14). The red sunset is a result of scattering. As the sun appears to move toward the horizon in the late afternoon, its rays must penetrate an increasing amount of atmosphere. Since molecules in the atmosphere scatter blue light best, the colors that penetrate the atmosphere along our line of sight are predominantly red. The phenomena we call twilight and dawn also result from the scattering of light by molecules in the atmosphere.

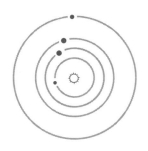

The moon lacks any atmosphere and its sky could best be described as black; when the sun "goes down" at any location on the moon, complete darkness comes on immediately with a severe drop in temperature.

THE EARTH'S MAGNETIC FIELD

It is very evident that the earth possesses a magnetic field. The fact that a compass aligns itself in virtually the same direction on every occasion demonstrates this fact. If we traveled in a northerly direction as indicated by a compass, we would eventually arrive at a point called the *north magnetic pole*. We would find ourselves, however, approximately 1600 km (994 mi) from the *geographic north pole* which is determined by the rotation of the earth. The geographic north pole is a point on the surface of the earth that remains stationary as the earth rotates. (These terms apply to the southern polar region as well.)

In our attempt to explain the cause of the earth's magnetic field we must keep the following observations in mind. Although the north magnetic pole is now located in northeastern Canada, it has not always been located there; it has wandered in a random fashion all around the north geographic pole. Observed changes in the location of the magnetic poles measured in hundreds of kilometers in a period of only a few hundred years rule out the theory of "lodestone" buried in northeastern Canada. The alignment of magnetic minerals in unearthed rocks indicates that the earth's magnetic field has also reversed itself many times in each of the nine epochs of the last 3.6 million years. When magnetic minerals in rocks crystallize from a liquid state, they align their magnetic components with the earth's magnetic field at the time of crystallization and then retain that magnetism until unearthed.

There is one generalized concept that provides at least a partial explanation of the earth's magnetic field. Whenever charged particles are made to flow in a circular motion, a magnetic field is generated. For example, visualize a disk as shown in Figure 3.15. If a charge is placed on each metal disk shown and the larger disk rotated between the hands, a nearby compass will align itself with the magnetic field created. If charged particles exist within the earth, they would be set into circular motion by the rotation of the earth. Perhaps eddy currents (little whirlpools) are set up

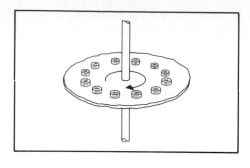

FIGURE 3.15 The magnetic property of a rotating charged plate.

within the liquid outer core producing individual magnetic fields, the total field being the algebraic sum of the individual fields.

Alternately, different layers of the earth and/or different latitudes may rotate at slightly different rates, producing relative motion among the charged particles and corresponding magnetic fields. (See the discussion of the Coriolis effect on page 130 for an explanation of the basis of this theory.) As we continue to observe the other planets with fly-by probes, orbiters, and landers, we will want to compare their magnetic fields with that of the earth. Such a comparison may yield a better explanation of the magnetic field of the earth.

FIGURE 3.16 The earth's magnetic field, distorted by the solar wind. The two Van Allen belts, somewhat doughnut-shaped, are regions of high-energy charged particles trapped by the earth's magnetic field.

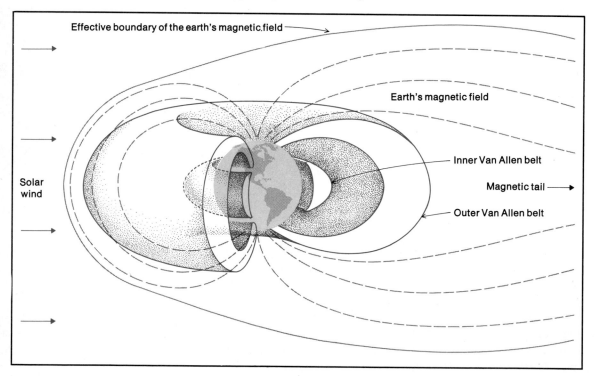

The magnetic field of the earth may be visualized by lines that run from the north magnetic pole to the south magnetic pole. These imaginary lines indicate the direction a compass would point if placed on a line. If a charged particle moved into the magnetic field, a force that acts at right angles to the lines would cause it to move in a circular path around the earth in an east-west circulation. Thus the earth's magnetic field traps some of the rapidly moving charged particles (electrons, protons, and others) that flow from the sun. Two doughnut-shaped regions of these highly energized particles called the *Van Allen belts*, are located at elevations of 3200 km (2000 mi) and 16,000 km (10,000 mi) above the equator; regions over the poles are relatively free of these trapped particles (Figure 3.16).

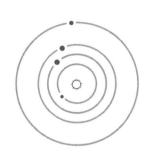

The magnetic field of the earth is sometimes greatly disturbed by activities on the sun. Associated with a solar flare, for instance, is a tremendous release of energy that often disrupts radio communications and produces changes in the magnetic field of the earth. Even when the sun is relatively quiet, there is a constant outflow of charged particles called the *solar wind* that distorts the magnetic field of the earth. The record of satellites that have orbited the earth passing in and out of the regions of the magnetic field, shows that this field is somewhat flattened on the side toward the sun and that a magnetic "tail" exists on the opposite side. Figure 3.16 shows the approximate configuration of the magnetic lines and also shows a cutaway view of the Van Allen belts.

The earth's atmosphere often puts on a real "show" called the *northern lights (aurora borealis) or southern lights (aurora australis)*. On such occasions, regions of the sky near the north or south magnetic poles glow with a light that may take several forms: a general glow, a fanlike shape, curtains hanging in the sky, or spikes shooting upward. Such a display originates from the solar wind. Some of the charged particles which have become entrapped within the Van Allen belts, find their way along the magnetic field lines to the north or south polar region. At this point the charged particles are capable of ionizing atoms of the earth's atmosphere. Such ions recombine with free electrons and downward electron transitions occur within the atom. As you will recall from Chapter 2, downward electron transitions that stop at least temporarily at energy level 2 produce visible light.

ROTATION OF THE EARTH

Early observers saw the sun, moon, planets, and stars rise in the east and set in the west, just as we do today. They assumed that the earth was stationary, however, and therefore that all these objects moved around the earth. This idea persisted as late as the sixteenth century. Even when theories regarding a rotating earth were introduced, no conclusive proof was available until the nineteenth century. In 1851 Jean Foucault, (1819-1868) a French physicist, demonstrated the fact of rotation. He suspended from the dome of the Pantheon in Paris a weighted pendulum and began to swing it toward one entrance (Figure 3.17). Later he found it swinging toward

FIGURE 3.17 The Foucault pendulum in the Pantheon, Paris. (Science Museum, London)

another entrance; the pendulum thus appeared to change the direction of its swing. The change of direction occurred at a given rate, like the hands of a clock. It was reasoned that the earth itself must be turning under the pendulum in order to produce the observed effect. Gravity is the only force that acts on a free-swinging pendulum, and gravity would not produce rotational motion for it acts only toward the center of the earth. If this experiment were performed at the north or south pole, the time required for one apparent rotation of the earth under the pendulum would be approximately 24 hours. As one moves from the poles toward the equator,

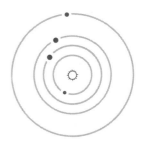

the apparent time for one rotation of the pendulum increases. At the latitude of Los Angeles (34°N) for example, the period is 43 hours. There is no apparent rotation at the equator.

Today astronauts have witnessed the earth's rotation from a vantage point in space. From that point, it is very obvious that this daily rotation of the earth is responsible for the apparent *diurnal* (daily) motion on the sun, the moon, the planets, and the stars. A photograph (time exposure) of the northern polar region of the sky taken over a period of several hours reveals circular star trails (Figure 3.18). Polaris, the North Star, can be seen in the photograph just off-center.

We know that the circumference of the earth at the equator is approximately 40,000 km (24,856 mi). Since the earth rotates once in 24 hours, the speed of a point on the equator is approximately 1600 km/hr (40,000 km / 24 hr) or roughly 1000 mi/hr. Farther to the north or south of the equator, say at latitude 34°, the speed of a point on the earth is closer to 1100 km/hr (700 mi/hr). At the poles, the speed due to rotation is zero. Because different latitudes have different speed of rotation, a rocket fired

FIGURE 3.18 Star trails in the region of the north celestial pole which result from the rotation of the earth. The trail made by the North Star, Polaris, is seen just off center. (Lick Observatory)

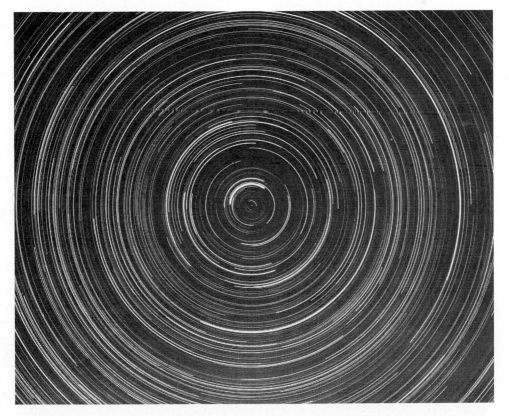

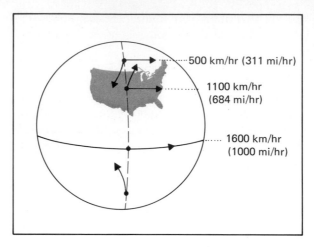

FIGURE 3.19 The Coriolis effect, due to the earth's rotation.

northward from the equator would possess an eastward component of velocity of 1600 km/hr. As it passed northward over more slowly rotating points, the rocket would appear to veer to the east, because its eastward component of velocity would be greater than that of the land over which it flew. If fired southward from a region near the north pole, a rocket would appear to veer to the west, because its eastward velocity would be less than that of the land over which it flew (Figure 3.19). This phenomenon is called the *Coriolis effect*. You can see that, since all moving objects experience this effect, the typical circulation of the wind and ocean currents in the northern hemisphere is clockwise. By the same reasoning you can see that objects moving in the southern hemisphere experience a veering to the left, relative to the earth's crust, and the result is a counterclockwise pattern of currents. The very fact that the Coriolis effect can be observed constitutes another proof of the earth's rotation. The Coriolis effect may also create currents within the earth; if it does, this may explain the earth's magnetic field.

REVOLUTION OF THE EARTH

Proof that the earth revolves around the sun did not come as part of Copernican heliocentric theory, nor as part of Galileo's work though his ideas were eventually generally accepted (see Chapter 1). The proof came, rather, in the nineteenth century, when techniques were developed to detect the apparent shift of nearby stars in relation to the background of more distant stars. If the earth were stationary, a given alignment of a nearby star and one more distant would not vary in a period of a few months. If the earth, however, did change its position in space, the alignment would also change. This apparent shifting of nearby stars against the background of more distant ones has in fact been observed, and the phenomenon is called *stellar parallax*. Since it is a periodic kind of change where a given star apparently shifts first one way and then another during

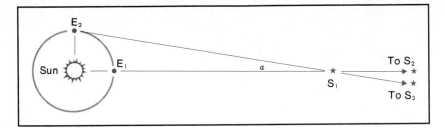

FIGURE 3.20 Stellar parallax, due to the earth's revolution.

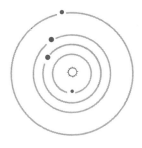

the course of one year, it must be due to the fact that the earth is revolving around the sun. The change is small—only 1/4800 of one degree for the nearest star. In Figure 3.20 the parallax angle (α) is greatly exaggerated.

As a consequence of the earth's revolution, the sun appears to move against the background of stars, aligning itself with different stars each day. The particular constellations through which the sun appears to move are called the signs ("houses") of the zodiac. In Figure 3.21 the sun appears to be aligned with Pisces (the fish), but as the earth revolves about the sun, it will appear next in Aries, and so on; it will spend about one month moving through each of the 12 signs of the zodiac. The time required for the apparent motion of the sun through all the signs of the zodiac, returning to

FIGURE 3.21 As the earth revolves about the sun, the sun appears to align itself with each of the signs of the zodiac. The sun is shown aligned with Pisces, but in one month's time it will align with Aries, etc. In one year, the sun will again align with Pisces.

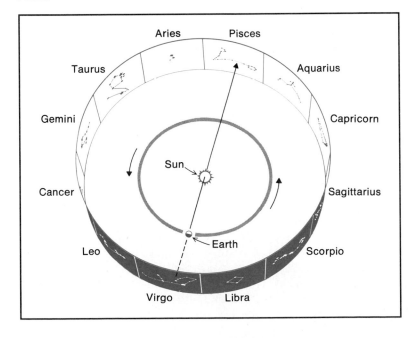

a given starting point is equal to 365 days plus a fraction of a day. This fraction is not obvious from one year's observation. However, when this cycle is observed over several hundred years and an average is taken, the period of the earth's revolution is found to be 365.2564 days. This is called the *sidereal year*—the year as measured by the stars.

THE SEASONS

As the earth revolves around the sun, its path defines a plane (flat surface) called the *ecliptic*. The word "ecliptic" derives from the word "eclipse," for it is within this plane that eclipses of the sun and the moon occur. The ecliptic plane "cuts" through the signs of the zodiac; the circle described by that intersection is the apparent path of the sun from our point of view on earth.

If the earth's axis of rotation were not tilted—that is, if it made an angle of 90° with the ecliptic plane—then the sun's rays would strike the equator of the earth directly at all times and we would experience no seasonal changes. However, the axis of the earth is tilted 23.5° away from a normal (perpendicular) to the ecliptic plane, and this tilt, together with the revolution of the earth around the sun, produces seasonal changes.

In Figure 3.22, position *A* shows the sun's rays striking directly over a point 23.5° north of the equator. Since direct solar radiation is the most intense, *A* represents summer in the northern hemisphere. Position *B*

FIGURE 3.22 The seasons result from the fact that the earth's equator is inclined 23.5° to its plane of orbit (the ecliptic plane).

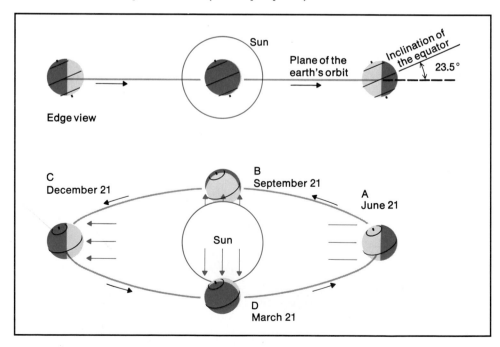

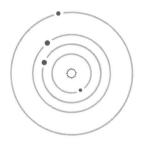

shows the sun's rays directly over the equator and represents autumn in the northern hemisphere. Position *C* shows the rays striking directly over a point 23.5°south of the equator and so represents winter in the northern hemisphere. The extreme northerly position of the sun (*A*) at 23.5°north of the equator determines the Tropic of Cancer; similarly, the extreme southerly sun (*C*) at 23.5° south of the equator determines the Tropic of Capricorn. The Arctic and Antarctic Circles are determined 23.5° from the North and South Poles, respectively (Figure 3.22). Note that someone living just inside the Arctic Circle experiences continual daylight in the later part of June, which is why the Arctic has been called the "land of the midnight sun." This region, however, is in continual darkness in the later part of December.

TIME: HOW LONG IS A DAY?

The most fundamental unit of time to which we relate our patterns of behavior and our routine is the day, yet there are several ways to measure the length of a day. Imagine a line in the sky, running north and south through a point directly over your head. This called your "local meridian." If you counted the time from when a given star crossed your meridian until it crossed again, you would have measured the time required for an earth rotation of exactly 360°. The time equals 23hr 56min and is called a *sidereal* day—a day based on a star reference. If, on the other hand, you counted the time from when the sun crossed your meridian until it crossed again, you would find that the average is 24 hours. This is called a *mean (average) solar day*—a day based on the sun as a reference. Figure 3.23 shows that the sun requires approximately four minutes longer to return to the observer's meridian than does the star. Why is this true?

The answer lies in the fact that while the earth is rotating it also moves along its orbit around the sun approximately 1° per day (360°/365.25 day = 1°/day; therefore, for an observer to see the sun cross his local meridian again requires one extra degree of rotation, a total of

FIGURE 3.23 Observing a sidereal day.

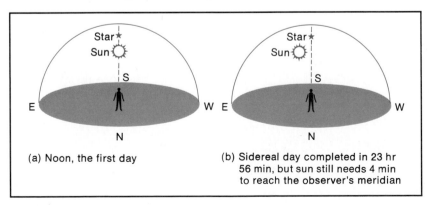

(a) Noon, the first day

(b) Sidereal day completed in 23 hr 56 min, but sun still needs 4 min to reach the observer's meridian

361°. The earth rotates 361° in 24 hours, approximately 15° per hour (361°/24 hr = 15°/hr) or 1° in four minutes, therefore a solar day is four minutes longer than a sidereal day (Figure 3.24). Since most of our daily routine is organized according to the sun's position in the sky—we eat breakfast soon after sunrise, lunch when the sun is high, and dinner soon after sunset—our clocks are built to agree with solar time, not sidereal time. As a consequence, a given star appears to rise and set four minutes earlier each night. Thus, if a star is on your meridian at 9:00 P.M. tonight, you may expect the same star to be on your meridian at approximately 8:56 P.M. the following night. At this rate of four min/day, a given star rises one hour earlier in 15 days and approximately two hours earlier each month. After one year the star will again be on your meridian at 9:00 P.M. Likewise, each constellation appears to shift a little night after night, and gradually new constellations dominate the early evening sky. This is why we speak of some constellations as being summer constellations and others as autumn, winter, or spring constellations (see Figure 1.3, pages 8 and 9).

In determining a mean solar day, the time period that clocks reflect, we presume that the earth is moving through the same angle of revolution each day. In reality, this is not the case. Although the earth's rate of rotation is very nearly constant, its rate of revolution changes; the earth moves fastest when closest to the sun and slowest when farthest from the sun, as expressed by Kepler's second law (see Chapter 1). The earth thus revolves through an angle of more than 1° per day in the winter and less than 1° per day in the summer. For this reason, the sun may actually appear on your

FIGURE 3.24 The sidereal and the solar day.

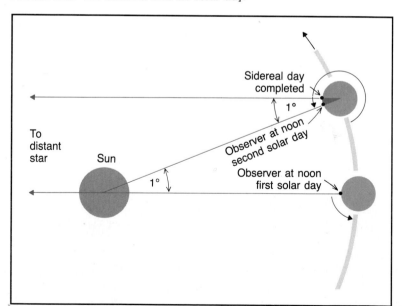

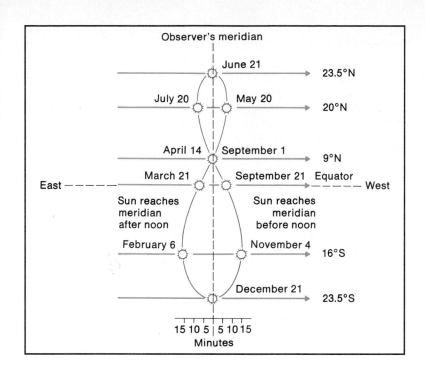

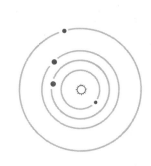

FIGURE 3.25 The analemma.

local meridian as much as 15 minutes before or after noon according to your clock (mean solar time). The difference between the apparent solar time and mean solar time is called the *equation of time*. The *analemma*, found on most globes of the earth, is a graphic representation of the sun's yearly cycle of changing latitudes and of its arrival time at a given local meridian. In Figure 3.25 we can read out these two facts for any day of the year. For instance, on June 21 the sun is 23.5° north of the equator (traveling over the Tropic of Cancer) and it will cross the observer's meridian at noon (no correction needed). On February 6 the sun is 16° south of the equator and will cross the observer's meridian 15 minutes after noon. Other corrections may be needed if you do not live in the center of a time zone or if you are on daylight savings time.

LATITUDE AND LONGITUDE

The location of any point on the earth's surface may be indicated by specifying two angles, one called *latitude* and the other *longitude*. How was this system devised? The north and south poles are two very special points on the earth, for the rotation (spinning) of the earth does not move them. Halfway between the poles we imagine a circle called the equator. The angle which a point makes with the equator is called its latitude. In Figure 3.26, angle beta is the latitude of point *P*. Latitude, specified as north (N) or south (S) of the equator, may be read off a chart showing a system of

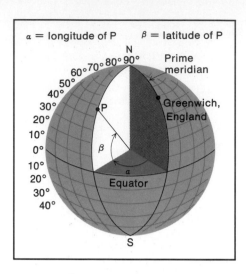

FIGURE 3.26 A system of longitude and latitude.

circles drawn parallel to the equator called *parallels of latitude*. These parallels are usually shown for each 10° of latitude.

In measuring east or west (longitude) there was no natural starting point, therefore an arbitrary choice was necessary. Because of the large number of observations of star positions that had been made at the Royal Greenwich Observatory just outside London, this location was selected as the starting point for measuring longitude. The line running from the North Pole through Greenwich, England, to the South Pole is called the *prime meridian*. Additional meridians of longitude may be visualized through any point; however, on a globe map we usually see one meridian for each hour of the earth's rotation, or 24 meridians in all. Each meridian makes an angle of 15°(360°/24 = 15°) with its neighbor. The smaller angle between the prime meridian and a meridian through point *P* is called the longitude of *P* (angle alpha in Figure 3.26). The longitude of a point is specified as east (E) or west (W) of Greenwich, the prime meridian. Table 3.2 gives the latitude and longitude of some cities around the world.

The meridians may also be thought of as hour lines, for when it is noon at Greenwich, it is 11:00 A.M. at the first meridian (15°W), 10:00 A.M.

TABLE 3.2

LONGITUDE AND LATITUDE FOR SEVERAL WELL-KNOWN CITIES		
City	Longitude	Latitude
Washington, D.C.	W 77°04′	N 38°55′
Los Angeles	W 118°20′	N 34°10′
Honolulu	W 157°45′	N 21°22′
Rome	E 12°32′	N 41°50′
Rio de Janeiro	W 43°10′	S 22°40′

at the second meridian (30°W), and so on. Thus in Los Angeles, which is near the eighth meridian (120°W), it is 4:00 A.M., eight hours earlier than Greenwich. Celestial events are usually given in Greenwich Time, or Universal Time. In order to convert to local time, subtract the number of hours equal to the number of standard meridians from Greenwich west to your city. Observers near Los Angeles, for example, would subtract eight hours to get Pacific Standard Time (or seven hours to get Pacific Daylight Time). Table 3.3 gives the correction for US and Canadian time zones.

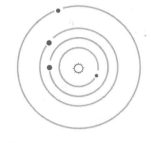

TABLE 3.3

FROM UNIVERSAL TIME, SUBTRACT TO GET LOCAL TIME.	
Pacific Standard Time—8 hr	Central Standard Time—6 hr
Pacific Daylight Time—7 hr	Central Daylight Time—5 hr
Mountain Standard Time—7 hr	Eastern Standard Time—5 hr
Mountain Daylight Time—6 hr	Eastern Daylight Time—4 hr

Likewise, cities which lie to the east of Greenwich must add one hour for each meridian. These same meridians serve as approximate centers for each time zone (Figure 3.27). The rather zig-zag boundaries of time zones are made so that cities will not be divided into two different time zones.

FIGURE 3.27   Time zones. The solid lines represent standard meridians usually shown on a globe of the earth. The shaded areas show the approximate boundaries of time zones in the United States.

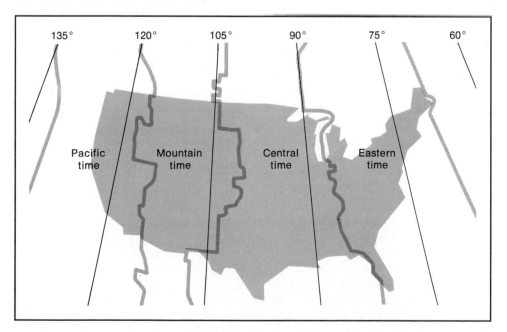

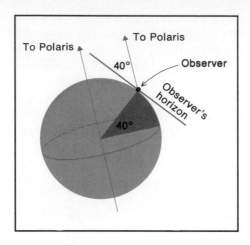

FIGURE 3.28 The angle which Polaris makes with an observer's northern horizon marks his or her latitude.

The latitude of a given point in the northern hemisphere may be approximated by noting the angle that the pole star, Polaris, makes with the observer's horizon. The position of the pole star is determined by extending the axis of the earth's rotation until it penetrates the sky. If you lived at the North Pole, you would expect to see Polaris directly overhead, near your *zenith*. It would make an angle of 90° with your horizon, an angle equal to the latitude of the North Pole. If you lived on the equator, you would expect to see Polaris near your northern horizon, making an angle of 0°, which is equal to the latitude of the equator. Taking an intermediate point—for example, 40° north of the equator—we find that Polaris indeed makes an angle of about 40° with the northern horizon (Figure 3.28). This method of approximating latitude works because the earth is essentially a sphere.

RIGHT ASCENSION AND DECLINATION

Astronomers specify the location of stars by a system which is very similar to longitude and latitude. Let us imagine that the stars are located on a transparent sphere, called the *celestial sphere*, at a given distance from the earth. When the earth's axis is extended until it intersects the celestial sphere, it determines the *celestial poles*, and the projection of the earth's equator onto the celestial sphere produces the *celestial equator*. In the same way, the projection of the meridians and parallels of latitude complete the coordinate system in the sky. As was true for longitude measure on earth, we must define a starting point (and a starting circle) from which to measure eastward in the sky. We cannot relate such a point to Greenwich, for no point in the sky remains over Greenwich. Let's look to circles already defined in the sky; the ecliptic (the plane of the earth's orbit and apparent path of the sun) and the celestial equator (an extention of the earth's equator). Visualize the sun moving along the ecliptic from its more southerly position in the winter and just crossing the celestial equator on

TABLE 3.4

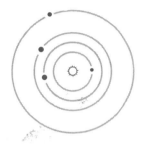

THE TERRESTRIAL SYSTEM AND THE CELESTIAL SYSTEM	
Terrestrial	Celestial
North and South Geographic Poles	North and South Celestial Poles
Geographic equator	Celestial equator
Parallels of latitude:	Declination circles:
Latitude (N or S)	Declination (+ or −)
Greenwich, England	Vernal Equinox
Prime meridian	Zero hour circle
Meridians of longitude:	Hour circles:
Longitude (E or W)	Right ascension
	(measured eastward only)

its way northward. The date is March 21 and the point of crossing in the sky is called the *vernal equinox*. Now visualize a circle running from the north celestial pole, through the vernal equinox, and on to the south celestial pole. This line is called the *zero hour circle*, and it will play the same role in the sky as the prime meridian played on earth. From this zero hour circle, place another hour circle every 15° to the east, naming them in succession—the first hour circle, the second hour circle, and so on—until you have numbered through the twenty-third hour circle. The twenty-fourth hour circle is then identical to the zero hour circle. Table 3.4 lists the parallels between the terrestrial (earth) system and the celestial (sky) system.

The angular measure corresponding to eastward longitude on earth is called the *right ascension* of a star and is measured in units of time. If a star has a right ascension of 5hr 30min, we know that is halfway between the fifth-and sixth-hour circles. Since we have 360° divided into 24 equal segments, each hour circle must make an angle of 15° (360°/24hr = 15°/hr) with its neighbor. If a star has a right ascension of 5hr 30min, then it makes an angle of 82.5° with the zero hour circle (5.5 × 15° = 82.5°).

The north-south angular measure corresponding to latitude on earth is called the *declination* of a star and is measured by the angle a star makes with the celestial equator. Declination is specified as positive (+) if north of the equator and negative (−) if south of it. The position of any star thus may be given by two numbers. The star Capella, for example, has a right ascension (R.A.) of 5hr 13min and a declination (Dec.) of +45° 57′ (Figure 3.29). To test your understanding of the system of right ascension and declination, practice reading out the R.A. and Dec. for the brightest stars shown on your star maps (Appendix 13) and check your work by referring to these same coordinates as listed in Appendix 8 (The Twenty Brightest Stars).

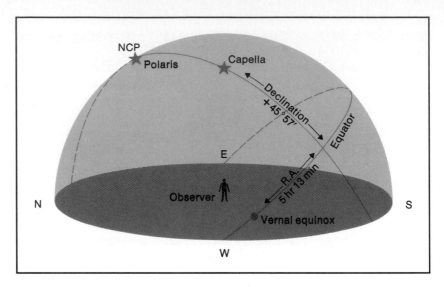

FIGURE 3.29 Circles in the sky, showing the right ascension and declination of the star Capella. NCP refers to the north celestial pole (Polaris).

APPARENT MOTIONS OF THE SKY

Although the sky seems to move overhead at the rate of one hour circle for each sidereal clock hour, we should remember that the earth's rotation produces this illusion. Since a given star appears to cross your local meridian about four minutes earlier each night, we should expect the hour circles also to shift by an equivalent amount each night. Table 3.5 presents

TABLE 3.5

	HOUR CIRCLE OVERHEAD AT 9:00 P.M. LOCAL TIME				
Date	Hour Circle	Date	Hour Circle	Date	Hour Circle
Nov. 6	0	Mar. 21	9	Aug. 6	18
Nov. 21	1	Ap. 6	10	Aug. 21	19
Dec. 6	2	Ap. 21	11	Sept. 6	20
Dec. 21	3	May 6	12	Sept. 21	21
Jan. 6	4	May 21	13	Oct. 6	22
Jan. 21	5	June 6	14	Oct. 21	23
Feb. 6	6	June 21	15	Nov. 6	24$_a$
Feb. 21	7	July 6	16		
Mar. 6	8	July 21	17		

aSame as zero.

the hour circle that will be overhead at 9:00 P.M. (local time) for various dates throughout the year. *Local time* refers to the standard clock time shown for any observer in a particular time zone.

Note that on March 21 the sun is located at the point of the vernal equinox, in the sky, and thus the zero hour circle will be overhead at noon (local solar time), the first hour circle is overhead at 1:00 P.M., the second hour circle is overhead at 2:00 P.M., and so on. By 9:00 P.M. the ninth hour circle is overhead. By remembering these facts, together with the fact that any given hour circle reaches your local meridian two hours earlier each month, you could create this Table 3.5 from memory.

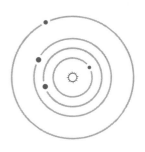

ALTITUDE-AZIMUTH SYSTEM

A coordinate system which may seem more natural to a specific observer is the altazimuth system. The observer's horizon, zenith, and the direction toward the north star become the reference points from which a star's location is specified. The *altitude* of the star is merely the angle measured along a vertical circle through the zenith from the horizon (Figure 3.30). The *azimuth* of the star is the angle measured clockwise from north along the horizon to the vertical circle. Many modern telescopes have altazimuth mounts and must be driven in both directions in order to follow a star's diurnal motion. Only the advent of the computer-controlled drive system has made this a useful design.

FIGURE 3.30 The position of a star, relative to an observer, may be specified by two angles—azimuth and altitude.

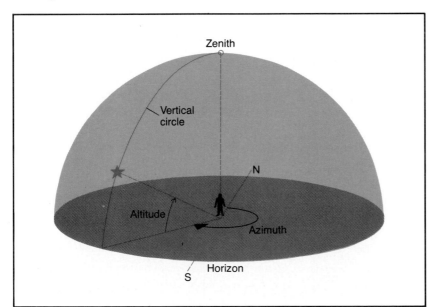

PRECESSION OF THE EARTH

Rotation and revolution are very familiar motions of the earth. Another periodic but not so familiar motion of the earth is *precession*. The completion of one cycle requires not 24 hours or 365.25 days but about 26,000 years; its effect is detectable only over a long period of time. During this period, the earth's axis will align itself with new points in the sky other than the location of Polaris. We can compare this motion to that of a toy top which, while it is spinning, slowly describes a circle at the upper end of its axis (Figure 3.31). Like the toy top, the earth bulges at its equator and is rotating on a tilted axis. The gravitational force that the sun and moon exert on the earth tends to straighten that tilt of axis; however, as with any rotating body, there is a tendency to maintain a given tilt of axis called the gyroscopic effect. Since the earth resists having its axis of rotation straightened, the gravitational energy produces the precessional motion of the earth (Figure 3.32). Because of precession, the axis of the earth over a 26,000-year period describes a circle in the sky with a radius of 23.5°. The axis will in time therefore no longer point to Polaris. In the year 14,000 A.D., the bright star Vega will be within a few degrees of the north celestial pole. This precessional motion causes the vernal equinox to move westward in the sky approximately 50 seconds of arc per year, and for this reason the full name of this motion is *precession of the equinoxes*.

Since one complete cycle of precessional motion requires approximately 26,000 years, the vernal equinox remains in each of the 12 houses of the zodiac for just over 2000 years. The vernal equinox has been passing through the constellation of Pisces since about the start of the Christian era. On the fall star map in Appendix 13, you may note that the vernal equinox now falls near the western part of Pisces; because of its westward motion it will move into the constellation of Aquarius early in the twenty-first century, ushering in the "Age of Aquarius." This is a happening of special interest to astrologers. Astronomers must also be aware of this

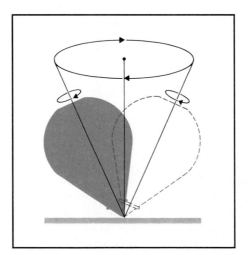

FIGURE 3.31 Precession of a spinning top.

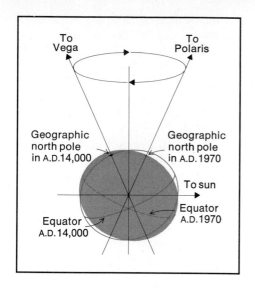

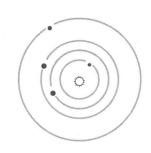

FIGURE 3.32 Precession of the earth.

gradual change; any shift in the position of the vernal equinox changes the right ascension and declination of all stars since the vernal equinox defines the zero hour circle in the sky. Right ascension and declination values given in Appendices 8 and 9 reflect the coordinates for 1950.

THE CALENDAR

The ideal calendar is one that reflects seasonal changes as accurately as possible. It should contain an average of 365.2422 days. This period, called the *tropical* year (the year of the seasons), is the time between successive passing of the sun through the vernal equinox. It is shorter than the sidereal year of 365.2564 days by 0.0142 days due to precession. In 46 B.C. Julius Caesar decreed a calendar that reflected the tropical year to a fair degree of accuracy. He believed that the year contained $365\frac{1}{4}$ days, thus added one day every fourth year and called it a leap year. This left an error amounting to the difference between 365.2500 and 365.2422, or 0.0078 days/year. This seems insignificant, yet in a thousand years it amounts to 7.8 days. Incidentally, Julius Caesar was partially responsible for placing more days in one month than in another. His month, July, contains 31 days. His successor, Augustus Caesar, had to have at least as many days in his month, August. In the process of calendar reform Augustus robbed February, leaving it with only 28 days and in leap years with 29 days.

In 1582 Pope Gregory brought further reform, first by correcting the date, which had fallen out of synchronization with the seasons since the adoption of the Julian calendar. He did this by removing 10 days from the month of October of that year. The Gregorian calendar, which we use today, more closely reflects the tropical year by the following correction to Julius Caesar's leap year formula: three times in a period of 400 years, the extra day normally placed in a leap year is deleted. For example, in the century years of 1700, 1800, 1900, 2100, 2200, and 2300, the extra day is deleted. In the century years of 1600, 2000, and 2400 (century years divisible by 400) it is retained. This plan yields an average length of 365.2425 days—very nearly the 365.2422 days of the tropical year. The Gregorian calendar is accurate within a 1-day error in 3500 years.

OTHER MOTIONS OF THE EARTH

We have studied three basic motions of the earth: rotation, revolution, and precession. The earth has additional motions that are due to the movement of the entire solar system as a unit. The sun, along with some 100 billion other stars, participates in the general rotation of the Milky Way galaxy. The sun is located approximately two-thirds of the way from the center of the galaxy to its edge; the speed of stars in this region is approximately 240 km/sec (860,000 km/hr or 534,400 mi/hr). The Milky Way Galaxy, furthermore, is moving through space in relation to other galaxies; the sun shares this additional motion and "carries" the earth with it. These motions are discussed in more detail in later chapters.

QUESTIONS

1. What is the name given to the lines on the celestial sphere that correspond to the meridians on earth?
2. What are the two fixed points on the earth's surface, in relation to its rotation?
3. What motion of the earth is primarily responsible for creating the seasons?
4. A spherical model of the earth (a globe) usually shows 24 meridians. How many degrees separate each meridian?
5. What is the length of time required for the earth to rotate from one meridian to the next, assuming a total of 24 meridians on the earth?
6. If the earth were tilted more than 23.5°, say 35°, would the seasons be longer? Would the winter season be more severe? Why?
7. At the present time, the sun is in the constellation of Gemini during the month of July. In ancient times it was in Gemini during the month of May. What phenomenon has caused this change?
8. If you lived on the equator, what percentage of the sky could you see if you stayed up all night?
9. How is the vernal equinox point determined?
10. Through how many hour circles does the sun appear to move each month?
11. What property of an object is measured by each of the following: volume, mass, and density?
12. In going from the crust of the earth to its core, might we expect to find increasing or decreasing density?
13. Eratosthenes observed that the sun's rays struck one city in Egypt at a different angle than in another city 800 km to the north at noon on a given day. How did this observation prove that the earth is not flat?

14. What is the most abundant element in the earth's atmosphere?

15. Describe the *solar wind* and discuss several effects it may have on the earth.

16. Describe an experiment by which the daily rotation of the earth is proved.

17. Describe an experiment by which the yearly revolution of the earth is proved.

18. What basic motion of the earth causes the sidereal day to differ from the solar day?

19. Twice during the year, the sun appears to move directly over the equator of the earth. List the approximate dates of these occurrences.

20. An observer is located 34° north of the equator. Where will Polaris (the North Star) appear from his point of view?

21. Which basic motion of the earth causes different constellations to be seen on a winter evening than on a spring, summer, or fall evening?

22. What properties of the earth cause it to precess?

23. How do stars seen by a person in North Africa (say in Casablanca, Morocco, latitude 34° N) compare with the stars seen from Mt. Wilson in California (latitude 34° N) in the evening of the same day?

24. As compared to residents of Florida, New York residents experience which of the following (several answers may be true): (a) longer summer days, (b) longer winter days, (c) same-length days all year, (d) shorter summer days, (e) shorter winter days. (*Note*: Here the word "days" refers to the length of daylight hours.)

25. If you place a stick in the ground at noon on the longest day of the year and then mark the tip of its shadow on the ground at precisely noon each day, what shape would the marks on the ground make in one year?

26. When it is 11:00 A.M. in New York City (approximately 75° W longitude), what is the time in a city of 105° longitude?

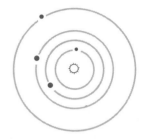

The Moon

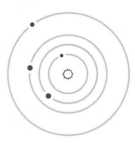

Four

We often think of the motions of the earth and moon separately; we picture the earth in its annual motion around the sun, and the moon in its monthly motion around the earth. In reality, the earth and moon are better thought of as a unit—a binary (dual) system orbiting the sun. The term *binary system* may refer to any two objects, but it usually refers to two objects that have an observable effect on each other's motion due to their mutual gravitational force. With the earth-moon system, the effect of the interaction is very obvious. If you flip through the left-hand flip pages beginning on page 180, you will see that the earth deviates from a smooth path around the sun by a significant amount, approximately 4800 km (2983 mi) on either side of an elliptical path marked by the dashed line. The moon also orbits the sun but deviates approximately 384,000 km (238,600 mi) on either side of that same dashed path. The only point which moves along the smooth path around the sun is the *barycenter* of the system, the point at which the earth and moon would balance if placed at either end of an imaginary rod 388,800 km (241,600 mi) long.

Since we already know the mass of the earth, this balancing idea allows us to determine the mass of the moon. Suppose a father tries to seesaw with his young daughter and they find they can balance only if the daughter sits three

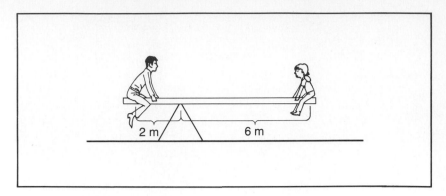

FIGURE 4.1 Father and daughter balance on a teeter-totter.

FIGURE 4.2 If the earth and moon were placed at opposite ends of a rod, the system would balance at the barycenter.

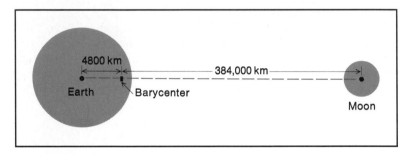

times as far from the pivot, or *fulcrum*, as the father (Figure 4.1). If we were to conclude from this observation that the daughter weighs only one third as much as the father, we would be correct. Based on this relationship, given that the barycenter of the earth-moon system is 4800 km (2983 mi) from the center of the earth and 384,000 km (238,600 mi) from the center of the moon (Figure 4.2), we conclude that the moon has only 1/80 the mass of the earth (4800/384,000 = 1/80), or

$$1/80 \times 5.98 \times 10^{27} \, g = 7.47 \times 10^{25} \, g$$

BOUNCING A LASER BEAM OFF THE MOON

We may determine the size of the moon if we know its distance from earth and the apparent angle its diameter makes with our eye. Today, to measure the moon's distance we merely bounce a laser beam off the moon and record the signal's round-trip time. Given that laser light travels at 300,000 km/sec (186,000 mi/sec), if it requires 2.5627 seconds for the signal to travel to and from the moon on a certain night, the round trip distance is equal to

$$2.5627 \text{ sec} \times 300{,}000 \frac{\text{km}}{\text{sec}} = 768{,}800 \text{ km} \ (447{,}700 \text{ mi})$$

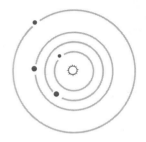

The one-way distance then is 384,400 km (238,900 mi). If on that same night the diameter of the moon appeared to make an angle of 0.518° with your eye, then we might picture these measurements as in Figure 4.3 and use the following reasoning to determine the size of the moon. The angle of 0.518° compares to the full circle of 360° in the same way that d, an arc of the circle, compares to c, the circumference of the circle recalling that the circumference is equal to 2π multiplied by the radius of the circle ($C = 2\pi r$), the diameter is calculated as follows:

$$\frac{d}{C} = \frac{0.518°}{360°}$$

$$\frac{d}{2\pi(384{,}400 \text{ km})} = \frac{0.518°}{360°}$$

$$d = 3476 \text{ km} \ (2160 \text{ mi})$$

The moon's diameter is thus about one-quarter that of the earth, yielding a volume of $2.3 \times 10^{25} \text{ cm}^3$, or about 1/50 the earth's volume. This is a useful comparison when we try to visualize just how large the moon really is. When both the mass of the moon and its volume are considered, its average density may be computed as follows:

$$\text{density} = \frac{\text{mass}}{\text{volume}} = \frac{7.5 \times 10^{25}}{2.3 \times 10^{25}} = 3.3 \frac{\text{g}}{\text{cm}^3}$$

Many of the moon rocks have proven to be almost this dense. Before humans landed on the moon, observers thought the moon was not *differentiated* (having layers of increasing density toward the center) but was

FIGURE 4.3 Finding the diameter of the moon.

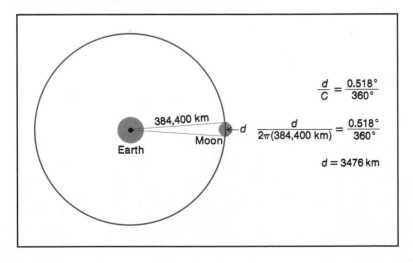

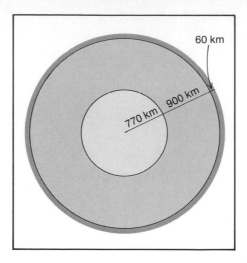

FIGURE 4.4 Layers of the moon.

homogenous throughout. They reasoned that a body as small as the moon would not experience pressures sufficient to produce differentiation. Several seismic (moonquake) stations left on the moon by Apollo crews, however, reveal quite a different picture. Although the typical moonquake is rather mild, registering 1 or 2 on the Richter scale, changes have been measured in the velocity of seismic waves at varying depths within the moon. One such change occurs at a depth of about 60 km (37 mi) and another at a depth of about 1000 km (621 mi). Because most moonquakes originate at these latter depths, we conclude that the moon probably has a crust about 60 km thick, a solid (rigid) mantle underneath, and then a

FIGURE 4.5 Eleven of the twelve known concentrations of mass on the moon. The mascons seem to be associated with a definite type of mare. (JPL-NASA)

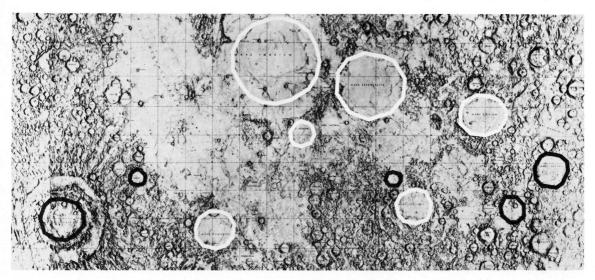

central core (Figure 4.4). Seismographs of meteor impact on the far side of the moon yield a clue as to the nature of the core. Although both shear and pressure waves are generated by such an impact, in passing through the moon's core the shear waves are lost. This indicates that the core may be molten. Estimates of the temperature of this core range upward to 1500°C.

Another interesting method of studying the subsurface of the moon has been developed: The Apollo command module, orbiting around the moon, sends radar signals of various wavelengths toward the moon and times their return, or echo. Signals of 115-cm wavelength penetrate the lunar soil approximately 20 m (65 ft) and signals of 13-cm wavelength penetrate as much as 200 m (650 ft). Scientists conclude that what now appear to be fairly smooth lowland regions of the moon represent valleys and basins that were filled in by molten lava at some stage in the moon's evolution. When a spacecraft flies over certain of these *maria* (lunar "seas"), it experiences an increased gravitational attraction, which indicates the presence of higher concentrations of mass. These concentrations are called *mascons*; they are the circled areas in Figure 4.5. Most observers believe that the original basins in which the maria formed were blasted out by the impact of huge objects in the early stages of the moon's history. Whether those objects were buried in the basins, thus accounting for the mascons, is not known. It seems more likely that the basins were filled by lava more dense than the surrounding material. It is interesting to note that mascons do not exist on the far side of the moon, where large maria are not filled with dark lava (See figure 4.14, page 159). Mascons are known to also exist on earth, where they are usually referred to as gravitational anomalies.

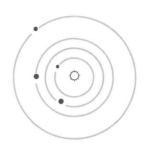

SURFACE FEATURES

The most obvious distinction that can be made in the surface features of the moon is between the flat lowlands called maria (seas), and the *highlands* (mountain ranges), which appear both lighter in color and brighter than the maria. Although the lowlands are called "seas," they contain no water. Because it lacks an atmosphere, the moon cannot maintain liquid water on its surface; liquid water exists on the earth's surface as a result of the pressure created by the earth's atmosphere. Furthermore, Apollo rock samples, unlike earth rocks, do not show water chemically bonded in the minerals. The circular maria associated with mascons such as those illustrated in Figure 4.5 range in diameter up to 1130 km (702 mi); irregular maria such as Oceanus Procellarum are significantly larger. The highlands extend for hundreds of kilometers and reach heights of over 6 kilometers above the level of the maria. These lunar highlands bear names of terrestrial mountain ranges such as the Alps and Pyrénées. Both maria and mountain ranges are identified in Figures 4.7 and 4.9, and their photographic counterparts are shown in Figures 4.6 and 4.8.

Craters represent still another major surface feature of the moon. Ranging in size up to 240 km (150 mi), craters are named for persons of

FIGURE 4.6 First-quarter moon (7 days old). North is at top, as would be seen by the naked eye. (Lick Observatory)

science and politics. Note that the highlands are dominated by craters, whereas relatively few occur within the maria; lava flows may have obliterated craters there.

Many of the surface features shown in Figures 4.6 through 4.9 can be seen with a pair of binoculars or a small telescope. If binoculars are used, the moon will appear as illustrated with its first-quarter phase apparently

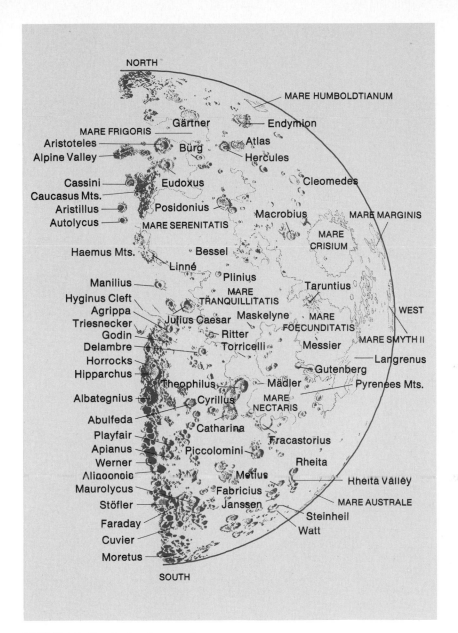

NORTH

MARE HUMBOLDTIANUM

Gärtner · Endymion

MARE FRIGORIS

Aristoteles · Atlas

Alpine Valley · Bürg · Hercules

Cassini · Eudoxus · Cleomedes

Caucasus Mts.

Aristillus · Posidonius · Macrobius · MARE MARGINIS

Autolycus · MARE SERENITATIS · MARE CRISIUM

Haemus Mts. · Bessel

Linné

Plinius

Manilius · MARE TRANQUILLITATIS · Taruntius

Hyginus Cleft

Agrippa · Julius Caesar · Maskelyne · WEST

Triesnecker · MARE FOECUNDITATIS

Godin · Ritter · MARE SMYTH II

Delambre · Torricelli · Messier

Horrocks · Langrenus

Hipparchus · Gutenberg

Theophilus · Mädler · Pyrenees Mts.

Albategnius · Cyrillus · MARE NECTARIS

Abulfeda · Catharina

Playfair · Fracastorius

Apianus · Piccolomini

Werner · Rheita

Aliacensis · Metius · Rheita Valley

Maurolycus · Fabricius · MARE AUSTRALE

Stöfler · Janssen

Faraday · Steinheil

Cuvier · Watt

Moretus

SOUTH

FIGURE 4.7 Key to the first-quarter moon.

lighted from the right. This is called the "naked-eye" view. If a simple telescope is used, however, the image will appear upside-down and backward, right to left. A simple rotation of the figure in your book will allow for comparison. See how many features you can identify. Because shadows tend to clarify surface details, the best time to view the moon is in its early or waxing phases (crescent, first quarter, and gibbous) and in its later or

FIGURE 4.8 Third-quarter moon. North is at top, as would be seen by the naked eye. (Lick Observatory)

waning phases (gibbous, third quarter, and crescent) (Figure 4.20, page 164). During these phases a portion of the moon is lighted and a portion is dark. The boundary between these two portions of the moon is called the *terminator*. Each night, as the phase of the moon changes, the terminator moves to a new location, revealing added detail. By contrast, when the moon is full, no shadows exist and little detail is seen. During early or late

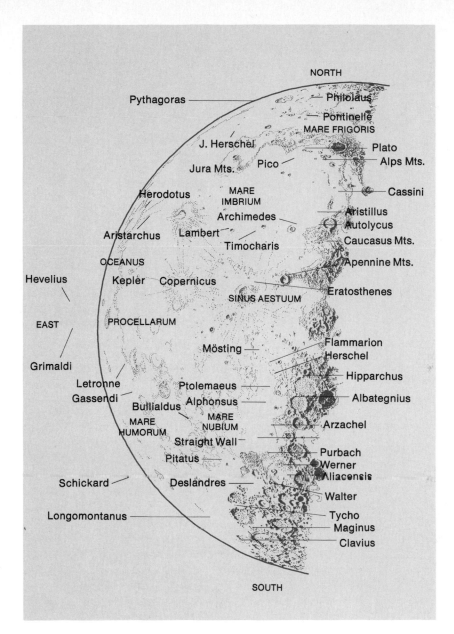

FIGURE 4.9   Key to the third-quarter moon.

phases, shadow lengths give very accurate evidence of the heights of features above the surface (Figure 4.10).

Two kinds of valleys exist on the moon: the type that one would expect in any mountainous region (Figure 4.11), and the type represented by the Prinz Valleys, shown in Figure 4.12. They flow downhill, however, this is not sufficient evidence to conclude that they were formed by water

FIGURE 4.10 The Crater Theophilus. The lengths of the shadows reveal the height of the rim of this crater to be approximately 4.2 km and the central peak to be 2.25 km. The diameter is approximately 104 km (65 miles). (Yerkes Observatory)

FIGURE 4.11 Scientist-astronaut Harrison H. Schmitt standing next to a huge lunar boulder during the third Apollo 17 extravehicular activity—EVA-3 at the Taurus-Littrow landing site. (NASA)

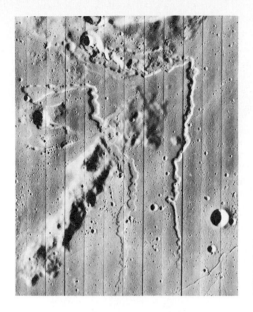

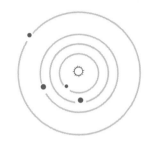

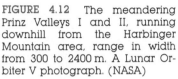

FIGURE 4.12 The meandering Prinz Valleys I and II, running downhill from the Harbinger Mountain area, range in width from 300 to 2400 m. A Lunar Orbiter V photograph. (NASA)

flow. Rather they may be the result of faulting (crustal movement) and slumping of the lunar soil. The *rilles* that mark much of the moon's surface resemble faultlike features on earth and may have originated from moonquake activity [Figure 4.13(a) and (b)].

One side of the moon is perpetually turned toward the earth; the far side had never been seen until a probe was placed in lunar orbit. As Figure 4.14 shows, the appearance of the far side is quite different from the near side; it is cratered rather uniformly and is almost devoid of dark maria. The explanation of this difference in terms of the origin and evolution of the moon poses a significant challenge.

MOON ROCKS

The samples of the moon brought back by the Apollo astronauts have an exciting story to tell. Their composition and age are important clues in the unraveling of the moon's history. There are three basic classes of moon rock:

1. *Basalt,*—an igneous rock much like the lava we find on earth. (An igneous rock is a rock that was once melted.) (Figure 4.15).
2. *Anorthosite,*—another igneous rock but one that cooled slowly beneath the surface of the moon (Figure 4.16).
3. *Breccia,*—a rock formed by the continual bombardment of the moon by meteorites and micrometeorites. This process involves first the pulverization of surface material then its compaction by the shock of meteorite impact. Breccia thus most closely ressembles sedimentary rock found on earth (Figure 4.17).

The basalts are common in the maria. Radioactive dating methods have shown them to be between 3.1 and 3.8 billion years old. The

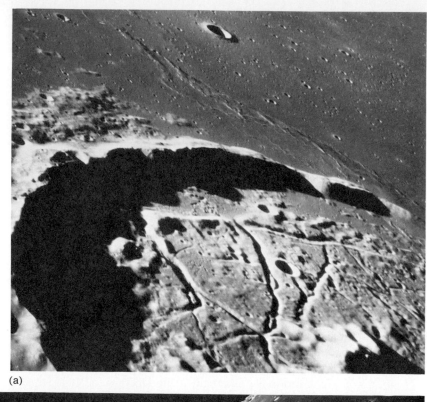

(a)

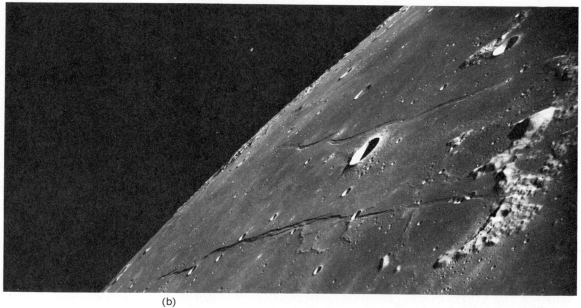

(b)

FIGURE 4.13 (a) A portion of Gassendi Crater on the near side of the moon, as photographed from the Apollo 16 spacecraft in lunar orbit. This view is looking southerly into the Sea of Moisture. (b) The Sea of Tranquility, including the crater Cauchy and two rilles. (NASA)

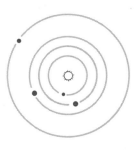

FIGURE 4.14 The far side of the moon as seen by Lunar Orbiter III. This photograph was taken 1400 km above the lunar surface and shows the prominent crater Tsiokovsky filled with dark material. This crater is about 225 km in diameter. (NASA)

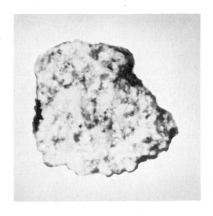

FIGURE 4.15 Moon rock: a coarse basalt. (Smithsonian Observatory)

FIGURE 4.16 A thin section of anorthosite, enlarged 22 times. (Smithsonian Astrophysical Observatory)

FIGURE 4.17 Moon rock: soil breccia. (Smithsonian Observatory)

anorthosites, which are common in the highlands, are between 3.9 and 4.4 billion years old. Although the breccias are found more typically in the highlands, some exist in the maria. The oldest grains of breccia, found in the highlands, are between 4.3 and 4.6 billion years old. How do these facts reveal the history of the moon?

HISTORY OF THE MOON

Before the Apollo program and the dating of moon rocks, many observers assumed that the moon was cratered at a rather uniform rate since its formation. We now conclude that definite periods of cratering and melting occurred periodically up to 3 billion years ago. Evidence indicates that the moon has been essentially frozen in time since then. Let us examine the evidence and the theories.

Most lunar surface material shows signs of having been originally crystallized from a molten state. Since the oldest material dates to 4.6 billion years, we presume that this dates the formation of the moon's surface. The source of heat for the overall melting of the lunar surface was most likely energy associated with the *accretion*, a gravitational gathering-in of material from the surrounding region, which initially formed the moon. Energy is given up by each particle as it joins the larger whole and a significant part of that energy is converted to heat.

Gravity tends to make larger bodies out of a multitude of smaller ones. Even after enough material was gathered together to make the moon, the surrounding space was cluttered with other chunks of material, some of which were very large. We believe that bombardment of the moon by these chunks of material during the final stages of accretion created its cratered appearance. All of the inner planets and most moons of the solar system show similar signs of early bombardment.

Thus the first billion years of the moon's existence can be characterized by general cratering due to the impact of very large chunks of material. Very large chunks may have blasted out the huge basins, which were later filled with lava to form the maria we see today. This filling of the

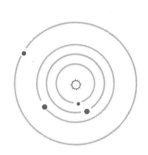

basins—some of the ringed nature, such as those identified with the mascons, and others of an irregular shape, such as Oceanus Procelarum—represent a second major heating of the moon. The source of heat on this occasion may have been the decay of radioactive elements in a deeper layer of the moon, similar to that which we believe takes place in the upper mantle of the earth. This type of heating would cause volcanic activity and associated flooding of basins with lava; thus the maria may have been formed. The dating of the maria basalt indicates that this flooding occurred between 3.1 and 3.8 billion years ago and might have extended over several hundred million years. The moon has been quiet since that time, with the exception of a few impacts of craters in the maria and elsewhere; these impacts can be identified by the bright rays which appear to emanate from them (see the Crater Copernicus in Figures 4.8 and 4.9, pages 154 and 155).

One process, however, continues to this day: the bombardment of the lunar surface by micrometeorites. This action occurring continuously over the billions of years of the moon's existence has pulverized lunar rock and produced a layer of lunar soil. Sharp crater rims are eventually rounded by such bombardment and, in addition, by the erosive effect of pulverized material tumbling down slopes due to gravity.

No water exists on the moon, not even in chemically-bonded form within the rocks, hence we do not see the erosion normally seen on earth. If the rocks did contain water, they would be broken by the freezing and thawing that would accompany the drastic temperature changes typical of the moon. Temperatures on the moon range from 102°C (215°F), when it is directly under the sun, to −176°C (−285°F), when it is at its coldest point. This wide variation in temperature can occur only on a body devoid of atmosphere. The earth's atmosphere provides a moderating blanket that limits the difference between daytime and nighttime temperatures to 30° or 40°C. Furthermore, the phenomena we know as twilight and dawn do not occur on the moon. As the sun goes down, darkness is immediate; an exception might be reflections from nearby peaks above the observer's head, but the percentage of light reflected is very, very small—only about 12 percent. The percentage of light which is reflected from a body is called its *albedo*. Whereas bodies without atmospheres typically have very low albedos, those with a cloudy atmosphere, such as Venus, reflect about 76 percent of the light received.

MAGNETIC FIELD

The moon today has no detectable magnetic field, yet some of the moon rocks possess a locked-in magnetism. This would indicate that at some earlier time, most likely 3 billion years ago, the moon possessed a strong magnetic field. This would be difficult to explain on the basis of its slow rotation (27⅓ days), however the moon may have been spinning more rapidly in its early history.

ORIGIN OF THE MOON

The major theories concerning the moon's origin are as follows:

1. The moon was formed far from the earth and was later captured by the earth.
2. The moon was "torn" from the earth leaving a great depression, the Pacific Ocean.
3. The moon formed out of the same nebula of gas and dust and at the same time as the sun, earth, and other planets.

Theory 1 seems unlikely. It is nearly impossible for one object to capture another unless at least one other object of the proper mass and motion is present. On the other hand, the possibility cannot be ruled out.

Theories 2 and 3 are based on the fact that the composition of the moon (both the highlands and maria) and the crust of the earth are similar. If you had to choose between theory 2 and theory 3 on the basis of the information on compositions listed in Table 4.1, which would be your choice? It would seem that if the moon had been torn from the earth, the percentage composition of the earth and moon should be almost identical. On the other hand, they seem close enough in composition to have been formed from the same nebula; differences may have resulted from layering (differentiation) of elements as the body cooled. Most experts choose theory 3 as most plausible.

TABLE 4.1

PERCENTAGES OF VARIOUS ELEMENTS IN THE CRUST OF THE MOON AND THE EARTH					
Oxygen (%)	Silicon (%)	Iron (%)	Magnesium (%)	Aluminum (%)	Calcium and Potassium (%)
Highlands 61.1	16.2	4.5	4.0	10.2	6.1
Maria 60.6	16.8	1.8	5.3	6.6	4.7
Earth 61.7	21.0	1.9	1.8	6.4	3.3

REVOLUTION OF THE MOON

Owing to the earth's rotation, the moon appears to move in a westward direction during one evening's observation. Closer study will show, however, that it is moving in an eastward direction against the background of stars. You may verify this by observing the moon when it is at the western limit of a bright star or cluster, such as the Plieades. If you observe the moon for several hours, its eastward motion will become apparent as it moves to occult (cover up) the stars one after another.

A more dramatic eastward change takes place when we observe the moon on successive nights, noting its position relative to the stars. If we continue this practice for approximately 27 days, we find that the moon returns finally to the same portion of the sky. To be exact, the moon's *sidereal period of revolution* is 27.322 days. At this rate, it will shift in the sky approximately 13° per day ($360°/27.322$ days $= 13°/$day)—a little less than one hour circle in a 24-hour period.

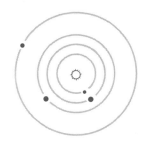

We can also observe that more than 27.322 days are required for the moon to complete its cycle of phases. Suppose we start counting time when the moon is full, shown in Figures 4.18 and 4.19 as position A. We note that the moon appears near the bright star Regulus. In 27.322 days the moon will return to align itself with Regulus, but in that length of time the earth will have moved through approximately 27° to position E_2, and the moon will then be in position B. Since the moon appears full only when the moon is on the side of the earth opposite from the sun, it will not be full until it moves through an additional angle of approximately 29° to position C. This additional revolution requires approximately 2.209 days, hence the moon's *synodic period*, or period of phasing, is 29.531 days ($27.322 + 2.209 = 29.531$). It is thus the earth's revolution that causes the moon's synodic month to the different from its sidereal month.

In passing from one full moon to the next, the moon goes through all its successive phases, as shown in Figure 4.20. We observe that the *new (all dark) moon* occurs when the moon is between the earth and the sun and that the *first-quarter* moon occurs when the moon makes a 90° angle with the sun as viewed from the earth. Between these two phases is the *waxing (growing) crescent,* the phase that most people incorrectly refer to as the "new" moon since this phase is the first we see when the moon is starting a new sequence of phases. The next phase is that of the *waxing gibbous,* when the face appears more than half lighted. The *full moon* occurs when the moon is on the opposite side of the earth from the sun.

FIGURE 4.18 The sidereal and the synodic month as seen in the sky. (a) A full moon, in conjunction with Regulus (start counting time here). (b) After 27.33 days–the moon, again in conjunction with Regulus, but not yet a full moon (this is a sidereal month). (c) After 29.5 days–the moon, again full, but now located approximately 30° east of Regulus (this is a synodic month).

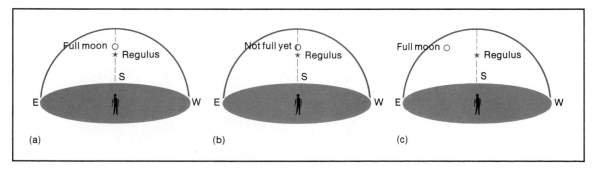

(a)   (b)   (c)

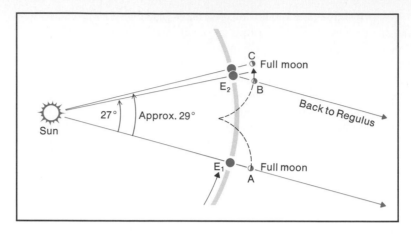

FIGURE 4.19 The synodic period of the moon.

FIGURE 4.20 The phases of the moon. North is at top, as would be seen by the naked eye. (Lick Observatory)

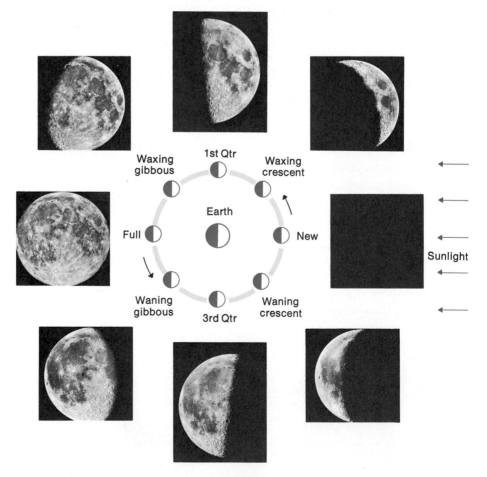

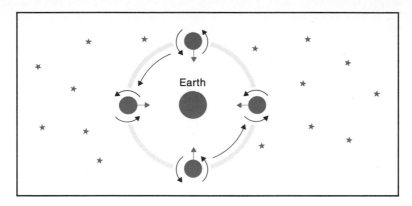

FIGURE 4.21 The synchronous rotation of the moon.

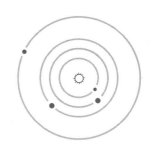

Then follows a series of *waning* phases in which the visible illuminated portion appears to shrink. Note that you can predict when and where a given phase of the moon will appear in the sky. The easiest phase to visualize is the full moon; since the moon is full only when it is directly opposite the sun, it will rise when the sun sets, will be overhead at midnight, and will set when the sun rises the next morning.

Throughout all its phases, the moon presents essentially the same side toward the earth. Does this mean that the moon does not rotate on its own axis? In order always to present the same side toward the earth, the moon rotates once with respect to the stars while revolving once around the earth (Figure 4.21). This motion, referred to as *synchronous rotation*, has a period of 27.322 days. Although the moon's rotation is uniform, its revolution is along an elliptical orbit and hence is not uniform. Changes in its speed of revolution result in the periodic exposure of small portions along its western and eastern edges, or limbs (Figure 4.22). When the

FIGURE 4.22 Two photographs of the moon showing libration. (Lick Observatory)

moon is at its greatest distance from the earth, it moves more slowly along its orbit (recall Kepler's second law, page 32) and its rotation gets ahead of its revolution. When the moon is closest to the earth, it moves more rapidly in its orbit and its revolution gets ahead of its rotation. We periodically view a little beyond the north and south polar regions as well since the moon's orbit is inclined 5° to that of the earth. Over a period of several months we are thus able to view a total of 59 percent of the moon's surface from the earth. These motions that result in the exposure of over one-half the moon's surface are called the *librations* of the moon.

In the next section we will answer the question, "Has the moon always pointed the same face toward the earth?"

TIDES PRODUCED BY THE MOON

Let us conduct an imaginary experiment in which we observe tides of the ocean on several successive evenings when the moon is also visible overhead. On the first night the moon is on your local meridian at 8:00 P.M. and high tide occurs a short time later. On the second night the moon is on your local meridian at 8:50 P.M. and again high tide occurs a short time later. On the third night the moon passes your local meridian at 9:40 P.M., and once more high tide occurs a short time later. Were this experiment conducted for several additional nights, you would conclude with confidence that the timing of the high tide coincides with the passage of the moon and must be due to its presence overhead. Why, though, are there two high tides in a period of 24 hours 50 minutes? Stated another way, how can a single moon create the double bulge in the earth's oceans, which is necessary to produce two high tides per day?

In order to explain the occurrence of two high tides, let us consider an imaginary earth: one that is perfectly smooth, with no mountains or valleys and alone in space, with no motion of its own. If water existed on the surface of this imaginary earth, it would be distributed evenly over the entire surface, since each molecule would experience the same force of gravity toward the center (Figure 4.23).

If this imaginary earth were then set into rotation, the layer of water would become deeper around the equator and shallower at the poles because the water near the equator would experience a greater tendency to be thrown off on a tangent (recall the concept of inertia, page 39). Rotation would only produce a constant bulge around the equator, however; it would not produce the fluctuating tides that we experience. The force that does produce the tides is the gravitational force of the moon on the earth and on the waters that cover it. The force of gravitation depends on the distance between the two objects in question: the greater the separation, the less the force. Suppose that a 1-gram mass is situated at three different positions as shown in Figure 4.24: in the water nearest the moon (*A*); in the center of the earth (*B*); and in the water farthest from the earth (*C*). Because of their positions relative to the moon, mass *A* will experience the greatest force and mass *C* the least force, the force on mass *B*

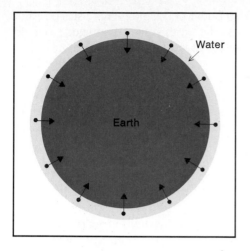

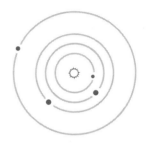

FIGURE 4.23 Water molecules (dots) distributed evenly over the surface of an imaginary earth, alone in space and without any motion of its own.

being intermediate between the other two. We may then expect the acceleration on each mass to be proportional to the force that is acting in it ($F = ma$). Therefore mass A will experience the greatest acceleration and mass C the least, with the acceleration on mass B being intermediate between the other two. If you were riding with mass B, what would the other two seem to do? Mass A would pull away from you and you would pull away from mass C. Likewise, the water nearest the moon pulls away from the earth and the earth pulls away from the water on the side away from the moon, creating a bulge on the side toward the moon and another bulge on the side away from the moon. In a very real sense, the tendency of tides is to tear a body apart.

The effect of the *differential* gravitational forces that the moon exerts on different parts of the earth can be illustrated by placing a system of springs and masses on a smooth table (Figure 4.25). As a force is applied to mass A, it will be accelerated to the right. While the spring to which it is attached is being stretched, it will create a lesser acceleration of mass B and likewise, while the second spring is being stretched, it will create yet a lesser acceleration of mass C. The result will be a separation of mass A from mass B and also mass B from mass C. We could describe the system as being "torn apart" and we see the resultant double bulge.

FIGURE 4.24 Tides on the earth.

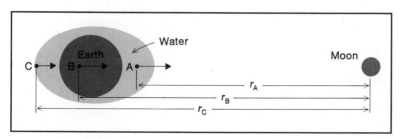

Our planet, in reality, is not smooth but has mountains, beaches, and deep trenches under the ocean. As the earth rotates within the bulged surface water, raised portions of the land literally run into the bulge and friction results, slowing the earth's rotation. Although this slowing trend only extends the length of a day by about 0.002 seconds in a period of 100 years, it is this phenomenon that changed the speed of the moon's rotation until it coincided with its revolution. Although the primal moon was devoid of oceans, it was flexible enough that the earth once caused land tides and the resulting friction changed the moon's rotation rate. Now the cool, semi-rigid moon has two earth-produced bulges permanently fixed, one pointing always toward the earth and the other always away from it.

The rate of slowing of the earth's rotation appears rather insignificant, yet in the time frame with which the astronomer deals, it is very significant. Could we turn time backward, say 500 million years, the length of a day would be only about 21 hours. This fact is confirmed by the fossil record of certain corals that show 416 daily growth rings in each yearly band. Assuming that the number of hours in a year is approximately the same today as it was 500 million years ago, a 416-day year would have had days of approximately 21 hours.

Because the sun is so much farther from the earth than the moon, it raises tides on the earth that are less than one-half the magnitude of those raised by the moon. When the sun and moon cause waters to bulge in the same direction, they produce extremely high and low tides, called *spring tides*. As shown in Figure 4.26(a), spring tides occur during new and full moons. When the forces of the sun and moon act at right angles to each other, the tides are less extreme and are called *neap tides*. Figure 4.26(b) shows that neap tides occur during first-quarter and third-quarter phases of the moon. Gravitational forces also act upon the rigid earth to produce land tides and sometimes trigger the release of stresses in the earth, producing earthquakes.

FIGURE 4.25 A system of springs and masses that illustrates the effect of the moon's gravitation force in producing tides on the earth. The circle attached to the center mass represents the earth.

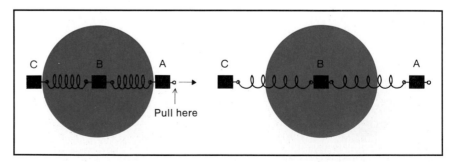

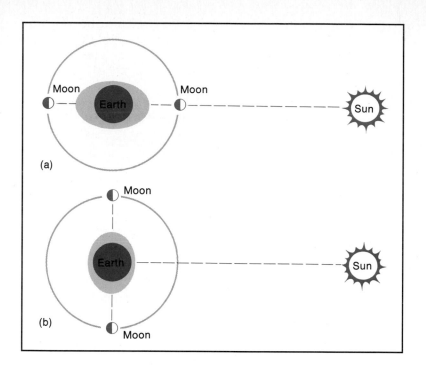

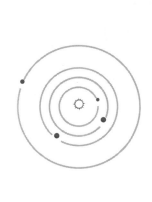

FIGURE 4.26 (a) Spring tides. (b) Neap tides.

ECLIPSES

Eclipses occur when one object passes within the shadow of another object. The structure of the shadow that is cast provides a key to understanding the eclipse. If the source of light is a point source, such as a star, the boundaries of the shadow cast by an object are simple and well-defined, as indicated in Figure 4.27. When the source of light is nearby and extended, the shadow that an object casts is more complex. The dense portion of the shadow is called the *umbra*; the source of light cannot be seen from this region. If you found yourself within the umbra caused by the passage of the moon between the sun and the earth, the sun would appear completely covered (Figure 4.28). The less dense portion of the shadow is called the *penumbra*; from within this region, the moon would appear to cover only part of the sun.

FIGURE 4.27 The shadow produced by a point source.

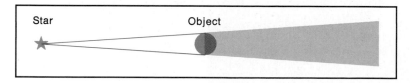

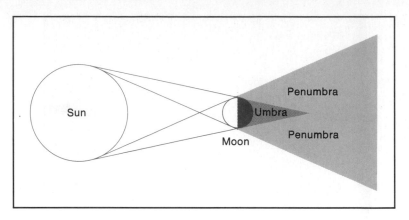

FIGURE 4.28   The shadow produced by an extended source.

SOLAR ECLIPSE

The length of the umbral portion of the moon's shadow is slightly more than 383,000 km (238,000 mi). The distance from the moon to the earth varies between 362,000 km (225,000 mi) and 406,600 km (252,700 mi). Thus we may conclude that under certain conditions the shadow of the

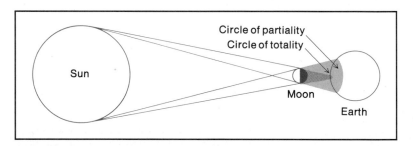

FIGURE 4.29   A solar eclipse. An observer within the circle of totality will see a total eclipse of the sun, while an observer in the circle of partiality will see only a partial eclipse.

FIGURE 4.30   The diamond ring effect. (Cliff Holmes)

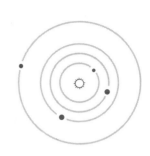

FIGURE 4.31 The corona of the sun, as seen during a total solar eclipse, March 7, 1970. (High Altitude Observatory)

moon will strike the earth but cover only a very limited region of it at any one time (Figure 4.29).

If you were standing within the region of the earth covered by the path of the dark umbral portion on March 7, 1970, you were able to see a truly spectacular event: a total eclipse of the sun. The combined motion of

FIGURE 4.32 The solar eclipse of March 7, 1970. An exposure was made every 6 min. Venus appears above the fifth image from the left. (Maurice E. Snook)

the earth and the moon caused this umbral portion of the shadow to sweep out a narrow path across the face of the earth, traveling at approximately 1600 km/hr (1000 mi/hr). The sun's disk was gradually covered by the disk of the moon.

For just an instant before total covering, a portion of the sun was visible shining through a depression in the moon's profile and producing the beautiful diamond-ring effect, as seen in Figure 4.30. When the sun was completely covered—the time of *totality*—the sky darkened significantly and the beautiful *corona*, or outer atmosphere of the sun, appeared (Figure 4.31). The corona of the sun is not usually visible owing to the brilliant glare of the *photosphere*, the visible disk of the sun. During the time of totality, many strange things happen: dogs howl, chickens go to roost, birds cease their singing, and some flowers close. Much of the surrounding landscape takes on strange hues of color. The total portion of a solar eclipse lasts no more than seven minutes and usually less; then as the moon continues to move, the entire disk of the sun is revealed (Figure 4.32).

The path of totality for the solar eclipse of March 7, 1970 is shown in Figure 4.33. The eclipse was witnessed by millions of people in the United States and Mexico. Those who lived within this narrow band, or who traveled to reach it, were rewarded by a spectacular sight. Almost everyone in Mexico, the United States, and Canada was in a position to see at least a

FIGURE 4.33 The path of totality for the solar eclipse of March 7, 1970. (Nautical Almanac Office, U.S. Naval Observatory, Washington, D.C.)

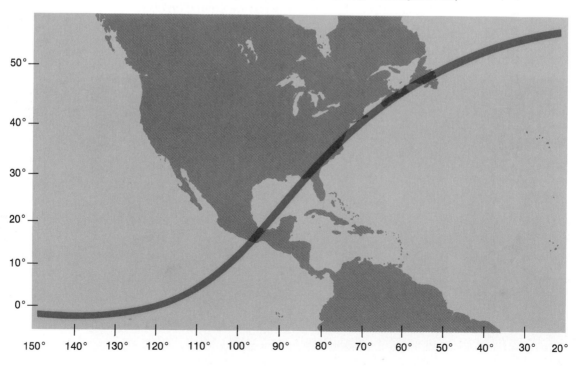

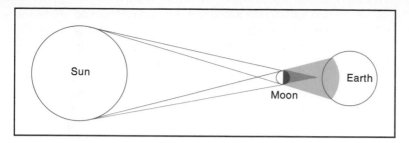

FIGURE 4.34 An annular eclipse of the sun.

FIGURE 4.35 The paths of total eclipses between July 1963 and May 1984. A solid line corresponds to a total eclipse; a dashed line, to an annular one. The beginning of the path is indicated by a small white circle; the end, by a black one. (From J. Meeus, K. Grosjean, and W. Vanderleen, Canon of solar eclipses. Elmsford, NY: Pergamon Press, 1966)

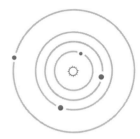

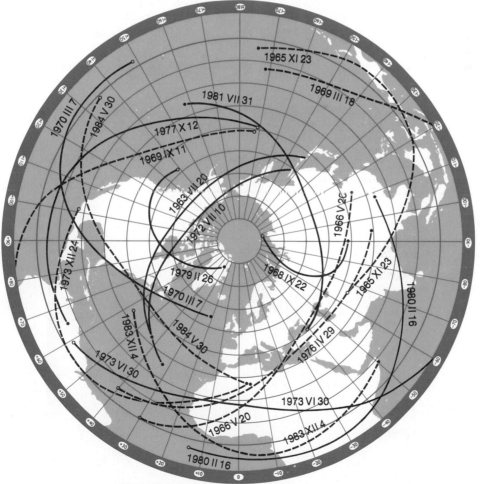

partial eclipse, since all countries were in the penumbral portion of the moon's shadow. Television via communication satellite played an unprecedented role in bringing this spectacular event to millions of people all over the world.

In addition to the partial and total eclipse, a third situation may arise. We know that the moon is often farther from the earth than 383,000 kilometers and that its shadow will therefore not reach the earth. If you were standing directly in line with the moon's shadow, you would see the central portion of the sun's disk covered, leaving a ring of sunlight visible around the moon's disk. This is called an *annular or ring eclipse* (Figure 4.34). The paths of solar eclipses between July 1963 and May 1984 are mapped out in Figure 4.35. Can you find the path of an eclipse that you may have seen?

A WARNING. Never look at the sun directly, whether by naked eye, or through a telescope or binoculars, for your eye may be irreparably burned by the sun's radiation. No pain is associated with this experience, therefore a person who has observed the sun directly might not immediately be aware that he is being blinded. A solar eclipse may be safely viewed by holding a small piece of cardboard in which a pinhole has been made, allowing the sun's rays to pass through the pinhole and fall on a second piece of cardboard (screen). A larger image may be formed by substituting a small telescope for the pinhole, as illustrated in Figure 4.36.

FIGURE 4.36 Observation of a solar eclipse using (a) a pinhole and (b) a small telescope.

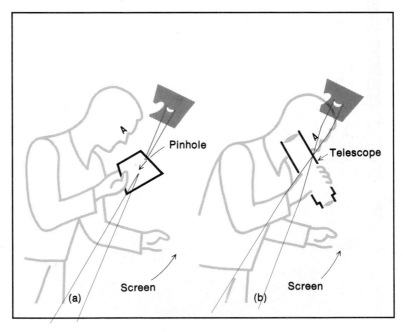

LUNAR ECLIPSE

A total lunar eclipse occurs when the moon passes completely within the umbral portion of the earth's shadow, which extends outward approximately 1,380,000 km (857,500 mi). At the moon's distance, the umbral por-

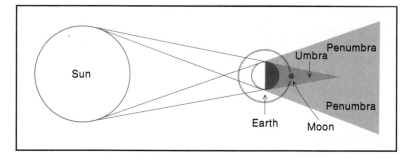

FIGURE 4.37 Lunar eclipses.

FIGURE 4.38 The moon is seen moving into the umbra of the earth's shadow. (R. T. Dixon)

tion is almost 9700 km (6028 mi) in diameter. The moon's diameter of 3476 km (2160 mi) allows it to fit easily within this shadow (Figure 4.37). The penumbral portion is approximately 16,000 km (10,000 mi) in diameter at the moon's distance. When a total eclipse occurs, the moon first moves into the penumbral portion of the earth's shadow and is only slight dimmed. As it moves into the umbral portion, the curvature of the earth's shadow may be seen projected on the moon (Figure 4.38). During totality the moon does not become darkened entirely but rather takes on a copper hue; illumination of the moon at this time is the result of sunlight refracted by the earth's atmosphere. Recall that the earth's atmosphere scatters blue light, allowing the red tones to pass more directly through it to the moon.

ECLIPSE SEASONS

It might seem that we should have a solar eclipse at each occasion of a new moon, when the moon passes between the earth and the sun, or that we should have a lunar eclipse on the occasion of each full moon, when the earth is between the sun and the moon. This is not the case, however, because the moon orbits the earth on an imaginary plane that is tilted 5° to the plane of the earth's orbit (Figure 4.39).

The points at which the two orbital planes intersect are called *node points*. Whenever the sun and moon are within a few degrees of a node simultaneously, an eclipse occurs. If both sun and moon are near the same node, then a solar eclipse occurs, and if they are near opposite nodes, a lunar eclipse occurs. Thus we may expect eclipses to occur approximately every six months during what may be called eclipse seasons. During any given season, one would expect a solar and a lunar eclipse about 14 or 15 days apart (see Figure 4.40).

However, eclipse seasons do not occur exactly at six-month intervals because there is a slow regression of the nodes along the orbital plane of the earth. A given node moves through 360° in a period of 18 years 10 days,

FIGURE 4.39 The plane of the moon's orbit is inclined 5° to the plane of the earth's orbit.

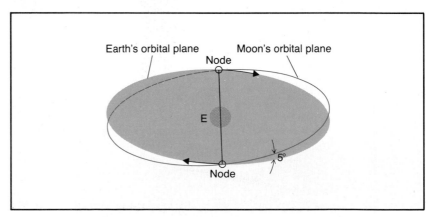

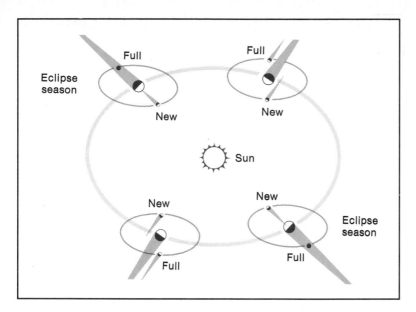

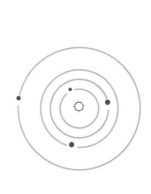

FIGURE 4.40 Eclipse seasons.

a cycle called the *saros*. Thus the circumstances which produce a given eclipse repeat themselves in that period. When both the phasing cycle of the moon and the seasonal changes are considered, yet another cycle— the *Metonic cycle*—is found with a period of 18.61 years. "Stonehengers" may have recognized this cycle with their 56 Aubrey holes, since $3 \times 18.61 = 55.83$ (almost 56) years.

A GENERALIZATION. What we have learned about the earth's moon will continue to form a basis for questions about other moons in the solar system.

QUESTIONS

1. In which direction does the moon appear to move against the background of stars?
2. Why does the sidereal month (27.322 days) differ from the synodic month (29.531 days)?
3. The moon presents almost the same "face" toward the earth at all times. Therefore, which of the following is true: (a) its period of rotation is 27.322 days; (b) its period of rotation is 29.531 days; (c) it does not rotate.
4. At what time of night would you expect to see a full moon overhead on your local meridian?
5. Do coastal ports usually experience one, two, or three high tides per day?
6. The moon appears to shift among the stars: (a) one hour circle (15°) per day; (b) less than one hour circle per day; (c) more than one hour circle per day.

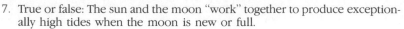

7. True or false: The sun and the moon "work" together to produce exceptionally high tides when the moon is new or full.

8. Where is the barycenter of the earth-moon system located?

9. How often does a point on the moon experience a cycle of changing tides?

10. Describe a mascon.

11. List five types of surface features on the moon.

12. Which major surface feature is most conspicuous by its absence on the far side of the moon?

13. Solar eclipses always occur during what phase of the moon?

14. An eclipse season usually lasts for about a month. How many eclipse seasons may occur in one year?

15. What part of the sun is visible during a total eclipse of the sun?

16. Describe two safe methods for viewing a partial eclipse of the sun.

17. Why don't you see an eclipse of the moon every month?

18. Describe the relative positions of the sun, moon, and earth when the moon is full.

19. If high tide occurs at 9 A.M. on a given day, when can you expect the high tide to occur the next morning? Why?

20. Why is the sun a less powerful factor than the moon in producing tides?

21. What is meant by land tides?

22. What does the term *earthshine* mean in relation to the moon?

23. If you lived on Mare Crisium (on the moon), you might expect to see a sunrise once every _____ earth days.

24. What factors of erosion on the earth are not found on the moon? Are there any factors of erosion which the earth and moon have in common?

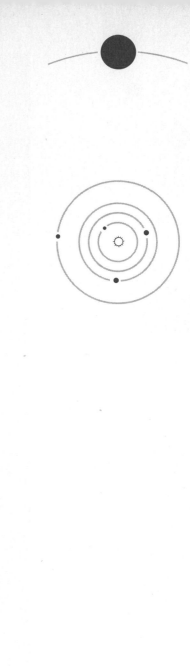

The Planets

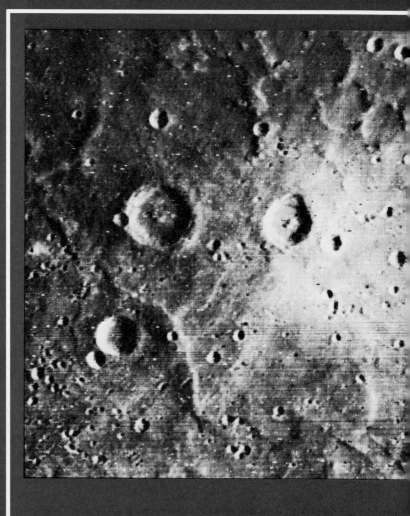

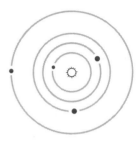

Five

As you open this chapter you can count on seeing our star, the sun, and its family of planets and moons in a way no generation before you has been able to see them, for during the 1960s and 1970s more information about the solar system was gathered than ever before. In fact, the deluge of information from space probes, orbiters and landers will probably take the decade of the 1980s to decipher. We have gone to the rocky worlds of Mercury and Mars and have seen them in detail. We have flown by, orbited, and dropped probes into Venus's cloudy atmosphere yet it still tends to hide many details from us. We have seen Jupiter and Saturn with unusual clarity, as well as many of their moons; and, we're on our way to Uranus and Neptune. In spite of the little hope that we will see Pluto close-up, astronomers have recently discovered that even this planet has a moon.

While committed to the primary task of describing the physical properties, appearance, and motion of the planets and their moons, the astronomer is ever tantalized by the question of their origin and evolution. What clues will unlock these secrets? As we study the nine planets and their many moons, and numerous smaller objects that revolve about the sun, we will evaluate several theories of their origin. One theory suggests that the planets were formed much farther from the sun than their present distances indicate

Moon   Earth

and later were captured at random by the sun; this may be referred to as the *random-capture* theory. Another theory proposes that the sun may once have passed very near another star, and due to their mutual gravity, a long streamer of material was pulled off. From this material the planets were formed: this could be called the *encounter theory*. A third theory suggests that as the sun condensed out of a huge gas cloud, which, rotating faster and faster, left a flattened disk of material around its nucleus out of which the planets formed. This theory of a common origin of the sun and its planets has had numerous variations. One suggests that rings of material were formed first and that each ring eventually became a planet. Another suggests that smaller condensations (knots) formed in the flattened disk and eventually many of these combined to form each planet. Such planets-in-the-making are referred to as *protoplanets*. Gravity is the fundamental force of attraction responsible for accretion, the process that tends to make small condensations grow into larger bodies by the gathering-in of new material. Disruptive forces that tend to oppose accretion include the random motion of various particles that will not yield to gravity and collisions that inevitably occur. Let us discuss some basic facts concerning the planets that may help us evaluate these theories.

THE PLANETS IN GENERAL

All of the planets revolve about the sun in the same direction and within almost the same plane of orbit as the earth; Their motions can thus be shown quite accurately on a flat piece of paper, as on the right-hand flip pages starting on page 115. The configuration of the solar system seems to rule out the randon-capture theory, for had the planets been captured at random, they would likely be oriented in various directions and on various planes (Figure 5.1). Their direction of revolution is counterclockwise as viewed from the "north side" of the solar system, the side toward which the north pole of the earth points. Because the planets move in virtually

FIGURE 5.1 (a) The orientation of planetary orbits that might be expected if they had been captured at random. (b) The actual disklike nature of planetary orbits.

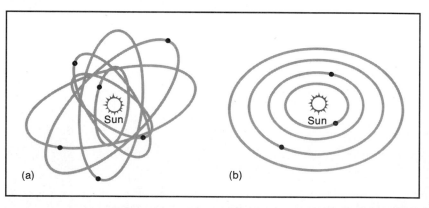

the same flattened plane, we on earth see the planets moving, against the background of stars, along almost the same ecliptic path. Thus, planets appear to move through the 12 constellations known as the signs of the zodiac. Planets are thus described as "being in" one of those signs at any given time; we might say, for instance, "Saturn is in Gemini."

The encounter theory of the solar system's origin is consistent with the fact that all the planets orbit the sun in the same direction and along the same plane. This idea, however, is discounted on a statistical basis; the chance of two stars passing close enough to pull streamers of material from each other is so low that it is hardly considered a possibility. Perhaps you can appreciate this judgement better if you realize that the average distance between stars in our part of the Galaxy is roughly 41,841,000,000,000 kilometers (26,000,000,000,000 miles).

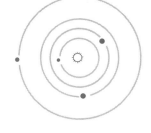

We are left, then, with the suggestion that the nebula which ultimately formed the sun was flattened due to its rotation and the planets formed from material in that flattened disk. According to this theory, the counterclockwise revolution of the planets, as we see them today, reflects the original direction of rotation of the solar nebula, and the plane of their revolution reflects the plane of that flattened disk. This theory, however, does not fully explain the direction of rotation that each planet possesses.

ROTATION OF THE PLANETS

Seven of the nine known planets rotate in a counterclockwise direction. The other two, Venus and Uranus, spin in a clockwise direction and we refer to their rotation as being *retrograde* (backward to that which appears usual). Variety among the planets is also manifest in the differences in the tilts of rotation of the various planets in relation to their planes of orbit. If we recognize that each particle that helped to form a planet originally had a motion of its own, it follows that the rotation and tilt of a resulting planet would reflect these motions.

SPACING OF THE PLANETS

At first glance, the distances between the planets and the sun may appear to have no pattern other than that the first four planets are relatively closely spaced and the remaining five are widely spaced. However, in 1766 Johann Daniel Titius of Wittenberg, Germany, pointed out a very interesting mathematical relationship. This relationship was later published by Johann Elert Bode (pronounced Bo-da) and is often erroneously called Bode's law. In the last column of Table 5.1 you will see the average spacing of the planets measured in astronomical units (A.U.). In the first three columns are the results of a series of simple steps by which these numbers may be closely approximated. The first column of figures begins with 0. Below that comes the numeral 3. Every number thereafter is twice the value of the one preceeding it. A second column is formed by adding 4 to the corresponding entries in the first column. A third column is then ob-

tained by dividing each entry in the second column by 10; the resulting numbers are very close to the actual distances measured in astronomical units listed in the last column. Although the discoveries of Uranus, Neptune, and Pluto were to come later, the Bode-Titius relationship is extended in Table 5.1 to include these planets for comparison. It is evident that this relationship does not fit the observed distances to Neptune and Pluto.

TABLE 5.1

		THE BODE-TITIUS RELATIONSHIP		
Planet	First Column	Second Column	Bode Numbers	Actual Distances (A.U.)
Mercury	0	4	0.4	0.39
Venus	3	7	0.7	0.72
Earth	6	10	1.0	1.00
Mars	12	16	1.6	1.52
	24	28	2.8	
Jupiter	48	52	5.2	5.20
Saturn	96	100	10.0	9.54
Uranus	192	196	19.6	19.18
Neptune	384	388	38.8	30.06
Pluto	768	772	77.2	39.44

Conspicuously absent from Table 5.1 is a planet corresponding to the 2.8 A.U. figure in the Bode-Titius progression of numbers. Astronomers thought that an unknown planet might be found at or near this distance, so they computed the period such a planet would have based on Kepler's third law and launched a search for it. In 1801, an object with the expected period was found and given the name Ceres. This discovery gave way to others and within a few years, many small objects were found having approximately the same period and orbital distance. These objects, called *asteroids*, will be discussed more fully in Chapter 7. Some observers have felt that because the Bode-Titius numbers so closely approximate the observed distances to the planets (except Neptune and Pluto), a force may be operative to produce this spacing. Others have felt that the spacing is coincidental.

The Bode-Titius relationship perhaps would be considered more significant had it led scientists to an explanation of why the planets formed at their respective distances from the sun. Numerous theories have been proposed that predict entirely different spacings of the planets. The Bode-Titius relationship thus should not be considered a physical law but should merely be treated as an interesting way to remember the spacing of the first seven planets and the primary asteroid belt.

EARTH-LIKE VERSUS JUPITER-LIKE PLANETS

The planets fit very naturally into two categories: (1) those that are earth-like, called the *terrestrial* planets; and (2) those which are Jupiter-like, called the *Jovian* planets. The terrestrial planets, besides the earth, include Mercury, Venus, and Mars. The Jovians include Jupiter, Saturn, Uranus, and Neptune. We will treat Pluto separately, for it is thought to be unlike either of the two categories.

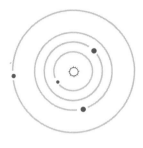

The distinction between the terrestrial and the Jovian planets are quite obvious. As indicated in Table 5.2, the terrestrials are small—earth size and smaller—and they have low mass but high densities. The higher densities of the terrestrials reflect a higher abundance of heavier elements and less significant atmospheres. With the exception of Venus, the atmospheres of the terrestrials are not even taken into consideration when computing their diameters and masses (Figure 5.2).

Why should the planets near the sun tend to have little or no atmosphere? Their relatively high temperatures and low masses tell part of the story. Molecules in the atmospheres which may once have surrounded these planets would have absorbed energy from the sun, thereby increasing their velocity until they were moving so fast that the gravitational pull

TABLE 5.2

PROPERTIES OF THE TERRESTRIAL PLANETS AS COMPARED TO THOSE OF THE JOVIANS[a]						
Planet	Diameter	Mass	Density	Rotation	Temperature	Atmosphere
Terrestrials						
Mercury	0.38	0.05	5.2	58d 15h	−173 to +427°C	None
Venus	0.95	0.82	5.3	−243d	−23 to +477°C	Significant
Earth	1.00	1.00	5.5	23h 56m	−74 to +38°C	Thin
Mars	0.53	0.11	3.8	24h 37m	−125 to +20°C	Very thin
Jovian						
Jupiter	11.23	317.9	1.3	9h 50m	−139°C	Very signif.
Saturn	9.41	95.2	0.7	10h 39m	−176°C	Very signif.
Uranus	3.98	14.6	1.3	12 to 24h	−215°C	Very signif.
Neptune	3.88	17.2	1.7	18 to 22h	−218°C	Very signif.
Pluto	0.23	0.002	1.1	6d 9h	−230°C	Unknown

[a] Diameters and masses are related to those of the earth (as 1). Densities are given in grams per cubic centimeter. Temperatures of the Jovian planets are not meant to be exact but to show a trend.

of the planet could no longer hold them; they would thus have escaped into space. The velocity required for molecules to escape from each planet is shown in Appendix 6 (page 535). Some molecules in the atmospheres of the terrestrials could certainly have been heated by the sun to those velocities. The less massive elements such as hydrogen and helium would be most easily accelerated to the escape velocity, thus although these elements are very abundant in the universe, only traces exist in the atmospheres of the terrestrial planets.

This explanation accounts for the absence of the least massive molecules; however, it does not account for the low percentage of other gases such as neon and argon. The terrestrials may at one time have lost virtually all atmosphere then may have developed another atmosphere from within the planet, through volcanic action.

The solid (crusty) nature of the terrestrial surfaces contrasts with the amorphous nature of the Jovian "surface." The Jovian planets might be characterized as balls of gas and liquid with only a small dense core at their centers. The Jovian atmospheres are believed to be composed of mainly hydrogen with lesser amounts of helium, and traces of other elements. These planets are able to retain the lighter atoms because of two

FIGURE 5.2 Relative size of the planets.

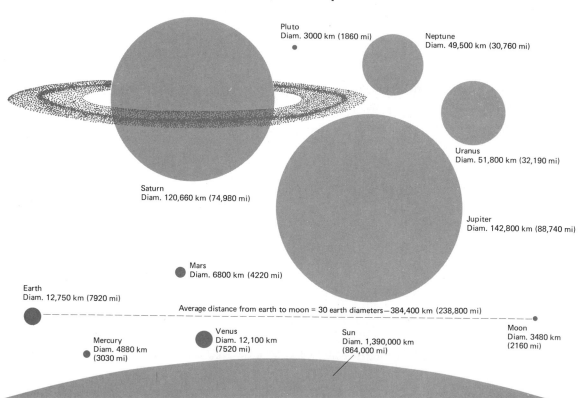

Pluto
Diam. 3000 km (1860 mi)

Neptune
Diam. 49,500 km (30,760 mi)

Saturn
Diam. 120,660 km (74,980 mi)

Uranus
Diam. 51,800 km (32,190 mi)

Jupiter
Diam. 142,800 km (88,740 mi)

Mars
Diam. 6800 km (4220 mi)

Earth
Diam. 12,750 km (7920 mi)

Average distance from earth to moon = 30 earth diameters—384,400 km (238,800 mi)

Mercury
Diam. 4880 km
(3030 mi)

Venus
Diam. 12,100 km
(7520 mi)

Sun
Diam. 1,390,000 km
(864,000 mi)

Moon
Diam. 3480 km
(2160 mi)

characteristics: (1) a large surface gravity (due to high mass), and (2) cool surface temperatures—individual molecules do not move fast enough to escape.

The planets can also be categorized according to their position in the solar system relative to the earth. The planets that lie within the earth's orbit (Mercury and Venus) are called the *inferior* planets; those planets which lie beyond the earth's orbit (Mars, Jupiter, Saturn, Uranus, Neptune, and Pluto) are called the *superior* planets. Let us describe the arrangements these planets can take on from our point of view on the earth.

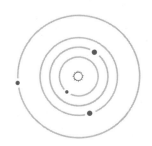

CONFIGURATIONS OF THE INFERIOR PLANETS

Whenever a planet is aligned with the sun so that both may be viewed in the same direction from the earth, the planet is said to be in *conjunction* with the sun. In Figure 5.3 we note two possible positions of alignment: one between the earth and the sun, called *inferior conjunction*, and one on the far side of the sun, called *superior conjunction*. During such configurations the planet is lost to our view due to the glare of the sun. We may describe the position of a planet by the angle it makes with the sun as seen from earth. This angle is called the *elongation* of the planet; when this angle increases to its largest value during a given revolution, the planet is said to be at *maximum elongation*. Since this may occur at either of two different locations, we specify one as a western elongation if the planet appears to the west of the sun in the sky, and the other as an eastern elongation if the planet appears to the east of the sun in the sky. The times of maximum elongation are obviously good times to view Mercury and Venus. Note that if Mercury or Venus were at maximum elongation west,

FIGURE 5.3 Configurations of the inferior planets.

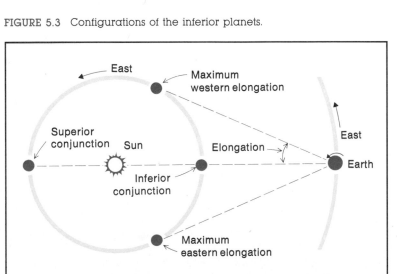

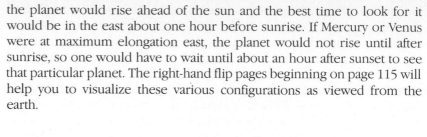

the planet would rise ahead of the sun and the best time to look for it would be in the east about one hour before sunrise. If Mercury or Venus were at maximum elongation east, the planet would not rise until after sunrise, so one would have to wait until about an hour after sunset to see that particular planet. The right-hand flip pages beginning on page 115 will help you to visualize these various configurations as viewed from the earth.

CONFIGURATIONS OF THE SUPERIOR PLANETS

When a superior planet is viewed from the earth, there is only one way it may align itself with the sun to produce a conjunction. A superior planet, however, may make any angle up to 180° with the sun as viewed from the earth; when exactly 180° from the sun, a planet is said to be at *opposition*—on the opposite side of the earth from the sun (Figure 5.4). This is a particularly advantageous time to observe the planet, for it is in this configuration that the planet makes its closest approach to the earth, is fully illuminated by the sun, and so appears at its brightest. It is visible all night because it appears to rise at sunset and set at sunrise. Careful observation will also reveal its apparent retrograde motion against the background of stars, for the earth is actually passing that planet at this time. The

FIGURE 5.4 Configurations of the superior planets.

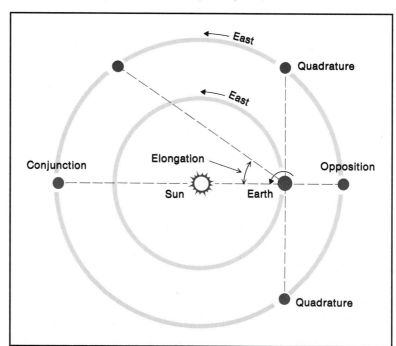

configuration of a planet (or moon) as seen 90° from the sun is called *quadrature*.

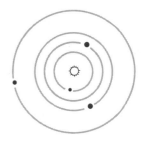

Let us now consider some of the distinctive characteristics of each planet. You will find the physical properties and motions of the planets and their moons summarized in Appendices 5, 6, and 7. Not all of these properties will be discussed in detail in the following text; therefore these appendices should be studied as an integral part of this chapter.

MERCURY

Mercury is the most elusive of all the planets visible with the naked eye. It appears sometimes as a morning object, rising just before sunrise, and at other times as an evening object, setting just after sunset. This cycle repeats itself just over three times during the year. Ancient observers did not realize that they were seeing the same object alternately in the morning and evening. The Greeks named the planet *Mercury* when it appeared in the evening and *Apollo* when it appeared in the morning. Mercury's elusiveness is explained by the fact that it revolves around the sun at an average distance of only 57 million kilometers (35.4 million miles), or 0.4 A.U., less than half the distance from the sun to the earth. As a consequence, it never appears more than 28° from the sun. Since the earth rotates at the rate of 15°/hr, Mercury appears to rise less than two hours before the sun when it is a morning object and appears to set less than two hours after the sun when it is an evening object. At best, Mercury is seen in the light of dawn or twilight, and under these conditions surface details cannot be seen clearly.

The plane in which Mercury travels is inclined 7° to the ecliptic plane, so the planet may appear several degrees above the ecliptic at some times and several degrees below it at others. Mercury's orbit has an eccentricity of 0.206, the second most eccentric orbit in the solar system. This eccentricity is shown in Figure 5.5, which indicates that Mercury's distance from the sun varies between 69 million kilometers at *aphelion*, its most distant point, and 46 million kilometers at *perihelion*, its nearest point. If we apply Kepler's law of equal area in equal time, we see that Mercury has a wide range of orbital speeds. At aphelion, Mercury travels at 40 km/sec; at perihelion, its speed increases to 60 km/sec. Mercury's average speed in orbit is 48 km/sec (172,800 km/hr, or 107,380 mi/hr). A careful study of Figure 5.5 will help you develop a feeling for the motion of this planet and of all planets in general. Each successive planet you study will be farther from the sun and will orbit more slowly; for example, the earth's average speed is 107,280 km/hr (66,660 mi/hr).

The *sidereal* period of revolution for Mercury, the time needed for one exact revolution (360°) relative to the stars, is 87.96 days. However, since we view the planets from a moving platform, the earth, a more meaningful reference is the *synodic* period. This is the time needed for a planet to move from any given configuration, such as inferior conjunction, back to the same configuration, as seen from a moving earth. Referring to

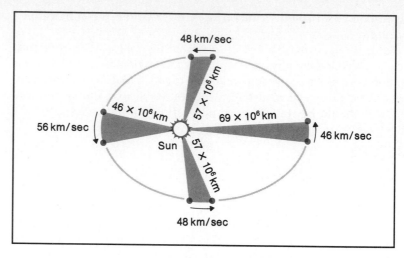

FIGURE 5.5 Mercury's orbital speeds and distances.

Figure 5.6, suppose that we begin counting time when Mercury and the earth are in position A, an inferior conjunction. Points B, C, and D show the corresponding positions of the planets after 44, 88, and 116 days. After 116 days, Mercury is again at inferior conjunction as viewed from the earth. The flip pages beginning on page 115 illustrate this 116-day synodic period of Mercury.

One of the exciting discoveries of the 1960s came in early 1965, when R. B. Dyce and G. H. Pettengil, using the 304-m (1000-ft) Arecibo antenna (see Figure 2.53, page 96), succeeded in bouncing radar signals off the surface of Mercury. The reflected signals, although very weak, indicated that the period of rotation of the planet is approximately 59 days. This came as quite a shock, since in 1890 the Italian astronomer Giovanni Schiaparelli had announced an 88-day rotation and this period was apparently confirmed by almost every visual observation made from that date until the year 1965. This new discovery was confirmed and refined by

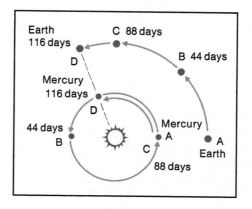

FIGURE 5.6 The synodic period of Mercury.

subsequent observations yielding a figure of 58.64 days, which is almost exactly two-thirds of the revolutionary period of the planet. This means that for every two revolutions of Mercury ($2 \times 87.96 = 175.92$), the planet makes three rotations ($3 \times 58.64 = 175.92$); the combined effect of these two motions produces a sunrise every 176 earth days. Thus there is a certain form of synchronization of the rotation of Mercury and its revolution with respect to the sun, as there is with the moon with respect to the earth. Mercury points first one face and then the opposite face toward the sun on successive times of *perihelion passage* (times when the planet makes its closest approach to the sun).

On March 29, 1974, Mariner 10 flew within 800 km (500mi) of Mercury and revealed for the first time its densely cratered surface (see the photograph on page 180 and Figure 5.7). At first glance, Mercury resembles the moon in that it has large, relatively smooth areas somewhat like the moon's maria, and it has a variety of crater forms ranging from those which appear to be very old with rounded rims to those which appear very young with sharp rims. We see some deeper craters with central peaks and some large shallow ones with evidence of repeated impacting.

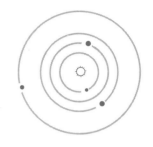

FIGURE 5.7 (a) A dark, smooth, relatively uncratered area on Mercury which resembles certain areas on the moon—lava flows being suggested. The largest craters shown are approximately 35 km (21.7 mi) in diameter. (b) Several fresh (sharp-rimmed) craters are seen in older basins. (c) A heavily cratered area of Mercury, showing many low, hill-like structures. The large valley at the bottom is approximately 100 km (62 mi) long. (NASA-JPL)

(a)   (b)   (c)

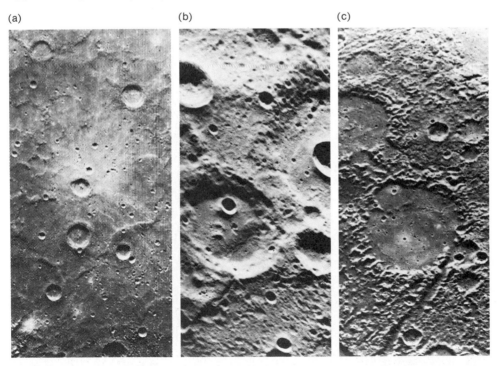

Measurements of surface temperatures using the infrared radiometer on board Mariner 10 indicated that Mercury has a high temperature of from 327°C to 427°C, depending upon the planet's distance from the sun. A drastic drop to−123°C was noted as the spacecraft began to traverse the shadowed portion. The coldest temperature has been estimated to be just under−173°C. A variation of approximately 600°C thus exists between the hottest and coldest points. These observations were not surprising since the planet is so close to the sun and has virtually no moderating "blanket" of atmosphere. The synchronization of rotation and revolution also contributes to the development of extreme temperatures; points on Mercury's equator experience 88 earth days of heating and then 88 earth days of cooling.

The discovery made by Mariner 10 that was least expected was a magnetic field strong enough to deflect the solar wind rather significantly; one does not usually associate a magnetic field with such a slowly turning planet. Although the strength of Mercury's magnetic field at the equator is at most one-fiftieth that of the earth, it does provide a definite bow shock and prevents particles from the sun from impinging directly on the surface

FIGURE 5.8   Mercury, as photographed by Mariner 10: (a) A photomosaic of Mercury, constructed from 18 photos, taken 6 hr after Mariner 10 flew past the planet on March 29, 1974 (note the distinct bright rayed craters in the upper-right portion of the photograph); (b) Fractured and ridged plains of the Caloris Basin as seen on Mariner's third encounter with Mercury; (c) The cratered terrain shown in this view appears very similar to parts of the moon. The large crater at the top shows conspicuous hills on its floor. (NASA-JPL)

(a)   (b)   (c)

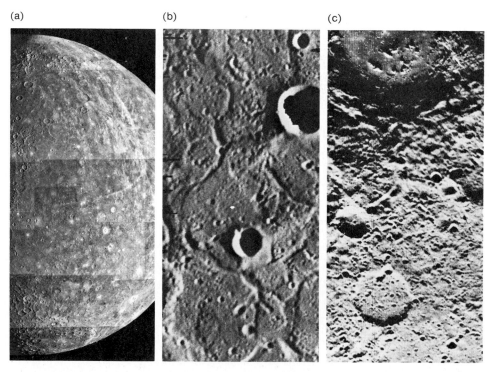

of the planet. If, as we have suggested for the earth, magnetic fields arise from electric currents in a molten core, we must explain how Mercury could still have a molten core after 4 billion years of cooling. First we must assume that Mercury was once molten to the extent that its more dense components, such as iron, settled toward the core, leaving a silicon-dominant crust. Although heat would flow very slowly from the core to the surface of Mercury, as with the earth, Mercury is much smaller than the earth and thus should have cooled completely, leaving no molten core. The fact that the axis of Mercury's magnetic field is tilted 7° to its axis of rotation seems to suggest that the magnetism is not simply locked into its crustal material. Perhaps the answer lies in the fact that Mercury has about 80 percent of its total mass in the core, a much higher percentage than in the earth. For a small planet to achieve such a high density (see page 185 and Appendix 6), its iron content in the core must be relatively higher as well.

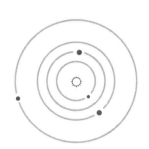

Observers searching for additional clues to the history of Mercury are particularly interested in the views of Mercury taken by Mariner 10 on its second and third passes (September 21, 1974, and March 16, 1975, respectively). The spacecraft had been placed into a solar orbit with a period of 176 days so that with each two revolutions of Mercury, the spacecraft encountered the planet again. This required no additional expenditure of fuel, other than for simple pointing procedures. Figure 5.8(a) is a mosaic of 18 photos showing distinct bright-rayed craters in the upper right portion of the photographs. Figure 5.8(b) shows the Caloris Basin, a very old major feature on Mercury's surface; its surface is cracked and wrinkled—evidence of a period of shrinking crust. Another surface feature thought to result from a time of shrinking crust is the scarp, a ridge of raised material (up to 2 km high and 1500 km long). Scarps sometimes cut right through craters, indicating that they may have been formed after the time of primary cratering. A surface feature which makes Mercury distinctly different from the moon is its *intercrater plains*, shown in Figure 5.8(c). If you compare this illustration with views of the moon in Chapter 4, you will see that it is unlike either the moon's heavily cratered highlands or its sparsely cratered maria. The intercrater plains may represent the oldest surface feature on Mercury.

The very low surface gravity and very hot daytime temperatures associated with Mercury have allowed virtually all atmosphere to escape. The light by which we see the planet is reflected directly from its surface. Like the moon, Mercury is not a very efficient reflector. Its albedo, the percentage of reflected light, ranges from about 13 percent in darker areas to 35 percent in limited regions where bright material appears to have been ejected during impact [see Figure 5.7(a)].

VENUS

Even before Galileo turned his telescope toward Venus to see it go through phases like the moon (Figure 5.9), humans were intrigued by this bright object that they often saw in the western sky just after sunset, thus

calling it the evening star. Because of its very dense cloud cover, Venus is the best reflector of sunlight in the solar system and at its brightest, it appears 16 times brighter than the brightest star. Some recognized this planet to be the same object that appeared alternately in the eastern (morning) sky; as Venus swings around the sun in an orbit not much smaller than the earth's, it appears first on one side of the sun and then on the other (Figure 5.10). Because the earth is also revolving, one complete synodic cycle requires approximately 19.5 months (584 days).

Early observations revealed Venus to be very nearly a twin of the earth in terms of its diameter and mass, and therefore its density. Yet Venus is actually vastly different than the earth. Its cloud cover is so dense that no person has ever seen its surface. In fact, it is the only terrestrial planet with a significant atmosphere. Astronomers who have tried to measure the temperature of the planet by analyzing the light it reflects have gotten a reading for only the top of the clouds, for it is at that level that the light is reflected; the temperature at that level was slightly less than 0°C. When radio astronomy came into its own in the 1950s, radiation from nearer the surface was measured in radio wavelengths, and temperatures in the order of 327°C (621°F) were recorded. With this discovery, we began to see Venus as a desert hot enough to melt lead.

In 1962, using the Goldstone antenna (see Figure 2.52, page 94), Roland Carpenter and Richard Goldstein succeeded in bouncing radar signals off Venus. To everyone's amazement, they discovered that Venus rotates backwards (clockwise) in a period of 243.16 days. Before this finding, the planet was thought to have a direct 225-day rotation period synchronized with its revolution. The evidence of the period of rotation consisted of a broadening in the returning wavelengths—a Doppler shift due to rotation. A certain wavelength was sent to the planet but a variety of wavelengths returned; those that were reflected from the approaching limb (edge) were shortened and those reflected from the receding limb were lengthened (Figure 5.11). Based on the amounts by which the wavelengths were changed, it was possible to compute the planet's velocity of rotation, hence its period. If you lived on this planet and were still dividing time into earth days, Venus's retrograde rotation in 243.16 days and its direct revolution in 225 days would produce a sunrise in the west every

FIGURE 5.9 The phases of Venus. (Lowell Observatory)

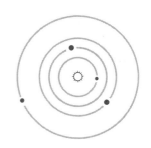

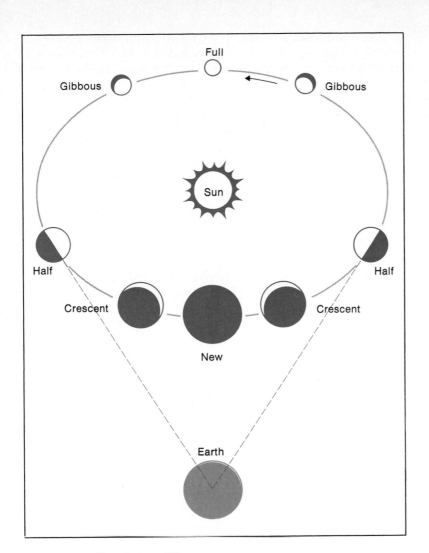

FIGURE 5.10   The phases of Venus.

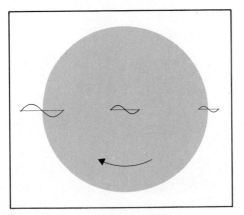

FIGURE 5.11   The Doppler effect upon radio signals reflected from Venus.

117 earth days. An interesting kind of resonance occurs between Venus and the earth. At each occasion of an inferior conjunction, Venus presents the same face toward the earth, even though it has rotated five times with respect to the sun. While the cause of this resonance is not fully understood, a tidal effect somewhat like that which synchronized the rotation and revolution of the moon with respect to the earth is suspected.

Radar astronomers are continually improving their "picture" of the surface of Venus. Of course, radio waves do not produce a photograph directly. Yet the returning signals may be interpreted to produce a map which is almost as instructive as a photograph. A recent upgrading of the radio telescope at Arecibo, Puerto Rico (see Figure 2.53, page 96) has permitted the use of wavelengths as short as 1 centimeter and has resulted in a 50-fold increase in the resolution of details on Venus. Features with dimensions as small as 20 km (12.4 mi) can be detected on the planet's surface. Figure 5.12 is a radar view of Venus that was made using the Arecibo antenna. In this view rough surfaces show as bright regions and smooth surfaces show as darker areas. This view is rather crude in comparison to current views from orbiting spacecraft, as will be shown later (see Color Plate 2).

Another phase in the study of Venus also began in 1962 when the first successful space probe was sent to survey the planet at close range. This probe, called Mariner 2, typified the fly-by approach taken by the United States in the study of Venus. A year earlier the USSR had begun a long series of attempts to land probes on the planet's surface. The first Soviet spacecraft are believed to have been disabled by the tremendous atmospheric pressure and high temperature near the surface; communications from the probes ceased before they reached the surface. That problem was

FIGURE 5.12 Radar view of Venus. The "Alpha" region (bright region in lower right corner) on the surface of Venus is 1350 km (837 mi) across, and it corresponds to rather rough terrain. The darker areas are relatively smooth. Note the bright craterlike structures scattered in the smooth (dark) region of the lower left area. (Courtesy of D. B. Campbell, Arecibo Observatory)

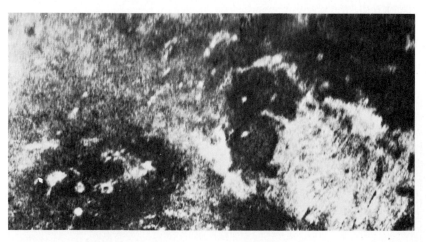

(a)

(b)

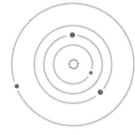

FIGURE 5.13 These two views represent the first direct photographs taken of the surface of Venus. This feat was accomplished by two probes of the USSR which were lowered onto the surface by parachute and survived the very high pressure and temperature of Venus long enough to transmit these television pictures back to earth. Views (a) and (b) were transmitted from Venera 9 and 10 respectively. (TASS, USSR)

overcome, however, and later attempts produced useful communications for as long as 110 minutes after the craft landed on the surface. Two such landers called Venera 9 and Venera 10 transmitted back to earth the television pictures you see as Figure 5.13(a) and (b). Venera 9 landed among numerous rocks in the order of 30 centimeters across that geologists interpret as indicating a rather young region. Venera 10 landed in a smoother, more mature region. Each view was made under existing light scattered by the dense atmosphere, with shadows lending detail.

The (U.S.) fly-by and the (Soviet) landing approaches have been equally effective in supplying data on many characteristic of Venus and have complemented each other in confirming various findings. For example, the atmospheric pressure at the surface is 90 times that on earth, a pressure equivalent to that which would be experienced by a diver 1 km (3280 ft) beneath the surface of the ocean. This pressure indicates directly that Venus' atmosphere has a mass of approximately 90 times that of earth.

A United States probe, Mariner 10, passed within 6000 km (3720 mi) of Venus on February 1974, and photographed the planet in ultraviolet light. Never before had specific patterns in the clouds of Venus been seen (Figure 5.14). They were revealed in ultraviolet light because of their composition—the clouds are composed of sulfuric acid droplets. This has also been confirmed by spectroscopic analysis in the infrared portion of the spectrum. A totally unexpected fact was revealed by this new view of Venus—the upper clouds (over the planet's equator) travel at very high speeds (360 km/hr or 224 mi/hr) and are able to circle the planet in four days. It was unheard of in the solar system for an atmosphere to rotate some 60 times faster than the solid body of the planet itself. Before we can attempt an explanation of the *superrotation* of Venus' atmosphere, we must answer two other questions: 1) why is it so hot on the surface of Venus, and 2) why does Venus have 90 times as much atmosphere as the earth?

The surface temperature of Venus is now known to be almost the same day or night, ranging between 430 and 480°C (810 and 900°F). This high temperature is not due to the direct radiation of the sun. The clouds of Venus reflect 76% of the radiant energy of the sun, so less energy reaches the surface of Venus than reaches the surface of the earth. The high surface temperature is attributed to the presence in the planet's atmosphere of CO_2 and H_2O, which together create a "greenhouse" effect. You have probably experienced the heating effect that occurs inside an automobile on a sunny day when the windows are closed. The sun's rays enter the car through the windows, are absorbed by the interior, and are then reemitted at a longer wavelength, primarily in the infrared range. These longer wavelenghts are not transmitted out through the windows as readily as the shorter wavelenghts are received, and hence there is a build-up of heat inside the car. When a higher temperature has been reached, the energy is radiated at the same rate it is received and that higher temperature is maintained. This is the same effect that florists use to maintain higher temperatures inside their greenhouses.

The carbon dioxide (97 percent) and water vapor (less than 1 percent) in the Venusian atmosphere act like the glass in a greenhouse; they absorb much of the infrared radiation from the planet's surface, thus preventing its radiation into space. Although, like any planet, Venus must eventually radiate all of the sun's energy that falls on it, the absorption by these compounds causes the surface and lower atmosphere to be heated to a high temperature before the outflow of energy balances the inflow.

FIGURE 5.14 A series of ultraviolet photographs of Venus, taken two days after Mariner 10 flew past the planet on February 5, 1974, during a 14-hr period, showing the rapid rotation rate of the upper cloud deck. The feature indicated by the arrows is about 100 km (62 mi) across. (NASA)

(a)   (b)   (c)

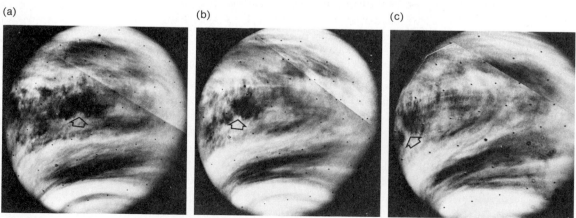

The earth's atmosphere contains only trace amounts of CO_2 and H_2O, however these gases serve to raise the surface temperature of the earth by approximately 40°C (72°F).

If one could imagine traveling from the surface of Venus up through its cloud layer which lies at an elevation of about 60 km, one would experience a steady decrease in temperature to a level considerably below the freezing point of water. This cooler temperature is also typical of the higher elevations. Molecules near the surface would certainly be moving at velocities that would allow them to escape the gravitational pull of the planet; but in the process of traveling to higher elevations, these molecules would experience cooling and their velocities would drop below that of escape. This cooling trend at higher elevations, then, is a significant part of the explanation of how Venus has been able to retain 90 times as much atmosphere as the earth.

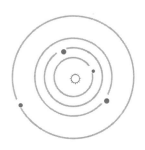

Whereas most of the heating of the atmosphere results from the infrared radiation of the surface of Venus, there is some direct heating of the middle and upper atmosphere by the sun itself. This heating is most effective near the equator, where the sun's radiation strikes the planet more directly. This heating triggers a movement of air to the north and to the south from the equator. When the gases have cooled, they descend to a lower elevation and return toward the equator. Such a north-south flow of atmosphere is called a *Hadley cell*. Venus has a very slow westward (retrograde) rotation; therefore the spot heated most effectively by direct sunlight gradually moves around the planet. Laboratory experiments with moving heat sources have produced superrotating liquids, so we look to this kind of an explanation for the superrotation of the Venusian atmosphere. Probes that have penetrated the atmosphere have measure winds that are virtually zero on the surface, but that increase with elevation until they are traveling 360 km/hr at the cloud tops. It is the superposition of the north-south Hadley cell motion upon these westward winds that is thought to produce the chevron appearance in the clouds (Figure 5.14).

The month of December 1978 was the most active period of exploration of Venus. On December 5, the orbiter Pioneer Venus 1 arrived at the planet and began an orbital reconnaissance that lasted approximately one year. One of its notable discoveries made using radar mapping was a canyon that dwarfs all others known in the solar system; it is 1300 km (810 mi) long, 230 km (145 mi) wide, and up to 4.5 km (2.8 mi) deep. Because of the probe's highly eccentric orbit, it actually samples various layers of the planet's upper atmosphere (ionosphere), dipping to within 378 km (235 mi) of the planet's surface on each pass.

On December 9 Pioneer Venus 2 arrived, along with its four miniprobes that were sent ahead to explore both day and night portions of the planet (Figure 5.15). Although not programmed to function on the surface, the "Day" probe actually survived impact and transmitted information from the surface for 67 minutes. The findings reported in the following sections are those primarily from the United States probes, but they are also confirmed by two additional probes of the Soviet Union, Veneras 11

FIGURE 5.15 The Venus Multiprobe Spacecraft, each component descending into the Venusian atmosphere to perform its own special function. (NASA)

and 12, which plunged through the dense Venusian atmosphere during the same month. The findings of the Soviet Union craft include continuous lightning below the 32-km (20-mi) level; this might explain the steady glow observed by the United States night probes.

As the Pioneer Venus 2 multiple probes descended through the atmosphere, they found three distinct layers of clouds beginning at the 60 km elevation, each composed of sulfuric acid droplets. Only the lower of the three was dense enough to appear as clouds in the usual visual sense; these lower clouds were 45 km (27 mi) deep.

The composition of the Venusian atmosphere is as follows: 97 percent CO_2 (carbon dioxide), 2 percent N (nitrogen), 0.1 to 0.4 percent H_2O (water vapor), 250 ppm (parts per million) He (helium), 6 to 250 ppm Ne (neon), 20 to 200 ppm Ar (argon), 240 ppm SO_2 (sulfur dioxide), and 60 ppm O (oxygen). The approximate percentages of CO_2 and N were known from earlier probes, but the discovery of primordial argon and neon in abundance levels several hundred times those in the earth's atmosphere was totally unexpected and may cause a total reevaluation of theories concerning the origin of the solar system. Most theories assume that the lighter primordial elements, up through argon and including neon, would have escaped the innermost planets, driven away by the heat of the sun. The most abundant form of argon in the universe is argon-36; according to the above theory Venus would have a lesser proportion of argon-36 than the earth or Mars. The findings of Pioneer Venus 1 and 2, however, indicate a reverse pattern, with the highest abundance on Venus, second highest on earth, and least on Mars. Could Venus have been formed under cooler circumstances than the earth or at a different time? These new findings will keep theorists busy speculating for years.

Why is the atmosphere of Venus so different from that of the earth when they are virtual twins in mass and size and Venus is only slightly closer to the sun? If we assume that the hydrogen and helium atmospheres present on the terrestrial planets at the time of their formation were driven off by solar radiation, then we would conclude that their secondary atmospheres must have formed from the outgassing of other elements, perhaps from volcanos. On earth, carbon dioxide is known to be released by volcanic action. The same processes may have occurred on Venus, except that the Venusian atmosphere retained a much higher concentration of CO_2. Perhaps the earth once released an equivalent supply of CO_2 but most of it was absorbed by its rocky crust. The earth is, in fact, thought to contain as much CO_2 as Venus, but instead of being in the atmosphere, it is chemically bound in the rocks of the earth. The CO_2 is combined chemically with calcium (Ca) and oxygen to form calcium carbonate ($CaCO_3$), a basic compound of much of the earth's present surface material. Why did the CO_2 in Venus's atmosphere not combine with its surface material? The answer may be that Venus, being closer to the sun, had a slightly higher original temperature than the earth. Laboratory experiments demonstrate that rocks at a higher temperature than on earth will not absorb as much CO_2.

Life itself plays a vital role in maintaining a rather constant low level of CO_2 in the earth's atmosphere. Plants, through the process of photosynthesis, convert CO_2 to oxygen. Marine organisms remove CO_2 from the ocean waters in the process of building their shells, allowing these waters to absorb more from the air above. Even a slight change in the rate at which CO_2 is used by living organisms might change the earth's ability to support life; a slight increase in the CO_2 content of the atmosphere would increase the greenhouse effect and raise the temperature of the earth. If such a process got out of hand, conditions on earth might one day resemble those on Venus.

The radar observations of the Venus Pioneer Orbiter over a period of many months have produced quite a detailed map of this planet's surface (see Color Plate 2). The planet appears very smooth, with 60 percent of its surface elevations within 500 m (1640 ft) of its average radius (6,051.4 km or 3760.3 mi). Three regions of the planet are significantly raised and may be thought of as "continents" for the sake of comparison to the earth. A region called Ishtar Terra (Figure 5.16) consists of a raised plateau almost the size of the United States, with an average elevation of 5 km (16,404 ft). Rising another 6 km (19,685 ft) above this plateau is a mountain called Maxwell; its height above the plains of Venus is 2 km (6,562 ft) higher than Mt. Everest. Another large region, about one-half the size of the African continent, is called Aphrodite Terra: its elevation ranges between 2 km and 5 km above the Venusian plains. Near Aphrodite's eastern

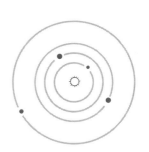

FIGURE 5.16 Radar information from the Pioneer Venus Orbiter has revealed the highest and most dramatic continent-sized highland region on Venus, a feature named Ishtar Terra, shown here by way of an artist's conception. The outline of the continental United States is provided for scale. The highest mountain peak rises almost 6 km above the surrounding level. (Ames-NASA)

end is a set of canyons which may indicate a rift in the surface of the planet. They extend for approximately 5000 km (3100 mi), and range in depth to almost 3 km (2 mi). A region called Beta Regio suggests the shape of double shield volcanos. This feature transcends in size any volcano on earth and exceeds in diameter the largest mountain on Mars, namely Olympus Mons; it has a base more than 1000 km (620 mi) in diameter and reaches to an elevation of 6 km (19,685 ft).

The landing of Veneras 13 and 14 in 1982 provided us with a close-up view of two more regions of Venus. In preparation for their encounter with the hottest planetary surface in the solar system, these probes were cooled to −73°C (−100°F). The fact that they were well insulated gave them a lifetime of about two hours on the surface. They were released from their respective fly-by "buses" two days before arrival at the planet. Their initial descent into the Venusian atmosphere was slowed and stabilized by parachute. During the last 45 km the probes experienced so much friction from interacting with the very dense atmosphere that they settled gently onto the surface as a marble would onto the bottom of an aquarium.

Each probe retrieved a thimble-full of Venusian soil and placed that soil into an internal laboratory. Here the sample was cooled to less than 38°C (100°F) and the pressure reduced to a fraction of one atmosphere. An X-ray spectroscope revealed the presence of calcium, sodium, silicon, iron, magnesium, and potassium. Venera 13 landed in a volcanic highland region called Phoebe Regio, marked by potassium-rich basalts somewhat like those associated with a Hawaiian volcano. Its sapphire lens recorded the surrounding terrain, as shown in Figure 5.17(a). Venera 14 landed five

(a)

(b)

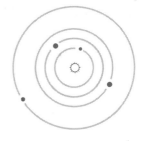

FIGURE 5.17 (a) Venera 13 view of surface of Venus; (b) Venera 14 view of surface of Venus. (Tass, Sovfoto)

days later in a lowland region composed of basalts similar to oceanic basalts on earth [Figure 5.17(b)]. The mapping efforts of the United States orbiter were used to select the most useful sites for Venera landings.

EARTH

It is interesting to see how our perception of the earth has changed in the last few years because we have been able to view it from space. The dominant colors are not so much the brown and green tones associated with the changing seasons on land, but the blues and whites of the sky, the oceans, and clouds. Shown on page 110 is a view of the earth as seen from a distance of 178,000 km (110,600 mi). What kind of detail can you see from such a distance? Certainly nothing with a dimension smaller than 80 km (50 mi) can be detected. With such limited resolution, we can barely make out larger geological features like mountain ranges. Certainly no indication of intelligent life is detectable—no freeways, waterways, or farm lands can be identified. The only possibility of detecting a city may be from the glow of its lights in a clear, night-time sky. Given the difficulty of identifying features on the earth from a distance of 178,000 km, one can easily see why it has been impossible for ground-based observers to tell whether life exists on Mars; Mars never approaches the earth closer than about 56 million km (35 million mi). But for a moment, let's imagine a form of intelligent being on that planet—"Martians"—with telescopes as good as ours, trying to determine if intelligent life exists on earth. Their job would be as challenging as ours. It is interesting to contemplate what technological developments might allow a Martian to discover life on earth.

In this brief discussion of the earth, we have directed our usual outward gaze back toward ourselves. This is what happens in all space exploration—we ask questions of each planet based on our present understanding of the earth, and the answers we get always prompt both more questions and a deeper understanding of the earth.

MARS

Mars seems to fire people's imagination more than any other planet. Many observers have speculated that some form of life, perhaps as intelligent as our own, might exist on this planet. This idea is perhaps due in part to the similarity of Mars to the earth, for Mars does exhibit polar caps and certain seasonal color variations similar to those we find on earth. As early as 1877, Giovanni Schiaparelli, director of the Milan Observatory, asserted that he had seen a network of fine straight lines that seemed to interconnect larger features on the Martian surface. Percival Lowell, an American astronomer, suggested that these lines represented canals, actual waterways built by the Martian inhabitants to irrigate their crops and to transport their goods on barges. The notion of intelligent life on Mars was long ago seized by science-fiction writers. Let us look at the evidence offered by recent detailed, close-up observations of this planet.

FIGURE 5.18 A high-resolution photovisual map of Mars showing the seasonal aspects of late Martian summer, 1969. The excellent quality allowed surface details of less than 30 km (18.6 mi) to be resolved. This map was produced from measurements of 20 photographs, 13 visual drawings, and 15 telescopic micrometer observations. South is at the top as seen in a telescope. (C Capen, JPL-Table Mountain Observatory)

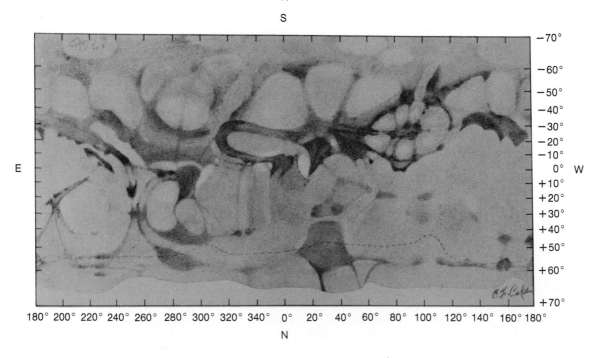

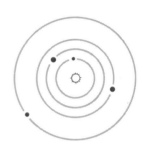

Before 1965 our only observations of Mars were made from earth-based observatories. Because of the obscuring effect of the earth's atmosphere, details of the Martian surface features, atmospheric composition, magnetic field, and the like were very sketchy. The best photo record of Mars gives only a hint at the presence of polar caps and large areas that are somewhat darker than the planet in general. A far more detailed view is shown in Figure 5.18 which represents the result of hundreds of hours spent by astronomers at the eyepiece of telescopes, sketching areas of momentary clearing. The space age opened up yet better methods for observation. Anticipating the close approach of Mars in 1965 at a time of opposition, scientists prepared a space probe that flew to within 9600 km of the planet, photographed its surface, and relayed, those photos together with other information back to the earth. The photographs which this probe, Mariner 4, returned to earth were far superior to any produced by earth-based telescopes.

Spurred by this success, scientists refined their equipment and prepared for another opposition of Mars in 1969, when the planet passed within 72 million km (45 million mi) of the earth (Figure 5.19). Because of the combined motion of the earth and Mars, such oppositions occur at 26-month intervals, Mars' synodic period. Because of the eccentricity of the two orbits, some approaches are much closer than others, as shown in Figure 5.20. Although much was learned during the Mariner 6 and 7 flights in 1969, much more information was gathered during the 1971 opposition.

In November 1971 Mariner 9 became the first artificial satellite to orbit another planet. For almost one year, this spacecraft radioed back to earth a continual flow of pictures and data about Mars; this information changed our concept of the planet significantly. The cameras on Mariner 9

FIGURE 5.19 Mariner 7 photographs of Mars, frames 73 and 74, showing the bright southern cap and the circle of Nix Olympica. The two frames were taken 47 minutes apart, showing the rotations of the planet. (JPL-NASA)

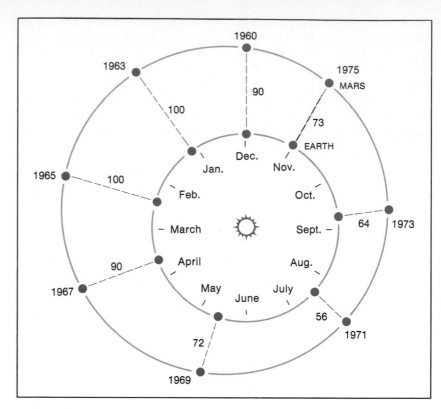

FIGURE 5.20 Mars-earth oppositions, with distance in millions of kilometers shown by numbers next to the dashed lines.

at first revealed only a huge dust storm that obscured most of the surface features of the planet, but as the storm subsided, four large volcanic mountains emerged through the dust. As the Martian atmosphere cleared further, the full extent of the largest volcanic mountain, Olympus Mons, was revealed; it is 600 km across at the base (covering an area equal to that of Arizona) and approximately 26 km (16 mi) high, far higher than any mountain on the earth (Figure 5.21). Three additional volcanic features of major

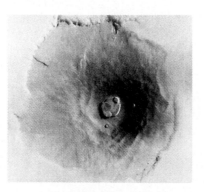

FIGURE 5.21 Mariner 9 photo of Nix Olympica, the Martian volcanic mountain: longitude 133°, latitude +18°. (JPL-NASA)

proportions exist in the same vicinity, indicating a region on the planet that has been active more recently than the cratered portions. If the lava-type material which poured forth to produce these volcanos was similar to that on earth, large amounts of water would have been released into the Martian atmosphere. Thus, these volcanos indicate not only the molten nature of at least a portion of the planet, but may also be relevant to the question of life on Mars. The presence of volcanos raises another question. Volcanic action on earth is often associated with the colliding of crustal plates, so it is reasonable to ask if this planet also has crustal plates that move. Figure 5.22 shows a map of Mars based on Mariner 9 observations. Stretching eastward from the region of the volcanos is a huge valley, approximately 4000 km (2500 mi) long, 100 km (62 mi) wide, and in several places up to 6 km (3.7 mi) deep. Such a valley, if superimposed upon the earth, would stretch across the entire United States. Details of the valley revealed by close-up views (Figures 5.23 and 5.24), have raised many questions about the process of its formation. The general form of the chasm suggests subsidence of the Martian soil due to faulting or plate tectonic motion. Note the fingerlike pit chains in the right-hand portion of Figure 5.24. These probably represent the beginning stages of a larger valley that will develop with continued subsidence. The scalloped edges along deeper portions of the valley suggest the enlargement of such pits. The tributary-like complex shown in the upper left-hand portion of Figure 5.23 seems to suggest erosion due to fluid action. This immediately raises the question, does water exist on Mars? It is well-established that liquid water could not persist on the surface of the planet with its present lack of atmosphere; the atmospheric pressure on the surface of Mars is less than one-hundredth that on earth. Under such low pressure, surface water would vaporize in all but the coldest parts of the planet. If the assumption is made that liquid water caused the erosion, a theory must be offered that explains how liquid water could have persisted long enough to erode a canyon or produce a river bed; one such theory is that a much denser atmosphere was once present.

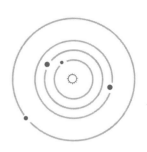

An alternative theory of the origin of the tributaries is that water on the surface of the planet exists not as a liquid, but in a frozen state beneath the surface. The theory suggests that the subsidence of soil exposed this frozen water resulting in evaporation and/or seepage, and that such seepage loosened material to produce the tributary-like structure. Some geologists believe that the side canyons of the Grand Canyon of Arizona were formed as a result of the seepage of ground water.

As we have noted earlier, one of the ways to determine the age of a given surface feature is to study the cratering that has occurred in that feature as compared to the cratering in the surrounding area. The craters that now exist within the Mariner Valley must have been formed after the valley itself was formed or else the formation of the valley would have obliterated them, regardless of its mode of formation. Because the count of craters within the valley is almost as high as in the surrounding area, the valley is considered to be quite old—in the order of 3 to 4 billion years.

A feature on Mars that represents a more convincing example of water erosion on Mars is the Amazonis channel, shown in Figure 5.25. Whereas the bottom of the Mariner Valley does not fall off in one direction, the Amazonis channel does appear to grade downward to allow a flow from south to north. Furthermore, the bars and braiding of the channel in

FIGURE 5.22 A shaded relief map of the entire equatorial region of Mars, based upon the photographic survey by Mariner 10. The principal features include volcanoes, craters, basins, and smooth-mantled areas. (JPL-NASA)

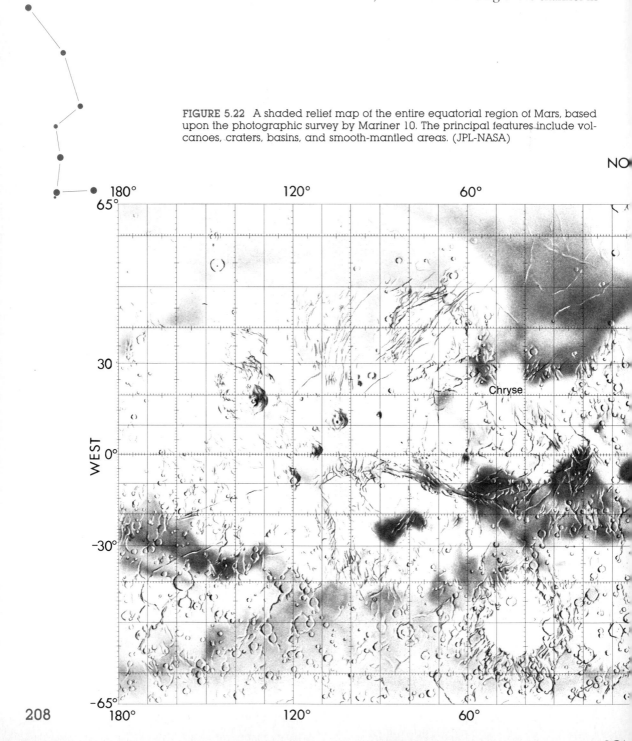

its northerly portion are very typical of river beds on earth. The entire Amazonis channel is 350 km (270 mi) long and 100 km (62 mi) wide (at its widest), which would require a river the size of the Amazon in South America to produce. Another way a channel-like structure on Mars might have formed is by an ice flow, the way glaciers carve out channels on earth.

In the summer of 1975, the United States launched one of its most ambitious and suspenseful space efforts: the search for life on an alien planet. Two identical probes, Viking 1 and Viking 2, went into orbit around

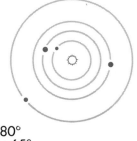

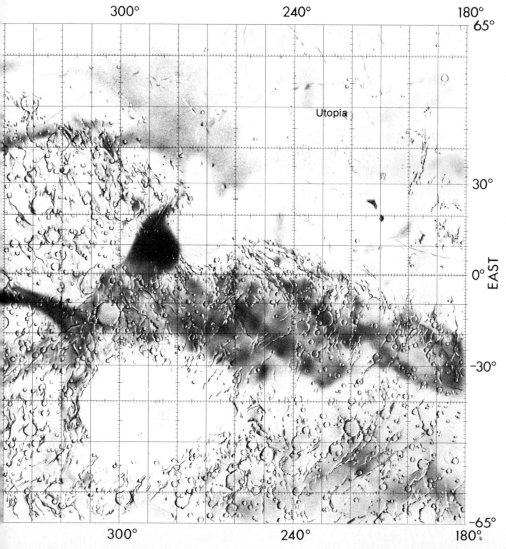

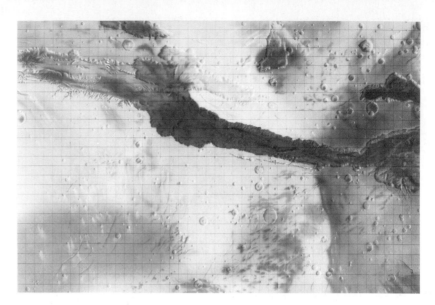

FIGURE 5.23 The Coprates region of the great chasm of Mars, based on Mariner 9 photographs together with large-scale albedo markings as seen from earth. (Prepared by James Roth and Dr. G. de Voucouleurs for JPL-NASA)

FIGURE 5.24 A close-up and profile of a portion of the Coprates region centered on longitude 75°, latitude −10°. (JPL-NASA)

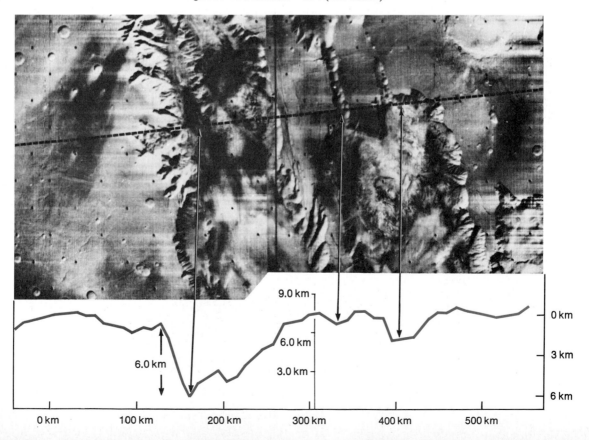

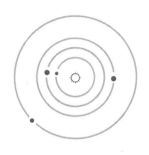

FIGURE 5.25   The Amazonis channel, thought to have been formed by running water. The section shown is approximately 100 km long, and the flow direction is thought to be toward the north (upper right). (JPL-NASA)

Mars after traveling just under one year to reach the planet. The first order of business was to search for suitable landing sites for their respective landers. In that process the Viking 1 orbiter photographed large portions of the planet in greater detail than ever before. The home of the Viking 1 lander, a river sand bar in the region of the Chryse Basin, is shown in Figure 5.26(a); the crater at the head of the teardrop erosion pattern appar-

FIGURE 5.26   (a) Braided channels record the flow of water in the Chryse basin, near the Viking 1 landing site. The shore of the channel is shown to the far right. (b) The crater Yuty, showing a flow pattern in the material that was ejected. (c) A Martian cyclone, as seen by Viking Orbiter 1, near the northern polar region. A frost-filled crater and patches of frost associated with the northern polar region can also be seen. (JPL-NASA)

(a)

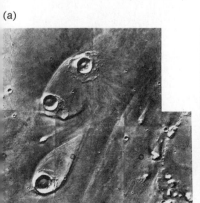

(b)

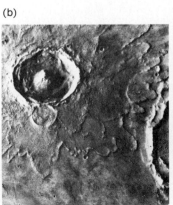

(c)

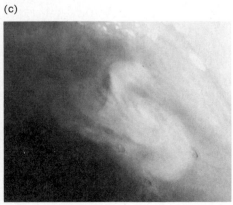

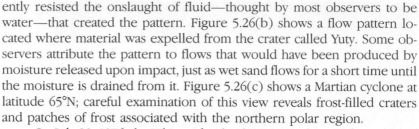

ently resisted the onslaught of fluid—thought by most observers to be water—that created the pattern. Figure 5.26(b) shows a flow pattern located where material was expelled from the crater called Yuty. Some observers attribute the pattern to flows that would have been produced by moisture released upon impact, just as wet sand flows for a short time until the moisture is drained from it. Figure 5.26(c) shows a Martian cyclone at latitude 65°N; careful examination of this view reveals frost-filled craters and patches of frost associated with the northern polar region.

On July 20, 1976, the Viking 1 lander (Figure 5.27) was released from its Orbiter and descended to Chryse Basin. On September 3 the Viking 2 lander descended to the Utopia Plain on the opposite side of the planet toward the northern polar region; there it experienced the extreme cold of a Martian winter a year later (see Figure 5.22 for these locations). Because the primary purpose of the probes was to search for life and because life, as we know it, is so closely linked with water, the two sites chosen were locations where water might have been present at some point. The first site was in a basin where sedimentation might have contained fossil remains of life and the second site was in a plain where atmospheric water concentrations appeared higher. A secondary constraint was that the sites had to be in a rather large smooth area so that a safe landing could be

FIGURE 5.27 The Viking lander. (NASA)

made. Before we discuss the life search program let us first look at what these landers saw.

Mars has always appeared reddish in the night sky, whether seen by naked eye or telescope. Now the true color was recorded at close range; it was found to be on the reddish side modified with tones of yellow and brown, with the sky a paler shade of the same color (see Color Plate 3). The color is probably due to iron oxides (rust) on the surface and to dust suspended in the atmosphere. Mars is thought to have high-velocity winds [exceeding 200 km/hr (124 mi/hr) at times] which cause the dust to be held suspended in the atmosphere to an elevation of 40 km (25 mi). Because the density of the Martian atmosphere is only one-hundredth that of the earth's atmosphere, the wind would be rather ineffective in moving anything but fine dust particles. Over a long period of time, however, the Martian winds play a significant role in creating surface features like the dunes shown in Figure 5.28.

The atmospheric composition of Mars is of critical interest because it contains every gaseous component necessary for life as we know it (see Table 5.3); it contains some poisonous gases as well. Although the abundances of surface material cannot be accurately determined yet, X-ray spectrometers on board the Viking landers have identified the components and their relative percentages, as shown in Table 5.4.

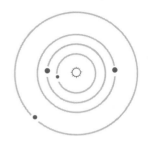

TABLE 5.3

COMPOSITION OF THE MARTIAN ATMOSPHERE	
Carbon dioxide (CO_2)	95%
Molecular nitrogen (N_2)	2%–3%
Argon (Ar)	1%–2%
Molecular oxygen (O_2)	0.3%–0.4%
Xenon (Xe)	Trace
Krypton (Kr)	Trace
Water vapor (near poles)	Trace

TABLE 5.4

COMPOSITION OF THE MARTIAN SOIL	
Iron	16%
Silicon	15%–30%
Calcium	3%–8%
Aluminum	2%–7%
Titanium	0.25%–1.5%

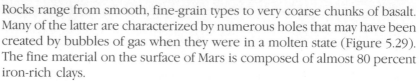

Rocks range from smooth, fine-grain types to very coarse chunks of basalt. Many of the latter are characterized by numerous holes that may have been created by bubbles of gas when they were in a molten state (Figure 5.29). The fine material on the surface of Mars is composed of almost 80 percent iron-rich clays.

Both probes contained seismometers for the detection of Marsquakes, however one detector failed, so the records of several quakes are a bit sketchy. In November 1976 a vibration was recorded that had all the characteristics of a quake of magnitude 6 on the Richter scale. Without the additional information that the inoperative seismometer would have provided, however, it was impossible to pinpoint the epicenter of the quake and to determine the nature of the planet's interior. Based on the many features which show Mars to be geologically active, such as the volcanic mountains, the Mariner Valley, and the Marsquakes, we may surmise that the planet must still have a source of internal heat and can be thought of as layered (differentiated) with a denser core.

As mentioned earlier, there is a pattern of changing seasons on Mars caused by its tilted axis and its revolution around the sun. Polar caps form in the north when the planet is tipped away from the sun and disappear as spring returns. For many years scientists surmised that these caps were simply CO_2 frost (dry ice), forming when the temperature dropped to $-123°C$ ($-190°F$) and sublimating (turning back to a vapor) at any warmer temperature; this theory was confirmed. An additional feature of these caps discovered by the Viking 2 lander situated nearer the north polar region is the existence of layers of water ice—permanent polar caps which lie beneath the changing frost caps (see Figure 5.30). Alternate layers of

FIGURE 5.28 A spectacular view of the Martian landscape, as seen by Viking 1 lander, shows sharp dune crests, which indicate recent wind storms. The small deposits downwind of rocks also indicate wind direction. The large boulder at left measures about 1 m by 3 m. The sun rose two hours before this photograph was taken. (JPL-NASA)

dust show periods when the deposition of ice waned, representing several major changes in climate. The upper dust covering tends to hide the ice caps from earth-based observatories; however, their terraced structure is shown quite clearly in maps based on Mariner 9 images (see Figure 5.31). Some observers believe that sufficient water is locked in these polar caps to cover the entire planet to a depth of 1 to 10 meters, if melted. Water may also exist as permafrost in the soil; temperatures in the polar regions range down to $-73°C$ ($-100°F$) in the summer season and down to $-125°C$ ($-193°F$) in the winter season.

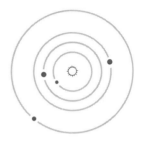

Over long periods, measured in thousands and millions of years, a variation in the tilt of Mars' axis and in the eccentricity of its orbit causes a change in the temperature range of the planet. Recall that the tilt of axis is measured from a perpendicular to the ecliptic plane. The tilt of its axis is not fixed but rather varies between 15° and 35°; this is primarily a result of the gravitational effect of Jupiter. When the angle of tilt reaches its maximum value of 35°, the seasonal variations would be much more extreme than at the present time. The variation in the eccentricity of Mars' orbit between 0.009 and 0.140 means a variation in how close to and how far from the sun Mars gets in its orbit; the higher the eccentricity value, the closer Mars is to the sun at its closest approach (perihelion) and the farther it is from the sun at its farthest point (aphelion). The layered effect in the polar caps shown in Figure 5.30 may be the result of extremes in temperatures associated with these long-term variations in axis tilt and eccentricity. Currently, Mars' tilt is 25° and eccentricity is 0.0934.

What motivates humans to search for life elsewhere in the universe, and what basic assumptions are made about the nature of life? Life as we know it is based on the element carbon. Almost an endless chain of molecules can be formed with carbon as the primary bonding agent. Since no other element can play the same role as efficiently as carbon, scientists assume that if life exists elsewhere in the universe, it would probably be based on carbon. The fact that organic (carbon-based) molecules have

FIGURE 5.29 The angular blocks in the foreground of this view are only about 4 meters from Viking 1 lander, and they range in size from a few centimeters to a few meters across. (JPL-NASA)

been found in meteorites and in gases between the stars supports this assumption. A fundamental question that arises from such considerations is whether life is something that is very likely to occur (in many places in the universe) or whether it is too unlikely to occur anywhere but on earth. What we have found on Mars should help us answer that question.

Two of the four experiments designed to detect life on Mars, gas-exchange and labeled-release experiments, were based on the fact that gases are always given off as a by-product of all life processes, plant and animal. In both experiments a nutrient solution was added to the Martian soil sample, whereupon a very significant release of gas immediately oc-cured followed by a period during which the release rate slowly dimin-ished. Although the first reaction was positive, the conclusion drawn by many observers is that a chemical rather than a biological reaction took place. A graph typical of each type of reaction is presented in Figure 5.32; the graph associated with the first two Viking experiments resembles that of the chemical reaction. A third experiment (pyrolytic-release) conducted

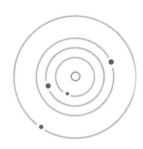

FIGURE 5.30 Mars's northern polar ice cap as seen by Viking Orbiter 2. The alternate layers of ice and Martian soil are particularly evident along the slopes that are exposed in the upper left portion of this figure. (JPL-NASA)

FIGURE 5.31 The polar regions of Mars: (a) northern region, October 12, 1972; (b) southern region, February 28, 1972. Note the distinctively different pattern of cratering in the two regions. (JPL-NASA)

(a)   (b)

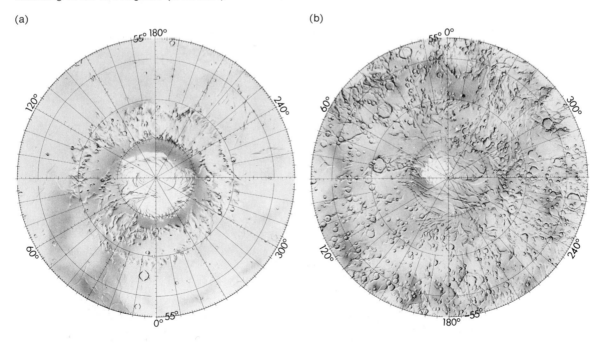

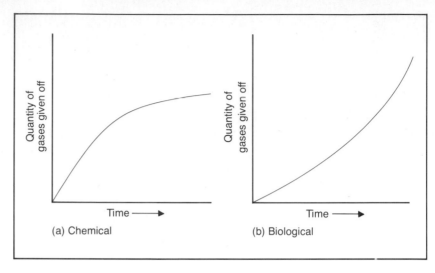

FIGURE 5.32 Reaction curves.

FIGURE 5.33 In the search for life, the Viking soil sampler arm was directed to move a rock and sample soil from beneath its original position, with the thought in mind that a living form might have sought refuge there from the sun's ultraviolet radiation. (JPL-NASA)

was based on the fact that living organisms assimilate carbon-14 as part of the life process. After proper incubation, the soil sample was pyrolyzed (vaporized by heat). Samples tested indicated positive results; however, because some of the samples were heated as high as 175°C (347°F) and still produced a positive result, it is assumed that the reaction was chemical since living microorganisms would not have survived such heat.

Although the results of these first three experiments seem to rule out life on Mars, we should hasten to say that some scientists give a more positive interpretation; they say that the experiments neither proved nor disproved the existence of life on Mars. A heavier blow was delivered, however, by the results of the final experiment, the gas chromatograph-mass spectrometer (GCMS) experiment, which is based on the fact that organic molecules generally are very large, heavy molecules. If a soil sample is pyrolyzed and passed through a gas chromatograph, the molecules will separate according to their weight and they can be identified using the mass spectrometer. Using such an instrument, organic components have been found in soil samples taken from places in Antarctica where it is very unlikely for organic molecules to exist. No evidence of organic molecules, however, was found in Martian soil. One would have thought that organic molecules derived from meteorite impact would have been found. One theory is that they were destroyed by the ultraviolet radiation of the sun (Figure 5.33).

Our search for life on Mars may not be over—approaches a new exploration might take are being discussed: a rover vehicle might cover a

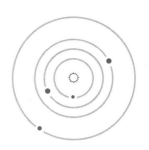

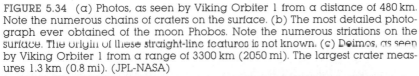

FIGURE 5.34 (a) Photos, as seen by Viking Orbiter 1 from a distance of 480 km. Note the numerous chains of craters on the surface. (b) The most detailed photograph ever obtained of the moon Phobos. Note the numerous striations on the surface. The origin of these straight-line features is not known. (c) Deimos, as seen by Viking Orbiter 1 from a range of 3300 km (2050 mi). The largest crater measures 1.3 km (0.8 mi). (JPL-NASA)

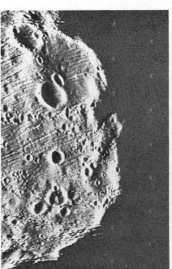
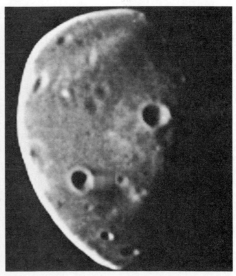

significant amount of surface, a rocket vehicle might return samples to earth, or perhaps a mission might land astronauts to explore the surface.

We have known since 1877 that Mars has two natural satellites, Phobos and Deimos; however, not until Mariner 9 turned its video eye their way did we have any idea of their surface appearance [Figure 5.34(a, b, c)]. Phobos measures 27 km (17 mi) in its greatest dimension and Deimos only 12 km (7.5 mi). Both moons have a dark surface that reflects only a small percent of the sunlight that falls on it. The moons are subject to impacts continually since they are devoid of atmosphere. The impacted surface of Phobos is shown very clearly in Figure 5.34(c), a view from the Viking orbiter. Note the *striations* (grooves) on the moon's surface—some pass through older craters and others are interrupted by yet younger craters. The history of this moon, including bombardment and perhaps shrinkage, can be read in its surface.

QUESTIONS

1. What basic properties distinguish the terrestrial from the Jovian planets?

2. The rotation of Mercury was once thought to be 88 days, a period equal to that of its revolution. It is now known to be _____ days, as recently measured by radar.

3. What properties of Mercury cause it to be one of the hottest and at the same time one of the coldest planets?

4. Why is the albedo of Venus so much greater than that of Mercury or of the earth?

5. How many hours after sunset is Venus visible when at maximum elongation east?

6. During what phase does Venus appear brightest? Why?

7. The maximum surface temperature of Venus is probably greater than that of Mercury. How is this possible when Venus is farther from the sun?

8. Which configuration of Mars brings it closest to the earth: conjunction, quadrature, or opposition?

9. Where and when is Jupiter seen when at opposition?

10. Which method of observation yields the most detailed information about Mars?

11. Are the straight-line figures referred to as "canals" actually visible at times on Mars?

12. Show by a drawing how Galileo's discovery that Venus goes through all possible phases from new to full and back to new again disproves the Ptolemaic model of the planetary system.

13. Find the semimajor axis of a planet's orbit that has an aphelion distance of 160 million kilometers (100 million miles) and a perihelion distance of 100 million kilometers (62 million miles).

14. Why do observers believe that Venus lost its original atmosphere and then developed a secondary atmosphere? How could it develop a secondary atmosphere?

15. What characteristics of Venus were revealed or confirmed by probes that either landed or flew by the planets?

16. What are the possible dangers of releasing ever-increasing amounts of carbon compounds into the earth's atmosphere by the burning of fossil fuels?

17. Group the terrestrial planets into categories according to the predominant element or compound in their atmospheres: (a) hydrogen, (b) nitrogen, (c) carbon dioxide, (d) other elements.

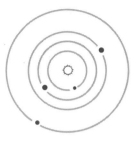

The Jovian Planets

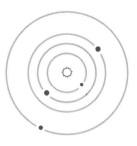

Six

The next four planets—Jupiter, Saturn, Uranus, and Neptune—are similar in many ways. Because they are very much like Jupiter, we call them the *Jovian* planets. Astronomers have been particularly fascinated by the Jovian planets because their composition is thought to represent the composition of the original nebula from which the solar system was formed more closely than the terrestrial planets. They formed in the cooler (outer) regions of that nebula far from the heat of the sun, and therefore probably were able to hang onto all their constituents; the heat of the sun caused at least the lighter atoms of the terrestrial atmospheres to escape.

Probes that are sent to get a close-up view of the Jovian planets must survive a passage through the asteroid belt that lies between Mars and Jupiter. Some 50,000 rocky chunks of material with an average size of a few kilometers orbit the sun in this belt. These objects are spread out in an orbit whose average radius is 2.8 A.U.; thus the belt consists mostly of empty space. In fact, the chance of a space probe being hit by one of these sizable asteroids is almost nil. Before a probe was sent successfully through this region, however, we thought that it might be hit by at least a pebble-sized particle and thus possibly be destroyed. We were also uncertain of the dust content of the asteroid belt.

Near Jupiter the probes must survive high-energy particle radiation more than 100 times that necessary to kill a human. The Pioneers were followed by Voyager 1 and 2. To date Saturn has also been visited, and we should reach Uranus by 1986. The most recent analysis of the information gathered by these probes is discussed below.

JUPITER

From the time Galileo turned his telescope toward Jupiter and discovered its system of four large moons, humans have been fascinated by the planet's appearance. Using only a modest telescope, one can see Jupiter's banded clouds (its alternating bright *zones* and dark *belts*), its great red spot which rotates with the planet in approximately ten hours (the shortest rotation period in the solar system), and its slightly flattened appearance (Figure 6.1). By watching the changing configuration of four of Jupiter's moons each night, one can determine the distinct period of each moon. Only slightly more sophisticated observations will yield Jupiter's mass (318 times the mass of the earth), its diameter (11 times the diameter of the earth), and its volume (1331 times the volume of the earth). Given only this information—the mass, diameter, and volume—could you determine whether this planet is solid, liquid, or gaseous? Recalling the way in which we computed the average density of the earth, let us use the figures just given to compute the density of Jupiter:

$$\text{density} = \frac{\text{mass}}{\text{volume}} = \frac{318 \times M_E}{1331 \times V_E} = 0.239 \times D_E$$

FIGURE 6.1 Late photo of Jupiter showing red spot. (NASA)

where M_E, V_E, and D_E represent the mass, volume, and density of the earth respectively.

This tells us that Jupiter's density is only approximately 24 percent of the earth's, or 1.3 g/cm^3 (0.239 × 5.5 g/cm^3 = 1.3 g/cm^3). Certainly this comparatively low density, which is not much greater than the density of water, points to a predominantly liquid form. This figure is consistent with a significant amount of atmosphere surrounding the planet and a small solid core. This low density also helps explain how the planet may be slightly flattened through rotation—a gaseous-liquid body can be deformed more easily than a solid.

There are many characteristics of distant planets, such as atmospheric composition, that can not be determined from earth-based observatories. The composition of Jupiter's atmosphere has been confirmed by several probes to be 88 percent hydrogen, 11 percent helium, and 1 percent other constituents, including traces of methane (CH_4), ammonia (NH_3), and other compounds of hydrogen, carbon, nitrogen and oxygen. The amount of helium present in Jupiter's atmosphere, which is relatively large, is close to estimates of both the amount of helium produced in the very early stages of the universe and the helium content of young stars. Some observers have suggested that Jupiter may be an aborted dwarf star, for it radiates almost three times as much thermal energy as it absorbs from the sun. What is the source of this surplus energy? Could there be a starlike process going on in the core of Jupiter? Even though Jupiter's mass is 318 times the mass of the earth, its self gravity is not sufficient to create the high temperature and pressure necessary to initiate the thermonuclear process characteristic of a star. However, thermal energy (heat) may be generated through continuing contraction. Whenever a body contracts, some of its gravitational energy is converted to thermal energy. In the case of Jupiter, a contraction of only one centimeter per year would release enough energy to explain the additional thermal radiation.

Spacecraft as far as 12 million kilometers from Jupiter have detected a magnetic field strong enough to deflect the solar wind (the outflow of charged particles from the sun). As the craft passed through the belts of high-energy electrons and protons entrapped in that magnetic field, the level of radiation was almost too high to be measured by the instruments on board. The integrated picture of Jupiter's magnetic field that emerges from the data shows the total amount of magnetic energy to be more than 400 million times that of the earth's magnetic field. The field is tilted 10° with respect to the planet's axis of rotation and is highly flattened and dimpled in toward the poles (Figure 6.2). As the planet rotates, its magnetic field appears to wobble up and down.

We know by observation from the earth that Jupiter is enveloped by a thick layer of dense clouds that perpetually obscure any hint of the planet's interior. Scientists, however, can derive a model of the interior of any planet with the aid of a modern electronic computer. First the computer is given information concerning the laws of nature, such as the law of gravity and the gas laws, and information gained from observations of the exterior

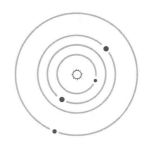

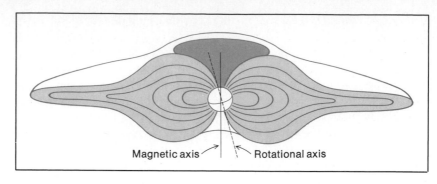

Magnetic axis — Rotational axis

FIGURE 6.2 The magnetic field of Jupiter as determined by the Pioneer 10 fly-by.

FIGURE 6.3 A model of Jupiter's interior. The planet is mainly liquid hydrogen. (NASA)

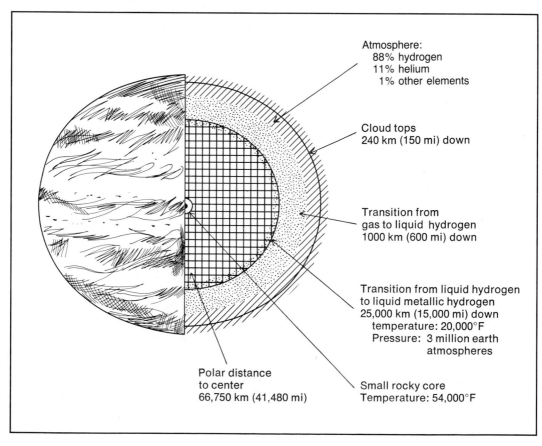

Atmosphere:
88% hydrogen
11% helium
1% other elements

Cloud tops
240 km (150 mi) down

Transition from
gas to liquid hydrogen
1000 km (600 mi) down

Transition from liquid hydrogen
to liquid metallic hydrogen
25,000 km (15,000 mi) down
temperature: 20,000°F
Pressure: 3 million earth
atmospheres

Small rocky core
Temperature: 54,000°F

Polar distance
to center
66,750 km (41,480 mi)

of the planet. Then the computer is instructed to make the millions upon millions of separate calculations necessary to derive a model of the planet's interior (Figure 6.3). According to the model, the portion we call gaseous, the atmosphere, occupies a layer at least a thousand kilometers deep. In the next major layer, hydrogen has been converted to a liquid state, due primarily to the very high pressure that exists even at that level. In a still deeper major layer, hydrogen is converted to a liquid metallic state due to ever-increasing pressures; the protons of atoms are stripped of their outer electrons, creating a fluid that is an excellent conductor of electricity. Jupiter's strong magnetic field may be the result of electric currents within this layer. The center of Jupiter is thought to be a relatively small silicate core several times the diameter of the earth created by further increases in pressure (of the order of 100 million times the earth's atmospheric pressure).

In 1979 Voyagers 1 and 2 flew by Jupiter to produce the most detailed picture of the planet ever made and to study four of its moons in close detail. The clarity of the pictures showing the motions of Jupiter's atmosphere exceeded the scientists, greatest expectations. The most obvious feature of the atmosphere is the banding pattern of bright zones known to be higher elevations, alternating with darker belts known to be regions where lower elevations are visible (Figure 6.4). The bright zones

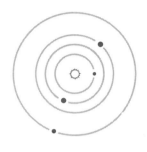

FIGURE 6.4 Motions in Jupiter's atmosphere. Gas, which would move toward the equator by convection, instead due to Coriolis force moves around the planet against the direction of rotation. Gas which would move toward the poles instead moves around Jupiter in the rotation direction.

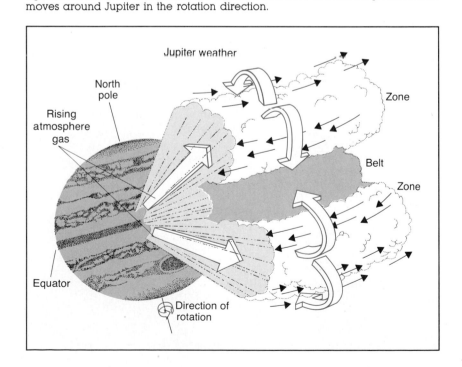

are probably composed of a cold strata of ammonia crystals, and the darker belts of reddish-brown hues probably consist of ammonium hydrosulfide and various polymers. This basic structure is thought to be a result of Jupiter's very rapid rotation—rotating in 9 hours 50 minutes. A point in the equatorial region must move at about 33,000 km/hr (20,500 mi/hr) due to rotation alone. Many astronomers believe that a Coriolis effect, such as the effect that causes wind and ocean currents on earth, has produced the banded structure in Jupiter's atmosphere. Also, the heat radiated by the planet causes the clouds to have a very turbulent vertical motion. The combination of these forces produces a cyclone that surrounds the entire planet with the higher zonal regions continually dumping their material back into the belts. The belts and zones have different velocities, so certain portions lag behind. Superimposed on these bands are a multitude of "swirls" and jet streams with speeds of up to 400 km/hr plus or minus the rotational velocity of the planet (Figure 6.5).

A major discontinuity in the banded pattern is the Great Red Spot, the largest "cyclonic" feature in the solar system; it measures 13,000 km × 25,000 km (8100 mi × 15,600 mi) (see Color Plate 4). The Voyager time-lapse photographs of the Red Spot leave no question as to

FIGURE 6.5 The Great Red Spot of Jupiter, together with one of the white ovals that can be seen in earth-based telescopes. The details of turbulence to the left of these two features had never been seen prior to these Voyager 1 views. (JPL-NASA)

the counterclockwise rotation of this feature. They showed bright white regions first drawn into the huge vortex, swirled around, stretched and deformed, and then ejected and reformed into another white region. The temperature of the Great Red Spot is about 3°C cooler than its surroundings and therefore is presumed to be of somewhat higher elevation.

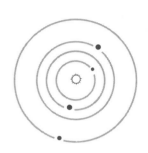

The Voyager probes have confirmed that some atmospheric phenomena on Jupiter are similar to those on earth, namely auroras and lightning. The auroras probably occur by the same process as on earth; charged particles flowing from the sun excite the atoms of Jupiter's atmosphere near the magnetic polar regions and as electrons within those excited atoms make downward transitions, several forms of those excited atoms make downward transitions, several forms of radiation may be produced. Bright auroras have been detected in ultraviolet light. Lightning seen on the darkened side of Jupiter results from the discharge between regions of differing electrical potential.

The 16 natural satellites of Jupiter fall into three obvious categories. If you look at the listing of these moons in Appendix 7, you will notice that the first eight are all within 2 million kilometers of the planet, and they orbit in the same direction as Jupiter very nearly in the equatorial plane of the planet. The next four orbit in the same direction as Jupiter at a distance of approximately 12 million kilometers and have inclinations of 25° to 30°. The last four moons orbit in a retrograde direction at distances of 21 to 24 million kilometers. This suggests that the first eight moons were formed from the same cloud of gas and dust as was Jupiter, their direction and plane of revolution reflecting the rotation of the original nebula. The remaining moons were probably captured, and their motions reflect the conditions under which this capture occurred. The four largest moons are called the *Galilean satellites* after Galileo, who first observed them with the aid of his early telescope, and can be easily viewed by the amateur with only a very modest telescope.

Why are the Galilean satellites of such intense interest to scientists trying to decipher the origin and history of the solar system? The Galilean satellites are very large, as moons go, ranging in diameters from that of the earth's moon to a size greater than Mercury (Table 6.1). Discovering these moons was like finding four more terrestrial planets formed in the cold part of the solar nebula. What features of these moons might serve as clues to their history? The extent to which a planet is cratered is an indication of its age. Major cratering probably took place early in the history of the solar system from the debris left over from the main condensations that became the planets and moons. We think that cratering still goes on but at a greatly reduced rate. Thus surface areas that show very abundant craters must be very old, and surface areas that show almost no cratering must be very young. Callisto and Io illustrate these extremes quite clearly (see Color Plates 5d and 5a). Many surface features are produced by the internal activity of a planet or moon. Volcanic activity, mountain building, spreading, the drifting of continental plates, and wrinkling due to contraction upon cooling are all caused by internal heat in a body.

Table 6.1

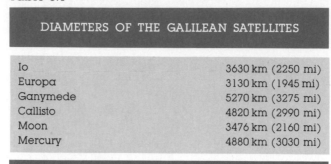

DIAMETERS OF THE GALILEAN SATELLITES	
Io	3630 km (2250 mi)
Europa	3130 km (1945 mi)
Ganymede	5270 km (3275 mi)
Callisto	4820 km (2990 mi)
Moon	3476 km (2160 mi)
Mercury	4880 km (3030 mi)

The most surprising fact about the Galilean moons is that no two of them are really similar. Just a glance at Color Plate 5 (a,b,c,d) will reveal this fact. Io is closest of the four to Jupiter and is literally bathed in the radiation from Jupiter. Fast-moving charged particles are continually striking the surface of Io and as a result, sulfur and sulfur dioxide (thought to be the main constituents of its surface) sputter off into the surrounding space. Figure 6.6 shows the sodium cloud around Io, and faintly the sodium and sulfur ring that surrounds Jupiter. These elements must continually be added to this cloud or it would not long survive. Voyager 1 showed eight volcanos erupting simultaneously on Io. It is estimated that each volcano may eject up to 10,000 metric tons of material, primarily sulfur and sulfur dioxide, per second. Plumes of such material were seen rising as much as 250 km above the surface of Io (Figure 6.7). By the time Voyager 2 passed this moon approximately four months later, only six of the eight

FIGURE 6.6 The sodium cloud surrounding Io and preceding it in orbit around Jupiter. (JPL-NASA)

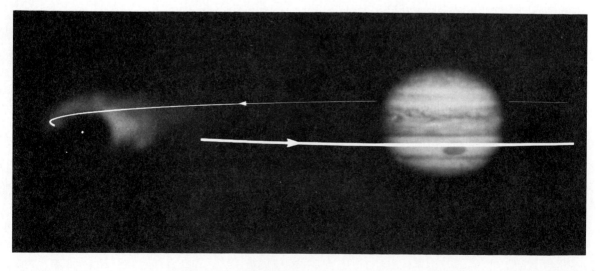

volcanos were still active; this shows the transitory nature of the volcanic activity. The splotchy appearance of Io, with areas of white, yellow, orange, and red color, makes it a glorious pizza baking and bubbling in a hot oven. Like a pizza, its surface is continually changing. Io has perhaps the youngest surface of any moon or planet in the solar system, so it seems logical to seek the source of the heat energy that must be present in this moon. Jupiter is the source of the radiation that strikes Io and it also creates an ever-changing tidal force on the planet. Io does not orbit its mother planet in a perfectly circular orbit but is continually perturbed by the other moons. As Io experiences changing deformation by the gravitational effect of Jupiter, internal friction is thought to produce the heat that drives the volcanos. This process can be illustrated by holding a sponge rubber ball in your hand and rapidly squeezing it one way and another; you would feel heat being generated. Io is the densest of the four Galilean moons; it has a density of 3.5 g/cm^3, not much greater than the density of our own moon. We thus expect Io to have a more rocky composition. Io reflects almost 60 percent of the light that falls on it.

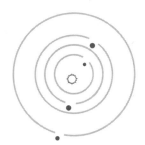

Europa, second of the four large Galilean satellites, also appears to be of rocky composition, with a density of 3.09 g/cm^3. Europa differs from Io in that its surface appears covered by a thin layer of ice. A network of crisscrossing lines, up to 130 km (81 mi) wide and thousands of kilometers long, have been seen. Very few craters have been found on Europa yet we would assume that it received numerous hits from debris accreting onto its surface in its early history. Perhaps the craters were not retained because its surface was warm from the decay of radioactive elements. Or, perhaps today Europa is covered by a layer of ice that hides such features. We wonder if the network of dark lines on the surface represents cracks that resulted from the expansion of the surface due to freezing, after this moon had evolved. The albedo of Europa is almost 70 percent; it thus ranks almost with Venus in its ability to reflect sunlight.

FIGURE 6.7 Two active volcanoes on the moon Io as seen by Voyager 1. The plumelike structures rise more than 100 km (62 mi) above the surface. (JPL-NASA)

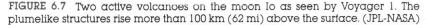

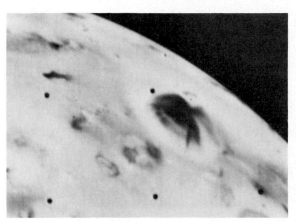

Next in order is Ganymede, the largest of the Galilean satellites. The density of this moon is only 1.9 g/cm³ which indicates a composition of about one-half water in the form of ice and one-half rocky material. At first glance, Ganymede takes on the appearance of our moon in that certain regions appear dark and other regions appear light. The darker regions are heavily cratered and are judged to be very old. The lighter regions show a variation in crater densities and in contrast to terrain seen on our moon, they are characterized by a grooved terrain. Grooved terrain can be described as a system of parallel ridges and valleys, each one from 5 to 15 kilometers wide and hundreds of kilometers long. One system of grooves may run into another system at any angle. These are thought to be younger features. The bright-rayed craters are also considered to be a young feature.

Callisto, last of the Galilean moons, has a density of 1.8 g/cm³, the least dense of the four. It may thus have a slightly higher water (ice) content than Ganymede. Callisto is cratered over its entire surface (see Color Plate 5d). Its early history of bombardment is literally frozen in its surface. The most spectacular feature on this moon is the series of concentric rings measuring 2600 km (1600 mi) across that seem to indicate the impact of a major body. Many craters are superimposed on this ring structure, hence the event that caused this structure must have occurred early in the history of the moon. We would also conclude that Callisto has little internal activity because surface features have not been erased. The record we see on Callisto's surface is probably 4 billion years old, the same age as the heavily cratered (dark) regions of Ganymede. The lack of high moun-

FIGURE 6.8   This striking view of Jupiter's ring was recorded by Voyager 2 on July 10, 1979. (JPL-NASA)

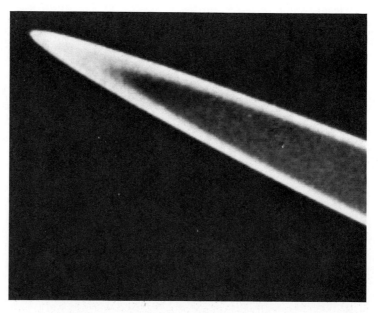

tains or very large craters on Callisto suggests a rather weak frozen crust, able only to retain the smaller features. Callisto appears darker than the other large moons of Jupiter because of its lower albedo (approximately 18%), yet this figure is still greater than that of the earth's moon.

Most unexpected was the discovery of a very thin ring around Jupiter. The outer edge of this ring is approximately 57,000 km (35,400 mi) above the planet's visible surface and the ring is approximately 30 km (18.6 mi) thick. The outer portion of the ring (see Figure 6.8) is about 5200 km (3230 mi) wide; the inner portion, which has fewer particles, may extend to the surface of Jupiter.

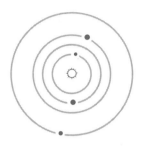

SATURN

The data about Saturn provided by the passage of Pioneer II, Voyager 1, and Voyager 2 will require many years to analyze. Before we look at some of the fascinating discoveries these probes made, let us review what was known about this planet before the probes were sent; these are facts still pertinent to your own observations of Saturn (Figure 6.9). Saturn is the most beautiful of all the planets because of its prominent ring structure. It requires almost 30 years to orbit the sun which means that its apparent motion through the signs of the zodiac is very slow. Viewed in an amateur's telescope over many years, Saturn reveals its tilted axis of rotation (29°); for a period of about 14 years we see the top side of its rings, then they seem to go "edge-on" for about a year, and then we see a view from beneath the rings. In 1966 and again in 1980 the rings were seen edge-on, and in an average-sized telescope they seemed to disappear—they are very thin (at the most only a few hundred kilometers thick).

FIGURE 6.9 Saturn, as photographed with the 2.6-m (100-in) Mount Wilson telescope. (Mount Wilson and Las Campanas Observatories, Carnegie Institute of Washington)

Three rather distinct sections of the rings can be recognized by amateur observers: the outer *gray ring*, the middle *bright ring*, and the inner *crepe ring*. Furthermore, even a small telescope will reveal a dark "gap" between the bright and gray rings. This is called *Cassini's division* and it was thought to be a region of the rings which lacked particles. Some observers suspected a fourth *subcrepe ring*, even closer to the body of the planet (Figure 6.10).

Through a telescope one can also see the flattening of Saturn near its poles and a hint of its banded appearance. Calculations of the planet's average density reveal the lowest in the entire solar system—0.7 g/cm^3. The low density is indicative of a planet largely composed of gas and liquid. The banded structure seen in the upper atmospheric clouds is also indicative of a gaseous composition. Horizontal banding together with flattening on this gaseous planet are probably closely tied to its very rapid rotation—a period just over 10 hours. The polar diameter is about 12,000 km (7500 mi) less than its equatorial diameter.

In 1972 the most distant radar bounce ever was recorded from the rings of Saturn. The radio signal was reflected with an efficiency of 60 percent, indicating that the particles in the rings were not merely ice crystals, as had been suspected, but ice-covered objects of some significant size—a few centimeters to a few meters. It had been known for some time that the rings were not a solid disk as they may appear in a telescope, but a muriad of particles that have varying orbital periods according to their distance from the planet. Whereas particles in the crepe ring have an average period of about 4 hours, those in the outer gray ring require on the order of 14 hours to complete one orbit. Such periods were predicted by Kepler's laws and confirmed by the Doppler shift in their spectral lines.

FIGURE 6.10 The rotation periods of the clouds of Saturn at various latitudes and of selected particles in the rings.

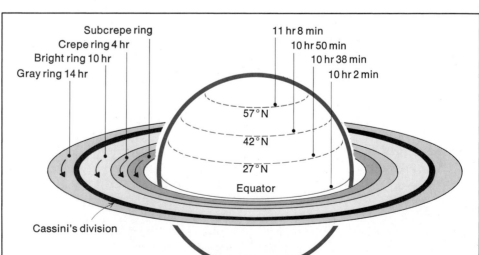

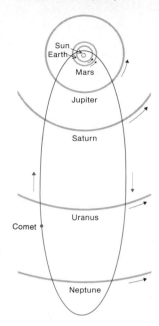

It is interesting to speculate why any planet has such a ring system. Astronomers believe that it would be impossible for a moon-size object to exist as close to Saturn as does its rings. Such an object would experience a much stronger gravitational force on its side nearest the planet than on its side away from the planet; these differential tidal forces would tend to tear the object apart or prevent it from forming in the first place. The distance within which a larger body cannot exist is called the *Roche limit*. The outer edge of the gray ring is thought to coincide with such a limit, hence only small bodies may exist inside that distance.

By the turn of the century, we knew of nine moons that orbit Saturn. Only one of these, called Titan, ranks in size with the Galilean satellites of Jupiter. It is the largest moon in the solar system and massive enough to hang onto an atmosphere. Outermost of Saturn's moons is Phoebe which travels at a distance of approximately 13 million km (8.1 million mi) in a highly inclined orbit and in a retrograde direction. We speculate that Phoebe was not one of Saturn's original moons but was captured at a later time. Its very low albedo, 5 percent, suggests a surface other than ice, perhaps that of a rocky asteroid.

As the probes Pioneer 11 (1979), Voyager 1 (1980), and Voyager 2 (1981) scrutinized Saturn at close range, they revealed many surprises concerning its magnetic field, atmosphere, moons, and other characteristics. Saturn's magnetic field is second only to Jupiter's in terms of its total energy. At the planet's surface, however, the strength is only about the strength of the earth's surface field and the polarity is opposite to the earth's; its north magnetic field is thus in what we would term its southern hemisphere. Unlike the earth and Jupiter, Saturn's magnetic axis coincides with its axis of rotation almost perfectly. Some theories of how a magnetic field is generated within a planet have assumed that an offset between these two axes is essential. The example of Saturn seems to invalidate this assumption. The model of Saturn's interior that has emerged from this close view includes a rock core no larger than the earth; it is substantially the same as that of Jupiter. Like Jupiter, Saturn radiates about three times as much thermal energy as it receives from the sun—probably also from continued contraction.

Voyager 2 was able to see the banded structure of Saturn's upper atmosphere better than any probe before it because the haze that often enshrouds the planet was unexpectedly absent. Jet streams (winds) were observed with speeds that differed from the general rotational velocity of the planet by as much as 1600 km/hr (1000 mi/hr). These activities extend further to the polar regions of Saturn than similar activities do on Jupiter. At no time was great contrast evident in Saturn's cloud colors as is true on Jupiter. Observations on the dark side of the planet revealed bright auroras in the polar regions.

Perhaps the most unexpected observations were of Saturn's ring (Figure 6.11). The detail seemed to be enhanced with each passage until the number of subdivisions in the rings exceeded a thousand. The three rings observable from the earth, the gray, bright, and crepe rings, were confirmed in their general form and were designated as the A, B, and C

The Jovian Planets   235

FIGURE 6.11 Details of Saturn's rings as seen by Voyager 2. (NASA)

rings respectively. The suspected D (subcrepe) ring was observed only briefly by a spacecraft. Newly discovered was the very narrow F ring, some 7000 km (4300 mi) beyond the outside edge of the A ring, a hint of a G ring at an additional 30,000 km (18,600 mi) distance, and an E ring perhaps as much as 100,000 km (62,000 mi) farther out. Only rings A, B, C, and F are considered thoroughly confirmed and only these rings show great amounts of detail (subdivisions). The probes all confirmed that Cassini's division is not empty as once thought, but is occupied by numerous ringlets of material. In fact, when Pioneer 11 viewed the rings from the side opposite the sun, Cassini's division appeared bright and the B ring (normally brightest) appeared dark. Totally unanticipated was the fact that the strands in the F ring appeared braided in Voyager 1 imagery. There also appeared to be "spokes" extending outward through (or over the top of) the B ring. Astronomers now suggest that fine dust is "lifted" out of the ring plane by the magnetic field of Saturn and that a given "spoke" may survive on the order of one hour. The braiding of the F ring has not yet been explained; no braiding was seen by Voyager 2 cameras.

Several new moons of Saturn have been discovered since 1979. At the outer edge of the A ring, perhaps keeping the edge of that ring sharply defined, is a moon designated as 1980S28. On either side of the F ring are what have been called the shephard moons, 1980S27 and 1980S26. Beyond the F ring are two moons which occupy essentially the same orbit. These *co-orbital* moons are designated 1980S3 and 1980S1 and one of them may represent what was reported in 1966 as the tenth moon of Saturn and called Janus. The question arises, How can two moons with different rates of revolution exist in essentially the same orbit without a collision resulting? Even though one moon moves slightly faster than the other, no collision has taken place. As the inner moon (the faster one) approaches, the two moons exchange energy (in the form of angular momentum) through gravitational attraction. The inner moon moves into a higher orbit and gains angular momentum, but it slows down as a result. The outer moon moves into a lower orbit and loses angular momentum, but it speeds up in the process. The two moons never pass, but rather they change roles. Because of their new orbits, one might say that S1 becomes S3 and vice versa.

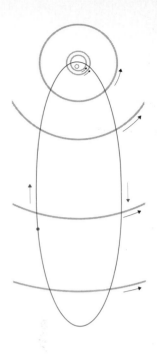

Two additional moons, designated as 1980S25 and 1980S13, are locked into a tri-orbital arrangement with Tethys (Saturn's fourth largest moon), one maintaining a position 60° ahead of Tethys and the other following at a position of 60° behind Tethys. These positions represent a stable arrangement, similar to the arrangement of the Trojan asteroids in relation to Jupiter (see Chapter 7, page 257). A second set of co-orbital moons consists of a previously known moon called Dione and its newly discovered mate (Dione B), designated as 1980S6. With the addition of four small moons identified from Voyager imagery, Saturn has a total of at least 21 moons (Figure 6.12).

Astronomers have been particularly fascinated with Saturn's largest moon, Titan, for it is the only moon in the solar system which is both massive enough and cool enough to retain an atmosphere. It is the composition of that atmosphere that has been so intriguing; scientists speculated that it would contain elements associated with living organisms, namely hydrogen, oxygen, carbon, and nitrogen. Molecules composed of such elements that were found on Titan include CH_4 (methane), HCN (cyanoacetylene), and C_2H_2 (cyanogen); these molecules are known to play a role in synthesizing amino acids, components of yet more complex organic molecules. Because Titan's atmosphere is more dense than the earth's and its surface temperature is −180°C (−292°F), some observers thought that prebiotic organic molecules might have "rained" onto Titan's surface and some sign of life might be evident. It is these same complex molecules that make up the layer of orange-colored clouds 250 km (155 mi) deep that obscure the surface of Titan [Figure 6.13(a)].

Saturn's larger moons could be characterized as spheres composed primarily of water ices with 20 to 50 percent rock and densities between 1.1 and 1.9 g/cm^3. Many of these show evidence both of numerous impacts by small objects and forms of internal activity. Mimas (Figure 6.13b) is not only cratered over its entire surface, it bears a crater scar that ranks with any in the solar system. The crater's diameter is 130 km (81 mi), its depth is over 10 km (6.2 mi), and its central peak ranks with the higher mountains on earth. This feature is thought to be billions of years old.

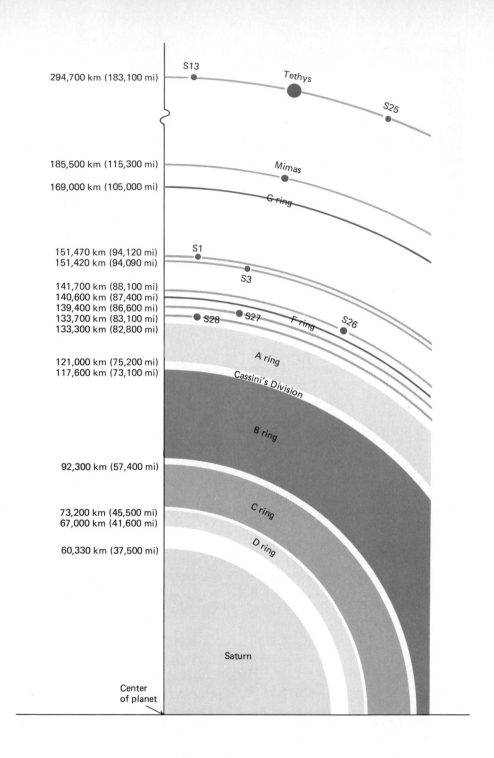

FIGURE 6.12 Spacing of rings and moons of Saturn.

(a)

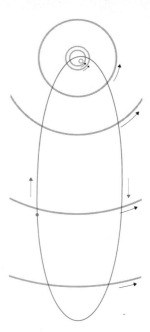

FIGURE 6.13 (a) Titan, Saturn's largest satellite, is shrouded by a thick atmosphere. Only a hint of the cloud band which is parallel to the planet's equator can be seen. (b) Mimas, one of Saturn's icy satellites. The huge crater (130 km or 81 mi in diameter) is thought to have almost split the satellite apart. (JPL-NASA)

(b)

Enceladus, by contrast, shows evidence of recent activities that have produced a very diverse surface (Figure 6.14). The smoother parts are thought to have originally been cratered, then resurfaced. The flow patterns and wrinkled surface suggest a continuing state of geologic-type activity; the intensity suggests something as unstable as a liquid mantle beneath a thin crust, with an unknown heat source.

The landscape of Tethys (Figure 6.15), is dominated by ancient craters, but they do not stand forth as if frozen in ice. We are suggesting that some mechanism altered these craters to provide a smoother overall (spherical) shape to the moon. A dominant feature is the 2500 km (1550 mi) long system of valleys called Ithaca Chasma that in their size resemble those of the Mariner Valley on Mars. Thus although Tethys is not thought to be active now, it must have once had an active period.

The remaining moons will not be described in detail. Note, however, the composite photograph of the eight small satellites shown in Figure

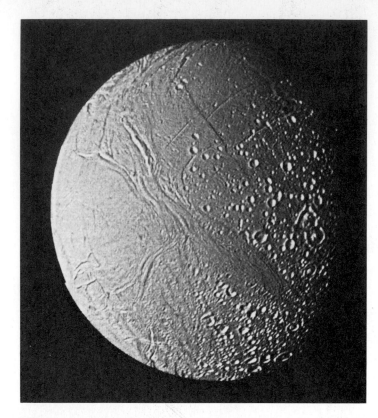

FIGURE 6.14 Enceladus has one of the youngest surfaces of Saturn's moons. Of its many different features, the straight line features represent fault lines associated with the movement of the crust. (JPL-NASA)

FIGURE 6.15 Tethys, dominated by a huge crater 400 km (250 mi) in diameter, shows evidence that its crust could not support the high walls that once may have resembled the huge crater on Mimas. (JPL-NASA)

6.16. Because these moons are small and have less mass, their self-gravity is unable to form them into a spherical shape—they look more like fragments of a larger body. Diameters shown in Appendix 7 are an indication only of their largest dimension.

A question that was first asked when Cassini discovered the apparent gap between the A and B rings and that is now more pertinent than ever is this: Is there any relationship between the locations of the moons of Saturn and the gaps in its rings? With the thousand or more newly discovered gaps, the question will be a troublesome one. The early thinking went like this: If a given particle in the rings orbits around the planet in exactly one-half the period of a given moon's orbit, that particle would experience its greatest gravitational tug from the planet at the same place in its orbit, and the cumulative effect would be to move it out of that location—leaving a gap in the rings. This has long been the explanation for Cassini's division, for particles near the inside of that division orbit in one-half the time required for the moon Mimas. A similar relationship may hold for particles with other periods such as $\frac{1}{3}$, $\frac{1}{4}$, or $\frac{1}{5}$ that of a given moon. Even with twenty-one moons and several simple fractional relationships for each moon, we could not begin to explain the thousand or more gaps between the ringlets. A search was made for small moons thought to be actually imbedded in each gap sweeping out particles, but none were found. So we leave Saturn with a number of unanswered questions.

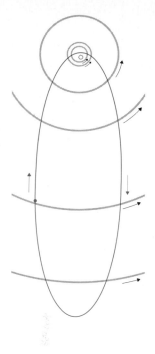

FIGURE 6.16 Saturn and seven of its major satellites. Clockwise from upper right are: Titan, Iapetus, Tethys, Mimas, Euceladus, Dione, and Rhea. (JPL-NASA)

URANUS

Uranus is the first planet of our study that was not known to the ancients. Only under ideal conditions can this planet be seen by the naked eye. It was on March 13, 1781, that Sir William Herschel, while viewing the skies through a self-made 16 cm (6.3 in) Newtonian reflector telescope, thought he spotted a nebulous (fuzzy) star or comet. The fact that this new object appeared larger as he increased the power of his telescope identified the object as something other than a star; stars are so far away that they appear as mere points of light regardless of the power used. Within four nights, Herschel was able to detect the motion of this object in relation to the stars, thus confirming its association with the solar system. Herschel's work had been that of an amateur until in 1782 King George III made him a court astronomer; this gave him the opportunity for a much more professional and full-time approach to the development of the telescope and to the study of the heavens. Herschel's emphasis was upon the gathering of more light in order to see dim nebulae and the attainment of a high degree of resolution in order to see detail of structure. His efforts culminated in the 1.25 meter (48 inch) telescope shown in Figure 6.17. By 1789, Herschel had discovered two large moons of Uranus, and two of the smaller moons of Saturn.

Uranus orbits the sun at an average distance just under 20 A.U. and in a period of 84 years. If a human could spend an entire lifetime on Uranus,

FIGURE 6.17 Sir William Herschel's 1.25-m (48-in) reflecting telescope, which he built himself. (Yerkes Observatory)

he or she would die at the age of only one Uranus year. At this great distance, the planet reveals no details in its cloud tops. Some information can be derived through ground-based spectroscopy; this technique reveals a predominance of molecular hydrogen (H_2), and enough methane (CH_4) to give its greenish color. Ammonia may also be present but in a frozen form due to Uranus' low temperature. There appears to be no additional source of energy within Uranus similar to that which we saw on Jupiter and Saturn. If we were to model its internal state based on an average density of 1.3 g/cm^3, we would think of a small rocky core no larger than the size of the earth, a thick layer of frozen ices, and a relatively thick layer of molecular hydrogen clouds.

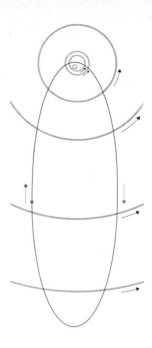

One of the most distinctive characteristics of this planet is the tilt of its axis, which is 98°. To understand this orientation, imagine a planet whose axis is not tilted at all—its equator is in the same plane as its orbit. Imagine that this same planet rotates in a counterclockwise direction, a point on its equator traveling in an eastward direction. Now tilt the planet's axis of rotation by 90° so that the axis lies in the plane of its orbit. Continue to tilt the axis of the planet 8° more, making a total of 98°. In what direction will it now be rotating? To answer this question, use your finger to represent the planet rotating in a counterclockwise direction with no tilt at all, then gradually tilt your finger until you have reached the 98° position. You will see that your finger is now rotating in a clockwise direction. Likewise, the rotation of Uranus is clockwise, or retrograde (Figure 6.18). It is cus-

FIGURE 6.18 Uranus, with three moons. (Lick Observatory)

tomary to define the tilt of any planet that rotates in retrograde by an angle greater than 90°.

Uranus has five moons, all of which orbit in the same plane as the planet's equator and in the same direction as the planet's rotation. Because of the planet's extreme tilt, the plane in which the moons are seen circling it is sometimes perpendicular to and at other times aligned with our line of sight.

On March 10, 1977, a most unexpected observation was made. A group of observers, flying in NASA's Kuiper Airborne Observatory, were preparing to witness Uranus as it passed in front of a faint star. This is called an *occultation* observation, and its value lies in the fact that as a planet passes in front of a star, details concerning the planet's atmosphere and size may be determined. A planet with no atmosphere will occult the star abruptly, whereas a planet with an atmosphere will create a slower dimming depending on the density of its atmosphere. As the group was observing Uranus before the occultation began, there were several dips in the intensity of the light coming from the star. The first was thought to be a momentary malfunction of the equipment, but by the time five dips had been recorded they were no longer attributed to malfunctions. The planet then occulted the star for 25 minutes and, as Uranus passed on by, five distinct dips again occurred. The comparative frequency of the dips on either side of the planet correlated very well. The interpretation that emerged is that Uranus has at least five rings. Because each dip in light output lasted only for one to four seconds, the rings are thought to be very thin, less than 100 km thick (Figure 6.19). These rings have more recently been confirmed by their infrared radiation which is brighter than that of the planet itself (in the 2.2 micron wavelength); using this method, nine

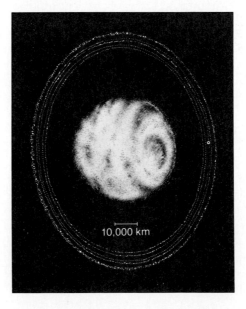

FIGURE 6.19 The rings of Uranus.

10,000 km

rings have in fact been identified. The Space Telescope described in Chapter 2 and/or the Voyager 2 probe which is programmed to pass Uranus in 1986 should provide further confirmation. Certainly this new discovery creates more anticipation concerning the Voyager fly-by.

NEPTUNE

The discovery of Uranus stimulated observation of that planet and its orbital motion. If Uranus were the only planet moving around the sun, its orbit would consist of an ellipse along which it moved so as to sweep out equal areas in equal times, as Kepler's laws predict. However, Uranus is not alone in the solar system, and its motion is perturbed (gravitationally disturbed) by every other planet. The algebraic sum of these perturbations shows up as deviations in Uranus' actual position as compared to that which would have been predicted with no other planets present. The total (net) deviation can be predicted by computing the individual deviations using Newton's laws of gravity.

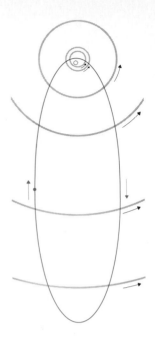

When the perturbations caused by all known planets were computed and other necessary corrections made, the degree to which Uranus deviated from the predicted position was large enough to raise suspicion that another perturbing object must be present. Two astronomers, John Adams of England and Joseph Leverrier of France, independently asked the question, Where would an object have to be located and what would be its mass in order to create the unexplained perturbations seen in Uranus? Assuming that the same law of gravity would hold for an unknown planet, Adams and Leverrier calculated its most probable location and their predictions were almost identical. Each man tried to persuade an observational astronomer to look for the new planet. Adams was put off at Cambridge, but Leverrier's request was met by J.G. Galle of the Berlin Observatory; he found the new planet in his first evening of searching, less than 1° from the location that both Adams and Leverrier had predicted. The date was September 23, 1846. This discovery was one of the most significant triumphs in astronomy to that date, for it further verified Newton's description of universal gravitation and demonstrated its predictive value.

The planet was named Neptune after the mythological god of the ocean—"king of the deep." Although there had been a valid basis for calling Leverrier its discoverer, history usually records both Adams and Leverrier as the codiscoverers of Neptune. In less than one month's time, the larger of Neptune's two moons (diameter 4000 km) had been discovered, and it was called Triton after Neptune's mythological son (Figure 6.20). The unusual feature of Triton is its retrograde motion—it is the only moon of significant size that has retrograde motion in relation to the rotation of the parent planet. All of the moons of Uranus orbit in retrograde, but that would be expected because the planet itself rotates in retrograde.

A quick look at the statistics in Appendix 6 will show Neptune to be roughly the twin of Uranus in size, mass, albedo, and density (Neptune's density is a little greater). One interesting difference is the evidence of an

FIGURE 6.20 Neptune and its largest moon, Triton. (Lick Observatory)

internal heat source in Neptune. At 30 A.U., the heat of the sun would not be very effective, and would leave a planet in unimaginable cold, perhaps −228°C (−378°F). Infrared measurements, however, show the cloudtops of Neptune to be at least 10°C warmer than this, or −218°C (−360°F). The planet itself, like Jupiter and Saturn, must therefore radiate heat equal to 2 or 3 times that received from the sun. Additional observations using charged-coupled devices (ccd detectors) at infrared wavelengths at the Catalina Observatory in Arizona have revealed some detail in the cloudtops of Neptune. This may be due to differences in absorption by methane in the planet's atmosphere.

The very slow progress of Neptune through the stars on its 165-year orbit has not permitted an exact determination of its orbit, but an interesting discovery has recently been made that could yield helpful information in this regard. There is the distinct possibility that Galileo recorded the location of Neptune as early as 1612–13, when that planet appeared in close proximity to Jupiter and its moons. This was 234 years before its discovery as a planet. Galileo had simply passed it off as a star in several of his drawings of Jupiter and its moons. This record has become a significant contribution to our understanding of Neptune's motion; it may force a slight correction in the plot of Neptune's orbit.

PLUTO

The orbital motions of both Uranus and Neptune were followed with much interest by numerous observers. In the early 1900s Percival Lowell, founder of Lowell Observatory in Flagstaff, Arizona, concluded that Neptune could not account for all the perturbations in the orbit of Uranus. By mathematical calculations he predicted that still another planet might be found in one of two possible locations. In 1930 a planet was found by Clyde Tombaugh, an assistant at the observatory, who had analyzed photographs of these regions in a device called a *blink microscope*. Pluto was found within 6° of one of the predicted positions by this technique. Figure 6.21 shows the discovery plates taken six days apart. Note how an object that had moved in those six days would appear to shift back and forth if first one plate is projected and then the other. We now know that the mass of Pluto is too small to produce detectable variations in the orbits of Uranus and Neptune, therefore most observers suggest that the discovery of Pluto was accidental. They attribute the discovery to the very patient search procedures of Clyde Tombaugh. In keeping with tradition, Pluto was named for the mythological god of the underworld. Note though the "coincidence" that the first two letters of the name are also the initials of Percival Lowell and that the symbol for Pluto (♇) is constructed from these initials.

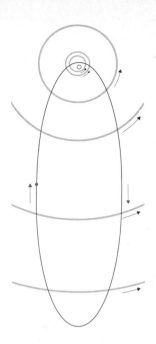

Pluto has the most unusual orbit of all the planets. Besides being the most eccentric, its plane of orbit is inclined 17° to the plane of the earth's orbit (the ecliptic). Pluto orbits the sun at an average distance of 40 A.U., or 5.5 billion kilometers, but it travels to within less than 30 A.U. at perihelion and out to almost 50 A.U. at aphelion. The planet was at aphelion in 1865 and will be at this point again in 2113. It will be at perihelion in 1989. The

FIGURE 6.21 Discovery of Pluto. These plates, taken 6 days apart, show the motion of a new planet among the stars (see arrows): (a) January 23, 1930; (b) January 29, 1930. (Lowell Observatory)

(a)   (b)

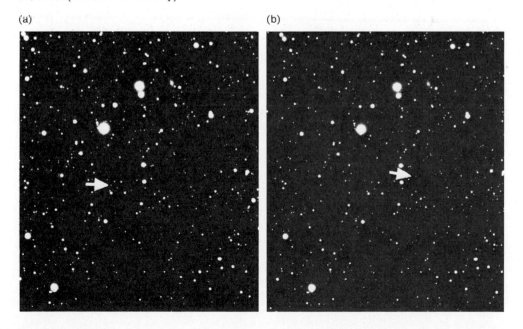

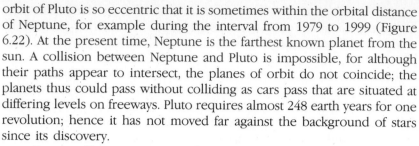

orbit of Pluto is so eccentric that it is sometimes within the orbital distance of Neptune, for example during the interval from 1979 to 1999 (Figure 6.22). At the present time, Neptune is the farthest known planet from the sun. A collision between Neptune and Pluto is impossible, for although their paths appear to intersect, the planes of orbit do not coincide; the planets thus could pass without colliding as cars pass that are situated at differing levels on freeways. Pluto requires almost 248 earth years for one revolution; hence it has not moved far against the background of stars since its discovery.

Because Pluto appears as only a tiny speck in even the largest telescopes, astronomers have had to guess what its physical properties might be, such as diameter, mass, density, and surface composition. Until very recently, many observers thought Pluto was basically somewhat like the earth's moon, having a hard crust, no atmosphere, low albedo, a diameter about half that of earth, and a density of 4 to 5 g/cm³. Recent photometric measurements of Pluto's spectrum in the infrared portion of the spectrum, however, reveal a surface covering of frozen methane at a temperature of −230°C (−382°F). When this condition is duplicated in the laboratory, an albedo of 40 to 60 percent is observed. Based on this finding, astronomers would conclude that Pluto is roughly the size of the earth's moon, which is significantly smaller than previously thought. The reasoning behind this assumption is that if Pluto is a better reflector than previously thought, it need not be so large in order to be as bright as it appears. If not so large,

FIGURE 6.22   The orbits of Pluto and Neptune.

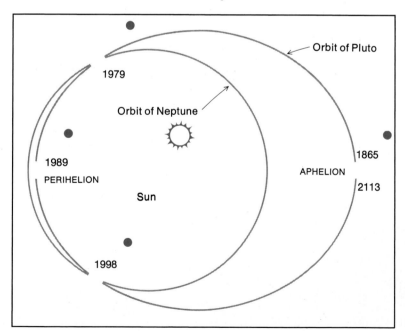

then the planet is not as massive as thought. Further, if it is characterized by frozen gases, then it is not as dense as thought—perhaps only slightly more dense than water, with a mass about 1/500 that of the earth. The size of Pluto has recently been confirmed by a process of *speckle interferometry* to be roughly the size of the earth's moon. The process of speckle interferometry, which is fundamental to the determination of star size, will be detailed in Chapter 9.

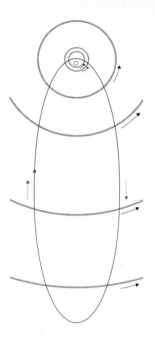

Because we cannot see details of Pluto's surface, we cannot determine its rotation directly. A study of its periodic changes in brightness suggested that there was one rotation in a period of 6 days 9 hr 17 min. One of the most surprising discoveries about Pluto is that it has a moon. When first photographed, the moon appeared as just a bump on the side of the planet's image; the observer was not sure what it was (Figure 6.23). The confirmation came by inspecting plates that had been taken earlier. These revealed the same bump on different sides of the planet, and the revolution period was determined to be 6 days 9 hr 17 min. Thus the revolution of the moon is now thought to be the cause of the brightness variations noted above. The moon is called *Charon* after the mythological ferryman of souls across the river Styx into Hades.

When we put all known facts together, we now see Pluto as being similar to a moon of Saturn—perhaps composed largely of frozen water ice. We wonder about its origin. Was it once a moon of Neptune that was perturbed into its present eccentric and inclined orbit by some passing object? Or, was it captured after the formation of the sun and other plan-

FIGURE 6.23 The discovery plate of Charon, moon of Pluto, as indicated by the "bulge" on the upper left side of the planet's image. The discovery of this moon has led to a complete reassessment of the planet's diameter, mass, and density. (U.S. Naval Observatory)

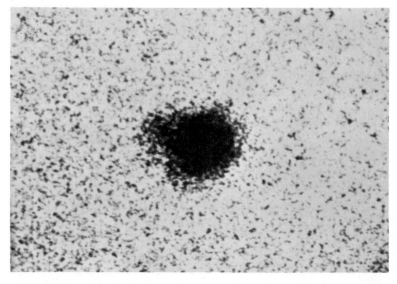

ets? Take note of its similarity to the nucleus of a huge comet as we study comets in the next chapter.

PLANET X

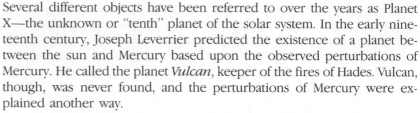

Several different objects have been referred to over the years as Planet X—the unknown or "tenth" planet of the solar system. In the early nineteenth century, Joseph Leverrier predicted the existence of a planet between the sun and Mercury based upon the observed perturbations of Mercury. He called the planet *Vulcan*, keeper of the fires of Hades. Vulcan, though, was never found, and the perturbations of Mercury were explained another way.

In 1971 Joseph L. Brady of the University of California did an extensive computer study of the perturbations of Halley's comet and concluded that a planet more massive than Saturn must orbit the sun beyond the orbit of Pluto. He predicted Planet X's period of revolution to be 464 years, and he predicted it would be located in the constellation of Cassiopeia. Extensive photographic searches, however, have revealed no such planet. It is considered likely that the irregularities in the flight of Halley's comet are due to nongravitational effects, which will be explained in Chapter 7.

You may recall reading in the newspaper in November of 1977 about still another Planet X. Although its discoverer, Charles Kowal, had not intended his newly found object to be labelled as the tenth planet, the press proclaimed its discovery far and wide. The speed with which even the casual thought of an astronomer is disseminated today is a striking contrast to the almost total lack of communication a century or two earlier. As a group, scientists communicate with each other primarily through scientific journals. An astronomer reports research results as soon as any assurance of validity exists, and within a very short time other astronomers are reacting and checking the original researcher's work. This line of communication is vital to the advancement of science. In the case of Kowal's Object, as it is called, other modes of communication came into play. The newspapers turned his object into "Planet X," and most failed to identify the highly speculative nature of his discovery.

What do we actually know about the Kowal Object, now called Chiron after the mythological Centaur, son of Saturn? Once it had been discovered by means of the blink microscope, the instrument used in Pluto's discovery, Charles Kowal and other astronomers began a search for it on photographic plates taken before its discovery. They succeeded in identifying it on numerous plates, like that shown in Figure 6.24. Its flight has been traced backward in time approximately 80 years, which is equivalent to $1\frac{1}{2}$ revolutions around the sun since its period is 50.68 years. Chiron orbits the sun at an average distance of 13.7 A.U., but has a more eccentric orbit than Pluto (its eccentricity is 0.3786). At aphelion, Chiron moves almost as far from the sun as Uranus, but at perihelion it moves within Saturn's orbit.

Nothing we have related thus far would keep Kowal's Object from

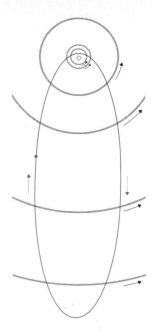

FIGURE 6.24 The motion of Chiron, as recorded during a 75-min. exposure on the Palomar 1.25 m (48-in.) Schmidt telescope. (Palomar Observatory Photograph)

being called a planet. The crucial factor, however, is its size. Judging from its faintness on the photographic plates, its size would be similar to an asteroid or a comet, 50 to 600 km (31 to 372 m) in diameter. The latest thinking favors its description as an asteroid. As you will see in Chapter 7, most asteroids orbit between Mars and Jupiter or in highly eccentric orbits near the sun. An interesting speculation is that Chiron may thus be the first of another class of asteroids. Or, it may simply represent an asteroid that was perturbed by Jupiter into this more distant orbit. Some of the outer moons of Jupiter and Saturn have motions that suggest capture by these giant planets—they may have been asteroids as well.

ORIGIN AND EVOLUTION OF THE PLANETS

We opened the last chapter with the question of how the solar system came into being. We suggested that the random-capture theory does not fit the orderly revolution of all nine planets in almost the same plane. We are also inclined to eliminate the encounter theory, not because it is inconsistent with planetary motion, but rather because of the low probability that the sun passed close enough to another star to have pulled material off from it. Due to the great distances between stars, it is estimated that no more than ten such encounters have occurred in the life of the Milky Way galaxy (15 to 18 billion years), and astronomers believe that many solar systems exist in the Galaxy.

By 1950 several variations of the *protoplanet* theory had been suggested. This theory may not be the "last word" on the subject, but in many respects it is more plausible than its forerunners. It suggests that the planets were formed at about the same time as the sun and out of the same nebular material. The process is thought to have begun when a huge cloud of gas and dust, many times larger than the diameter of the entire solar system, became unstable. Due to gravitational attraction, a condensation (knot) occurred at one point within the gas cloud. This condensation we would term the *protosun*—the sun in the making. Any such cloud is thought to have some component of rotation and, much like a spinning ice skater pulling in his or her arms, the cloud rotated faster and faster as it contracted. As a result, it left some of its gas and dust in a flattened disk surrounding the protosun, and within this disk other condensations formed several protoplanets, which continued to move in the same direction and in the same plane as the original disk.

At the same time, the planets assumed a rotation and tilt of axis that reflected the combined motion of the gas and dust particles from which they were formed. The rotation and tilt of the planets show great variation, yet small angles of tilt and counterclockwise rotation are more prevalent, just as we might expect within a disk revolving in a counterclockwise direction. The same general process of condensation surrounding each protoplanet may have acted on a smaller scale to form the moons. While there is no absolute correlation between the mass of a planet and its number of moons, a tendency does exist for the more massive planets to have more moons. Another correlation that has recently become evident is between the presence of a regular system of moons around a planet and the presence of rings. Jupiter, Saturn and Uranus all have both.

According to this protoplanet theory, as the material that composed the disk became more concentrated in the protoplanets, the space between the sun and these objects cleared, allowing the sunlight to reach them. This heat was naturally most intense on the inner planets, driving off the lighter elements and perhaps most of the atmosphere they had, leaving only a more dense core of heavier elements. At the distance of Jupiter or beyond, the effect of heating would be greatly reduced, allowing the lighter elements such as hydrogen and helium to remain as part of their atmospheres. This model seems to fit with the facts concerning the atmospheres. Evidence indicates that the terrestrial planets went through a period of numerous impacts with objects in space and that they had active volcanic periods, during which time they developed a secondary atmosphere composed of gases expelled from the planet itself.

One of the most fundamental unanswered questions is: Why does the sun rotate so slowly? If the sun and the planets were formed from the same rotating cloud of gas and dust, we would expect the sun to rotate faster and faster as it condensed (like the spinning figure skater); this relationship is called *conservation of angular momentum*. If the sun had followed this pattern, it would rotate in a period of only a few hours, not in its present period of 24.6 days. One theory suggests that the planets were magnetically coupled to the sun and that through such coupling the planets "robbed" the sun of some of its rotational energy with that energy now represented by the revolutional energy of the planets. It is also thought that a significant amount of gas and dust, equivalent to almost 1 percent of the sun's mass, was lost to the solar system in its early stages. This material could have carried with it a significant part of the sun's original angular momentum. It is a fact that the sun contains over 99 percent of the mass of the solar system, yet it has less than 3 percent of its angular momentum.

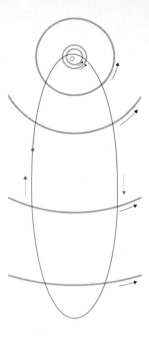

QUESTIONS

1. True or false: The volume of Jupiter is greater than the combined volume of all the remaining planets.

2. List the Jovian planets.

3. How many moons does Jupiter have? How many are larger than the earth's moon?

4. What property of Saturn is demonstrated by the fact that it would float in water?

5. What problems would be encountered in maintaining human life on any of the Jovian planets?

6. What principal characteristics distinguish Uranus from other planets?

7. What kind of observations led to the discovery of Neptune?

8. For a period of 20 years beginning in 1979, Pluto orbits nearer to the sun than does Neptune, yet these plants will never collide. Why?

9. What ideas about the origin of planets are suggested by the fact that they orbit the sun in almost the same plane as does the earth?

10. Suppose you could stand on the visible surface of Saturn at a point near its equator. Describe the apparent motion of the particles in each ring from your point of view.

11. Find the period of a hypothetical planet that orbits the sun at an average distance of 8 A.U. (You may wish to refer to Chapter 1.)

12. Speculate on how life on earth might have been affected if Jupiter had been a dim star which together with the sun formed a binary system of stars.

13. What is the fundamental difference between stars and planets?

14. What is the predominant element in the atmosphere of the Jovians?

Asteroids, Comets, and Meteors

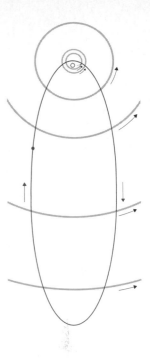

Seven

The characteristic that makes asteroids, comets, and meteors obvious members of the solar system is the fact that they orbit the sun. Kepler's laws apply to these objects as well as the planets: they have elliptical orbits, they move so that a line joining any one of them to the sun would sweep out equal area in equal time, and if the periods of two such objects are compared, the squares of those periods will equal the cubes of their semi-major axes.

When the Bode-Titius relationship was first published in 1772, no object was known to orbit the sun at 2.8 A.U. (see Table 5.1, page 184). The very fact that the other numbers in this relationship so nearly agreed with the actual distances to the known planets suggested the existence of an object at that distance. Searches were initiated but remained fruitless until 1801, when quite by accident Giuseppe Piazzi, director of the observatory at Palermo, noticed a small starlike object that was not recorded on his charts. After several nights he noticed a slight shift in its position among the stars. At first he thought that the object was a comet; however, after continued observation and computation, it was identified as a small planet orbiting the sun at an average distance of 2.76 A.U. Thinking he had found the lost planet, Piazzi called this small object Ceres after the goddess of the harvest. Within a year the German astronomer, Heinrich Olbers, found another object in a very similar orbit and named it Pallas. This second discovery gave a clue that there might be many more such objects and in fact marked the beginning of a long series of discoveries of asteroids (Table 7.1).

TABLE 7.1

DISCOVERY OF ASTEROIDS		
Year	Name	Diameter
1801	Ceres	1000 km (620 mi)
1802	Pallas	600 km (370 mi)
1804	Juno	250 km (160 mi)
1807	Vesta	540 km (340 mi)
1845	6 known	
1890	300 known	

By the late nineteenth century important new photographic techniques had been developed. If you have ever taken a time exposure of the city lights, you probably noticed the streaks of light made by moving cars. As a time exposure is made of a given star field, the telescope must move with that star field in order to produce sharp images of the stars; if an object in the sky is moving in relation to the stars however, it will make a streak on the film. Dr. Eleanor Helin (Figure 7.1), a modern asteroid searcher, uses a special technique to identify her prey: namely, she takes a twenty minute exposure, covers the film for 10 minutes and then takes another 10 minute exposure, tracking the stars throughout this 40-minutes interval. The characteristic streak that is obtained is shown in Figure 7.2.

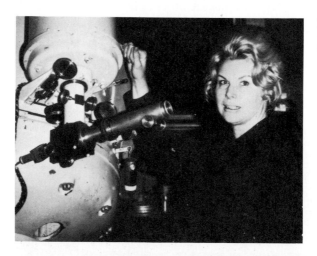

Figure 7.1 Dr. Eleanor Helin. (World Space Foundation)

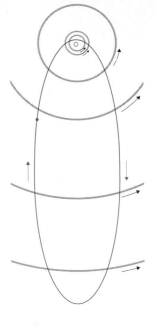

Figure 7.2 The discovery photo of asteroid 1982TA, using a 20-minute exposure, a 10-minute break and a second exposure of 10 minutes. (E. Helin, IPL and Palomar Observatory)

The angular movement of an asteroid can be identified and also its direction; it moves toward the shorter dash. Dr. Helin is particularly interested in asteroids with orbits that bring them near the earth. One of her most unique discoveries is an object called 1976AA that has a diameter of about 1 kilometer and a period of revolution just under 1 year. Asteroid 1976AA comes closer to the earth than any other object except the moon.

The number of asteroids identified to date is in excess of 100,000; thus recently it has not been practical to name each of them. As new discoveries are made, they are numbered in the order of discovery: (1) Ceres, (2) Pallas, . . . , (1566) Icarus, and so on. Although most asteroids orbit the sun between Mars and Jupiter, a number have rather eccentric orbits. Of particular interest are asteroids such as Hermes, Apollo, Adonis, Icarus, Geographos, and Eros, because their orbits cross the earth's orbit bringing them at times relatively close to the earth. For instance, in 1937 Hermes came within 0.8 million kilometers of the earth, about twice the distance to the moon. It is thus possible that the earth was hit by asteroids in the past and that it will be hit in the future (see Fig. 7.3).

Asteroid Icarus is of special interest because it passes very close to the sun—within 27 million kilometers—and then moves out beyond the orbit of Mars. Every 19 years it comes within a few million kilometers of the earth. Icarus is only 0.5 kilometer in its largest dimension, according to radar measurements. It was discovered by Walter Baade when he photographed a star field using the 1.22-m Schmidt telescope on Mt. Palomar. Because of its sun-grazing orbit, it was called Icarus after the mythological young man who flew too near the sun. The story is that with his father, Daedalus, he attempted to escape from an island prison by flying away on artificial wings. Daedalus, a master craftsman, had fashioned two sets of wings and attached them to their bodies with wax. He warned his son not to fly too high because the wax would be melted by the sun's heat. The boy was so thrilled at being free and able to fly; however, that he soared higher and higher. The wax melted, and Icarus fell to his death in the sea.

Asteroids fall into four categories based on their orbits: (1) those whose orbits are nearly circular, with orbital radii of about 2.8 A.U. (these form the traditional asteroid "belt"); (2) the Apollo group, those with very

eccentric orbits, with semi-major axes of 1 A.U. or less (many of which cross the earth's orbit); (3) the Trojan group, those which appear to be trapped in Jupiter's orbit, 60° ahead and 60° behind Jupiter itself (these bear the names of Homer's Trojan heroes, such as Hector and Nestor); and (4) a possible group of asteroids with very eccentric orbits and semi-major axes of several A.U. (see discussion of Chiron, page 250). Asteroids are so spread out that they seldom come within a million kilometers of each other, thus even within the asteroid "belt," they should not be thought of as "filling" the space. Pioneer and Voyager space craft passed through the belt without detecting a single asteroid.

If the surface detail of an asteroid has never been seen through a telescope or with a space probe, then how do we know of its appearance? The shapes of some asteroids have been determined by analyzing the variation in their brightness. Figure 7.4 shows the light curve of Geo-

Figure 7.3 The orbits of selected asteroids.

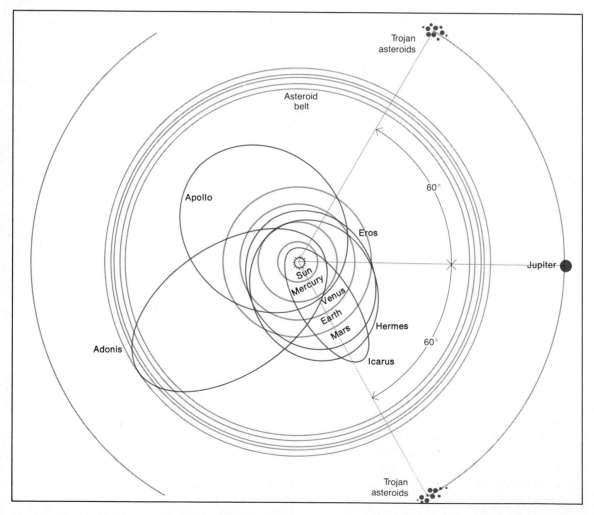

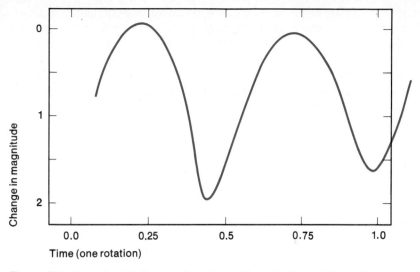

Figure 7.4 The asteroid Geographos shows this periodic variation of two magnitudes, which is interpreted as resulting from one rotation of the object.

graphos. It is interpreted as follows: The object is longer in one dimension that in another and as the asteroid tumbles through space it sometimes presents a broadside view (the time of greatest brightness) and at other times it presents an end view (the time of lesser brightness). This light curve is thus repeated over and over again. Astronomers believe that most of the smaller asteroids resemble the moons of Mars (see Fig. 5.34, page 219); such bodies do not have enough self-gravity to pull themselves into a spherical shape. The large asteroids, such as those listed in Table 7.1, are thought to be spherical. Since 1970 astronomers have been applying new techniques to determine the physical properties of asteroids. These techniques include polarization, infrared detection, and spectroscopic studies. The combination of these observations provides measures of albedo, diameter, and surface composition. The great majority of asteroids fall into two classes: (1) those that appear to be covered with carbon compounds and appear dark—their albedos range in the order of 5 percent; and (2) those that are light in color and have higher albedos, ranging from 10 to 30 percent, with a few as high as 37.5 percent. The latter could be characterized as stony or silicate types. A majority of the asteroids fall in the carbonaceous category. There seems to be a pattern of distribution of types within the asteroid belt: the inner particles tend to be the lighter, stony type and the outer (cooler) particles tend to be the darker, carbonaceous type. This may be a clue as to the nature of the original nebula from which the solar system formed.

ORIGIN OF ASTEROIDS

Some observers have suggested that asteroids were produced when a single planet exploded; however, if this were the case, we would expect a

more random arrangement of types. It seems more reasonable to expect that numerous smaller bodies formed throughout the solar nebula and then had various fates: some of these bodies may have gravitationally joined to others to form planets; some may have remained separate— these would be the larger asteroids, thought to be more nearly spherical in shape; others may have been torn apart, forming the smaller, irregularly shaped asteroids; and finally, some of these early bodies must have collided with the planets to produce the cratering we see today. As discussed in the following section, nuclei of comets are thought to be composed of frozen gasses and rock and when the comet has lost most of its gas, the rocky core appears like an asteroid.

A fundamental question follows from the identification of asteroid types: Will the meteorite types that we will study at the end of this chapter (see page 276) correspond in some way to the asteroid types? Are asteroids the source of some meteorites? The reflectivity curves of many asteroids match those of certain meteorites quite closely, which is suggestive.

COMETS

A comet presents one of the most spectacular sights in the sky. Often unheralded, it appears first as a small source of light, somewhat fuzzy and often resembling a nebula or distant galaxy. As the comet moves closer to the sun, however, a vaporous tail is driven from its head; this tail always extends away from the sun, growing as it approaches the sun and diminishing as it recedes from the sun. Comets presented a fearsome sight to ancient observers and were interpreted as foretelling terrible events to come, such as the death of a king. The tail was thought to contain poisonous gases that might endanger earthly inhabitants. There is a tale of some enterprising person that at a gathering of comet watchers provided viewers with pills to protect them from such dangerous gases—for a fee.

If a comet actually moved within the earth's atmosphere, it would appear to streak across the sky in a matter of seconds, like a so-called "shooting star." It is, however, millions of miles from the earth even at its closest approach, so it shows only the slightest apparent motion. Whenever a bright comet appears, therefore, it will likely be seen for several weeks. Most comets are not bright enough to be seen with the naked eye so they go unnoticed by the casual observer. Even the brightest may go unnoticed by city dwellers. When you hear of a comet being visible, remember to view it from a dark location, away from city lights.

Comets are distinguished from other members of the solar system by their very eccentric orbits. A comparison between the earth's orbital shape and that of the comet Kohoutek is shown in Figure 7.5. Some comets have very long periods, ranging from 10,000 to 80,000 years or longer, and their eccentric orbits take them to distances of several thousand astronomical units from the sun. Other comets have short periods, ranging from 3 to 100 years; their orbits are less eccentric and are generally confined to distances less than that of Pluto's orbit. If a comet with a long period

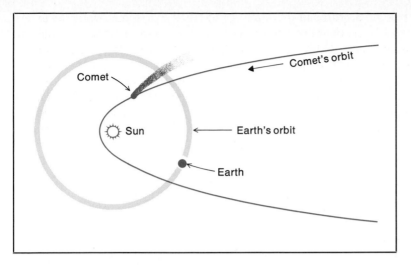

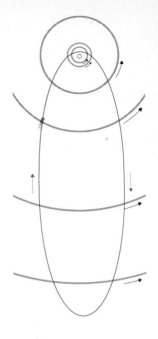

Figure 7.5 The highly eccentric orbit of the Comet Kohoutek (1973f) compared to the almost circular orbit of the earth. The shape of the comet's orbit is almost like that of a parabola, typical of long-period comets.

approaches the sun and is perturbed by the gravitational field of a planet that it happens to encounter, the shape of its orbit may be altered so that its period becomes short (Figure 7.6). Jupiter has perturbed the orbits of 45 to 50 comets, called the *Jovian group*, that now orbit the sun in a period of from 5 to 10 years. A comet may appear at any point in the sky and move in direct or retrograde motion. They seem to be randomly oriented in the solar system.

It was not realized at first that a comet might become visible over and over again. In 1705 the British astronomer Edmond Halley published the orbits of 24 comets and noted that the time interval between the comets of 1531, 1607, and 1682, was 75 or 76 years. He reasoned that if those dates

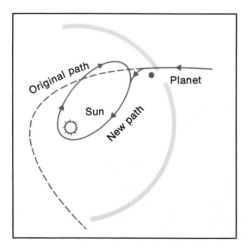

Figure 7.6 A comet may be perturbed by a planet so that the shape of its orbit is changed.

represented returns of the same comet, then that comet would again appear in 1758. The comet did in fact reappear on Christmas night of that year. It has since been called Halley's comet (Figure 7.7). The period of the comet is known to vary between 74 and 79 years, due to gravitational disturbances by the Jovian planets and to the expulsion of gases from its nucleus as it rounds the sun. It is interesting to note that Halley's comet appeared in the year of Mark Twain's birth (1835) and reappeared in the year of his death (1910), just as he had prophesied. Because such significance was associated with the appearance of comets, the records of such events were preserved and now we are able to trace the returns of Halley's comet back to 240 B.C. The comet's next passage near the sun is expected very early in 1986, however the circumstances which made its 1910 appearance so spectacular, namely the favorable position of the earth with respect to the comet, will not be duplicated. Viewers will have to seek out a dark location, away from city lights, in order to have a good view of Halley's comet this time around. The flip pages beginning on page 235 depict the flight of Halley's comet.

Figure 7.7 Halley's Comet of 1910. The two photos were taken a day apart; by examining the position of the comet relative to the background of stars, you can detect the motion of the comet. (Lick Observatory)

(a)

(b)

ORIGIN OF COMETS

The comets that become visible are probably not just interstellar wanderers that enter the gravitational influence of the sun by chance. They are more likely members of a spherical "comet cloud" surrounding the solar system at some distance beyond 50,000 A.U. This cloud may extend halfway to the nearest star, Alpha Centauri, located at a distance of 260,000 A.U. Examples of periodic comets that orbit at these great distances include the Pons-Brooks comet, the Griggs-Mellish comet, and others with aphelion distances of 30,000 to 60,000 A.U. Members of the great comet cloud are thought to be perturbed by major planets or passing stars and thus brought closer to the sun. Further perturbations may then entrap them in a smaller orbit, which makes their return more predictable. The theory of a comet cloud consisting of up to 100 billion members was set forth in 1950 by J.H. Oort of Leiden Observatory in the Netherlands. The comet Kohoutek viewed in 1973 tends to confirm Oort's theory; calculations based on the almost parabolic orbit of the comet show that it came near the sun from a great distance—almost 4000 A.U. On December 28, 1973, it passed within 21 million kilometers of the sun; it will not return for approximately 80,000 years (Figure 7.8).

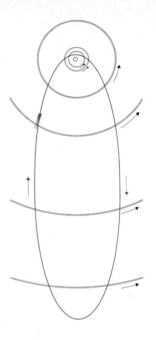

Figure 7.8 The Comet Kohoutek (1973f), as observed on January 12, 1974. (Palomar Observatory Photograph)

Many astronomers believe that comets were among some of the earliest members of the solar system. They think that comets may have formed before the sun and planets approximately 4.6 billion years ago, while the solar nebula was still a very large, roughly spherical cloud of gas and dust. Our study of comets may thus give us additional clues to the origin of the solar system. What are comets like and of what are they made?

PHYSICAL NATURE OF COMETS

When a comet is far from the sun, it consists only of a nucleus resembling a dirty snowball, or a swarm of dirty snowballs. It is composed of a mixture of rocky and metallic particles bound together by water ice and frozen gases such as carbon dioxide, methane, and ammonia. As the nucleus approaches to within about 3 A.U. of the sun, the solar radiation causes the frozen material to sublimate (vaporize), thus producing a large gaseous region surrounding the nucleus, called the *coma*. The thinness of the coma is revealed by the stars seen shining through it. Although the nucleus is very small—often only a few kilometers in diameter—the coma usually grows to a diameter several times that of the earth. When a comet passes quite near the sun, its surface temperature may rise to several thousand degrees. At this time, the absorption spectrum of reflected sunlight can be seen as well as a bright-line spectrum from the comet itself. A comet's emission spectrum results primarily from the ultraviolet radiation of the sun which excites the gases to fluorescence, revealing the presence of

Figure 7.9 The hydrogen halo of the Comet Kohoutek, as seen by Skylab 4 in the far ultraviolet wavelengths. (NASA)

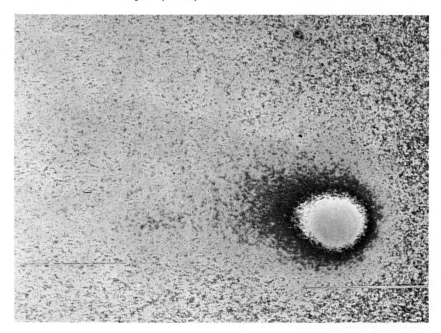

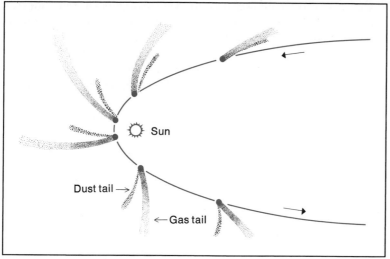
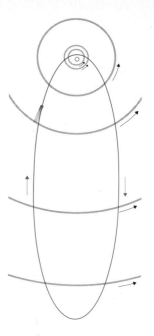

Figure 7.10 The tail of a comet, showing the separation of the gas and dust tails.

carbon, nitrogen, oxygen, and hydrogen in the following radical forms: CH, NH, CN, NH_2, OH, C_2, and C_3. Observers suspect that these derive from the breakdown of ammonia (NH_3), cyanogen (C_2N_2), methane (CH_4), and water (H_2O) ices which compose the nucleus of the comet, along with rocky and metallic particles, including iron, nickel, sodium, calcium, and carbon. The International Ultraviolet Explorer Satellite has confirmed cer-

Figure 7.11 Comet Arend-Roland of 1957, showing an antitail, which certain comets develop. (Palomar Observatory Photograph)

| April 26 | April 27 | April 29 | April 30 | May 1 |

tain molecules (or ions) and has identified additional ones, namely: H, O, CO, C, S, CS, OH, CO^+, and CO_2^+. The Skylab crews had a particularly good opportunity to photograph comet Kohoutek and to analyze its composition spectroscopically because of Skylab's freedom from atmospheric interference. They found a hydrogen "halo" engulfing the entire comet as shown in Figure 7.9. Radio observation of this comet revealed rather complex molecules such as CH_3CN (methyl cyanide) and HCN (hydrogen cyanide). Since comets contain materials that closely resemble molecules associated with living organisms, some observers have suggested that comets may play a role in making life possible in diverse locations.

Most comets produce a tail only after approaching to within about 2 A.U.'s of the sun. The gas and dust liberated by the sun's heat experience two fundamental forces that ultimately determine their direction of motion. One force is due to the sun's electromagnetic radiation, the other to the solar wind. The gas molecules, atoms, or ions (charged particles) in the tail appear to be most directly influenced by the solar wind. The reason may be in part that the solar wind has a magnetic field that would carry off

Figure 7.12 Halley's Comet of 1910, showing the growth of the tail as the comet approached the sun in April and May and then the decline of the tail as the comet receded from the sun in June. (Mount Wilson and Las Campanas Observatories, Carnegie Institute of Washington)

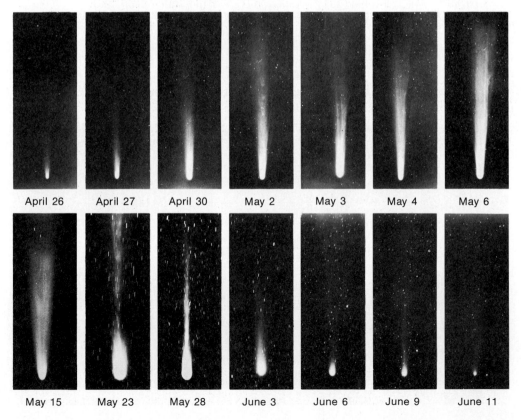

| April 26 | April 27 | April 30 | May 2 | May 3 | May 4 | May 6 |

| May 15 | May 23 | May 28 | June 3 | June 6 | June 9 | June 11 |

ions from the coma, forming a tail. Dust particles, on the other hand, are larger and appear to be influenced primarily by the sun's radiation. Therefore, comets typically develop two tails: a gas tail that tends to point more directly away from the sun, and a dust tail that tends to curve and lag behind the gas tail. The separation of these two tails becomes more apparent as the comet recedes from the sun, for at that time the gas tail precedes the head of the comet and the dust tail tends to follow the head (see the lower portion of Figure 7.10 and Color Plate 7). In the case of the comet Arend-Roland (Figure 7.11), the dust tail appears to point toward the sun. This illusion, however, is created by our angle of view; the comet is receding from the sun with its gas tail leading the way, and its dust tail merely curves backward enough to appear as an antitail. Note the change in structure of the gas tail in these views.

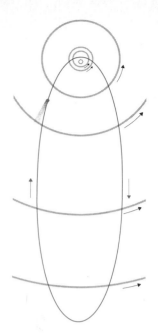

The tails of some comets have reached the tremendous length of more than 160 million kilometers (100 million miles), a distance greater than that from the sun to the earth. An average tail, however, is only a few million kilometers long. The tenuous nature of the tail is known by the fact that stars may easily be seen through it (Figure 7.12). Only when we see the tail of a comet at a right angle to its length do we see its full length. If we view the comet from a direction in line with its tail, then no tail is apparent at all.

LIFETIME OF A COMET

We know from direct observation that Halley's comet has been returning to perihelion about every 76 years since the year 240 B.C. when it was first recorded. With every such passage near the sun a comet loses thousands to millions of tons of material through sublimation, yet this may represent less than one percent of its mass. How long can it last? Some comets should be good for more than 100 passes near the sun. In a few cases though we have failed to find a comet when it was expected, and in its place have experienced meteor showers produced by the disintegrated particles that once composed its nucleus. Such a shower perhaps indicates the death of a comet. We think that comets continually inject particles into

Figure 7.13 Comet West broke into four parts, over a 10-day period, as it passed near the sun. (New Mexico State University)

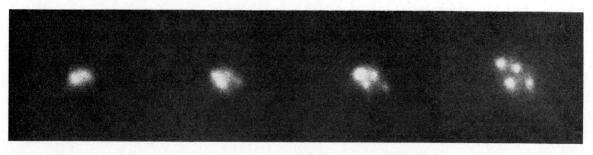

Asteroids, Comets, and Meteors   267

Figure 7.14 This spectacular photograph of Comet West shows a great deal of structure in both gas and dust tails. (Lick Observatory)

their orbits, which result in meteor showers on earth (a full discussion follows on page 273).

As a comet approaches perihelion, it experiences such tremendous tidal forces that astronomers are hard pressed to explain how the nucleus "hangs" together at all. Indeed, a number of comets have broken into two, three, or four parts when close to the sun. The recent splitting of comet West into four parts as it passed the sun is shown in Figure 7.13. This will certainly shorten its lifetime because more of its nucleus is exposed to sublimation by the sun's heat. Figure 7.14 shows comet West before the break-up of its nucleus.

COMET SPIN AND PERTURBATION

As a comet approaches the sun, the side of its nucleus that faces the sun experiences heating and as material in the comet sublimates, a jetlike force results and tends to push the comet away from the sun. If the nuclei of some comets are rotating, as Fred L. Whipple of the Smithsonian Astrophysical Observatory has suggested, the jet-like force will either accelerate or retard the comet: if the comet's direction of rotation is the same as its direction of revolution, it will accelerate; if the comet's direction of rotation is opposite its direction of revolution, it will retard (Figure 7.15). In either case the orbit of the comet will be perturbed (changed) by a force that is non-gravitational and a given comet will return to the sun earlier or later than its last orbital period would predict. We now seem to have an explanation for the perturbations of Halley's comet that Brady thought indicated yet another planet in the solar system (see page 250).

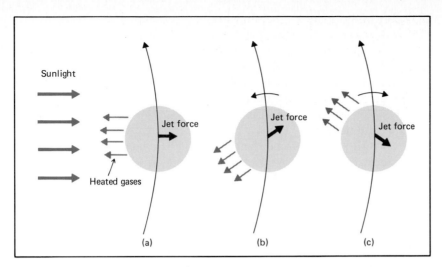

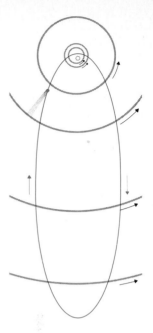

Figure 7.15 If rotating, a comet may produce its own perturbing force: (a) non-rotating, (b) rotating counter-clockwise, (c) rotating clockwise.

HOW ARE COMETS DISCOVERED AND RECORDED?

The record of comets for 1967 presented in Table 7.2 is typical of almost any year. It is evident from this list that the names of many persons have been immortalized by their discovery of a comet. Comets are assigned labels that indicate the order of discovery. The label consists of the year of

TABLE 7.2

COMETS SIGHTED IN 1967		
Date	Designation	Name
Jan. 3	1967a	p/Tuttle
Feb. 4	1967b	Seki
Feb. 11	1967c	Wild
Feb. 12	1967d	p/Tempel II
June 5	1967e	p/Reinmuth II
June 29	1967f	Mitchell-Jones-Gerber
Aug. 7	1967g	p/Finlay
Aug. 3	1967h	p/Encke
Aug. 8	1967i	p/Schwassmann-Wachmann II
Oct. 5	1967j	p/Wolf
Oct. 5	1967k	p/Wirtanen
Oct. 5	1967l	p/Arend
Oct. 5	1967m	p/Borrelly
Dec. 28	1967n	Ikeya-Seki

discovery and a letter of the alphabet indicating the order of discovery during that year; a comet listed as 1967d, for example, is the fourth discovered in 1967. Comets may be further identified; the designation p/Tempel II, for example, indicates that this comet is a periodic comet, merely returning to the sun, and that it was originally discovered by Tempel as his second comet (II). Those comets lacking the "p/" designation represent comets newly discovered in 1967.

Are the names of the discoverers those of professional astronomers alone? No! There is no such criterion for the naming of comets. The only requirement is that the name be that of the person who first identified the object as a comet. Lubos Kohoutek, a professional astronomer, found comet 1973f while searching photographic plates for asteroids. Most amateurs do not have access to such plates, however. Comets are often found by visual observation, especially in dark regions of the sky where photographs are not usually taken. Remember that comets may appear at any point in the sky—they generally do not follow the ecliptic as did comet Kohoutek. If one finds a dark location away from city lights, and then adopts a systematic approach to slowly scanning regions of the sky, one will undoubtedly be rewarded by finding several comets.

The tool used for a comet search can be either a telescope or a pair of binoculars with a rather large aperture. If a telescope is used, it should have a short focal length; a focal ratio of f/4 would be appropriate, where the focal length is only four times the aperture. The telescope should also have a low-power eyepiece. Either of these instruments will provide a large, bright field of view, allowing the observer to recognize the rather faint, diffuse image that a comet presents in its early stages of visibility. Since the image of a comet often resembles that of a nebula or distant galaxy, an adequate sky map is essential; known diffuse objects, such as nebulae, clusters, and galaxies, will be shown on such a map. Charles Messier, a French observer whose primary concern was comet searching, compiled a catalogue of known nebulae, clusters, and galaxies that kept getting in his way as he sought comets. In his catalogue objects are designated by numbers; for example, M1, M2, etc. The Great Galaxy in Andromeda, for example, bears his catalogue number M31. Although Messier found numerous comets, he is probably better known for his catalogue of nebulae. His list together with the more recent *New General Catalogue (NGC)* and the *Index Catalogue (IC)* provides an essential tool for comet searchers today. (The *Messier Catalogue* is given in Appendix 9.)

If a certain diffuse object is seen that does not appear on the sky map, check the object for apparent motion against the background of stars; motion may not be evident for hours or even days. Make drawings and notations of its position among nearby stars. Note its right ascension and declination from a sky chart. If in fact the object does display movement and is diffuse in appearance (asteroids also display movement but are not diffuse), it probably is a comet. What does one do next? To record your discovery, send a telegram to the International Astronomical Union's Central Bureau for Astronomical Telegrams, Cambridge, Massachusetts (Western Union address TWX 710-320-68442: ASTROGRAM CAM), stating: your name, the date and time of observation, its diffuse nature, its right ascension and declination, the direction of motion, and the estimated magnitude, or brightness. Your observation will then be verified and the orbit calculated. If it is a periodic comet that you have merely rediscovered, it will of course not bear your name. If it is indeed a new comet and your telegram stating the facts of your observation arrives first, then your name will be given to that comet.

Comets are often seen after they have moved close to the sun, when they have a tail. It is not easy to spot the dim image of a comet that is just becoming visible at a distance of more than 500 million kilometers from the sun. To increase your chances of discovery, find a location as far away from city lights as possible and direct your attention to the darker portions of the sky. Good hunting!

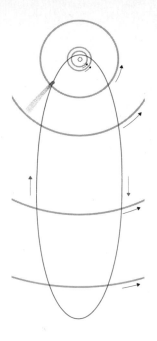

METEORS

Unlike the comet, which seems almost fixed among the stars throughout an evening's viewing session, the *meteor* is a phenomenon that lasts only a few seconds. The word phenomenon is used here to stress that what you see shooting across the sky is not the object itself, but rather a streak of light caused by the passage of a small solid body into the earth's atmosphere. The object itself is called a *meteoroid* while still in flight. It experiences the heating effect of friction between itself and atoms that make up the earth's atmosphere. As a result of this heat, the air is ionized (atoms lose electrons). As the ionized atoms "find" free electrons with which to recombine, downward transitions of these electrons occur and light is produced by transitions that stop at least temporarily at energy level 2.

Occasionally the object that produces a meteor survives its flight and falls to earth. Such an object is called a *meteorite*. Since most meteoroids are only the size of a grain of sand or a pebble at most, they are destroyed in flight. The rare meteoroid of larger size often produces a spectacular display called a *fireball*, or *bolide meteor*. The streak of light persists, and often explosions and other loud noises accompany the event. Objects that exhibit this rapid motion and that sometimes deposit material on the earth are obviously very close; in fact, they are usually within 160 km (100 mi) of the observer during their visible flight.

The meteorite itself is of great interest to the astronomer, for until recently it represented our only sample of extraterrestrial material. These objects may carry clues to the origin of the solar system and to life elsewhere within it. Analysis of certain meteorites has shown that they contain hydrocarbons and as many as 18 different amino acids. Six of the 18 amino acids—aspartic acid, glycine, alanine, glutamic acid, valine, and proline—are of the type normally found in the proteins of living cells. Figure 7.16 shows a photomicrograph of a section of meteorite in which investigators at the University of Chicago have found inclusions (white spots) containing a form of oxygen (isotope ^{16}O) so pure as to suggest its origin in interstel-

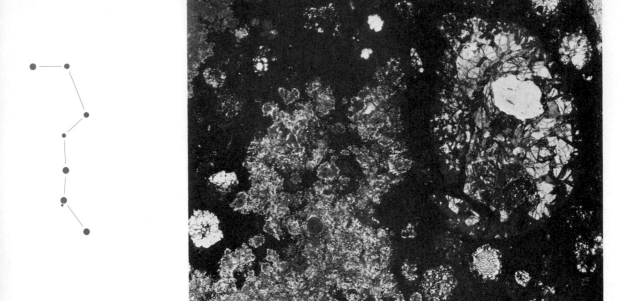

Figure 7.16 A photomicrograph of a section of the Allende meteorite showing inclusions (white) of pure oxygen 16. (Courtesy of Richard J. Kjarval, Graphic Arts Facility, Physical Science Division, University of Chicago)

lar material before the birth of the solar system; on earth, other isotopes of oxygen (^{17}O and ^{18}O) are usually found together with ^{16}O.

OBSERVING METEORS

Are meteors predictable? Where should one look to see a meteor or meteor shower? It is usually possible to see a few meteors on almost any night by simply assuming a reclining position and staring at the sky. Of course a dark location away from city lights is better for viewing meteors. Under these conditions, your eyes will pick out streaks of light quite easily, even if you do not know exactly where to look in the sky (Figure 7.17). However, there are some guidelines for viewing meteors. In general, the best time is during the early morning before dawn (that is, between midnight and dawn). The brilliance of a meteor depends largely on the speed with which it enters the earth's atmosphere. In the evening hours, the meteors you see would be overtaking the earth, as it moves around the sun, and hence would be entering the atmosphere quite slowly—approximately 16 km/sec (10 mi/sec). In the early morning hours, the meteors you see are meeting the earth head-on, in effect adding the earth's speed to their speed, resulting in relative speeds as high as 70 km/sec or 43 mi/sec (Figure 7.18). This corresponds to the effect one would experience driving in

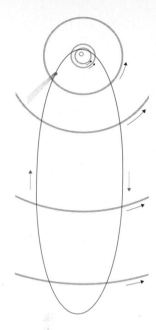

Figure 7.17   Meteor trail near Pleiades. (Yerkes Observatory)

Figure 7.18   Viewing meteors.

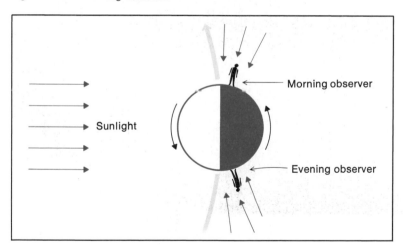

a rain storm—the rain would hit the front windshield with much higher velocity than it hits the rear window.

METEOR SHOWERS

On certain nights, more than the usual number of meteors will be evident. Streaks of light may appear in many parts of the sky, yet if we extend imaginary lines back toward their points of origin these lines will appear

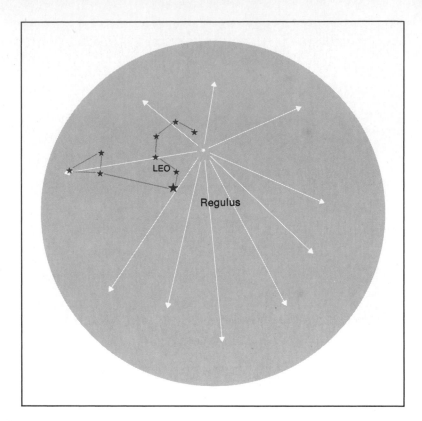

Figure 7.19 The radiant point of the Leonids.

to intersect in a definite point called the *radiant*. We are seeing a *meteor shower*, where many particles enter the earth's atmosphere within a short period of time. Although the meteors appear to radiate from a given point (Figure 7.19), this is merely an optical illusion; the individual particles that produce the shower are actually travelling in parallel paths. Just as parallel railroad tracks appear to converge at a point in the distance, so these parallel streaks of light seem to emerge from a single point. An analysis of the orbits of these particles reveals that they are associated not only with each other, but also with a comet that passed along the same orbital path at an earlier time; the particles are probably debris left from the comet itself. Since the earth crosses the orbit of a given comet at about the same time each year, these meteor showers are predictable, and the partial list of such showers provided in Table 7.3 can be used for any year. Each shower is named for the constellation in which the radiant is located.

A given shower may be better one year than the next because the particles that produce it tend to travel in swarms rather than being evenly distributed along the orbit of the comet (Figure 7.20). When the earth passes through a swarm, a spectacular display results. On November 13, 1833, the earth passed through a swarm that was later associated with the

TABLE 7.3

METEOR SHOWERS				
			Radiant	
Approximate Date	Name of Shower	Associated Comet	R.A.	Dec.
Jan. 1–4	Quadrantids	—	15h 20m	+52°
April 19–23	Lyrids	1861 I	18h 4m	+33°
May 1–6	May Aquarids	Halley	22h 16m	− 2°
July 26–31	Delta Aquarids	—	22h 36m	−11°
Aug. 10–14	Perseids	1862 II	3h 8m	+58°
Oct. 9–11	Draconids	Giacobini-Zinner	17h 40m	+55°
Oct. 18–23	Orionids	Halley	6h 8m	+15°
Nov. 1–15	Taurids	Encke	3h 40m	+17°
Nov. 14–18	Leonids	1866 I	10h 00m	+22°
Dec. 10–16	Geminids	—	7h 32m	+32°
Dec. 21–23	Ursids	Tuttle	14h 28m	+75°

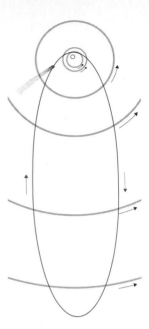

comet 1866 I, and it is estimated that more than 30,000 meteors per hour were visible. An outstanding display of this shower, the *Leonids*, occurs in a cycle of 33 years, the same as the comet with which it is associated. Further verification of the 33-year cycle came when spectacular showers were observed in 1932 and 1965, when up to 50 meteors/hour were seen.

More prevalent than the shower meteors are the sporatic meteors that enter the earth's atmosphere almost continuously. They seem to bear no specific relationship to the comets and may appear in any part of the sky at any time.

Figure 7.20 Orbit of the swarm of particles which produces the Leonids.

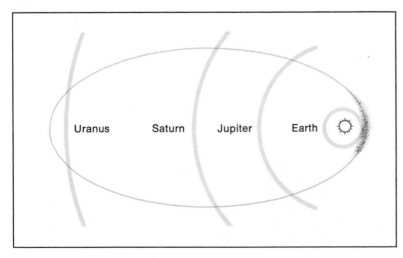

PHYSICAL PROPERTIES OF METEORITES

A typical meteoroid is a small, porous, easily fragmented body that disintegrates in flight. The average meteoroid recorded by the camera has a computed density of about 0.26 g/cm^3, which is somewhat like that of pumice stone. The exceptional meteoroid that reaches the earth is usually more dense, ranging from 3 to 8 g/cm^3. Such meteorites fall into one of three main categories. The stones (aerolites or chondrites) constitute about 92 percent of all meteorites; they are very much like ordinary stones, containing iron, silicon, carbon, magnesium, aluminum, and other metals (Figure 7.21). The stony-irons (siderolites) represent 1 to 2 percent of all meteorites; a polished slice of this type shows a matrix of stone and iron (Figure 7.22). The *irons* (siderites), representing 5 to 6 percent of all meteorites, are composed largely of iron with varying amounts of nickel (1 to 20 percent). Why, one may wonder, do museums and planetariums usually display mostly irons and stony-iron meteorites? The answer is that these two types are much easier to recognize. The "stones" are so similar to ordinary rocks that they often go unnoticed. The irons are quite easily identified as possible meteorites because of their higher density, however this is not proof by itself. The irons, upon cutting, polishing, and etching with a mild acid, reveal a very interesting crystaline structure that is a unique feature (Figure 7.23). These are called *Widmannstätten lines*; no natural rocklike mineral found on earth possesses such a pattern. The size of the crystals shown in such iron meteorites indicates a very slow cooling process. If the irons had cooled as small separate objects, such large crys-

Figure 7.21 The world's largest stony meteorite fell in the Jiling province of northeastern China in 1976. The large fragment shown weighs 1770 kg (almost two tons). The fusion crust shows the effects of heating due to friction as it passed through the earth's atmosphere. (Thomas Tsung)

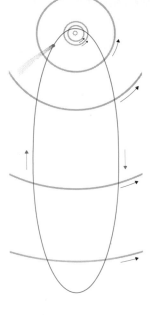

Figure 7.22 A polished slice of a stony-iron meteorite from Brenham County, Kansas. The dark material is olivine. (Ronald Oriti, Griffith Observatory)

tals would not have had time to form. They may thus represent the core of a much larger object. The irons contain 1 to 20 percent nickel as well. Nickel does not appear in this proportion along with iron in terrestrial materials.

One particular subdivision of the stones called carbonaceous chondrites represents objects approximately 4.6 billion years old, some of the oldest in the solar system. The polished section in Figure 7.16 on page 272

Figure 7.23 The polished section of an iron meteorite, showing Widmannstätten lines. (Ronald Oriti, Griffith Observatory)

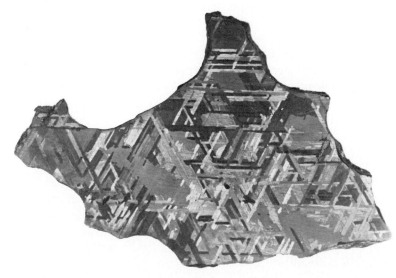

shows the round chondrules for which the chondrites are named. The existence of such distinct granules indicates that this particular type has never been heated above 500°K.

ORBIT AND ORIGIN OF METEORITES

The facts just mentioned suggest that meteorites originate outside the earth. The temperatures that the irons must have experienced to produce the separation of materials of different densities must have been higher than anything they could have experienced during their flight to earth or on its surface. Our discussion of meteor showers indicated the close relationship between meteors and comets. Of the more than 400 meteors for which the Harvard Observatory computed the original orbits, the large majority were shown to have a cometlike orbit; most probably originated from comets. Several had orbits suggesting asteroidal origin, and radioactive dating of meteorites indicates an age of 4 to 5 billion years. Some observers have suggested that before or during the creation of the planets, objects of lesser size formed. The series of events suggested by these observers is as follows. The centers of these objects experienced very high temperatures, perhaps from short-lived radioactivity, that created a molten core. Materials of higher density gravitated to the center and, upon gradual cooling remained in that separated state. The object later experienced a collision and broke into many small chunks. The iron meteorites originated from the core, the stones from the outer portions, and stony irons from the intermediate layers.

METEORITE CRATERS

If one stands on the rim of the Barringer Crater near Winslow, Arizona, one cannot help but ask several questions. What cataclysmic event could have caused this great natural depression in the earth's surface? What evidence do we have of its origin? Will an answer to this question shed any light on the origin of similar features on the moon? If we assume that the Barringer Crater is the result of the high-speed impact of a massive object from outside the earth, then we would expect that the object itself would suffer some damage upon impact. Its kinetic energy must have been converted to other forms of energy in a matter of seconds. Some of that energy would have been transformed into heat, some would have been used in blasting out the hole (throwing material outward), and some might have been spent in fracturing and possibly vaporizing some portion of the object itself. While it is estimated that the weight of the object must have been in the order of 100,000 metric tons, only about 30 metric tons of meteoric material have been found around the crater. Such an object might have been an asteroid that hit the earth; however, no central object has been found. To have blasted out such a hole using TNT would have required more than 10 million tons buried about 100 m (330 ft) underground (Figure 7.24).

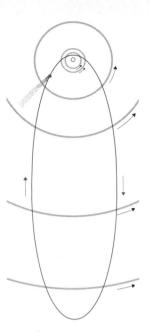

Figure 7.24   The Barringer Crater near Winslow, Arizona. (Meteor Crater Museum)

We would expect to see evidence of the high temperature and high pressure that would have accompanied such an impact. A number of craters are surrounded by a very dense form of silica called *coesite*. Silica of this density is not generally found, and it is believed to result from the tremendous shock wave that accompanies the impact of a large object on the earth. Under the sudden increase in pressure, rocks do not always shatter along their normal fracture lines, and the shatter structure of the crustal materials surrounding the impact area very often shows a special form. Features associated with high temperature include glasslike objects formed from constituents of the soil. Tektites, which are discussed in the next section, may be a by-product of such an impact.

With these theories about the features meteorites should produce in mind, the scientist sets out to look for such features. Where does he look? Unlike the moon, the earth has only a very limited number of sites that resemble a crater structure. The most obvious sites have been tested, and many have at least two of the distinguishing features mentioned above. Less obvious sites are sometimes revealed by satellite photography or by a very careful study of topographic maps that show the contours of a given region. Although a certain site may look almost flat today, its crater form may have been very obvious at an earlier time. The change would have taken place gradually as a result of erosion. Atmospheric factors such as wind, rain, and temperature variations effectively obliterate the once obvious surface features. The moon craters, on the other hand, are subject to almost no erosion because an atmosphere is lacking, and therefore remain almost unchanged for thousands of years.

The age of a crater can be roughly inferred from its appearance. The Barringer Crater, the result of a relatively recent event, is probably no

Asteroids, Comets, and Meteors   **279**

more than 30,000 years old. On the other hand, the Vredefort Crater of South Africa, almost indiscernable now, is probably 250 million years old. The impact force that produced the Vredefort Crater is estimated to be 500,000 times the force that produced the Barringer Crater. The Vredefort Crater has a diameter of almost 10 km (6.2 mi).

In this century there have been two large falls. Both occured in Siberia—one in 1908, the other in 1947. At the rate of one large fall for every 1000 years, the earth would have experienced 1000 such events in the past million years, yet no such frequency is indicated by the number of known craters. Perhaps many eroded craters lie undiscovered.

THE SIBERIAN EVENT

On the morning of June 30, 1908, an event took place in the Tunguska Valley of Siberia that has scientists still speculating. Witnesses several hundred kilometers away saw a fireball that seemed to "fill much of the sky" and gave off a heat which they clearly felt. A terrifying sound accompanied the fireball and then an explosion occurred with such force that even at that distance the ground shook and windows broke. When a Soviet biologist visited the site in 1925, he found no crater. Instead, the blast had simply leveled everything for several tens of kilometers in every direction, charring all the trees with the exception of a small stand in the center of the devastated area. Such a pattern indicates an aerial detonation with the power of several hundred atomic bombs.

This destruction does not appear to be the result of a huge iron meteorite or asteroid hitting the earth; in that case the evidence would most likely be a huge crater larger than the Barringer. In fact, no meteorite fragments have been found in the vicinity of this event to this day. An alternate suggestion is that the object was very massive, yet far less dense than the iron meteorite and able to vaporize readily without leaving a distinct remnant. A body composed of loosely packed snow (water ice) could meet all these requirements. If the Tunguska object was such a body, it might have resembled the nucleus of a small comet—too small to have been recognized in the usual visual sense. Several unusually luminous nights were observed in Europe just prior to the event, thought to have been caused by the dust in the atmosphere from a comet passing near the sun. An atmospheric pressure (shock) wave was detected at numerous meteorological stations around the earth, and there was some indication that the wave traveled around the earth more than once. As such a body approached the earth, it would set up a shock wave traveling faster than the speed of sound and creating a very high temperature in the atmosphere ahead of itself. Such high temperature could account for the charring of the trees.

Others have suggested a nuclear blast or the collision of an antimatter body with the matter body of the earth. When antimatter and matter come into contact, both are annihilated resulting in a great release of energy. Events of either of these types, however, would have also produced telltale signs of radiation; no such signs have been found. Still others have suggested that an extremely dense black hole passed through the earth. Can you add your own explanation?

TEKTITES

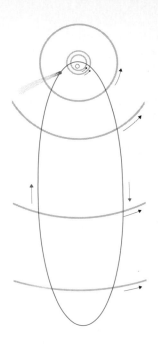

No subject in astronomy has aroused more diverse opinions than the origin of tektites. Before we present two of these different views, let us consider the factors which seem to be fairly well agreed upon. Tektites are small, blackish, glassy objects that have been found only in restricted land and sea areas (Figure 7.25). The largest number have been found in the Southwest Pacific area, including Australia, Indochina, Indonesia, and the Philippines. Smaller numbers have been discovered along the Ivory Coast of Africa, in Germany, Czechoslovakia, the United States, and the USSR. All tektites show evidence of having been heated to a molten state. This heating may have occurred at their time of origin and/or as they passed through the earth's atmosphere. When sliced in a thin section, a tektite appears translucent green or brown. Typically, tektites are homogeneous (crystal-free) structures that appear to have cooled very rapidly. They lack water and have unusually high silicon content, as would be true of ordinary window glass. Note the typical teardrop or dumbbell shapes in the samples from Thailand called *indochrinites* (Figure 7.26). These are sometimes called "splash" forms, suggesting that they cooled while in flight, having been thrown from a meteorite impact or a volcano. The surface markings on these samples are evidence of the frictional heating that resulted from their passage through the earth's atmosphere. Part of the surface material was undoubtedly lost in flight—a process called *ablation*. All surfaces of the indochrinites from Thailand show this effect. Contrast the

Figure 7.25 Tektites are found only in a limited number of locations on the earth's surface. Each dot represents a field over which tektites were strewn.

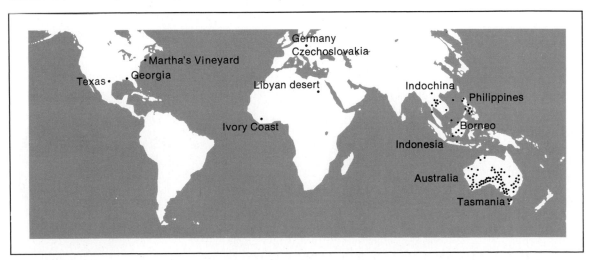

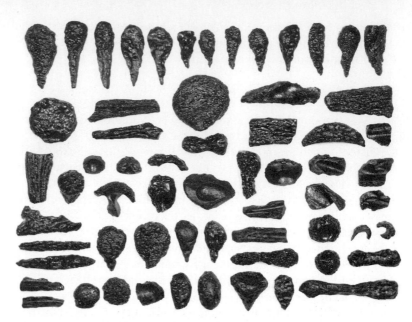

Figure 7.26 Some tektites from Thailand. (Ronald Oriti, Griffith Observatory)

shape of the tektites typical of Australia called *australites* (Figure 7.27). These "buttons," as they are often called, show a molten origin as all tektites do, but their appearance suggests a second period of heating during a controlled, tumble-free flight through the earth's atmosphere. The ablation effects show up on one side only, the side that approached the earth. The models shown in Figure 7.27 actually were artificially manufactured by NASA scientists by placing a sphere of glass in a wind tunnel and applying heat to one side with a blow torch. This was done to determine the shape a space reentry vehicle should have to survive a flight through the earth's atmosphere. The artificial models so closely resemble the true australites that it is not easy to distinguish them.

The dates of the various tektite types according to radioactive dating

Figure 7.27 Australites-tektites from Australia. (NASA)

are as follows: the general group called Australasian tektites (from Australia, Indochina, Indonesia, the Philippines, and nearby) are approximately 700,000 years old; the Ivory Coast tektites are approximately 1.2 million years old; the Moldavites (from Czechoslovakia) are approximately 15 million years old; and the North American tektites are approximately 35 million years old. The tektites thus did not originate in just a single event.

Opinion diverges when we try to determine the place of origin of the tektites and the mechanism whereby they were placed into flight. Researchers have suggested that they may have come from as far away as the moon, but probably no farther. Had they come to earth from the distance of Mars, it is reasoned, their dispersal on the surface of the earth would have been greater. Tektites may have been thrown from the moon as a result of either meteorite impact or volcanic activity. They may have on the other hand been thrown by either of the two same forces from earth. Let us consider these four distinctly different hypothetical origins. You may think of still further possibilities.

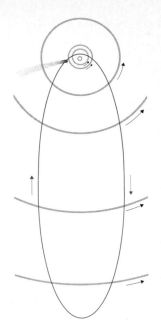

One group of researchers considers the most difficult problem posed by the two theories of terrestrial origin to be imagining a mechanism that would impart sufficient velocity to these blobs of glass to sustain them in flight long enough to produce all the observed effects, including the original melting and/or ablation effects on their surfaces. Because of the lesser surface gravity of the moon, this question is less problematic in theories of lunar origin; the accreted blobs could more easily escape the moon's gravitational field. If tektites originated by impact of a meteorite on earth, we must ask where the craters are that would be left by the impact. For the youngest tektites from Australia there are no signs of remnant craters. Craters have been found, however, near the tektite falls in Germany (the Ries Basin) and the Ivory Coast (Bosumtwi Crater in Ghana). The dating of these craters shows fairly close agreement with the ages of their respective tektites. This seems to support theories of terrestrial origin, but further study is called for.

We have compared the composition of tektites with that of earth rocks and earth soil and have found significant differences in silicon content and water levels. Terrestrial rock is typically lower in silicon and higher in water content than tektites. The water could easily have been lost in the heating process. When lunar rock samples were returned by the Apollo crews, none seemed to correlate with tektite composition. The makeup of the glassy portions of the lunar soil, on the other hand, seemed very similar to that of tektites. One group of observers sees this as strong evidence for a lunar origin, whereas another group considers the craters found in Germany and the Ivory Coast to be equally strong evidence for a terrestrial origin. Certainly the question of origin cannot be easily settled.

MICROMETEORITES

Approximately 3000 metric tons of material other than meteorites and tektites fall on the earth each day from extraterrestrial sources. This mate-

rial consists mostly of *micrometeorites*, individual particles so small that they are not heated as they pass through the earth's atmosphere and therefore produce no visible streaks of light. This form of deposit may actually serve to enrich the soil in certain lean areas of the earth's surface, yet the percentage of material that is added to the earth's mass even in a million years is negligible.

QUESTIONS

1. The first asteroids discovered had orbits that lie between the orbits of which two planets?

2. What is the general range of size of the asteroids?

3. Why are astronomers particularly interested in the asteroids with very eccentric orbits that carry them close to the sun and across the orbits of several planets?

4. New asteroids are discovered most easily on photographic plates that have been exposed for a hour or more. During the exposure, the telescope is moved to follow the stars. How is an asteroid recognized on such a plate?

5. If all the material of the known asteroids could be combined into one "planet," how would that planet compare in size and mass to the earth?

6. Does a comet have a tail during its entire orbital motion around the sun? Explain.

7. The planets all orbit the sun in the same direction and along almost the same plane. Is this also true of comets? Explain.

8. Explain what is meant by the "Jovian" family of comets.

9. What is thought to be the primary nature of the nucleus of a comet?

10. Why does the gas tail of a comet always extend away from the sun?

11. May only professional astronomers have their names associated with a new comet? Explain.

12. Describe the usual appearance of a comet when it is first discovered.

13. When is the best time to observe meteors? Why?

14. What is the range of size of most meteoroids as they begin their fiery flight into the earth's atmosphere?

15. Meteor showers are thought to be associated with what other astronomical object?

16. Why is the occurrence of a meteor shower predictable?

17. List the different types of meteorites.

18. Most meteorites that have been dated by radioactive elements have been found to be: (a) much younger than the earth, (b) about the same age as the earth, or (c) much older than the earth?

19. Why have we not found a meteor crater for each large meteorite that has hit the earth in the last billion years?

20. True or false: The composition of tektites proves their origin on the moon. Explain.

21. What evidence does the astronomer have for concluding that most asteroids have irregular shapes?

22. Why is it true that comets often develop two tails? How would you recognize the composition of each tail?

23. In addition to the theory of a comet cloud surrounding the solar system, can you think of any other theory that might explain the origin of comets?

24. In what ways do Kepler's laws apply to comets? How is the period of a comet related to its average distance from the sun?

25. Why is it impossible to predict the approximate number of meteors that will be seen during a given meteor shower, even if that same shower has been observed many times before?

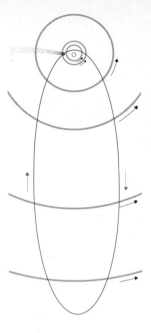

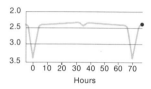

The Sun

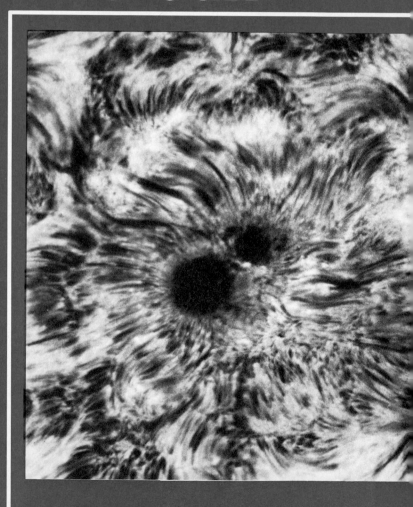

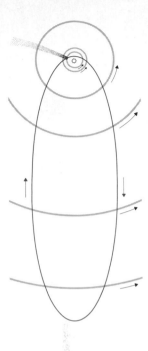

Eight

To refer to the sun as a star is to recognize its true nature, a nature drastically different from that of the planets or their moons. The sun is a tremendous seething inferno, generating its own energy from within. Imagine enjoying a sunny afternoon at the beach. You would be affected by three distinct forms of radiant energy: (1) infrared, or radiant heat, sensed by your body as warmth; (2) light, sensed by your eyes; and (3) ultraviolet, manifest perhaps by a tan or sunburn. It is the sun's energy that drives the winds that create swells of the ocean that eventually produce the breakers crashing in on the seashore. It is the sun's energy that lifts water (through evaporation) from the sea, only to dump it back onto the earth as rain. It is the sun's energy that converts elements such as carbon, hydrogen, and oxygen into food for our consumption and free oxygen for us to breathe through photosynthesis. In fact, most of the energy sources on earth can be traced back to the sun. Fossil fuels, such as coal and oil, are derived from organisms that were once dependent upon the sun for their existence. Life as we know it could not exist were it not for the energy received from the sun.

The earth is situated at a very strategic distance from the sun. The average distance from the sun to the earth is called the *astronomical unit*. Modern methods using radar

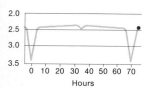

have greatly simplified the task of finding this distance. The method used does not involve sending radar signals toward the sun since the echo would be very difficult to differentiate from radio signals emitted by the sun itself; the method used involves an indirect approach. A radar signal is directed toward a given planet or asteroid, and the time required for the echo of that signal to be detected is noted. Recalling that a radar signal travels through empty space at the same speed as light, the observer may compute the distance to the object based on the time required for the two-way trip. Knowing both the distance and period of an object that orbits the sun, the astronomical unit can be calculated using Kepler's third law, which says that the square of a planet's period is proportional to the cube of its average distance from the sun. Suppose, for example, that the average distance between the earth and Mars at opposition is found by radar observation to be 78,389,294 kilometers; the computation of the astronomical unit would then take the following form (Figure 8.1):

$$\frac{P_1^2}{P_2^2} = \frac{r_1^3}{r_2^3}$$

$$\frac{(1.88)^2}{(1)^2} = \frac{(x+78,389,294)^3}{x^3}$$

$$x = 149,597,890 \text{ km}$$

(details of the solution have been omitted)

FIGURE 8.1 Average distance between the sun and earth, and between the sun and Mars. The period of Mars (P_1) is 1.88 years; the period of the earth (P_2) is 1 year.

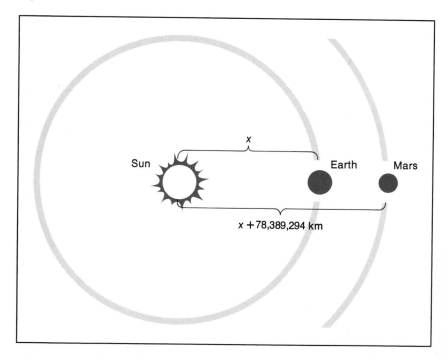

Here P_1 is the period of Mars, P_2 is that of the earth, and r_1 is the average distance of Mars from the sun. The result, 145,597,890 kilometers, is the internationally adopted value of the astronomical unit. The approximate value of this unit in the British system of measure is 93 million miles.

Because the sun is approximately 300,000 times closer than the next nearest star, we may determine its diameter rather easily.

PHYSICAL CHARACTERISTICS OF THE SUN

The determination of the diameter of the sun follows directly from our knowledge of its distance from the earth and the apparent angle that its diameter makes with our eye. Suppose that the sun's diameter makes an angle of 0.532° on a day when its distance is 149,598,000 kilometers. The calculations that follow exemplify a very useful technique requiring only simple arithmetic. For angles of less than 1°, this method actually provides greater accuracy than more advanced methods using trigonometry.

Visualize the earth at the center of a huge circle with a radius of 149,598,000 kilometers (Figure 8.2). Think of the sun's diameter as a portion of the circumference of that circle. Note that the angle of 0.532° compares to the entire angle at the center of the circle (360°) in the same way

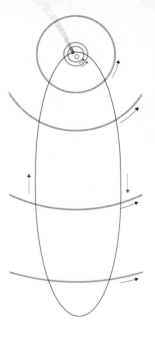

FIGURE 8.2 Determining the diameter of the sun.

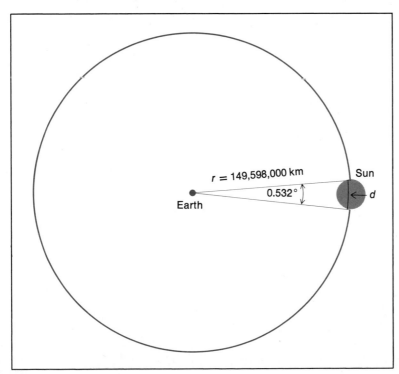

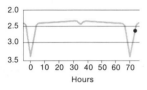

that the diameter of the sun (d) compares to the entire circumference of the circle ($2\pi r$):

$$\frac{0.532°}{360°} = \frac{d}{2\pi(149,598,000 \text{ km})}$$

Solving for d:

$$d = \frac{0.532(2\pi)(149,598,000)}{360} = 1,390,000 \text{ km}$$

This gives a diameter of approximately 1,390,000 kilometers, almost 110 times that of the earth. The immense size of the sun may be appreciated by knowing that a relatively small sunspot is equal in diameter to the earth (Figure 8.3).

The difference between the volumes of the sun and earth is even more dramatic. Recall that the volumes of two spheres compare as the cubes of their radii (or diameters); for example, if we double the diameter of a sphere, its volume grows by a factor of eight [$(2)^3 = 8$]. If we cube the

FIGURE 8.3 A white light photograph of the sun showing its comparitive size to that of the earth. (Mount Wilson and Las Campanas Observatories, Carnegie Institute of Washington)

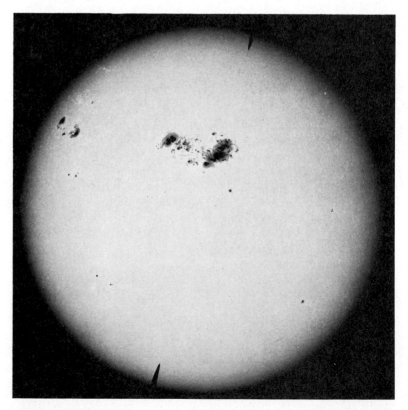

ratio of diameters just determined above, namely 110:1, then we find that the sun could contain 1,331,000 earths in its volume.

The mass of the sun can be found by studying its gravitational effect on the earth or any other planet. The earth, if left only to its own inertia, would fly off into space; however, it is accelerated because of its mutual attraction to the sun by 0.59 cm/sec². Assuming that the force which produces this acceleration is gravity, we set $F = ma$ equal to $F = Gm_1 \times m_2/r^2$, which gives us:

$$m_1 a = \frac{Gm_1 \times m_2}{r^2}$$

where m_1 is the mass of the earth and m_2 is the mass of the sun.

The mass of the earth cancels out, leaving:

$$a = \frac{Gm_2}{r^2}$$

Then solving for m_2 the mass of the sun is:

$$m_2 = \frac{ar^2}{G} = \frac{0.59 \times (1.496 \times 10^{13})^2}{6.67 \times 10^{-8}}$$
$$m_2 = 2 \times 10^{33} \text{ grams}$$

When compared to the earth's mass of 6×10^{27} grams, we see that the mass of the sun is approximately 300,000 times that of the earth. We are now prepared to compute the density of the sun; this will help us understand its nature. Density can be found by dividing the sun's mass in grams by its volume in cubic centimeters, however let's try a more instructive method. We know that the sun's mass is approximately 300,000 times that of the earth and its volume is approximately 1,300,000 times that of the earth. Let us use these comparisons as follows:

$$\text{Density}_{sun} = \frac{\text{mass}_{sun}}{\text{volume}_{sun}} = \frac{300,000 \times \text{mass}_{earth}}{1,300,000 \times \text{volume}_{earth}}$$

$$= \frac{1}{4} \times \text{density}_{earth} \quad \text{(approximate)}$$

$$= \frac{1}{4} \times 5.5 \text{ g/cm}^3$$

$$= 1.4 \text{ g/cm}^3$$

Thus the sun is approximately one-fourth as dense as the earth on the average. This figure is much closer to the density of water. The sun is probably like a big ball of gas that is more dense toward its middle due to pressures in the order of a billion atmospheres. Hydrogen "bombs" going off in the sun's core, however, keep it in a gaseous state throughout.

THE SUN: A HYDROGEN FUSION REACTOR

We often use the expression "The sun is burning." Were the sun merely burning, however, it could not begin to produce the vast amount of energy

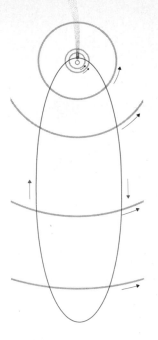

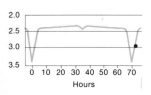

Hours

that radiates from its surface. The astronomer now realizes that the true source of the sun's energy is a nuclear reaction similar to that of the hydrogen bomb but taking place under controlled conditions. At the temperature of 15 million degrees Kelvin (and up) and the pressure at the core of the sun of one billion atmospheres, the atoms tend to fuse together and in the process release energy. The hydrogen atom itself supplies the "fuel" for this reaction. Hydrogen is the simplest atom in the universe, consisting of one *proton* (a positively charged particle) in the *nucleus* and one *electron* (a negatively charged particle) moving about the nucleus. For our present purpose, we may ignore the electron because we are concerned primarily with nuclear changes. Two positively charged protons will under ordinary conditions repel each other. However, under the conditions of high temperature and high pressure in the core of the sun, two protons may be forced so close together that a nuclear binding force becomes more effective in holding them together than the electrostatic repulsion that tends to drive them apart. When two or more nuclear particles "stick" together, a new isotope or element is formed. The first step in this reaction occurs when two hydrogen nuclei fuse to form a special type of hydrogen called *deuterium* (^2_1H). This step may be pictured as follows, where $\oplus$ represents a proton and Ⓝ represents a neutron:

$$\oplus + \oplus \longrightarrow \oplus\text{Ⓝ} + \text{positron} + \text{neutrino}$$

or

$$^1_1H + {}^1_1H \longrightarrow {}^2_1H + {}^0_1e + \text{neutrino}$$

A positron (^0_1e) is viewed as a particle that has a mass similar to an electron but charge that is positive. The neutrino is a particle thought to have little or no mass and no charge. Neutrinos are thus very hard to detect, and yet they carry off significant amounts of energy from the sun. The second step in this fusion reaction occurs when the deuterium atom fuses with still another hydrogen atom to form helium–3 (^3_2He) and a gamma ray (γ ray):

$$\oplus\text{Ⓝ} + \oplus \longrightarrow \begin{matrix}\oplus\oplus\\\text{Ⓝ}\end{matrix} + \text{gamma ray}$$
$$^2_1H + {}^1_1H \longrightarrow {}^3_2He + \gamma \text{ ray}$$

The gamma ray is an energetic form of electromagnetic radiation; it too carries off a certain amount of energy from the sun. The third step of the reaction occurs when two such helium–3 (^3_2He) atoms fuse to form an atom of helium–4 (^4_2He) and two hydrogen atoms:

$$\begin{matrix}\oplus\oplus\\\text{Ⓝ}\end{matrix} + \begin{matrix}\oplus\oplus\\\text{Ⓝ}\end{matrix} \longrightarrow \begin{matrix}\oplus\oplus\\\text{Ⓝ}\text{Ⓝ}\end{matrix} + \oplus + \oplus$$

or

$$^3_2He + {}^3_2He \longrightarrow {}^4_2He + {}^1_1H + {}^1_1H$$

Only four hydrogen atoms were used in forming the one helium–4 atom, since two hydrogen atoms are given back in the final reaction. However, the mass of the helium–4 atom is less than the total mass of the four hydrogen atoms—the difference in mass has been converted into energy. It is this energy that constitutes the primary source of energy for the sun and for stars in general.

In order to specify the mass of a given nucleus, we define a unit called 1 *amu* (atomic mass unit) as 1.66×10^{-27} kilograms. The mass of a hydrogen proton appears to be slightly greater than that of a single proton in the helium atom; it is equal to 1.008 amu. Four such hydrogen protons would have a mass of 4.032 amu, yet when these four are fused to form one helium-4 atom, its mass is only 4.003 amu—0.029 atomic mass units have been converted into energy. How much energy? The relationship between mass and energy was expressed by Albert Einstein in the equation $E = mc^2$, where E is the amount of energy (in ergs), m is the mass that is converted (in grams), and c is the velocity of light (in centimeters per second). The velocity of light is 3×10^{10} cm/sec; this number squared would equal 9×10^{20}. Therefore Einstein's equation, stated in words, says that the energy produced is equivalent to 9×10^{20} (that is 900,000,000,000,000,000,000) times the mass that is converted. The sun converts almost 4.5 million metric tons of mass per second; thus it produces energy at a rate of almost 4×10^{33} erg/sec (4×10^{26} watts). Based on the amount of hydrogen in the core of the sun, this reaction could be sustained for more than 10 billion years. This nuclear reaction called the *proton-proton cycle*, is initiated and sustained by a central core temperature of about 15 million degrees Kelvin. At higher temperatures a reaction called the *CNO* (carbon nitrogen oxygen) *cycle* may become more important, but the net result is almost the same; four hydrogen nuclei react to form one helium-4 nucleus, and energy is released equal to the difference in mass. Carbon, nitrogen, and oxygen merely serve as catalysts for the reaction. The CNO cycle is considered in detail in Chapter 13.

Thermonuclear fusion is thus different from the chemical reaction of burning; however, astronomers typically speak of this fusion process as *hydrogen burning*. We will adopt this convention for the remainder of the book. *Helium burning*, then, will mean helium fusion to form still heavier elements.

NEUTRINOS FROM THE SUN

The very first step of the proton-proton cycle involves the production of a neutrino, a particle which possesses little or no mass and no charge. Neutrinos, because of their virtually massless nature, pass directly from the core of the sun to the earth; they do not interact with the other material of the sun. Thus, if these particles could be detected, astronomers would in effect be looking into the core of the sun. However, because of their character, neutrinos are difficult to capture. A high percentage of neutrinos

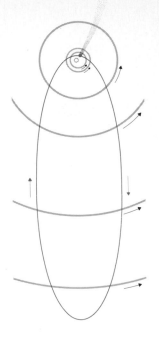

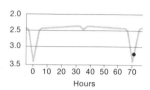

may pass through the entire earth without encountering a single atomic nucleus. A very interesting attempt to capture neutrinos has been made over the past ten years. In a rock cavern, some 1600 m below the surface of the Homestead Mine in Lead, South Dakota, a 378,500 liter (100,000 gal) tank of tetrachloroethylene (C_2Cl_4) was installed under the auspices of the Brookhaven National Laboratory (Figure 8.4). It is theorized by the researchers that neutrinos could be expected to react with the chlorine-37 atoms to produce a radioactive form of argon that is easily detectable. Because similar reactions could be initiated by other particles (that is, free protons or neutrons) that might penetrate the tank if placed at ground level, the underground site allows the earth to act as a shield in this regard. It was predicted that 100,000,000,000 neutrinos would penetrate every

FIGURE 8.4 The Brookhaven solar neutrino experiment, located 4850 feet underground in the Homestake Gold Mine in Lead, South Dakota. The tank contains 100,000 gallons of tetrachloroethylene (C_2Cl_4), with which neutrinos react to produce radioactive argon-37. Using this device, astronomers seek evidence of the thermonuclear process going on in the core of the sun. (Brookhaven National Laboratory, Upton, New York)

square centimeter of the earth's surface per second. The aim was to detect some interaction between the neutrinos and the chlorine atoms in the tank. Counts of radioactive argon atoms to date reveal that the capture rate is approximately one-third of the predicted rate. These experimental results pose a serious question for the solar astronomer. Could the well-established proton-proton theory be wrong? Could the sun's core be cooler than predicted? Or, do we fail to understand the neutrino? Some observers have suggested that the thermonuclear reaction in the core of the sun is not as constant as we had thought and may simply be at a low point in a long cycle. An alternate experiment at Brookhaven, using gallium-71 may help to answer some of these questions.

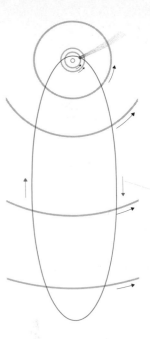

LAYERS OF THE SUN

The fusion of hydrogen into helium is continually building a helium core within the sun. The immediate layer surrounding that core is the site of the

FIGURE 8.5 Layers of the sun, shown with approximate temperature distribution.

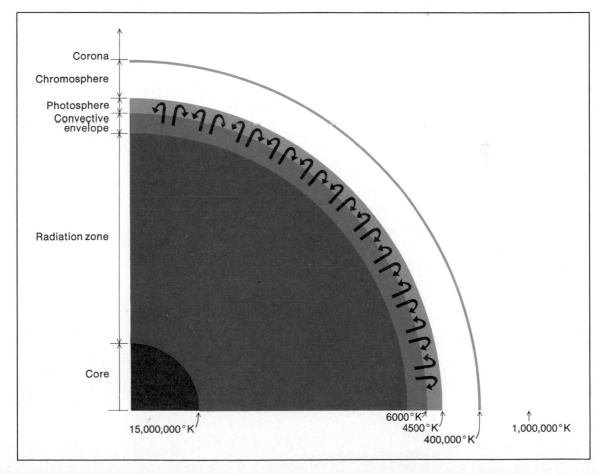

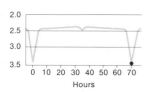

thermonuclear reactions (hydrogen "burning"). The tremendous quantities of energy produced must be transported to other layers and ultimately to space. Astronomers believe that the major transport mechanism within the body of the sun is radiation. Energy generated in the core of the sun travels through the *radiation zone* by means of a multiple absorption and reemission process. A photon which leaves the core does not radiate in a straight line to the surface of the sun but rather experiences multiple collisions, scatterings, absorptions, and reemissions. If you could follow a single photon through such a myriad of random reactions, you would find that a million years or more is required for that initial reaction to reach the visible surface of the sun and be radiated to the outside world. Thus, as we view the sun in visible light, we are not seeing the core but rather the outer layer from which the photons are finally emitted. Neutrinos by contrast come directly to the earth from the core of the sun, which is why astronomers are so interested in detecting them.

The sun's next layer, called the *convective envelope*, transports energy only by means of the motion of the material within it. Hot gases rise to deliver energy to the visible surface of the sun, then the cooled gases fall to be reheated (Figure 8.5). You are probably familiar with the phenomenon of convection in the heating of a room. If the heater is located on one side of the room, the air which it warms will tend to rise, flow across the ceiling, and fall on the far side of the room as it cools setting up a circulation pattern. We can imagine the violent motion which takes place in order to transport the vast quantities of energy being emitted from the radiation zone through the convective envelope. But what evidence do we have of this turbulence? When the sun is photographed in white light, its "surface" takes on a spotty appearance, referred to as the *granulation of the photosphere* (Figure 8.6). The bright spots are interpreted as rising columns of hot gases, and the darker regions as cooled gases returning to lower levels to be heated once again. This interpretation is borne out when we inspect

FIGURE 8.6 The granulation of the photosphere. The rising columns of gas cause a Doppler shift in the spectrum in one direction, whereas the falling gases cause a shift in the opposite direction, producing the waviness of the spectral lines. (Aerospace Corporation; Sacramento Peak Observatory)

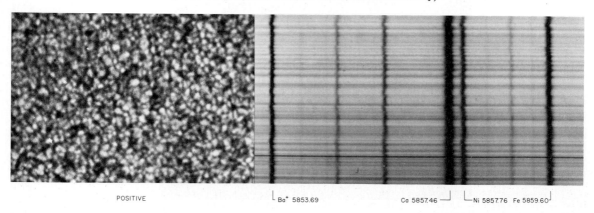

the spectrogram accompanying Figure 8.6. The wavy nature of the spectral lines is due to the Doppler shifts caused by the motion of the gases: blue shifts are associated with the rising spots that are approaching the observer; red shifts are associated with the cooler gases receding from the observer. The convective envelope is thought to represent about one-fourth of the sun's radius.

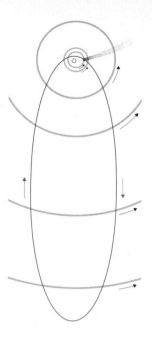

The layer immediately above the convective envelope is called the *photosphere*. This layer is the surface of the sun visible in a white-light photograph (Figure 8.7). Some variation in depth is shown by the darkening near the sun's limb (edge). When we look at the central portion of the solar image, we can see through the photosphere to its lower levels, where temperatures are of the order of 6000°K. When we look near the limb of the sun, we are seeing the higher levels of the photosphere, where temperatures are 1000°K to 1500°K cooler; the limb therefore appears darker to us. This temperature decrease explains the darker appearance there and, in addition, is an essential factor in the formation of the solar spectrum, which is an absorption spectrum (Figure 8.8). Recall from page 89 that an absorption spectrum is formed when light of a continuous nature

FIGURE 8.7 A white light photograph of the sun showing limb darking due to the cooling trend in the upper levels of the photosphere. (Mount Wilson and Las Campanas Observatories, Carnegie Institute of Washington)

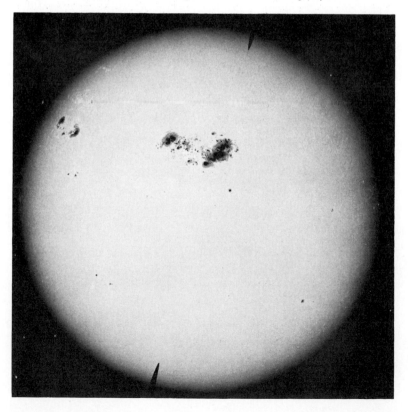

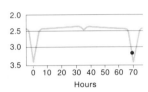

2.0
2.5
3.0
3.5

0 10 20 30 40 50 60 70
Hours

passes through a relatively cooler, low-pressure gas. It is within the photospheric layer that absorption takes place according to the elements that are present. In other words, if we could look at the sun's spectrum at the lower boundary of the photosphere, it would likely be of a continuous nature; Viewed from the upper boundary, it appears to be an absorption spectrum. As a result, the composition of the photosphere and the lower chromosphere is revealed, but not of the core or other interior layers.

The next outer layer of the solar atmosphere, the *chromosphere* (colored sphere), extends approximately 10,000 kilometers above the photosphere. The chromosphere is normally invisible due to the the brilliance of the photosphere; however, when the central disk of the sun is just covered by the moon during an eclipse, the reddish light of the chromosphere may be seen. When the light from this layer enters the spectrograph, its bright-line spectrum is revealed (Figure 8.9). The chromosphere is composed largely of hydrogen gas, as shown by a bright red line (the hydrogen-alpha line) in its spectrum and by its characteristic red hue. In addition to hydrogen, the bright lines reveal the presence of sodium, calcium, magnesium, and ionized helium. The astronomer no longer has to

FIGURE 8.8 The solar spectrum, with certain lines identified with known elements. (Mount Wilson and Las Campanas Observatories, Carnegie Institute of Washington)

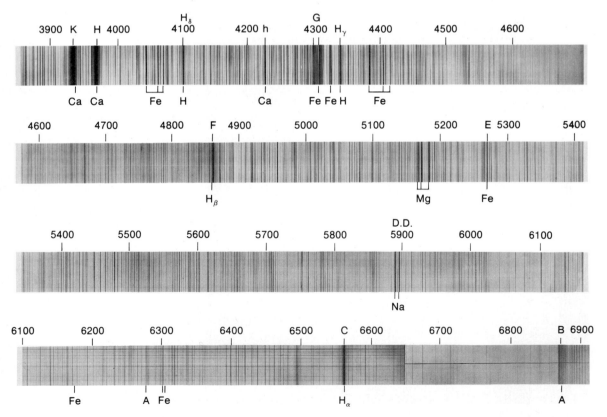

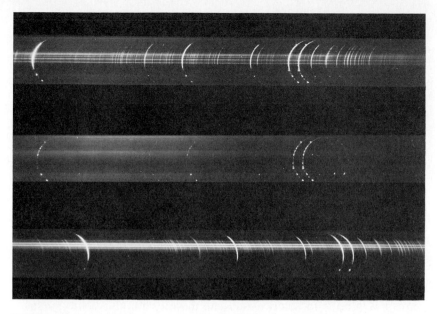
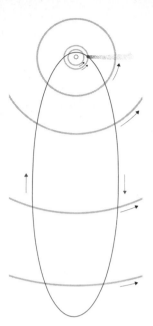

FIGURE 8.9 The bright line spectrum of the chromosphere, photographed during a total solar eclipse. (Mount Wilson and Las Campanas Observatories, Carnegie Institute of Washington)

wait for the occurance of a solar eclipse to study the chromosphere; it may be viewed by covering the solar disk near the focal plane of the telescope. Such a device is called a *coronagraph*. The spectrum of the chromosphere reveals a temperature gradation from 4500°K near the photosphere to 400,000°K in its upper levels.

The third principal layer of the solar atmosphere is the *corona* (Figure 8.10). Like the chromosphere, the corona is normally invisible, due to the brilliance of the photosphere; however, if it is viewed during a total solar eclipse or photographed using a coronagraph, its extended luminous region can be seen, largely as a result of the scattering of light by particles in the corona. The visible portion is only a very small part of the corona; the corona is now known to extend beyond the earth itself. In fact, most of the planets are continually bathed in its outflow of material. We have noted that the sun sends out in every direction many forms of electromagnetic disturbances, including light, radio, and infrared; however, this does not represent the full extent of its emission. There also appears to be a continual outflow of particles, including negatively charged electrons and positively charged protons of hydrogen gas, the basic constituents of the solar atmosphere. While the corona is more static near the sun, farther out near the planets it constitutes an expanding envelope and therefore produces a rapidly moving stream of charged particles called the *solar wind*. This stream is responsible for driving the tail of a comet away from the sun. At the earth's distance from the sun, the solar wind has a velocity of about 1,450,000 km/hr. The amount of material that is removed from the sun in

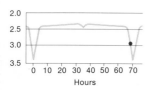

the form of this solar wind is approximately 1 million metric tons per second, which in terms of the sun's total mass represents a negligible amount of material. As these charged particles approach the earth's magnetic field, they are deflected into the Van Allen belts and eventually spiral toward the north and south polar regions, where they react with ions of the earth's upper atmosphere to produce the northern and southern lights.

The temperature of the corona can be determined by the degree to which elements in this layer are ionized. Iron, for instance, has 26 electrons in each atom under normal conditions of pressure and temperature. In the corona of the sun, the spectrum reveals that as many as 13 of these electrons have been stripped from the iron atoms, indicating a temperature of about 2 million degrees Kelvin. This very high temperature demands a special explanation, for none of the usual mechanisms of heat transfer will suffice. One theory suggests that acoustic (soundlike) waves are generated in the convective envelope of the sun, and that as they travel outward they produce shock waves whereby energy is transferred to the corona. While this theory is very plausible, no direct evidence of such waves has been found; hence alternate theories are also being pursued. For instance, it is possible for magnetic fields to heat the corona by electrical currents they develop, in the same way that an electric current flowing in a resistor is used to heat homes. This latter theory has definite merit in light of the discovery of corona "holes."

It is only recently that astronomers have recognized the significance of "holes" in the corona of the sun. Such gaps are especially evident when

FIGURE 8.10 The corona: (a) during a sunpot minimum; (b) during a sunspot maximum. (High Altitude Observatory; Yerkes Observatory)

(a)

(b)

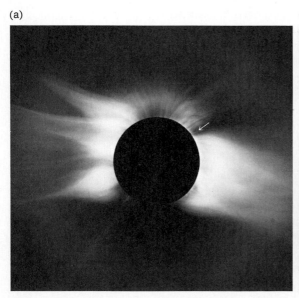

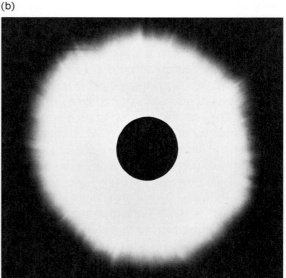

the sunspot cycle is near a minimum, when very few sunspots are evident. In Figure 8.10(a) note that virtually no coronal plumes extend from the polar regions, which is quite normal. However, note the dark region marked by the small white arrow. This is a "hole" in the corona. It is marked by relatively cool temperature, very low density, and open magnetic field lines. Magnetic regions of the sun are normally connected to regions of opposite polarity nearby. In the case of a coronal "hole," no region of opposite polarity exists nearby and its field lines remain open. The interesting result appears to be an unusual outpouring of the solar wind from that region, characterized by very high velocities. When such a stream of charged particles reaches the earth, the earth's magnetic field may be greatly disturbed. Such disturbances are called *geomagnetic storms*, and represent only one of many ways in which the earth may be affected by solar activities.

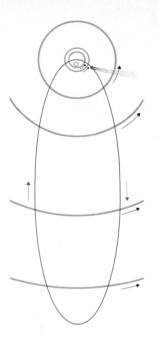

THE SOLAR SPECTRUM

The sun radiates energy in a wide range of wavelengths: radio energy (detected in the form of radio noise), infrared (radiant heat), visible light (to which our eyes are sensitive), and ultraviolet (which produces our suntan or sunburn). In addition, the sun emits X rays on the occasion of solar flares. Figure 8.11 shows a plot of the average output of the sun in these various wavelengths as measured at the top of the earth's atmosphere. The height of the curve, for any given wavelength, tells us the rate of energy production at that wavelength. When all of these energy rates are added for the whole spectrum, we get the total energy received at the top of the earth's atmosphere. This is called the *solar constant* and it amounts to 2 calories per square centimeter per minute. This means that at the distance of 1 A.U., the sun's energy could raise the temperature of 1 cm^3 of water by 2°C in one minute. The earth intercepts only about one-billionth of the sun's total energy. This fact may help us to appreciate the tremendous energy which the sun radiates in all direction.

As early as 1814, the German scientist Joseph Fraunhofer recognized that the rainbow-like spectrum of the sun was crossed by several dark bands, which he designated by capital letters. Several of these Fraunhofer lines, as they are called, are shown in Figure 8.8: the H and K lines of calcium, the F line of hydrogen, the D line of sodium (Na), and the C line of hydrogen. Fraunhofer's identification was only the beginning of one of the most productive tools of astronomy—spectral analysis.

Because of its high intensity, the visible spectrum of the sun is the easiest natural spectrum to work with. Solar spectra as long as 13m may be obtained. Such a spectrum contains more than 30,000 absorption lines. It is presumed that most of these lines are formed in the photosphere of the sun. The earth's atmosphere may create some of the lines, however, in which case they are called *telluric* lines; such lines are usually associated with such elements as nitrogen, oxygen, and water vapor, which are found in the earth's atmosphere. Telluric lines may be distinguished from the

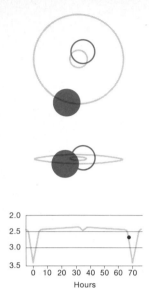

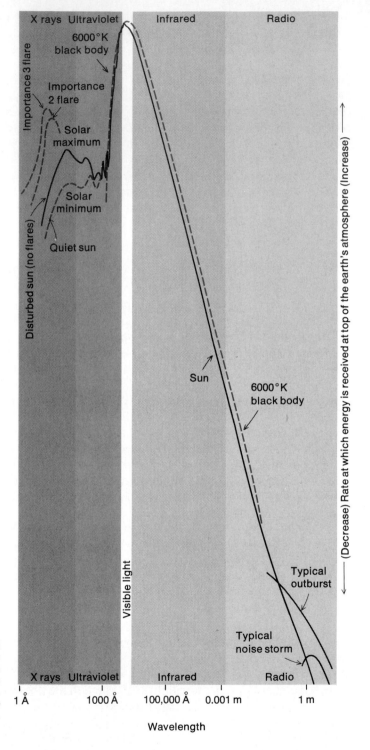

FIGURE 8.11 The solar spectrum. The total solar energy received at the top of the earth's atmosphere is represented by the total area under the curve. This is called the solarconstant and amounts to approximately 2 calories per square centimeter per minute. (Compiled by H. H. Malitson, NASA-Goddard Space Flight Center)

spectral lines of the sun in that they show no Doppler shift; such a shift is evident in the spectral lines of the limbs of the sun that advance and recede due to rotation.

How does the astronomer determine which elements are indicated by the some 30,000 lines in the solar spectrum? The laboratory spectrum of a known element is placed alongside the solar spectrum to see if all the lines of the given element have corresponding lines in the solar spectrum. In Figure 8.12 you can see that the lines of iron do in fact match a given set of lines in the solar spectrum, from which we conclude that iron is present—in a vaporized form—in the photosphere of the sun. Using this procedure, over 70 of the 92 natural elements that occur on the earth have been found to exist within the solar atmosphere. One should not assume that the heavier elements are very abundant, however, for hydrogen and helium still constitute the major portion of the solar atmosphere. The total of all other elements present represents less than 2 percent of that atmosphere (by mass), as indicated by the relative weakness of their spectral lines (see Figure 8.8, page 298). Table 8.1 lists the more abundant ele-

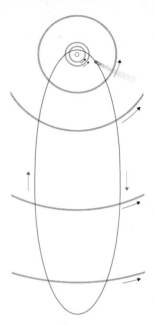

TABLE 8.1

ABUNDANCES OF SELECTED ELEMENTS IN THE SOLAR ATMOSPHERE	
Element	Percentage by Mass
Carbon	0.32
Nitrogen	0.10
Oxygen	0.84
Magnesium	0.07
Silicon	0.12
Sulfur	0.05
Iron	0.07

FIGURE 8.12 A small portion of the solar spectrum. The central bands with dark lines represent the spectrum of the sun. The bright lines above and below the solar spectrum represent the spectrum of iron (photographed for comparison). (Mount Wilson and Las Campanas Observatories, Carnegie Institute of Washington)

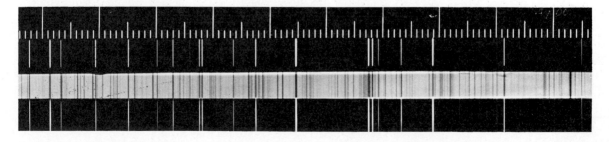

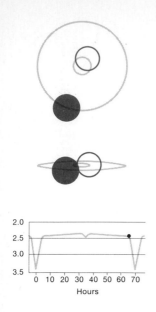

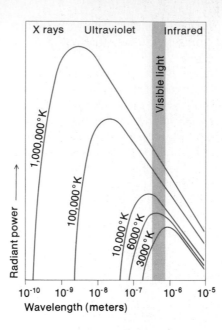

FIGURE 8.13 The energy distribution for black bodies at a number of different temperatures.

ments. We will see, as we study other stars, that the heavier elements are even less abundant in some cases.

The spectrum of the sun also indicates its temperature. The photosphere radiates maximum energy at a wavelength of approximately 4700 angstroms, indicating a surface temperature of 6000°K. This follows directly from a consideration of the energy radiated by a blackbody at various temperatures. A *blackbody* is any object that absorbs all the energy that falls on it and simultaneously radiates all of this energy. As the temperature of the blackbody increases, not only will the total energy radiated increase, but the wavelength at which maximum energy occurs becomes shorter (Figure 8.13). The sun's energy distribution best fits the blackbody curve corresponding to a temperature of 6000°K.

SPECTROHELIOGRAM

Solar activities cannot be seen easily in photographs taken in the ordinary (white) light of the sun [Figure 8.14(a)]. They can be seen better in a *spectroheliogram*, a photograph of the sun taken in the light of only one narrow portion of the spectrum. The word itself reveals its meaning, for if we take it apart in reverse order, "gram" means picture, "helio" means sun, and "spectro" indicates that we use the spectrum to obtain the picture. The *spectroheliograph* is the instrument by which such a picture is taken. The instrument uses either a filter that allows only the desired wavelength to pass, or a more complicated arrangement of a spectrograph and moving slit. If we choose to use the light in the first red line of the hydrogen spectrum—the hydrogen-alpha line—we obtain a picture of the sun that represents the upper photosphere [Figure 8.14(b)]. If we want to view yet a higher level (the lower chromosphere), we may use the light of the K line of ionized calcium (Ca II). The temperature within the bright regions, as viewed in the light of calcium, may reach 20,000°K, even though these regions are usually directly over a sunspot which appears dark in white light [Figure 8.14(c)].

With the increasing use of the Orbiting Astronomical Observatory, astronomers can now study the sun in the ultraviolet wavelengths. In order to interpret the results of such studies, as seen in Figure 8.14(d,e), we must first understand how they were obtained. We have just referred to Ca II, which stands for calcium atoms that have been ionized once, or lost one of their electrons; Ca I stands for neutral calcium atoms, atoms that have the same number of electrons as protons. The representation of the sun shown in Figure 8.14 (d) was derived from the information received from an ionized form of oxygen, namely O VI; this is oxygen that has been ionized five times. In such an atom, certain electron transitions are possible that are not possible in neutral oxygen. One such transition produces

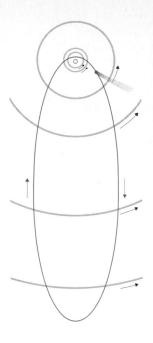

FIGURE 8.14 The sun: (a) in ordinary light; (b) in the light of Hα; (c) in the light of Ca II; (d) in ultraviolet radiation of O VI; (e) in ultraviolet radiation of Si XII; (f) X-ray photograph. (a–c, Mount Wilson and Las Campanas Observatories, Carnegie Institute of Washington; f, from L. P. VanSpeybroeck, A. S. Krieger, and G. S. Vaiana, Nature 227, 818–822 (1970))

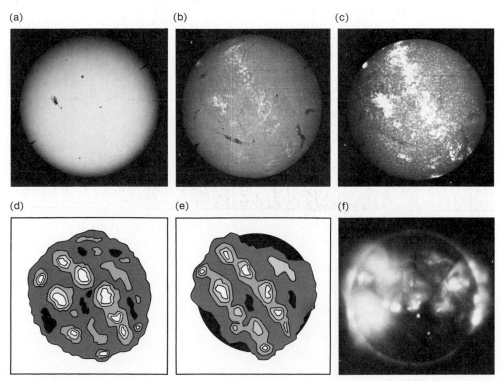

(a) (b) (c)

(d) (e) (f)

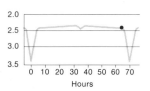

2.0
2.5
3.0
3.5
0 10 20 30 40 50 60 70
Hours

an ultraviolet wavelength of 1032 angstroms, and it was at this wavelength that the information was received. A temperature in the range of 350,000°K is necessary to ionize oxygen to this degree. Such a temperature exists in the upper chromosphere; hence we have a "picture" of the sun at that level. It is not a photograph in the usual sense, but rather a plot showing the hot spots (lighter areas) and cooler regions (dark areas).

Suppose we want to look at yet a higher level of the sun. We choose a wavelength produced by an ionized atom at still a higher temperature, say, 2,000,000°K. Silicon XII (Si XII) is such an atom, emitting an ultraviolet line at 499 angstroms. This line allows us to "see" the sun at a level well into the corona [Figure 8.14 (e)].

Rockets fired above the earth's atmosphere have given us still another view of the sun. X-ray wavelengths have revealed perhaps the most energetic regions of the sun, areas of the corona where very high temperatures prevail [Figure 8.14 (f)]. By using the entire spectrum, astronomers have thus gained a multilevel view of the sun.

MAGNETIC FIELDS OF THE SUN

Solar activities are largely controlled by magnetic fields evident on and near the sun's visible surface. A strong suggestion of such a magnet field, is seen in the spectroheliogram of Figure 8.15. The dark filaments surrounding a pair of sunspots resemble the pattern of iron filings near a magnet.

FIGURE 8.15 A spectroheliogram showing dark filaments in the vicinity of a sunspot group. Strong magnetic fields are suggested in this Big Bear Solar Observatory photo. (Mount Wilson and Las Campanas Observatories, Carnegie Institute of Washington)

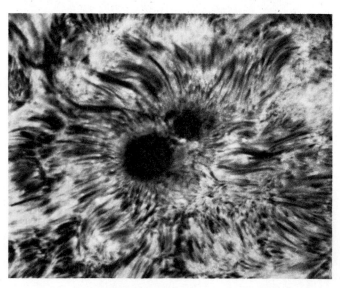

FIGURE 8.16 The Zeeman effect. The splitting of spectral lines results from the presence of a magnetic field on the region of radiation. (Yerkes Observatory)

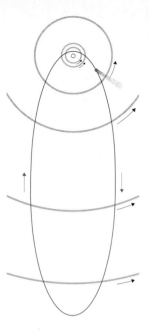

More direct evidence can be obtained, however, by the use of the *Zeeman effect*. When light is created in the presence of a magnetic field, the lines of its spectrum are broadened or split. The degree to which the lines are split indicates the strength of the magnetic field. The direction of optical polarization of the lines indicates the polarity of the field at any point (Figure 8.16).

We can understand the splitting of spectral lines in terms of a magnetic property of the atom itself. When an atom is within a magnetic field, the energy levels of the atom are split (Figure 8.17). Since the electron may exist at any one of the sub-orbitals (energy levels), it may make transitions from 2a or 2b or 2c to energy level 3, thus giving rise to multiple (split) lines.

If the sun is scanned on any given day, a plot of its magnetic properties may appear as in Figure 8.18. The bright areas represent the stronger fields and the slant of the region indicates its polarity. The strength of the magnetic field near sunspots is about a thousand times that of the general field of the sun. In fact, it is very likely that it is the strong magnetic field itself that creates the sunspot—it may impede the flow of energy through the convective envelope.

FIGURE 8.17 In the presence of a magnetic field, sub orbitals (energy levels) exist; hence additional electron transitions are possible. Certain of these transitions account for the multiple lines in Figure 8.16.

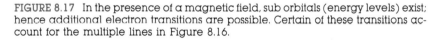

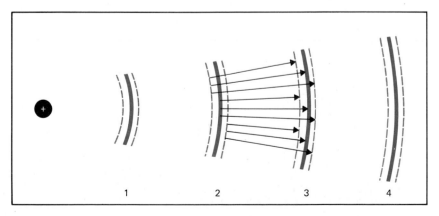

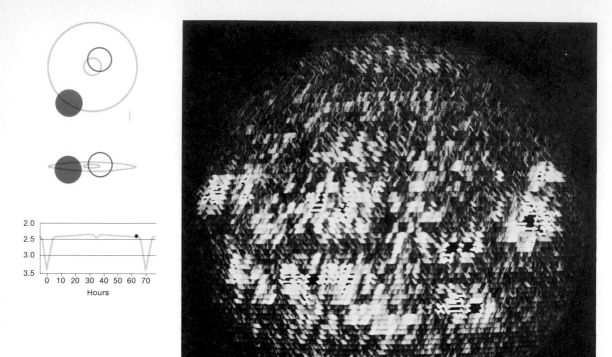

FIGURE 8.18 The magnetic fields of the sun. The brighter areas represent the stronger fields, and the slant of a region indicates the polarity of that region. (Mount Wilson and Las Campanas Observatories, Carnegie Institute of Washington)

SUNSPOTS

A *sunspot* is a region on the visible surface of the sun that consists of gases cooler than those surrounding it. The very dark central portion called the *umbra* has a temperature of approximately 4300°K. The gray *penumbral* region surrounding the umbra has a temperature of approximately 5000°K. These values contrast with a general surface temperature of 6000°K. If one could isolate a sunspot from its surroundings, it would appear bright; a sunspot appears dark only in contrast to its surroundings (Figure 8.19).

In 1842 an amateur astronomer, Heinrich Schwabe, suggested that sunspots increase and decrease in number in a somewhat regular cycle. Professional astronomers soon confirmed that a cycle does exist. Can you determine the period of that cycle as you inspect the record shown in Figure 8.20? Rather than looking at any one cycle, note that seven cycles are shown in a period of approximately 78 years (1876–1954). This yields an average period of 11.1 years. As obvious as this cycle appears to us, we

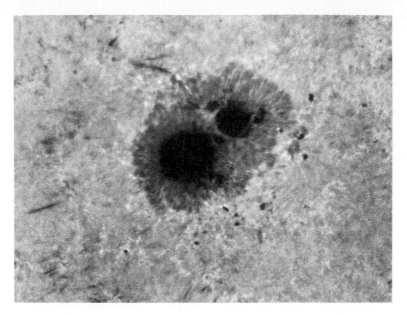

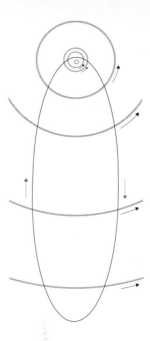

FIGURE 8.19 Details of a sunspot group photographed in white light. The darkest portion is called the umbra and the gray portion is called the penumbra. Sunspots represent cooler regions on the visible surface of the sun. This same sunspot group, photographed in the red light of Hα is seen in Figure 8.15. (Mount Wilson and Las Campanas Observatories, Carnegie Institute of Washington)

FIGURE 8.20 The butterfly pattern of sunspots from 1874 to 1953. The top view shows the plotting of the sunspots according to their latitude on the sun. The middle graph shows the number of sunspots. The lower graph shows the variations in the magnetic fields on the sun's surface. Note here the very close correlation between the sunspot cycle and the magnetic cycle. All three graphs reveal the 11-year cycle of these phenomena. (Royal Greenwich Observatory)

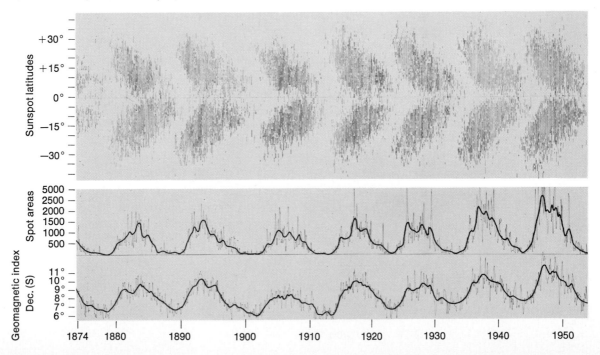

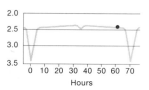

should not conclude that the sunspot phenomenon over a long period is quite as regular as it appears between 1874 and the present date.

MAUNDER MINIMUM

Between the years 1645 and 1715, almost no spots were observed on the sun. Because this seemed unbelievable, observers first concluded simply that no one was recording sunspots during that period. However, after an extensive search for records by W. Walter Maunder of the Royal Greenwich Observatory, in 1893 it was concluded that the spots really were missing during the above period. This conclusion has been more recently substantiated by the work of John A. Eddy of the High Altitude Observatory. You are probably aware that the growth rings of older trees allow the scientist to "read" the history of a local climate over many years. A Dutch scientist, Hessel DeVries, investigated the content of carbon-14, a radioactive form of carbon, within annual tree rings. Carbon-14 is created in the upper atmosphere of the earth by action of cosmic rays that strike the earth's atmosphere from all directions. Its formation correlates with sunspot activity in the following way: when the sun is active (possessing lots of sunspots), its extended magnetic field protects the earth from cosmic-ray bombardment and less carbon-14 is formed. DeVries studies show a sharp rise in carbon-14 content during the period we now identify as the Maunder Minimum, a period of low sunspot activity. During this period the total energy received from the sun was evidently also at a low, for glaciers appear to have grown during this time. The Maunder Minimum seems to correlate with what many observers would term the "Little Ice Age," a time of extreme cold, particularly in Europe and Greenland. This correlation is further confirmed by the study of ice cores taken from the polar caps of Antarctica. These cores can be read much like the rings of trees; such a reading revealed a much reduced rate of deposition of nitrates in the ice during the interval between 1645 and 1715. Deposits of approximately 25 parts per billion are the usual, however during this period deposits dropped below 10 parts per billion.

As many as 12 periods, like that of the Maunder Minimum, are indicated in tree rings. Today there is a tremendous interest in the relationship between solar activity and terrestrial weather, for even a change of 1% in the output energy of the sun could create a change of 1°C in the average temperature of the earth. A downward change could bring on another ice age and an upward change could melt much of the polar caps and inundate the coastal regions—what a delicate balance it is that maintains life! Only very long-term changes are produced by variation in solar activity, not short-term or localized variations. Our own lifetime may be too short to experience such change in any clear-cut fashion.

THE CYCLE OF SUNSPOTS

Let us trace the appearance of sunspots over a period of 11 years. First, consider a "quiet" sun devoid of spots, as it appeared in 1944–45. Next we

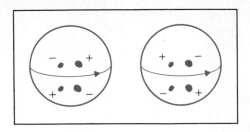

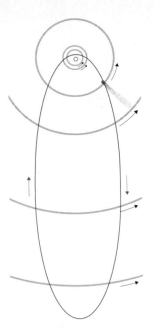

FIGURE 8.21 Successive 11-year cycles of magnetic polarity of sunspots.

see a few spots appearing in the northern and southern hemispheres at about 30° latitude. As the number of spots continues to increase, we observe that they also appear closer to the solar equator. At the end of the 11-year period, the spots diminish in number as they move near the equator. A plot of sunspots over time according to their latitude and number reveals the butterfly pattern shown in the upper band of Figure 8.20. Note the correlation between sunspot activity and magnetic activity as shown in the lower two bands of this illustration.

Sunspots often appear in pairs, one spot possessing a positive and the other a negative polarity. In any given 11-year cycle, the positive spot may lead the negative spot in the northern hemisphere, while the opposite is true in the southern hemisphere. In the next 11-year cycle, the order of polarity is reversed (Figure 8.21). We will try to explain how this may relate to the reversal of the sun's magnetic field.

ROTATION OF THE SUN

If you could provide yourself with a safe method of observing the sun, it would be an interesting experiment to watch sunspots move across the face of the sun. You would witness the rotation of the sun in a little less than one month's time (Figure 8.22). Careful study over several years would reveal that the apparent period is not constant but varies with the latitude of the sunspot. When near the equator, a given sunspot group seems to require only about 25 days for one rotation. At latitudes further north or south of the equator, longer periods are required; at plus or minus 30° and up to 27 days would be required nearer the poles, up to 33 days. This reminds us of the manner in which Jupiter rotates with each latitude having its own rate, and confirms the sun's gaseous nature.

Some observers have speculated that it is the differential nature of the sun's rotation that creates its magnetic field and produces strong fields in the sunspots.

SOURCE OF THE SUN'S MAGNETISM

If the sun were standing still or rotating as a solid, one would expect its magnetic field lines to appear as in Figure 8.23(a), with lines running north and south. Because the equator of the sun turns faster than higher latitudes, which is called differential rotation, the magnetic lines tend to "wind-up" as shown in Figure 8.23(b through e), thereby strengthening

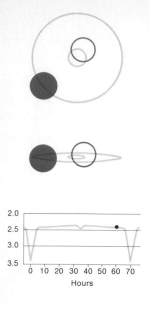

2.0
2.5
3.0
3.5

0 10 20 30 40 50 60 70
Hours

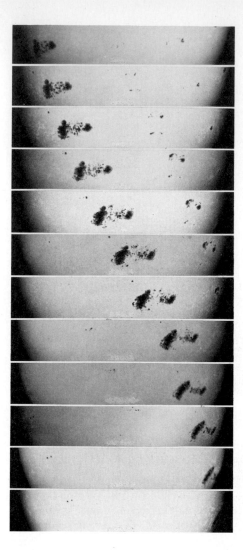

FIGURE 8.22 The sunspots show the rotation of the sun, photographed daily from April 2 to 13, 1947. North is at right. (Mount Wilson and Las Campanas Observatories, Carnegie Institute of Washington)

the field. Experimentation with plasmas in the laboratory seems to confirm this possibility. Adjacent lines are thought to exert a buoyant force on each other, causing them to burst through the surface. At this eruption point, a sunspot pair may occur, one spot exhibiting a positive polarity and the other a negative polarity. While this theory of the magnetism of the sun represents some degree of speculation, note how it correlates with other solar activities discussed below.

PLAGES

When the sun is photographed in the light of hydrogen-alpha or calcium II, bright regions called *Plages* often appear [see Figure 8.24 (b,c)]. The brightness of the region results from its higher density and temperature compared to the surrounding chromosphere. Plages are thought to be due to the focusing of energy into a given region of strong magnetic fields, often above a cooler sunspot. Note the bright region in Figure 8.24(b), right center, and the corresponding sunspot in the same region of Figure 8.24(a). The bright plage is yet more evident in the light of calcium II [Figure 8.24(c)]. There appears to be a direct connection between plages and sunspots. Bright plages may appear before their corresponding sunspot and therefore herald its appearance. The plage is often much larger and remains after its corresponding sunspot has disappeared.

SPICULES

In the upper chromosphere of the sun, spikelike columns of gas jet outward into the corona at speeds of up to 32 km/sec (Figure 8.25). These *spicules* may reach heights of 16,000 kilometers and any given one may last for 10 to 15 minutes.

PROMINENCES

For many years, activities of the solar atmosphere were observed only during a total solar eclipse when the moon covers the entire photosphere of the sun. Gigantic protrusions were observed at that time. Now, by using the spectroheliograph, these prominences may be studied at any time. Several types of prominences may be distinguished. One type, known as

FIGURE 8.23 Differential rotation of the sun may produce the magnetic fields associated with sunspots. This sequence shows how the faster rotation of the sun's equator may take north-south lines of magnetism and develop east-west lines and finally swell upward to form a loop over a sunspot.

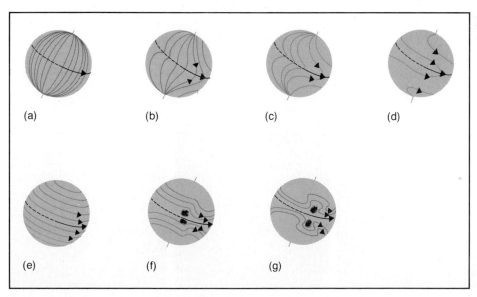

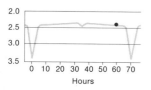

2.0
2.5
3.0
3.5

0 10 20 30 40 50 60 70
Hours

the quiescent type, may extend more than 320,000 kilometers into the solar atmosphere and remain relatively fixed. A second type, called a *loop prominence*, takes on the shape of an arch or loop and exhibits motion within the loop, suggesting the presence of a magnetic field (Figure 8.26). A third type of prominence is the *eruptive* type, which may send material outward for thousands of kilometers into the corona of the sun at speeds of up to 640 km/sec. The spectacular view shown in Figure 8.27 was taken from the solar observatory which stands near the rim of the Haleakala crater on the island of Maui, a particularly fine location for observing the sun.

The dark filaments, as seen in hydrogen-alpha light, represent prominences viewed from above (Figure 8.28). They tend to form along the boundary between the regions of opposite polarity associated with sunspot pairs.

THE SOLAR FLARE

The *solar flare* is the most dynamic activity associated with the surface or atmosphere of the sun. During a flare, a tremendous release of energy occurs in a very short time (Figure 8.29). The sudden brightening, usually in the vicinity of a sunspot group, accompanies a violent outrushing of material. Typically a flare occurs over an area with a diameter of 200,000 kilometers; the temperature associated with it may exceed 100 million degrees Kelvin. Flares often occur along the boundary between positively and negatively polarized regions of sunspot groups. Picture the concentration of energy in the tip of a bull whip when it is snapped—a flare is somewhat similar in that a great deal of magnetic energy is released in a very small area.

Solar flares have a direct influence on the earth and their X-ray emissions would kill humans if the earth did not have a protective atmosphere. Because the moon has no atmosphere to provide protection for astronauts walking there, a constant vigil was kept for solar flares. Had one occurred

FIGURE 8.24 The sun: (a) in ordinary light; (b) in Hα light; (c) in light of Ca II. (Mount Wilson and Las Campanas Observatories, Carnegie Institute of Washington)

(a)　　　　　　(b)　　　　　　(c)

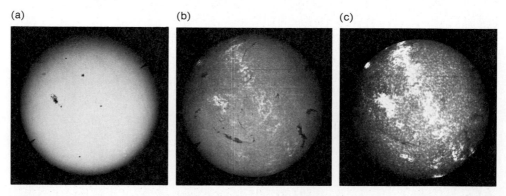

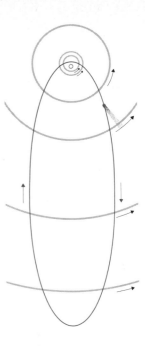

FIGURE 8.25 Spicules in the sun's chromosphere photographed in red light of Hα at Big Bear Solar Observatory. (Mount Wilson and Las Campanas Observatories, Carnegie Institute of Washington)

FIGURE 8.26 A solar prominence that reached a height of 330,000 km, photographed in the light of calcium (K line). (Mount Wilson and Las Campanas Observatories, Carnegie Institute of Washington)

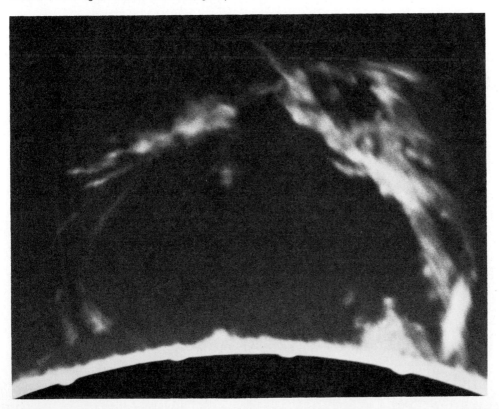

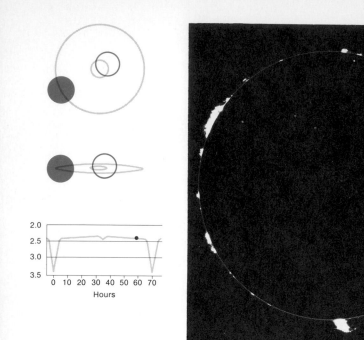

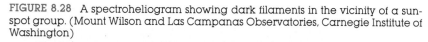

FIGURE 8.27 Prominences visible on March 1, 1969. The very large eruptive prominence extends approximately 645,000 km above the surface of the sun. (Haleakala Observatory, Institute for Astronomy, University of Hawaii)

FIGURE 8.28 A spectroheliogram showing dark filaments in the vicinity of a sunspot group. (Mount Wilson and Las Campanas Observatories, Carnegie Institute of Washington)

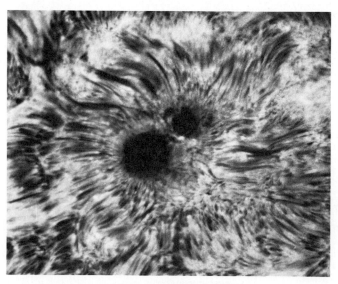

during their walk on the moon, they would have been immediately instructed to enter the spacecraft for protection. The ionospheric layers of the earth's atmosphere are altered by the radiation from a flare, which sometimes makes them fail to reflect radio signals, thereby causing a communication blackout on shortwave radio bands. Such a blackout may last for a few hours or up to several days. Auroral displays (the northern and southern lights) are usually greatly enhanced by the acceleration of particles in the solar wind following a flare activity. About a day after a flare occurs, the magnetic properties of the earth are usually disrupted, causing the needle of a compass to react in very strange ways.

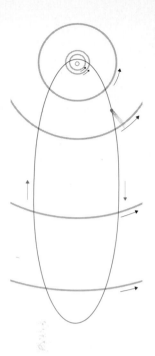

OUR NEAREST STAR

The sun is our nearest star and we have been able to examine it quite closely. Figure 8.30 is designed to remind one of the many features we have discussed. As we move out into space to examine the other stars of our Galaxy, we ask the same questions about them as we have asked about the sun: How far are they? How big? How massive? Do they rotate? What kind of atmosphere do they have? Do they have spots? Do they have magnetic fields? What is their temperature? What kind of energy do they radiate?

FIGURE 8.29 Solar flare. This sequence of spectroheliograms was taken over a period of 1 hr in the light of (a) Hα (Hydrogen-alpha); (b) Hα plus 0.6 Å; (c) Hα plus 0.9 Å. Each sequence shows the development of the flare at a different level and also its association with a sunspot. (Aerospace Corporation; San Fernando Observatory)

(a)

(b)

(c)

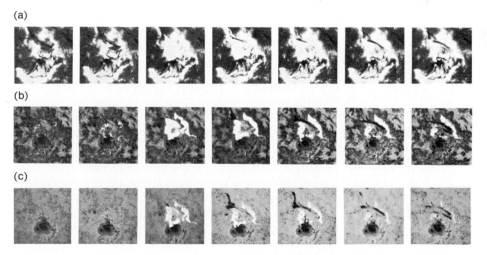

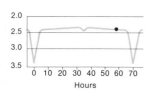

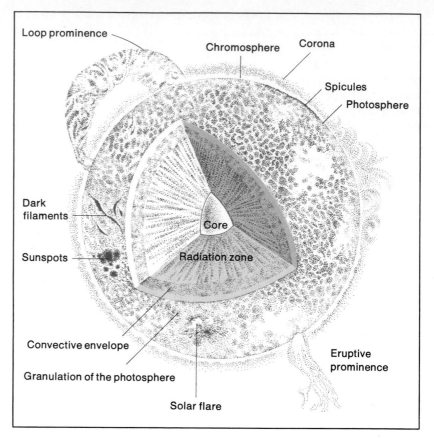

FIGURE 8.30 The active sun.

QUESTIONS

1. What basic characteristics identify the sun as a star rather than a planet?
2. What is the apparent diameter of the sun, as measured in degrees? How does this compare to the apparent diameter of the full moon?
3. List several means by which the rotation of the sun is measured?
4. Does the sun rotate as if it were a solid? Explain.
5. What does the granulation of the photosphere seem to suggest?
6. State the minimum temperature necessary to sustain the process whereby hydrogen is converted to helium in the core of the sun.
7. If hydrogen is considered to be the "fuel" of the thermonuclear process in the core of the sun, then what occurrence in the process explains the tremendous release of energy that results?
8. How much material (mass) of the sun is converted to energy every second?
9. At what level (or layer) of the sun is the dark-line spectrum of the sun produced?

10. List the things that an astronomer might deduce from the solar spectrum.

11. Of what does the solar wind consist?

12. The temperature of the corona of the sun has been measured by its ionization level and found to be _____ °K.

13. Define the solar constant.

14. List the kinds of electromagnetic radiation that the sun produces.

15. The number and location of sunspots seems to repeat in a cycle of how many years?

16. When the spectral lines of the sun appear to split, what condition is indicated?

17. When the sun is photographed in the light of hydrogen or of calcium, levels of the sun may be seen other than the photosphere. Explain what is seen in each of these lights.

18. The prominences of the sun are not seen in ordinary (white) light. How may they be viewed?

19. What is the most dynamic activity associated with the sun?

20. How do solar activities affect the earth?

21. The normal spectrum of the sun is an absorption spectrum (dark lines). Why does the sun show an emission spectrum (bright lines) during a total solar eclipse?

22. Why is the chromosphere red?

23. The sun is now being photographed in the light of oxygen VI (ultraviolet) and in X rays. What do these photographs reveal by contrast with those taken in white light or in hydrogen-alpha light?

24. What condition within a star or its atmosphere is necessary to ionize oxygen five times?

25. Assuming that the sun converts 5 million metric tons of mass into energy every second, find how much mass it has converted in the 5 billion years of its lifetime.

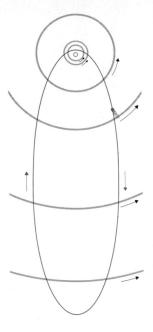

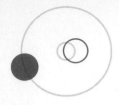

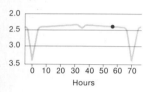

Hours

Stars
in General

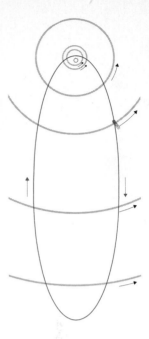

Nine

In order to determine factors such as the direction, distance, velocity, size, temperature, and luminosity of stars, astronomers must rely entirely upon their ability to interpret the radiation that they receive from them. Let us first consider the method used to determine distances to nearby stars.

When an inaccessible area on the earth must be measured, the surveyor uses a method of triangulation. For instance, suppose that we wish to know the distance from dock A on the bank of a river to dock C on an island in the middle of the river as in Figure 9.1. By laying off a base line AB along the bank of the river, say 100 m long, and then measuring the angles formed at A and B, one can solve for side AC, the desired distance, using trigonometry.

To find the distance to an object that is very far away, the object is viewed from two different locations (the ends of our base line) and the angle through which the object appears to shift as we move from one end of the base line to the other is measured. Such an angle is called the *parallax* of the star. To see how this works, try a simple experiment. Hold a pencil at arms length and view it first with one eye and then the other (Figure 9.2). Note how it appears to shift back and forth against whatever background you are viewing. The base line in your experiment is the distance between your eyes. In the first example that follows, the base

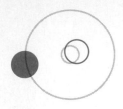

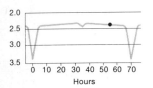

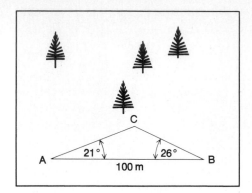

FIGURE 9.1 Triangulation.

line is the radius of the earth and the pencil is the moon. In the second example, the base line is the distance from sun to earth and the pencil is a nearby star.

To determine the distance from the earth to the moon, let us suppose that two observers are stationed at positions A and B on the earth (Figure 9.3) and that they are in radio communication with each other. The solution that follows is based on the fact that the two lines pointing to star D (see Figures 9.3 and 9.4), even if separated by 93 million miles, may be treated as parallel because the star itself is so far away; there is no measurable deviation from parallelism. Observer B radios that he sees the moon in alignment with star C. Observer A can measure the angle between stars C and D, and since the rays of light from star D may be considered parallel

FIGURE 9.2 Hold a pencil at arm's length and first close one eye and then the other. The pencil appears to shift against the background objects, due to the different position of your eyes—illustrating the concept of parallax.

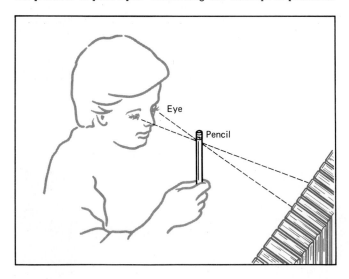

for both observers, we see that the angle between the stars is equal to the angle formed at the moon. This fact is expressed by the geometric theorem: *Alternate interior angles of parallel lines are equal.*

Since the base line AE, the radius of the earth, is a known distance, it is possible to calculate the distance from A or B to the moon. This method, called *geocentric parallax*, is of only limited usefulness because of the relatively short base line. However, the usefulness of the idea may be extended by using a much longer base line, such as the radius of the earth's orbit (see Figure 9.4). Suppose that when the earth is in position A, the nearby star S aligns with the more distance star D, but three months later, when the earth is at position B, star S aligns with star C. By measuring the angle between stars C and D, the astronomer then knows the angle formed at the star S. This angle (p), called the *heliocentric parallax* of the star, is always less than 1 second ($\frac{1}{3600}$ degree) in size. Triangles that have such a small angle can be solved by the following method. Consider a very large circle with center at star S and with radius equal to the distance r (Figure 9.5). The base line BF is a small part of the large circle and compares to the entire circumference of the circle as the angle of parallax (p) compares to the entire circle of 360°, giving:

$$\frac{BF}{2\pi r} = \frac{p°}{360°}$$

Solving for r, we get

$$r = \frac{360°(BF)}{2\pi p°}$$

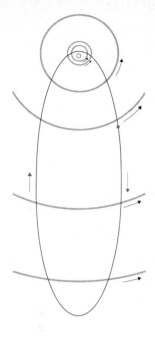

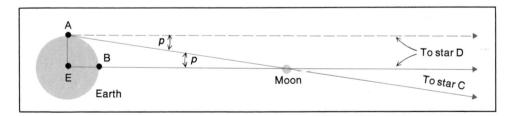

FIGURE 9.3   Geocentric parallax.

FIGURE 9.4   Heliocentric parallax.

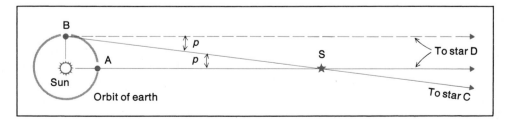

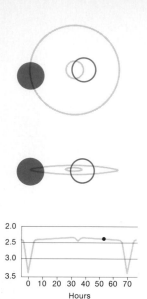

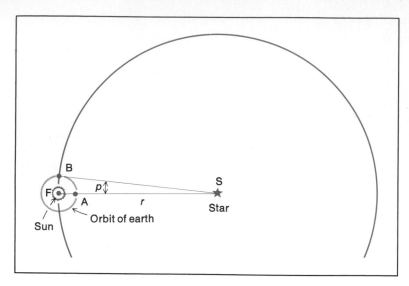

FIGURE 9.5   Properties of small angles.

Changing units so p is in seconds and clearing the factor 2 yields:

$$r = \frac{206{,}265(BF)}{p''}$$

Since BF is equal to 1 A.U., we can now find r in astronomical units:

$$r = \frac{206{,}265 \text{ A.U.}}{p''}$$

If we let 206,265 A.U. equal 1 parsec, we get a much simpler expression:

$$r = \frac{1}{p''}$$

A *parsec* is simply the distance at which a star must be situated in order to exhibit 1 second of heliocentric parallax, using 1 A.U. as a base line. The term "parsec" is a contraction of the two words *parallax* and *second*. Further calculations show that one parsec is equal to 3.26 light-years. Notice that the greater the distance r to the star, the smaller will be its parallax angle p. Our nearest star, Alpha Centauri, has a parallax of 0.75 second, hence a distance of $1/0.75 = 1.33$ parsecs (pc), or 4.3 light-years. For stars as distant as 100 pc (parsecs), the parallax angle is only $0.01''$ ($\frac{1}{360{,}000}°$). The measurement of an angle this small is generally fraught with errors which are large enough to nullify the effectiveness of the method beyond approximately 50 pc. The parallax of over 10,000 stars has been measured within this distance. In a subsequent section of this chapter, Distance of Moving Clusters, we will study a more accurate method. First, let us examine several concepts regarding the motion of stars.

MOTION OF STARS

Stars show very little evidence of motion, and yet most are traveling at phenomenal speeds. These motions are not readily apparent because the stars are so far from us. If we could watch the stars in the Big Dipper over a period of 100,000 years, however, we would realize that they are moving in various directions. Figure 9.6 shows the approximate location of the stars 100,000 years in the future; the direction of motion of the stars is indicated by small arrows. (The left-hand flip pages beginning on page 284 reveal the motion of the stars in the Big Dipper.)

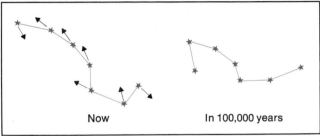

FIGURE 9.6   Motions of stars in the Big Dipper.

PROPER MOTION

The *proper motion* of a star, the rate at which its position in the sky changes, is measured in terms of seconds of arc per year. Many stars exhibit such small changes that they must be photographed over many years in order to measure that change. A star that shows a shift of 2 seconds of arc in a period of 10 years has a proper motion of 2 seconds divided by 10 years, or 0.2 seconds of arc per year. Barnard's Star has the largest observed proper motion, equal to 10.25 seconds of arc per year (Figure 9.7). This star shows relatively large proper motion because it is the second closest star to the sun. More than 300 stars have proper motions of at least 1.0 seconds of arc per year, but the average star seen with the naked eye has a proper motion of less than that. If the distance to a star is known, its proper motion can be translated into a velocity that is at right angles to the observer's line of sight, called the *tangential velocity* of the star. Suppose that star A has a proper motion of q seconds of arc per year and is located at a distance of r parsecs. Its tangential velocity (v) may be computed as follows. Visualize a circle of radius (r), with the observer at the center of the circle and star A on the circumference (Figure 9.8).

The length of the velocity vector (v) compares to the entire circumference of the circle as the angle of proper motion (q) compares to the

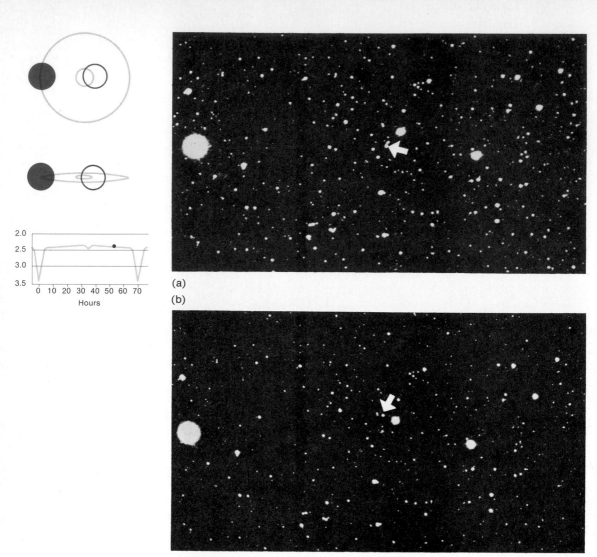

FIGURE 9.7 Barnard's Star, showing its change in position (proper motion) over a period of 22 years: (a) August 24, 1894; (b) May 30, 1916. (Yerkes Observatory)

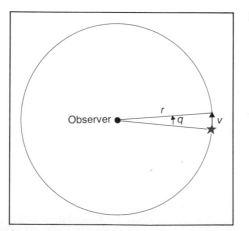

FIGURE 9.8 Tangential velocity (v).

entire circle of angles (360°), measured in seconds of arc (360° = 360 × 60 × 60 seconds).

$$\frac{v}{2\pi r} = \frac{q}{360 \times 60 \times 60}$$

Solving for v:

$$v = \frac{2\pi rq}{360 \times 60 \times 60}$$

Converting the units to kilometers per second (steps not shown):

$$v = 4.74 \; rq$$

For example, the tangential velocity of Barnard's Star, when $q = 10.25$ sec/yr and $r = 1.83$ pc, is:

$$v = 4.74 \, (10.25) \, (1.83)$$
$$= 88.91 \text{ km/sec } (320,076 \text{ km/hr})$$

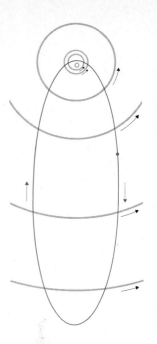

SPACE VELOCITY

Suppose that star A is traveling in a direction AB, as shown in Figure 9.9. Its tangential velocity is shown by AC and its radial velocity by AD. Recall from our discussion of the Doppler effect (page 91) that the radial velocity may be computed directly from the shift in a star's spectral lines, since this shift is due to its motion. We may then define the *space velocity* of star A by

$$(AB)^2 = (AC)^2 + (AD)^2$$

a relationship that is true for any right triangle. For example, if the tangential velocity of a given star is 60 km/sec and its radial velocity is 80 km/sec, then its space velocity is

$$(AB)^2 = (60)^2 + (80)^2$$
$$(AB)^2 = 3600 + 6400$$
$$(AB)^2 = 10,000$$
$$AB = 100 \text{ km/sec}$$

The space velocity of a star represents its motion with respect to the sun, not with respect to any absolute standard. The sun is moving at the rate of approximately 240 km/sec clockwise around the center of the Milky Way galaxy, and the Galaxy is moving with respect to all other galaxies

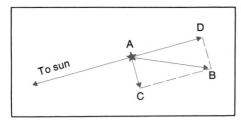

FIGURE 9.9 The motion of a star relative to the sun.

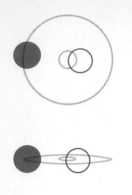

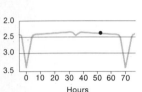

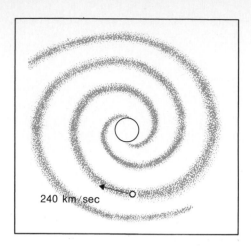

FIGURE 9.10 The sun's motion within the Milky Way galaxy.

(Figure 9.10). How then do we find an absolute standard? It seems that we must be content to express the motion of any object as it relates to some stated frame of reference.

PECULIAR VELOCITY

If we were to select a group of stars, say those within 100 parsecs of the sun, we could average their velocities and thus arrive at a velocity (speed and direction) that described the motion of this entire group of stars. To illustrate this idea, imagine yourself driving on a freeway. Some cars would pass you from time to time and likewise you would pass some cars. Some might also be changing lanes at a given moment and so their direction would be different from most of the other cars. Nevertheless, it would be possible to describe the general motion of the entire group of cars within one mile of yourself by directing a helicopter to fly overhead, and by telling the pilot to adjust his speed and direction in order to fly along with your group of cars. His speed and direction might then be used to describe the speed and direction of the entire group in general. In a similar manner, the local group of stars—those within 100 pc of the sun—has an average speed and direction that we call the *local standard of rest*. A star's velocity with respect to the local standard of rest is referred to as its *peculiar velocity*. Any star that is moving with the same speed and direction as the local group has a peculiar velocity of zero. The sun shows a peculiar velocity of approximately 20 km/sec toward the bright star Vega.

We are now prepared to examine one of the most fundamental methods of determining the distance to stars.

DISTANCE TO MOVING CLUSTERS

Clusters of stars are gravitationally bound to each other, therefore they move through space together. For this reason, we may expect all members of a given cluster to show the same peculiar velocity, and we can identify

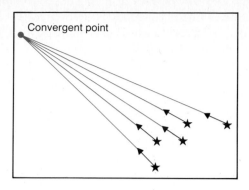

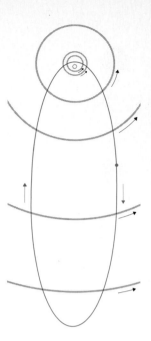

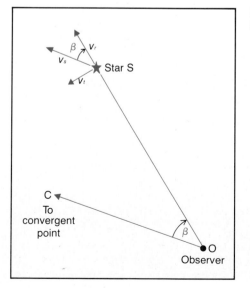

FIGURE 9.11 Convergent point of a moving cluster.

members by this fact. Consider, for instance, the Hyades cluster in Taurus. This cluster, which is moving away from us, is spread out over a sufficient angle in the sky so that the proper motions of the individual stars appear to converge toward a single point—as railroad tracks seem to converge in the distance (Figure 9.11).

Because the convergent point of such a cluster can be observed, it follows that the angle between a given member of the cluster and that convergent point can also be measured. Note in Figure 9.12 that the line OC is parallel to the space velocity (v_s), of the star; therefore the angle (β) between the convergent point (C) and the star (S) is equal to the angle between the space velocity (v_s) and the radial velocity (v_r) of the star.

Recalling again that the radial velocity of a star may be measured directly by observing the Doppler effect in the star's spectrum, it is possible to solve for the tangential velocity (v_t) using angle β as follows:

$$v_t = v_r \tan \beta$$

FIGURE 9.12 The moving-cluster method.

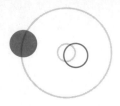

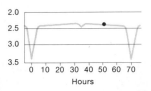

2.0
2.5
3.0
3.5

0 10 20 30 40 50 60 70
Hours

where tan β may be determined by reference to ordinary tables of trigonometric functions. Recall that we derived a relationship between tangential velocity (v_t), proper motion (q), and distance (r):

$$v_t = 4.74\ rq$$

Solving for r:

$$r = \frac{v_t}{4.74\ q}$$

Because we have calculated v_t above and have observed the star's proper motion, we can now compute the distance to the star. In the case of the Hyades cluster, this distance is approximately 46 pc. This method of determining distances involves smaller observational errors than that of trigonometric parallax. It has thus become the fundamental method by which the H-R diagram (discussed later in this chapter) is calibrated, and it serves as the foundation upon which all other distance determination rests.

FIGURE 9.13 The apparent magnitude of a star may be judged from the size of the image it makes on a photographic plate. (Palomar Observatory Photograph)

BRIGHTNESS OF STARS

In the second century B.C., Hipparchus grouped stars, according to their brightness as seen with the naked eye, into six magnitude classes; the brightest stars were classified as first magnitude and the faintest as sixth magnitude. He produced the first real catalogue of stars, showing both position and brightness. Of course, he was forced to rely upon his own ability to judge brightness without the aid of any instrument. Modern methods for measuring stellar brightness—techniques of *photometry*—are much more precise.

Photography has played a very important role in this area of astronomy. The apparent magnitude of a star may be judged by the size of the image it makes on a photographic plate, the brighter the star, the larger the image on the plate (Figure 9.13). A photographic plate is different from the human eye in that an image will continue to build up, or darken, as the exposure time is increased (Figure 9.14); the image formed in our eye by contrast fades and must be continually reinforced, so there is no build-up.

The photoelectric photometer provides a still more accurate method

FIGURE 9.14 Region of Orion Nebula, showing effect of aperture and time exposure on star images. (a) Small aperture and/or short time exposure. (b) Large aperture and/or long time exposure. (Lick Observatory)

(a)   (b)

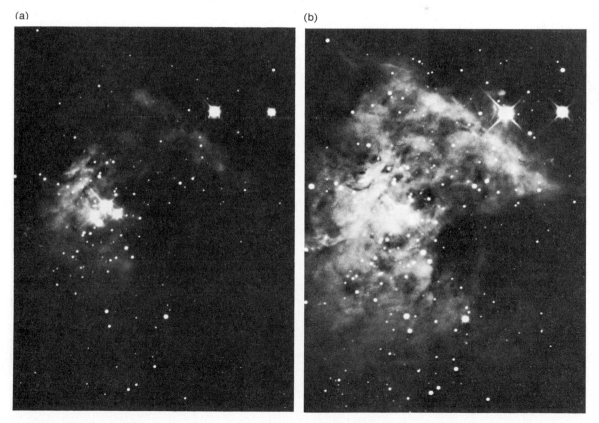

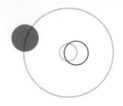

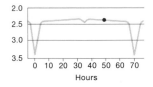

2.0
2.5
3.0
3.5

0 10 20 30 40 50 60 70

Hours

of brightness determination. You are probably familiar with exposure meters used in photography (many are now built into the camera). When light falls on the sensitive surface of the meter, a small electrical current is produced. This current then causes a needle to move, indicating the intensity of the light. With a similar but much more sensitive device, the light of a star is collected in the telescope and directed through a *photomultiplier tube* (Figure 9.15). The small current that the light produces in the photomultiplier is amplified to the point where it may be recorded. This method provides accuracy in the measurement of star magnitudes to the nearest one hundredth magnitude.

THE SCALE OF BRIGHTNESS

In an effort to preserve the familiar scale of Hipparchus, the modern astronomer has established the following classification scheme. The typical first magnitude star is 100 times as bright as the typical sixth magnitude star, so we seek a multiplying factor for each of the five steps between first and sixth magnitudes which will produce a 100-fold increase in brightness.

FIGURE 9.15 A photo multiplier tube showing (a) a cut-away view and (b) a functional diagram. The complex physical and electronic nature of this tube makes it sensitive to very small fluctuations in light intensity. (EMR Division of Weston Instruments, Inc.)

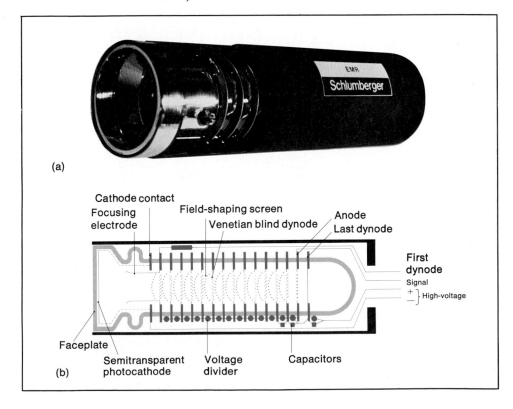

Letting n be that factor, we write:

$$n^5 = 100$$

$$n = \text{the fifth root of } 100 = 2.512 = \text{approx. } 2.5$$

This means that each step in the magnitude scale is 2.5 times as bright as the previous step, five such steps will produce the 100-fold increase:

$$(2.5)\ (2.5)\ (2.5)\ (2.5)\ (2.5) = 100 \text{ (approximately)}$$

Table 9.1 lists light ratios that correspond to given changes in magnitude. For example, a star that is three magnitudes brighter than another is 15.9 times as bright [(2.5) (2.5) (2.5) = 15.9].

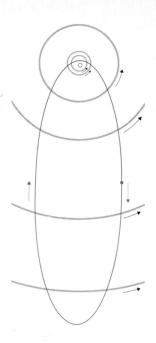

TABLE 9.1

LIGHT RATIOS FOR VARIOUS MAGNITUDE CHANGES	
Magnitude Change	Light Ratio
0.5	1.6/1
1.0	2.5/1
1.5	4/1
2.0	6.3/1
3.0	15.9/1
4.0	39.8/1
5.0	100.0/1
6.0	251.2/1
10.0	10,000.0/1
20.0	100,000,000.0/1
25.0	10,000,000,000.0/1

The system of magnitudes used today reflects two basic changes that have taken place since the time of Hipparchus: (1) sensitive detectors have been developed, and (2) a scale of brightness has been defined mathematically. Modern astronomers recognize that some stars which Hipparchus would have called first magnitude are significantly brighter than others that he placed in the same category. For instance, the bright star Sirius is approximately nine times brighter than Aldebaran. The scale of brightness has therefore been extended beyond 1, upward to 0, −1, −2, and so on. The apparent magnitude of Sirius is −1.4, whereas Aldebaran is +0.86. Some objects, such as certain planets, the moon, and sun, are still brighter, hence the extension of the magnitude scale to include these, as shown in Table 9.2. This table also shows an extension in the other direction to include dimmer objects that were unknown to Hipparchus because he lacked a telescope. Even an average pair of binoculars will reveal stars of magnitudes +7, +8, and +9. The larger telescopes with very sensitive

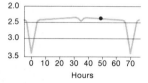

2.0
2.5
3.0
3.5

0 10 20 30 40 50 60 70
Hours

detectors permit the astronomer to record objects whose magnitudes are as low as +26.

TABLE 9.2

APPARENT MAGNITUDES OF FAMILIAR OBJECTS	
Object	Magnitude
Sun	−26.5
Moon (full)	−12.5
Venus (when brightest)	−4.4
Jupiter	−2.5
Sirius	−1.4
Rigel	+0.0
Aldebaran	+0.9
Polaris	+2.0
Naked-eye limit	+6.0
Binocular limit (average)	+9.0
15-cm telescope limit	+13.0
5-m telescope limit	
visual	+20.0
photographic	+23.5
electronic (C.C.D.'s)	+26.0

A magnitude change not found in Table 9.1, such as 17, may be calculated as follows. Since a change of 5 magnitudes corresponds to a factor of 100 in light ratio, and a change of 2 magnitudes corresponds to a factor of 6.3, then a total of 17 magnitude steps corresponds to a light ratio of 6,300,000/1, as shown below:

Magnitude change: $17 = 5 + 5 + 5 + 2$
Light ratio: $(100) \times (100) \times (100) \times (6.3) = 6,300,000$
$$= 6.3 \times 10^6$$

Note that whereas the magnitude changes are added, the corresponding light ratios are multiplied.

ABSOLUTE MAGNITUDE

Because stars are situated at widely different distances from us, it would be a serious error to assume that apparent magnitudes indicate their actual luminosity. We have all experienced the fact that as we move away from a source of light, its brightness seems to diminish. If we could accurately measure its brightness, we would find that when we were twice as far from the lamp, its brightness would be one-fourth what it was from the starting position. This can be understood by imagining a screen at distance d. (Figure 9.16) If this screen were moved to a distance of $2\,d$ a screen four times the size of the first would be needed to intersect all the energy directed at it; the intensity of that energy at twice the distance is thus one-fourth of the energy at distance d. We express this by saying that light falls off with the square of the distance and that distance is a critical element in determining the apparent brightness of a star, as compared to its actual brightness. In order that all stars may be compared on an equal basis, let us first imagine that all of them are at an arbitrary but definite distance, say 10 parsecs. Then let us determine their magnitudes, as seen from that distance. We define the apparent magnitude of a star, if placed at 10 pc from the sun, as its *absolute magnitude*. If our sun were placed at a distance of 10 pc, it would be very dim, just visible to the naked eye. It would have an absolute magnitude of +5. Rigel, on the other hand, if moved to 10 pc would brighten from its apparent magnitude of zero to a −7 magnitude. At a distance of 10 pc, Rigel would appear 630 times as bright as compared to the apparent brightness at its present distance.

Apparent magnitude, absolute magnitude, and distance for a given star are related by the following proportion:

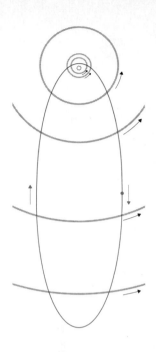

$$\frac{L(10)}{L(r)} = \left(\frac{r}{10}\right)^2$$

FIGURE 9.16 In traveling twice the distance, the light covers four times as much area. Therefore we would expect its intensity to be only one-fourth.

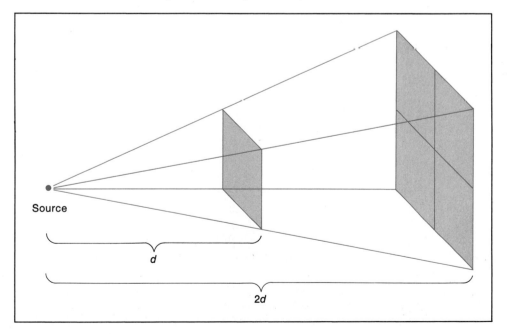

Source

d

$2d$

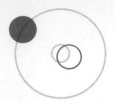

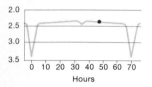

where $L(10)/L(r)$ is the light ratio of a star moved to 10 pc, compared to that star at its true distance of r pc. Letting $r = 1000$ in the equation above, we obtain the following light ratio:

$$\frac{L(10)}{L(1000)} = \left(\frac{1000}{10}\right)^2 = \left(\frac{100}{1}\right)^2 = \frac{10,000}{1}$$

The given star would thus appear 10,000 times brighter when moved to 10 pc. From Table 9.1 we see that this particular light ratio corresponds to a change of 10 magnitudes; thus if the apparent magnitude of the star was +7, its absolute magnitude will be −3 (7−10 = −3). (The absolute magnitude of many stars are shown in Appendix 8).

By similar calculations we may determine that a star situated at a distance of 16 pc will appear 1 magnitude dimmer than its absolute magnitude, that a star located at 25 pc will appear 2 magnitudes dimmer than its absolute magnitude, and so on. This relationship is shown in Table 9.3.

TABLE 9.3

DIFFERENCE IN APPARENT AND ABSOLUTE MAGNITUDES AT VARIOUS DISTANCES						
Distance (parsecs)	16	25	40	63	100	160
Difference between absolute and apparent magnitudes	1	2	3	4	5	6

TEMPERATURE OF STARS

Stars radiate energy in a broad range of wavelengths. The total energy emitted from a star in all wavelengths is called the star's *bolometric* magnitude. The astronomer, however, is usually only able to measure the magnitude of a star within a limited range of wavelengths. If a yellow filter is used in front of a photometer, what is recorded is called a *visual magnitude* (m_V). This magnitude most closely resembles the magnitude as seen by the human eye since the human eye is most sensitive to yellow light. If a blue filter is used, a different magnitude (m_B) will be recorded. A hot (blue-white) star will appear brighter when measured through the blue filter; a star that tends toward the cooler (reddish) color will appear brighter when measured through the yellow filter. For any given star, the difference between (m_B) and (m_V) is called the *color index* of the star:

Color index = $m_B - m_V$ (or simply spoken as $B-V$)

The color of a star may be determined by this difference, as shown in Table 9.4.

TABLE 9.4

THE COLOR INDEX OF A STAR[a]		
Color Index	Color	Surface Temperature
−0.6	Blue	50,000°K
−0.3	Blue-white	20,000°K
0.0	White	10,000°K
+0.2	Yellow-white	7,000°K
+0.6	Yellow	6,000°K
+0.8	Orange	5,000°K
+1.5	Red	3,500°K

[a]In reality, stars listed as yellow appear only slightly yellowish, stars listed as orange appear only slightly orangish and stars listed as red only slightly reddish, all deviating from white by very little.

We see that the color of a star is mainly a function of its surface temperature; stars range from the very hot (blue-white) with surface temperatures of approximately 50,000°K, to the very cool (reddish) with surface temperatures of approximately 3000°K. Invisible stars exist that have surface temperatures as low as 1000°K and radiate only in radio and infrared wavelengths.

As the light from a distant star travels through space, it may encoun-

FIGURE 9.17 Examples of different spectral types. (Mount Wilson and Las Campanas Observatories, Carnegie Institute of Washington)

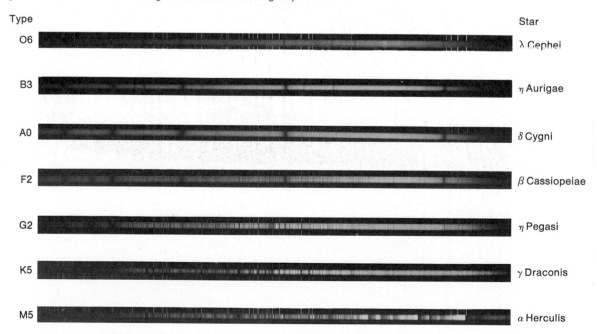

Type		Star
O6		λ Cephei
B3		η Aurigae
A0		δ Cygni
F2		β Cassiopeiae
G2		η Pegasi
K5		γ Draconis
M5		α Herculis

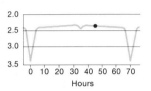

ter dust clouds that tend to redden the light, changing the apparent color of the star. Since the color index indicates only apparent colors, it is not a very reliable means of assessing a star's true color and its true surface temperature. The spectrum of a star gives us a better indication, since some spectral lines show up very clearly at certain temperatures and not at others (Figure 9.17). Stars were originally classified according to the strength of the hydrogen lines in their spectra, with A-type stars exhibiting the strongest hydrogen lines, B-type the next strongest, and so on to O-type. These same classes were also used to group stars according to surface temperature. The order of the classes based on hydrogen lines is not the same as the order of classes based on temperature, however. O-type stars are the hottest, B-type the second hottest, followed by A, F, G, K, and M. This order may be remembered rather easily by the following mnemonic device. Setting a word to each letter class, we get: "Oh Be A Fine Girl, Kiss Me." Each spectral class is further subdivided into ten subdivisions as follows: B0, B1, B2, B3, B4, B5, B6, B7, B8, B9, A0, A1, A2, and so on. How may these classes be recognized based on their spectra? At the extremely high temperature typical of the O-type star, the energetic collisions that occur among the atoms often knock electrons entirely free producing ions with their own characteristic spectra. At the other extreme of low temperatures, atoms may unite to form molecules such as titanium oxide that would be evident in the spectrum of the star. Table 9.5 gives examples of stars in each class, together with their typical spectra.

TABLE 9.5

STELLAR SPECTRA			
Class	Example	Spectra	Surface Temperature
O5		Ionized helium, nitrogen, oxygen	50,000°K
B3	Achernar	H, He strong	15,000°K
A1	Sirius	H lines at max. Ca lines weak	11,000°K
A3	Fomalhaut	H lines strong Ca lines stronger Metals weak	9,000°K
F0	Canopus	H lines weak Ca lines strong	7,600°K
F5	Procyon	Ca lines very strong Neutral metals	6,600°K
G0	Capella	Ca lines at max.	6,000°K
G2	Sun	Iron lines strong	
K2	Arcturus	H lines weak Molecular bands	5,000°K
M2	Betelgeuse	Neutral lines strong TiO molecules present	3,500°K

HERTZSPRUNG-RUSSELL DIAGRAM

Early in the twentieth century a very important study concerning the relationship between the surface temperature of a star and its luminosity was made by two astronomers, Ejner Hertzsprung of Denmark and Henry N. Russell of the United States. The results of their work are shown graphically in Figure 9.18; the diagram is known in their honor as the *H-R diagram*. Note that the spectral classes of stars have been arranged across the bottom of the diagram in order from hotter to cooler, and the absolute magnitudes of stars have been arranged upward along the side of the diagram in order from dimmer to brighter.

Stars of known distance have been plotted on the chart according to their spectral class and absolute magnitude. The stars are not evenly distributed over the chart, but rather are grouped in several areas. We shall try to understand the significance of this grouping, but first let us see where several familiar stars fall on this diagram. Our sun is of the spectral class G2 and has an absolute magnitude of +5, hence it is placed at point A on the diagram. Betelgeuse, a very cool star, is of spectral class M2. It has

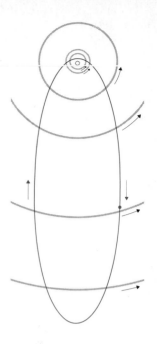

FIGURE 9.18 The H-R diagram. The points represent approximately 6700 stars of known absolute magnitude. (Yerkes Observatory)

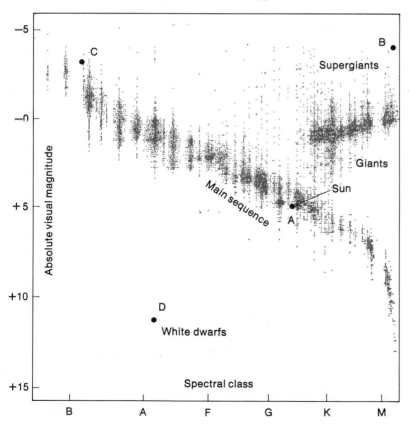

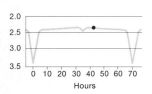

2.0
2.5
3.0
3.5

0 10 20 30 40 50 60 70
Hours

an absolute magnitude of −4, however, which places it at position B. We might ask ourselves how a cool star could possibly be so luminous. The answer is that it must be very large; its size makes up for the fact that it is so cool. All stars that are plotted in the upper right-hand portion of the diagram must likewise be very large—they are called *giants* and *super-giants*. Spica is of the spectral class B1 and has an absolute magnitude of −3, so it is placed at point C. The band of stars passing through C and A are called the *main-sequence stars*. They represent what you might presume to be the normal relationship between temperature and brightness, that is, the hottest stars being brightest and the coolest being dimmest. For average-size stars, this general relationship is true. However, there are some star types that plot near point D. What characteristics do they possess? They must be very hot (white) to be plotted toward the left of the diagram and they must be rather dim to be plotted toward the bottom of the diagram. The only explanation of how a star could be both hot and dim

FIGURE 9.19 Familiar stars plotted on the H-R diagram.

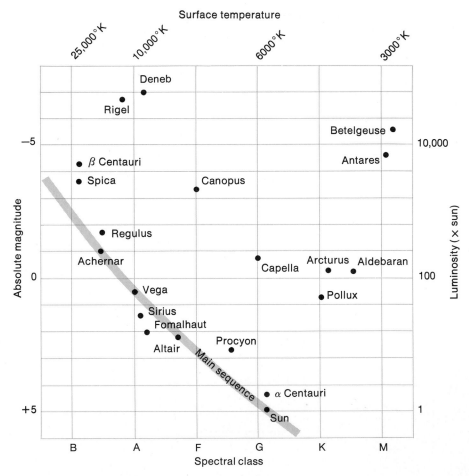

is that it must also be small. This lower-left portion, then, is where the *white dwarfs* are plotted. Some familiar stars are plotted on an H-R diagram in Figure 9.19.

Only a limited number of luminous stars are plotted close to the sun in the H-R diagram shown in Figure 9.18. It should not be inferred from this that the sun's temperature and magnitude are uncommonly low; If Figure 9.18 included all stars within a sphere of 500 light-years, we would see far more stars lying below and to the right of the sun. If stars in such a sphere could be counted according to their absolute magnitude (nearest whole number), a histogram which portrayed that count would look something like Figure 9.20; the sun is thus far above average among the nearby stars. In order to plot the additional stars, the H-R diagram would have to be extended downward and to the right, because most of the unplotted stars are both cooler and dimmer than those shown. The only red cool stars that can be seen with the naked eye are giants and supergiants. Stars that are both cool and dim are now being revealed by infrared techniques. Some familiar stars are plotted in Figure 9.19. As you identify some of these stars in the sky, think about where they plot on the H-R diagram and what that tells you about their color, temperature, absolute magnitude, and size.

A further subdivision of the giants and main-sequence stars may be made as follows (see Figure 9.21 for a graphical representation):

Ia Most luminous supergiants
Ib Less luminous supergiants
II Bright giants
III Normal giants
IV Subgiants
V Main-sequence stars

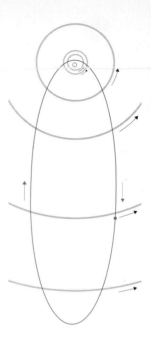

FIGURE 9.20 A histogram showing numbers of stars in absolute magnitude steps.

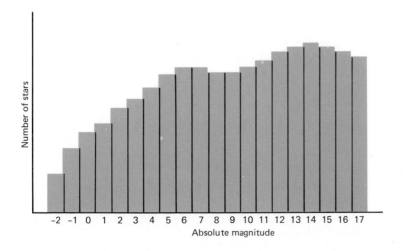

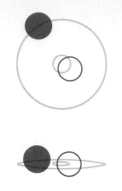

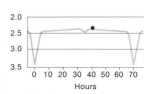

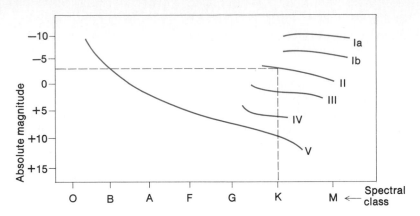

FIGURE 9.21 Subdivisions of H-R classifications.

These are often referred to as *pressure classes* because the very large supergiant stars typically exhibit the lowest pressure, the giants slightly greater pressure, and the subgiants and main-sequence stars having even greater pressures. This is exactly what one would expect given that a giant or supergiant star represents what was once a normal (main-sequence) star that expanded to several million times its original volume. Its individual atoms are much farther apart and, for a given temperature, this means a greatly reduced density. Furthermore, the reduced temperature generally associated with the red giant or supergiant also tends to reduce the pressure. How though does the pressure within a star's outer layers reveal itself to an earth-based observer? The answer lies in yet another characteristic of the spectrum, the width (shape) of its component lines. In what sense does a spectral line have shape? The shape of a spectral line is revealed only after we have portrayed its blackness graphically as shown in Figure 9.22. This is accomplished by scanning the photographic record of the spectrum, using a very narrow beam of light, and recording the amount of light transmitted at each wavelength. Because the spectrum of a star is composed of dark lines, the deepest dips occur at the darkest part of each line. Note that the deepest dip in Figure 9.22 has a definite width and a definite shape—it appears to flare out into a flattened "V" shape near the top. That flaring, or broadening, is due to pressure in the outer layer of the star. Atoms are close enough together to collide and distort each other's energy levels; therefore electron transitions are not as well-defined as they would be in a condition of lower pressure. By this line of reasoning, one would expect the stars with lowest pressure (the supergiants) to produce the narrowest spectral lines and those with higher pressure to produce lines that are more flared.

Pressure Class Ia and Ib represent the very largest supergiants and they show the narrowest spectral lines. Class II, III, and IV are characterized by increasing pressure levels and show increasingly broader spectral lines. Class V is by definition the main-sequence class—the state one might think of as normal for the larger portion of a star's life. These stars have yet

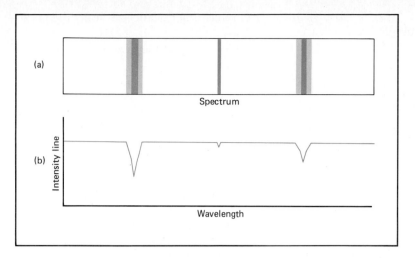

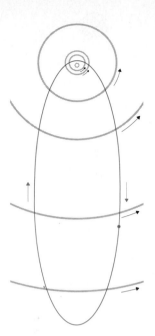

FIGURE 9.22 A great amount of information may be obtained by measuring the darkness (blackness) of spectral lines. Note the "wings" on the left-hand line.

higher pressures and therefore show even broader lines. The determination of the pressure class of a star is an essential part of the next method of distance determination called *spectroscopic parallax*.

SPECTROSCOPIC PARALLAX

The use of the method of heliocentric parallax for determining distances to stars is limited to nearby stars within 50 pc—the accurate measurement of parallax angles for stars farther than that becomes impossible. The method of spectroscopic parallax, on the other hand, is limited only by our ability to detect the spectrum of the star in question and to measure its apparent magnitude. We will be using the H-R diagram that shows the plots of numerous stars of known spectral type, absolute magnitude, and pressure class. We assume that any star of a given spectral type and pressure class will have the same absolute magnitude as any other star of like spectral type and pressure class. This is an expression of a fundamental assumption astronomers make all the time. Stars in our part of the universe are similar to stars in other parts of the universe. Step-by-step instructions for using the method of spectroscopic parallax are as follows:

Step 1. Classify the star by spectral type.

Step 2. Place the given star in its proper pressure category (Ia, Ib, II, III, IV, or V) on the H-R diagram.

Step 3. Extend a vertical line corresponding to the spectral type upward on the diagram until it intersects the proper pressure category line. For example, take a star of K0 type which is also in Pressure Class II and plot it as shown in Figure 9.21.

Step 4. From the point of intersection, draw a horizontal line to the left-hand side of the diagram and read out the corresponding absolute magnitude, which in the case of the example used above is −3.

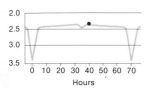

2.0
2.5
3.0
3.5
  0 10 20 30 40 50 60 70
         Hours

Step 5. Observe the apparent magnitude of the star, using the photo-multiplier tube, that is, +7.

Step 6. By comparing the absolute magnitude of a star to its apparent magnitude, determine its distance, using:

$$\frac{L(10)}{L(r)} = \left(\frac{r}{10}\right)^2$$

The difference in apparent and absolute magnitude is 10 [7 − (−3) = 10]. This corresponds to a light ratio of 10,000/1 (see Table 9.1, page 333).

$$\frac{L(10)}{L(r)} = \left(\frac{r}{10}\right)^2$$

$$\frac{10,000}{1} = \left(\frac{r}{10}\right)^2$$

$$10,000 = \frac{r^2}{100}$$

$$r^2 = 1,000,000$$

$$r = 1000 \text{ pc}$$

SIZE OF STARS

The sun is the only star that lends itself to easy size determination. Stars, in general, so far away that even the closest, Alpha Centauri, appears only as a point of light. Star images do appear as fuzzy disks on photographic plates; however, this is not a representation of their actual size—it is largely the result of turbulence in the earth's atmosphere. As a photographic plate is exposed over a period of several minutes to several hours, the earth's atmosphere refracts the starlight in various directions and a fuzzy disk results rather than a sharp point. To determine the diameter of a star, we must use special techniques.

The moon continually moves around the earth, occulting (covering up) one star after another as it goes. If the rate at which a star is occulted is measured very accurately, the size of the star can be determined. Suppose, for example, that you took a reading of the light received from a certain star every 0.001 second and your record showed that the moon took 0.01 second to occult the star completely. From a knowledge of how fast the moon moves (its angular rate is 0.54 second of arc per second), you could compute the diameter of the star. This method has been used very success-fully for over 50 stars that are occulted periodically by the moon. It should also be noted that it is primarily the larger stars (supergiants) that lend themselves to any method of diameter determination.

A very effective method for determining stellar size that overcomes the distortion created by the earth's turbulent atmosphere is called "speckle interferometry." Although the earth's atmosphere does spoil the sharper image that a star would make if no atmosphere were present, it is possible to produce a sharp image of a star by limiting the exposure time to approximately 1/100 second and combining multiple images. Such a

344   Chapter Nine

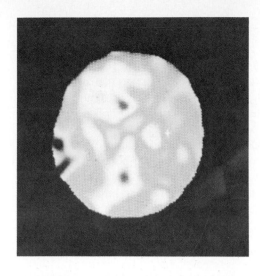

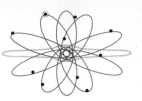

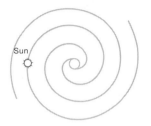

Sun

FIGURE 9.23 Speckle photograph of Betelgeuse. (Kitt Peak National Observatory)

"speckle" photograph of Betelgeuse is shown in Figure 9.23. Each speck is actually a photograph of the star in itself; there are many specks only because there are many different light paths through the turbulent atmosphere. Each speck contains much information that does not meet the eye. A special device called a *densitometer* passes over each speck, reading into a computer the variations in brightness it detects. After many specks have been read and compared, the computer can produce a single image of the star. This image not only reveals the size of the star but also variations in brightness on its surface thought to correlate with surface temperature variations.

For stars that do not lend themselves to any of the above methods, for example those that are very distant, yet another method is available. Suppose that the spectrum of a star indicates a surface temperature of 3000°K, half the temperature of the sun, and that the absolute magnitude of the star is zero, 5 magnitudes brighter than the sun (the sun's absolute magnitude is +5). A difference of 5 magnitudes indicates that the given star is 100 times more luminous than the sun. It is also known that a star whose temperature is one-half that of the sun radiates $\frac{1}{16}$ as much energy over each square centimeter of its surface. Taking these two factors into consideration, we see that the given star must have 1600 (16 × 100) times as much surface area as the sun. That is, it has to be larger to make up for being cool (the factor of 16) and for being brighter (the factor of 100). Since the surface area of a sphere is proportional to the square of its radius (surface area = $4\pi r^2$), we may compute the factor by which the star's radius is larger than the sun by taking the square root of 1600, which is 40—the radius of the given star is thus 40 times that of the sun. Among the largest known stars is Betelgeuse in the constellation of Orion, having an average diameter of 750 times that of the sun—a diameter greater than the diameter of Mar's orbit around the sun.

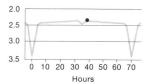

THE GREAT VARIETY OF STARS

We have discussed how stars vary greatly in size, surface temperature, color, and brightness; however, we have not yet conveyed the complete range of this variety. Our entire discussion has centered on an extremely narrow portion of the electromagnetic spectrum—visible light. It may surprise you to discover that of the number of stars in the Milky Way galaxy, more than half cannot be detected using visible light; we must turn to other wavelengths and other tools for detection of these objects.

First, let us consider a fundamental relationship between the surface temperature of a heated object and the wavelengths that it radiates. The curves in Figure 9.24 show that as an object is heated, it radiates a broader

FIGURE 9.24 The radiation curves for stars of different temperatures.

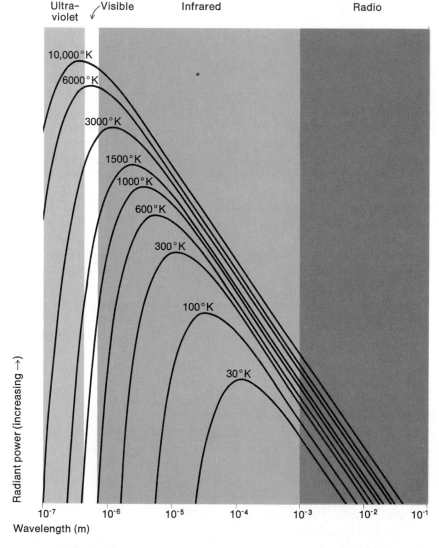

range of wavelengths. The wavelength at which the maximum radiation occurs is related to the surface temperature by a simple mathematical statement known as Wein's law:

$$\text{Wavelength}_{\max} = \frac{3000}{T°}$$

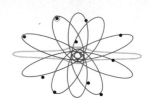

where wavelength is measured in microns (1 micron = 10^{-6} m = 1 millionth of a meter) and T is the surface temperature in degrees Kelvin. For example, the sun has a surface temperature of 6000°K and it radiates its peak energy at a wavelength of 0.5 micron, because 3000/6000 = 0.5 micron. Note that 0.5 micron is the same as 5000 Angstroms; it is near the center of the range of wavelengths called visible light. On the other hand, a star whose surface temperature is only 1500°K will radiate its maximum energy at 2 microns, which falls in the infrared range. Over 50 percent of the energy radiated at 2 microns will penetrate the earth's atmosphere, making this one of the better "windows" through which the astronomer may survey the infrared sky.

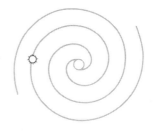

THE INFRARED SKY

If a map of the sky were constructed from information gathered only in the range of wavelengths from 2.0 to 2.4 microns (infrared) thousands of new objects which could never be detected by ordinary methods, such as astrophotography, would be revealed. You can see from the light curves shown in Figure 9.24 that a star like Betelgeuse, with a surface temperature of 3000°K, would radiate a very large proportion of its total radiation (area under the curve) in infrared, and therefore should appear very bright on an infrared map of the stars. You can also see that stars with surface temperatures of about 1500°K radiate almost all their energy in infrared; thus they will show up only on an infrared map.

Because the water vapor, carbon dioxide, and ozone in the earth's atmosphere tend to absorb most of the infrared that would otherwise fall on the earth's surface, we must go above that atmosphere to get a more complete infrared view of the sky. This was first accomplished by balloons, rockets, and high-flying aircraft, but these approaches provide only limited observing time. The advent of the orbiting satellite provided enough observing time to satisfy the hungry researcher. Some bodies have been identified that radiate thousands of times more total energy in the infrared range than the sun radiates in all its wavelengths put together. One such object is a star which is embedded in Eta Carinae (see Color Plate 12). Its infrared radiation is thought to come from a huge dust cloud that surrounds the star. The radiant energy from the star heats the dust to a temperature of roughly 250°K. It appears that this is one of many "cocoon" stars that are surrounded by shells of dust.

A second general type of infrared emitter is called a *protostar*—a star in the making. Stars are thought to condense out of huge clouds of gas and dust. Under the influence of gravity, the process of condensation always produces heat by giving up some potential energy. A protostar will thus

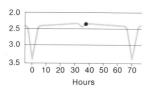

radiate in infrared, even if its temperature is only 100°K (such a source would have a peak output at 30 microns) and its presence will be revealed long before it begins to shine in visible light.

When we try to photograph the stars toward the center of our galaxy in visible light, we find it to be impossible because of the intervening clouds of dust that obscure their light. The infrared radiation of those stars, on the other hand, penetrates the dust clouds. Not only are stars near the center of the galaxy revealed in infrared but also galaxies that lie far beyond our own galaxy—those which happen to line up with the center of our Galaxy.

Stars are believed to cool off in their old age; thus old stars represent another class of infrared objects. At almost the other extreme from the cool, infrared objects we see the very energetic X-ray source.

X-RAY ASTRONOMY

By Wien's law, an object having a surface temperature of 30 million degrees Kelvin would radiate mostly at a wavelength of 0.0001 microns (equivalent to 1 Angstrom), in the X-ray portion of the spectrum.

X rays are not produced by hot stars alone, however. They have been detected from what is thought to be an extremely small but dense rotating star, called a *pulsar*, in the Crab Nebula (see Figure 14.5, page 428). This rapidly rotating object accelerates electrons to speeds near that of light and sends them hurtling through a magnetic field. The Crab pulsar can be detected "beeping" on and off at radio wavelengths and also pulsing in X-ray wavelengths. Astronomers could never have developed such a comprehensive picture of pulsars without using a wide range of wavelengths.

One of the most striking speculations of the past few years involving X-ray astronomy has concerned the "black hole." Very massive stars are thought to collapse in their old age and to develop very high surface gravities. Under such conditions, material may be drawn from a nearby star toward the black hole. Some of this material is thought to go into orbit around the very dense core, becoming so hot and turbulent as to generate X rays. Astronomers are quite confident that the X rays received from Cygnus X-1 represents the presence of a black hole. Black holes will be discussed more fully in Chapter 14.

The detection of X rays typifies what has been happening in all areas of astronomy in recent years—the opening of virtually the entire spectrum to our "view."

QUESTIONS

1. How long is the base line used in the method of heliocentric parallax?
2. The method of heliocentric parallax can be used to determine the distance to stars as far away as _____ parsecs.
3. Find the distance to a star whose heliocentric parallax is 0.50 second.

4. Which motion of stars appears to change the shape of the Big Dipper constellation?

5. The radial velocity of a star is indicated by what kind of observation?

6. Define the term "space velocity." Why cannot this be called the true velocity of a star?

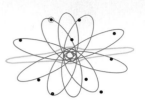

7. The sun moves within the Milky Way galaxy at a rate of _____ km/sec, owing to the rotation of the Galaxy.

8. What do we know about a star whose peculiar velocity is zero?

9. The computation of the tangential velocity of a star depends upon our knowledge of what two properties of the star's location and/or motion?

10. What was Hipparchus' system for categorizing stars by brightness?

11. List three means by which the apparent brightness of stars may be judged today?

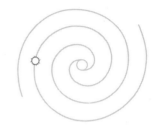

12. A change of 1 magnitude corresponds to a change of _____ in luminosity.

13. A star of −3 magnitude appears _____ times as bright as one of +2 magnitude.

14. True or false: Venus sometimes appears brighter than the brightest star.

15. True or false: The use of time exposures on film has extended the limiting magnitude of the Palomar 5-m telescope by several magnitudes.

16. The true brightness of two stars may best be compared by consideration of which of the following: (a) their apparent magnitude, (b) their absolute magnitude, (c) neither of these?

17. How is the apparent brightness of a star related to its absolute brightness and its distance from the observer?

18. Describe at least two ways by which the surface temperature of stars may be measured.

19. What is the general range of surface temperatures of stars? Can the surface temperature of a star be judged in even the roughest fashion by the naked eye? Explain.

20. What relationship is demonstrated by the fact that the majority of stars fall along the main sequence on the H-R diagram?

21. What advantage does the method of spectroscopic parallax have over that of heliocentric parallax for determining the distance to a star?

22. What is the range in star sizes as compared to the sun?

23. A 100-watt light bulb has a certain apparent brightness if viewed from a distance of 10 m. What will be the change in its apparent brightness if viewed from 20 m? from 50 m?

24. An object as cool as 300°K will radiate its maximum energy in what part of the spectrum?

25. The wavelength at which a star emits its maximum energy is inversely proportional to its surface temperature. If a star whose temperature is like that of the sun (6000°K) and has its maximum output at 0.5 micron, find the wavelength for the maximum output of a star whose temperature is 12,000°K. How will its color compare to that of the sun?

26. What increase in luminosity would accompany a change in temperature of a star from 3000°K to 6000°K, assuming alternately that (a) the diameter of the star remains the same, and (b) the diameter of the star is halved?

27. The sun appears to have a magnitude of −26.5 as seen from earth. What must be its magnitude as seen from Pluto, 40 times the distance from the earth to the sun?

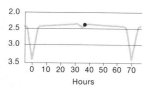

Multiple Star Systems

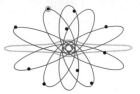

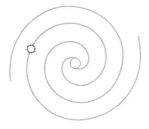

Ten

The casual observer would hardly suspect that more than one-half of the stars seen with the naked eye are in reality multiple systems, containing two, three, or more stars. When two stars appear close to each other, they may indeed be close to each other, or they may actually be remote from each other and just happen to line up from the observer's point of view. Sir William Herschel, a famous observer of *binary*, or double stars, theorized initially that such stars differed greatly in their distance from the earth since one star of the pair was often dimmer than the other. In trying to prove his theory, however, he found just the opposite to be true; he discovered that most pairs lie in each other's gravitational field—that is, they are close enough and massive enough to influence each other's motion in an obvious way. They literally revolve around a common point between them called their *barycenter*. In proving his original theory wrong, Herschel made a very important discovery. He continued his observations and by 1820, he had catalogued some 800 double stars.

OPTICAL AND VISUAL BINARIES

On the rare occasion that two stars do appear close together from the observer's point of view but are really distant from

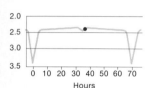

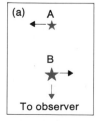

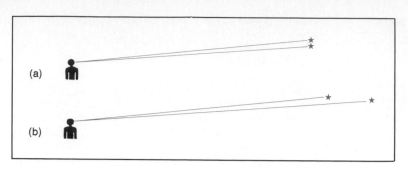

FIGURE 10.1 Stars may appear close together because (a) they are actually close, or (b) they merely line up from the observer's point of view.

each other, they are called *optical binaries*—their apparent closeness is only an optical illusion. Such a system is not a true binary system.

A much more common class of binary stars are those which lie in each other's gravitational field and are visible as doubles in a telescope. These are called *visual binaries* (Figure 10.1). Recall that the ability of a telescope to separate close stars is called its resolving power (see page 84.)

SPECTROSCOPIC BINARIES

If the components of a true binary system are so close that they cannot be *resolved* (separated) by a telescope, they may still be identified as a binary system by their spectrum. Consider a pair of binary stars, A and B, in mutual revolution around their barycenter as in Figure 10.2. When the stars are positioned as in (a) or (c), both have the same Doppler shift in their spectral lines. However, when they are positioned as in (b) or (d), one star is moving in a direction toward the observer and the other star is moving away; hence the spectral lines of one star will be slightly different from those of the other, and two lines will result (see Figure 10.3). A binary system, or more specifically a spectroscopic binary system, is thus revealed by the periodic separation of the spectral lines.

You can easily identify a system of stars that illustrates both the visual and the spectroscopic types of binaries. You can see with your naked eye

FIGURE 10.2 Spectroscopic determination of a binary system.

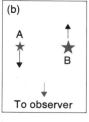

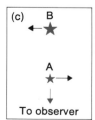

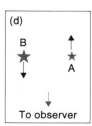

FIGURE 10.3 The double lines (B) as seen in the spectrum of a spectroscopic binary star system. The top band (A) shows no separation, since the two stars are in alignment with the observer. (Mount Wilson and Las Companas Observatories, Carnegie Institute of Washington)

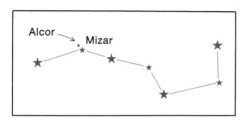

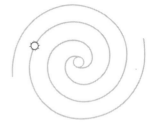

FIGURE 10.4 The Big Dipper.

the two most obvious members of the system, Mizar and Alcor, in the handle of the Big Dipper (Figure 10.4). Several minutes of arc separate these stars. If you look at Mizar through a small telescope, you will discover that Mizar is actually not a single star, but a visual binary; its brighter component is Mizar A and its dimmer component is Mizar B. If you were to analyze this system spectroscopically, you would further discover that Alcor, Mizar A, and Mizar B are each a pair of spectroscopic binaries. The system thus has six stars.

SPECTRUM BINARIES

A binary system can be identified by its spectrum in the following way. The spectrum of a hot star shows lines associated with ionized atoms, whereas the spectrum of a cool star shows lines associated with molecules. If the observer sees both kinds of lines in the spectrum of what appears to be a single star, the only possible explanation is that the system is multiple, that is it contains two or more stars; a single star could not be both hot and cool at the same time.

ASTROMETRIC BINARIES

Although the dimmer component of a binary system might go unnoticed by any of the above methods because it is so dim, its presence could be detected by the gravitational effect it has on the brighter star—the brighter star moves along a wavelike path in the sky. Such a system is called an *astrometric binary system*. Recall that the moon and the earth have such a gravitational effect on each other (see left-hand flip pages starting on page

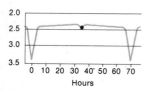

182). Note the wavelike path of the earth as a result of the moon's presence. If an observer were thus far enough from the earth that the moon was not visible, the earth's wavelike path would be evidence that the earth had a moon. The best chance of finding binary systems astrometrically is with the closer stars because they typically show greater proper motion and a wavelike path would be more obvious if an invisible partner star did exist. Look ahead to Figure 10.9 (page 357) and you will see the wavelike path of Sirius A. The binary nature of Sirius was first suspected based on its wavelike path and it was classified as an astrometric binary; modern techniques, however, reveal its white dwarf companion star as shown in Figure 10.9(b) and it may now be classified as a visual binary.

ECLIPSING BINARIES

Some binary systems, both visual and spectroscopic, are oriented so that one star passes in front of the other; that is, we see the edge of the system's plane of orbit. These are known as *eclipsing binary* systems. As one star is eclipsed by the other, we expect some variation in the total light reaching us from that system. If we plot the variations in light received as these eclipses occur, we may discover a number of facts. Consider a binary system composed of a large star and a small star where the small star is the

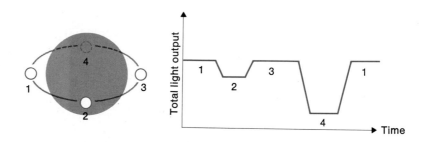

FIGURE 10.5 The light curve of a binary system.

FIGURE 10.6 Partially eclipsing binaries.

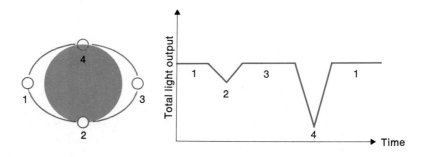

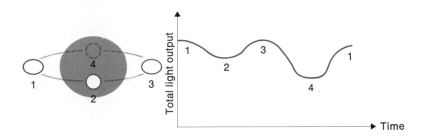

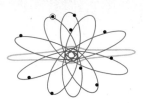

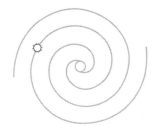

FIGURE 10.7 A binary system, distorted by gravitation.

hotter of the two (Figure 10.5). When the small star is in position 1 or 3, we see the combined output of light from both stars. When it is in position 2, we see part of the cooler star being covered with only slightly less light being received; when it is in position 4, we see a total eclipse of the hotter star, so that the least amount of light is received. If the orbit of this system is tipped slightly more in our direction, we observe only partial eclipses and the light curve appears as in Figure 10.6. The left-hand flip pages beginning on page 410 depict an eclipsing binary system in motion and the light curve being received.

Figure 10.7 shows the light curve for a pair of stars so close together that they distort each other's shape by their mutual attraction. They could be said to be continually in some partial state of eclipse.

It is important to realize that astronomers usually cannot see what is actually going on in an eclipsing binary system; they can only record the varying light output and represent it as a light curve. All the characteristics we have been discussing must be deduced from this curve.

The light curve gives an indication of the size of a star in relation to its orbital size. A smaller star is very rapidly eclipsed, making the sides of the curve very steep. A larger star requires more time for an eclipse and thus produces a less steep curve. This represents yet another way that the size of stars can be determined.

CLOSE (CONTACT) BINARIES

Some binary stars are so close to each other that their atmospheres bulge toward each other and they actually exchange material (mass). In order to understand some of the ramifications of this phenomenon, let us briefly preview Chapter 13. Stars are not static throughout their lifetime, but rather change (evolve), sometimes growing very large and sometimes shrinking to a very small size (the size of the earth, or smaller). Suppose that one star of a binary pair has three times as much mass as the sun, or 3 "solar masses," whereas the other star has only one solar mass. More massive stars typically go through their life cycle faster than the less massive stars; hence while the 1-solar mass star is still of the "main sequence"

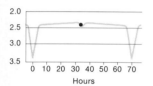

variety, the 3-solar mass star may swell to become a red giant. The outer edge of the 3-solar mass star may come so close to the 1-solar mass star as to experience a stronger gravitational attraction to it and, as a result, transfer some of its mass to that star. The mass of the two stars will be redistributed, changing the evolutionary rate of each star. Further evolution in the more massive star may produce a highly condense star, one that will pull material back from the low-mass star. The exchange of material can thus complicate the evolutionary pattern of the two stars involved. Furthermore, when material is pulled toward a very concentrated star, that material may be so highly agitated and heated it may produce X rays. Several tens of such systems have been discovered that show certain periodic variations in their X-ray radiation characteristic of eclipsing binaries.

DETERMINATION OF THE MASS OF BINARY SYSTEMS

A binary system lends itself very well to the determination of the mass of both the system and of the individual members. Three basic factors of the system are related: the mass of each star (m_1 and m_2), the separation of the stars (r), and their period of revolution (p). This relationship, first expressed by Kepler as $p^2 = r^3$, and later refined by Newton, is stated thus:

$$m_1 + m_2 = \frac{r^3}{p^2}$$

where (m_1) and (m_2) are expressed in solar masses (masses as compared to the sun's mass), (p) in years, and (r) in astronomical units. The sum of the masses (m_1) and (m_2) can be determined given p and r, which can be determined from observation. Suppose for example that two stars were orbiting each other at an average separation of 5 A.U. and that their period of revolution was 10 years; then

$$m_1 + m_2 = \frac{(5)^3}{(10)^2} = \frac{125}{100} = 1.25 \text{ solar masses}$$

This only gives us the total mass; we must determine how the mass of the system is distributed between the two stars. If the stars were equal in mass, they would orbit around a point midway between the two, called the *barycenter* (or center of mass). If one star were three times as massive as the other, then we would expect the barycenter to be nearer the greater mass (Figure 10.8). These factors are related by the proportion:

$$\frac{m_1}{m_2} = \frac{x_2}{x_1}$$

By observing the displacement of the stars in a binary system as compared with several nearby stars, the system's barycenter can be located and thus the distribution of mass can be calculated.

Consider the case of Sirius, a binary system with its "A" component the brightest star we see, and its tiny "B" component the first white dwarf ever discovered. As this system moves through space, its barycenter marks

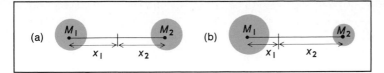

(a) and (b) diagrams showing M_1, M_2 with x_1, x_2 distances

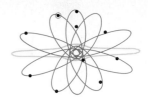

FIGURE 10.8 The barycenter of a binary system.

a uniform path indicated by the dashed line in Figure 10.9. The two components oscillate in varying degrees to either side of this path.

The approximate period of this system has been observed to be 50 years, and the separation of its members, 20 A.U. The total mass of the system is computed to be 3.2 solar masses, as follows:

$$m_1 + m_2 = \frac{(20)^3}{(50)^2} = \frac{8000}{2500} = 3.2 \text{ solar masses}$$

Sirius "B" is observed to orbit 2.2 times as far from the barycenter as does Sirius "A." The combined mass must therefore be distributed in the ratio of 1/2.2; the mass of Sirius "A" is approximately 2.2 times that of the sun, and the mass of Sirius "B" is equal to that of the sun. Think of the earth and moon, shown in the left-hand flip pages beginning on page 182, as two stars in a binary system.

FIGURE 10.9 (a) The path of a binary pair, Sirius A and Sirius B. (b) Three exposures that reveal the small companion of Sirius. (Lick Observatory)

(a)

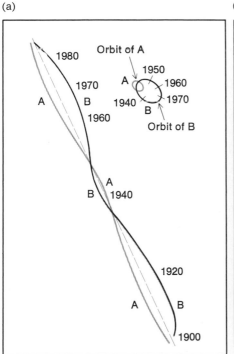

(b)

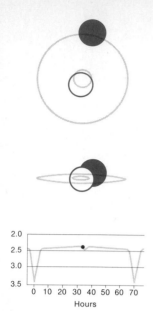

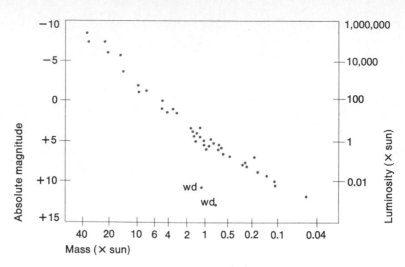

FIGURE 10.10 The mass-luminosity relationship.

MASS-LUMINOSITY RELATIONSHIP

One of the most significant results of our being able to determine the masses of binary stars becomes obvious when we plot the masses of these stars against their absolute magnitudes (Figure 10.10).

Most of these stars fall within a narrow band running from upper left to lower right. Stars low in mass are also low in absolute magnitude, that is, they are the dimmer stars, and high-mass stars are also high in absolute magnitude, that is, they are the brighter stars. This is called the *mass-luminosity relationship*. It is likely that all main-sequence stars satisfy this relationship. On the other hand, white dwarfs and red giants generally do not show this relationship. Figure 10.10 shows several white dwarfs that do not fit within the narrow band.

Once the mass-luminosity relationship was established, it was used to estimate the masses of main-sequence stars based on their absolute magnitudes, even though they were not part of a binary system. The assumption concerning the relationship between mass and luminosity derived for binaries applies equally to stars which do not happen to be members of a binary system. Note that stars of 40 solar masses have a luminosity factor of approximately 1 million times that of stars of 1 solar mass. This suggests that massive stars "burn" their fuel much faster and that, as a result, their lives are much shorter.

ORIGIN OF BINARIES

Binary systems are so numerous that any theory concerning the origin of stars must account for these special associations. Theories concerning the origin of binary systems run somewhat parallel to those of the solar system.

One theory suggests that two stars became entrapped within each other's gravitational field as a result of passing very close to each other. The chance of such an encounter is very small and this theory does not account very well for the multiplicity of binary systems.

A second theory suggests that a rapidly rotating star broke up into two or more parts. We might expect a rotating star to throw off some material from around its equator, but it is difficult to think of a star being torn in half by such a motion.

It is conceivable that, as a cloud of hydrogen gas began to condense to form a star, more than one center developed, eventually producing two or three stars within the same contraction. Such stars would possess a motion in common with the gas cloud from which they formed.

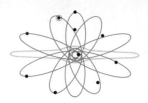

CLUSTERS

If seeing conditions are fair to good, even a pair of binoculars will usually reveal from a dozen to hundreds of stars within its field of view no matter

FIGURE 10.11 The Pleiades (Seven Sisters), a typical open cluster in Taurus. (Lick Observatory)

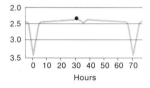

what part of the sky is observed. Certain regions, however, seem to "come alive" with stars when they are scanned. One such region is that of the *Pleiades cluster* (Figure 10.11), a group of stars visible to the naked eye but doubly spectacular in binoculars or in a low-power telescope—hundreds of stars can be seen. Is this more than merely an accidental association? The answer to this question can be found by studying motions of these stars. While each star differs slightly in speed and direction, there is a general trend toward the same direction and the same speed; they are thus a true cluster.

ORIGIN OF CLUSTERS

Given that stars form by means of condensations in huge nebulae which themselves contain enough material for hundreds or thousands of stars, it is reasonable to assume that most stars originally formed in clusters and that while some have lost their identity with a cluster, others have retained it. If the stars of an identifiable cluster did have a common origin, we may assume that they are all about the same distance from us, so that their apparent magnitude is proportional to their absolute magnitude. In other words, we assume that a star which appears to be brightest among the group is in fact brightest. Based on the apparent relationship between luminosity and the mass of a star, the more massive members of a cluster ought also to appear the brightest. When clusters are observed with these relationships in mind, it appears that the more massive members of the cluster tend toward the central region. Perhaps clusters were initially formed with this concentration of mass around the center. On the other hand, the more massive stars may gradually be moving toward this central position.

CLUSTER TYPES

There are two basic types of clusters. Examples of the *open cluster* type include the Pleiades (Figure 10.11), the double open cluster in Perseus (Figure 10.12), and the Beehive cluster in Cancer (M67) as seen on page 320. They are also known as *galactic clusters* because they occur primarily in the disk of our Galaxy. Such clusters usually contain young, hot (O- and B-type) stars numbering in the hundreds. They form a rather loosely bound system that may disperse in a matter of a few tens of millions to a few billions of years because of the gravitational forces within the Galaxy as a whole. By the time such stars age, they are already separated by distances which no longer suggest a common origin. Over 1000 open clusters are known in our Galaxy.

The second type of cluster is the *globular cluster*. A cluster of this type contains up to 100,000 stars that appear so tightly packed that the central regions cannot be resolved into individual stars, even with the largest telescopes (Figure 10.13). Just how close are these stars in the center of the globular cluster? Even in the most dense portion there are probably no more than 64 stars per cubic light-year (Figure 10.14). Within this region an average of three months would be required for light to travel from one star to its nearest neighbor, a distance of 2.4 trillion kilometers. By comparison, if we took a cubic light-year at random in the Milky Way Galaxy, our chances of finding even one star within that space would be only 1 in 64.

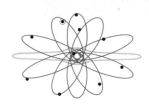

Globulars seem to form a halo around the nucleus of our Galaxy (Figure 10.15). They do not participate directly in the revolution of the Galaxy but rather have elliptical orbits with eccentricities of approximately 0.6. They typically move to within 4,000 pc of the galactic center. Their orbits and relationship to the Galaxy are somewhat similar to the orbits and relationship of comets to the solar system. The right-hand flip pages beginning on page 345 demonstrate the motion of the globulars in the halo of the Galaxy. The distance to these objects may be determined by the variable stars that they contain, called RR Lyrae, or cluster-type variables (see Chapter 11).

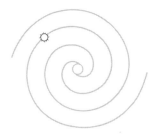

FIGURE 10.12   The double open cluster in Persei. (Lick Observatory)

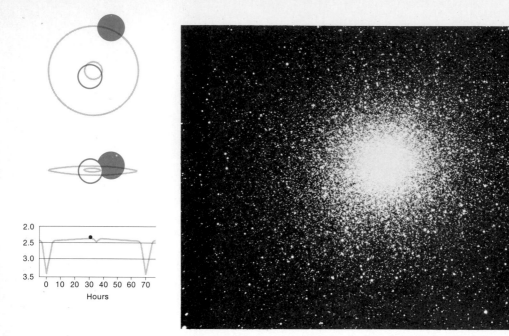

FIGURE 10.13   The globular star cluster in Hercules (M13). (Palomar Observatory Photograph.)

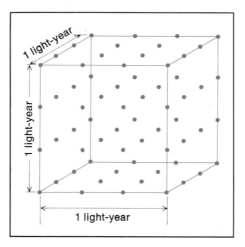

FIGURE 10.14   The density of stars near the center of a globular cluster.

1 light-year

1 light-year

1 light-year

FIGURE 10.15   The Milky Way galaxy, showing the halo, the nucleus, and the disk.

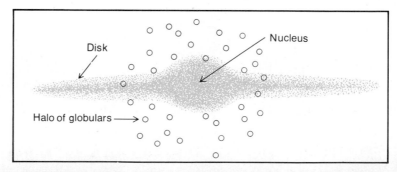

Disk

Nucleus

Halo of globulars

X-RAY BURSTERS

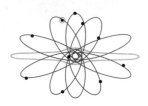

A very interesting phenomenon is unfolding that appears to be related to globular clusters. Sudden bursts of X rays and gamma rays seem to come from certain of these clusters. The discovery of these "bursters," as they are now called, is a fascinating story that is typical of today's astronomy.

The space age ushered in all sorts of new techniques for using the entire electromagnetic spectrum to study the universe. Recently launched X-ray satellites include: "Uhuru" (see page 105); the ANS (Astronomical Netherlands Satellite), built and operated by the Dutch but launched and partially tracked by the United States; and the HEAO (High Energy Astronomical Observatories) series. While scanning the computer output of the ANS satellite on September 25, 1975, an American astronomer, Jonathan Grindlay, noticed a sudden increase in the X-ray count. Upon obtaining a second-by-second record for this period (Figure 10.16), he was astonished by the sudden tenfold increase that occurred in a period of two seconds. He wondered if it was due to a quirk in the complicated system of detection and transmission. The same kind of record, however, was found in Uhuru data. Furthermore, a satellite called Vela, launched in 1973 to detect gamma rays as an indication of nuclear test ban treaty violations, revealed a corresponding burst at even higher energy levels. An inspection of Vela's earlier records showed that it had detected sudden bursts of energy soon after it had been launched. Although Vela can be credited with detecting and recording these X-ray "bursters," they were not actually discovered until Grindley, along with several other groups working independently, became aware of them and recognized their significance; without the human function of recognition, the next logical function, that of reasoning, would never follow.

The "burster" recorded in Figure 10.16 is associated with the globular cluster NGC6624, a known source of X rays (Figure 10.17). Since its discovery, at least 20 other bursters have been found; some of these are associated with globular clusters, and some may be associated with other objects. The 10-second burst represented by this innocent-looking graph in Figure 10.16 probably represents more X-ray energy than the sun emits in an entire year—it thus certainly demands an explanation.

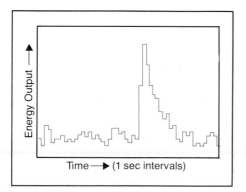

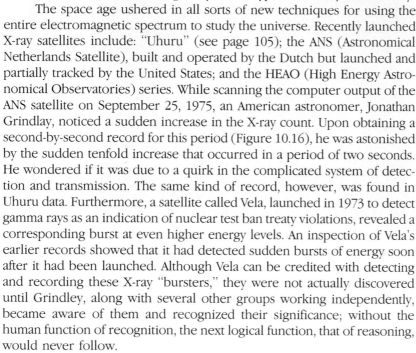

FIGURE 10.16 A typical X-ray burst, as recorded in NGC6624.

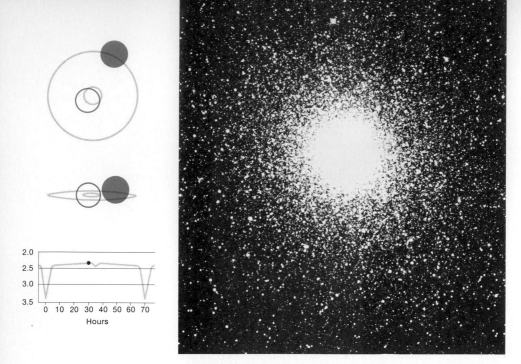

FIGURE 10.17 NGC6624, an X-ray burster.

The models astronomers have devised to explain the "burster" all assume that very dense stars are involved. Let us try to visualize such a star—one that has twice as much mass as the sun, but has shrunk to a diameter of 20 to 30 km. The density of such a star, called a *neutron star*, would be equivalent to the density of the neutron itself (on the order of 10^{14}g/cm^3). Let us further suppose that this star is a member of a binary pair of stars, and that because of its ultra-compact nature, it tends to gravitationally attract material from its mate. Although the flow of gases onto its surface would generate a background level of X-ray radiation, this would not account for the bursts. Stars typically have a magnetic field, and this field would tend to be strengthened by the contraction of the star. Such a field could hinder the flow of gases, but when a sufficient mass of gas accumulated, the field could collapse and a sudden outburst of X rays would occur as gas again fell onto the star's surface. As indicated by a number of "bursters," the whole process must then be repeated but at varying intervals of time. Variability seems to be characteristic of most X-ray radiation.

Grindlay's own model for bursters involves an even more massive and denser star called a black hole. Such a star is capable of pulling gas and objects toward its surface from greater distances than the neutron star. According to his theory, gas falling toward its surface would experience frictional heating and would generate X rays. The surrounding gases would be heated by such radiation which could trigger an outflow of gases that would at least temporarily halt the flow toward the surface. This deterrent could not be sustained very long, however, and a sudden collapse would trigger another outburst of X rays.

If you had asked astronomers what they expected to find with X-ray satellites, probably no one would have conceived of bursters. The unexpected is certainly a vital part of the fascination of science today.

ASSOCIATIONS

The loosest form of star clustering is called an *association*, a diffuse grouping of similar types of stars. Almost a hundred such associations are known in the Milky Way galaxy; an example is a group situated in the region of the Orion Nebula. These highly luminous stars are thought to be very young, lying close to the flattened disk portion of the Galaxy. We will see in Chapter 15 how these very loose, short-lived associations mark the spiral arms of the Milky Way much as bright lights mark the streets of a city as seen from an airplane at night. Other associations, called *T associations*, are composed of T Tauri variables which are thought to be very young, low-mass stars that have not yet stabilized to become main-sequence stars.

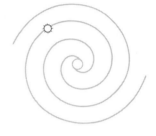

QUESTIONS

1. Most of the pairs we see as double stars are: (a) two stars that are many light-years apart but just happen to line up from our point of view, or (b) two stars that are close enough together to influence each other's motion in an obvious way. Which is true?

2. How may a true binary system be recognized, even if the two stars cannot be separated by the largest telescopes?

3. In the case of eclipsing binary stars, what properties of such a system can be determined from its light curve?

4. What is the most important characteristic of stars that can be determined in a binary system?

5. What does the location of the barycenter of a binary system tell the astronomer?

6. When the mass of a star is plotted against its absolute magnitude, what becomes evident from the graph?

7. What does the fact that many stars of a cluster are moving in the same direction and at the same speed suggest about their origin?

8. What is the most obvious difference between the open-type and the globular-type cluster?

9. Within the Milky Way galaxy, where are the globular clusters found?

10. Describe the motion of the globulars in relation to the Milky Way galaxy.

11. (a) Find the mass of both stars in a binary system if the total mass is 10 solar masses and their distances from the barycenter are 2 A.U. and 3 A.U., respectively. (b) Find their period.

12. Find the total mass in a binary system which has a separation of 3 A.U. and a period of 3 years.

13. What change would accompany a reduction of the distance between two binary stars if some unknown force shoved them closer together?

14. What assumption must be made about a star or star system if both the lines of ionized helium and the lines of molecules show up in the same spectrum?

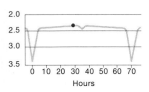

2.0
2.5
3.0
3.5
0 10 20 30 40 50 60 70
Hours

Variable Stars

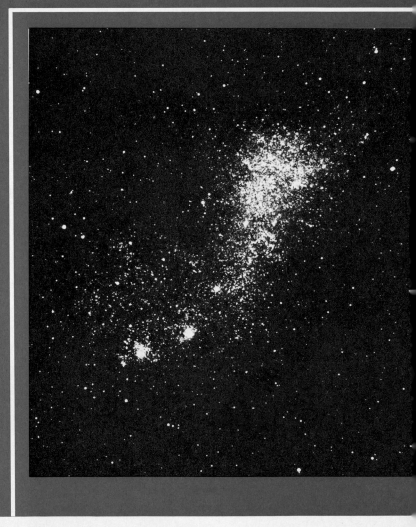

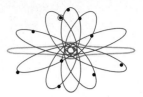

Eleven

Whereas eclipsing binary systems only appear to vary in brightness, others actually vary in their light output. The existence of such stars was recognized by Hipparchus in the first century B.C., and today thousands of such *variable stars*, as they are called, have been catalogued.

The *cepheids* represent perhaps the most important class of variable stars. They are named after a particular star, Delta Cephei, the delta star, or fourth brightest star, in the constellation of Cepheus (Figure 11.1). A system of brightness classification, using letters of the Greek alphabet, was initiated by Johann Bayer in his catalogue of stars published in 1603. With few exceptions, he assigned the Greek letter alpha to the brightest star in a given constellation, beta to the second brightest, gamma to the third brightest, delta to the fourth brightest, epsilon to the fifth, and so on. (See Appendix 10.)

The variable nature of Delta Cephei was discovered in 1784 by John Goodricke, a young British astronomer. The light curve of Delta Cephei, shown in Figure 11.2, indicates its period of 5.4 days. Note that this light curve has a distinctly different shape from that of an eclipsing binary system.

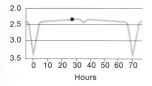

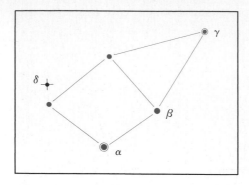

FIGURE 11.1 The constellation of Cepheus.

CEPHEID VARIABLES

We can see from Figure 11.2 that the light of Delta Cephei builds up rather rapidly from a magnitude of +4.3 to +3.6, a difference representing 0.7 magnitudes and a doubling of luminosity. This particular star appears twice as bright when at its maximum as at its minimum (Figure 11.3). Variations in Delta Cephei are quite evident with the aid of binoculars. Cepheids in general have periods of variation ranging between 1 and 50 days, and their light output varies over a range of 0.1 to 2 magnitudes; however, the amplitude and period of any one cepheid is unusually steady. Polaris, the North Star, is an example of a Cepheid that has a variation of only 0.1 magnitudes in a period of four days; this variation is not detectable by the naked eye.

We might wonder why certain stars fluctuate in their light output while others appear to be rather stable. What is happening to the star itself to cause these changes? The spectrograph helps to answer this question. When a variable star is subjected to spectral analysis, the pulsating of the star becomes apparent—it grows larger for a time, then smaller again. Let us suppose, for example, that a particular star is moving away from us and therefore has a red shift in its spectral lines due to its radial velocity. As this star grows, its spectral lines will be shifted slightly from their normal red shift position back toward the blue, since the star's motion due to expansion is basically toward us. On the other hand, when the star is decreasing in size, both its contraction and its radial velocity will produce a slightly

FIGURE 11.2 The light curve of Delta Cephei.

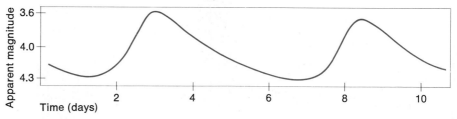

larger-than-usual red shift. If we compare a plot of the rate of change in size with the light curve, we find a distinct relationship between the two. The actual size of the star at any given time is shown in Figure 11.4 (bottom graph).

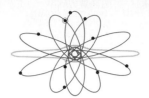

If a star were to behave simply like a "ball of gas," we might expect the time of its maximum brightness to correspond to its most condensed state; the star at that time would also be hottest, owing to its contraction. The point of greatest light output, however, follows shortly after this time of greatest contraction. The core of the star may in fact put out most of its energy at the time of greatest contraction, but this effect does not reach the outer layers until later, thus producing the apparent time delay. Can the variation in light be explained by the change in size alone? Delta Cephei has an average diameter of 40 million kilometers (it's a giant as stars go), and it varies by about 1.2 million kilometers on either side of that value—a total change of about 6 percent. This in itself could not account for the doubling in luminosity; the spectrum of the star reveals that accompanying a change in diameter is also a change in temperature. A relatively small change in temperature creates quite a drastic change in energy output per square unit of surface area. Even though the star may be shrinking during a given part of its cycle, it is heating up and this fact more than makes up for its smaller size. The variation in brightness of a Cepheid is thus more a function of temperature change than size change.

FIGURE 11.3 A double exposure with slight displacement, showing the variable star WW Cygni at maximum and minimum brightness. The variable is located in the center of the photograph. (Mount Wilson and Los Campanas Observatories, Carnegie Institute of Washington)

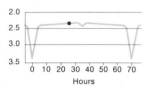

CEPHEIDS: INDICATORS OF DISTANCE

In 1912 Henrietta Leavitt of Harvard College Observatory became interested in a variable star that had been discovered in the Small Magellanic Cloud (Figure 11.5). She went to the southern station of the Harvard Observatory in South Africa and took many photographs of stars in this Small Cloud near the south pole of the sky. From these photographs she made a remarkable discovery—that there is a definite relationship between the apparent magnitude of the Cepheids in the Cloud and their period of variation. The longer period Cepheids were also the more luminous and the shorter period variables were the less luminous. This relationship is clearly shown in the plot of Figure 11.6.

The fact that these stars appear to lie along a diagonal line on this plot instead of being scattered at random over the whole chart shows the relationship between the period and the apparent magnitude of these stars. Does it necessarily follow that, because of this relationship, there is also a relationship between the period and the absolute magnitude of these same stars? Leavitt reasoned that because all these stars are in the Small Magellanic Cloud, they should all be approximately the same distance from the observer. This supposition was true, even though she did not know their distance. Whereas she thought the Small Magellanic Cloud was a cluster of stars in the Milky Way, it was later established as another

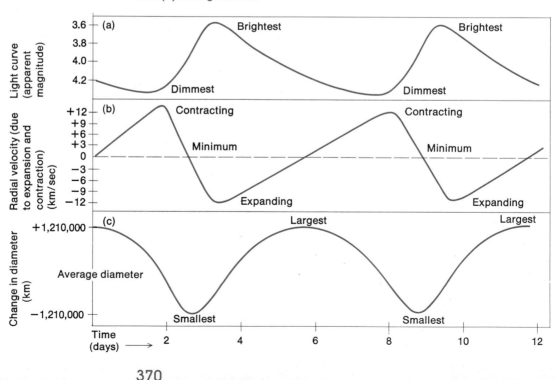

FIGURE 11.4 Graphs of a Cepheid variable. (a) light curve. (b) rate of change in size. (c) change in size.

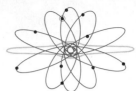

FIGURE 11.5 The Small Magellanic Cloud. (Mount Stromlo and Siding Spring Observatories, the Australian National University)

galaxy outside our own, at a distance of more than 150,000 light-years.

When a group of stars are found to lie at approximately the same distance from the observer, their absolute magnitudes compare in precisely the same way as their apparent magnitudes. If we assume that Ceph-

FIGURE 11.6 A plot of Cepheids in the Small Magellanic Cloud.

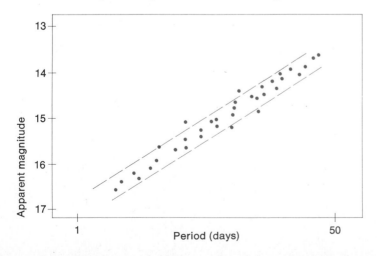

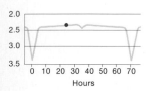

eids in one part of the universe behave like Cepheids in any other part of the universe, then it is only necessary to find the distance to a few Cepheid-type variables in order to convert apparent magnitudes to absolute magnitudes; no Cepheids could be found in the Cloud, however, given the limits of the method of heliocentric parallax (approximately 50 parsecs).

Recognizing the great (potential) significance of Leavitt's discovery, Harlow Shapley and others joined her in the search for cepheids in our own neighborhood of stars, and by using a method of *statistical parallax* they found the distance to a number of RR Lyrae stars (see below). Like RR Lyrae stars, Cepheids of a 1-day period were found to have an absolute magnitude of approximately zero, while those of a 50-day period were set at -4 in absolute magnitude; the latter were thus relatively luminous stars ranging from 100 to 4000 times the luminosity of the sun, again an indication of their giant nature since Cepheids are also rather cool as stars go.

RR LYRAE STARS

The naming of variable stars was not standardized until relatively late, thus many different types of names exist. When you see a double letter in front of a constellation name (as in RR Lyrae), you know the astronomer is designating a variable star. A more modern designation is of the form: V335 Herculis, the "V" meaning variable.

RR Lyrae stars are a class of variables with periods ranging from a few hours to one day, but with magnitudes of about zero, regardless of their period. Note the rather flat distribution of RR Lyrae stars on the graph shown in Figure 11.7 and note the sense in which they seem to be a continuation of 1-day Cepheids. Assuming that the calibration of this period-luminosity relationship was correct, Walter Baade, of Mt. Wilson and Palomar Observatories, conducted a search for variables in the Andromeda galaxy (see Color Plate 15). He was able to photograph a variety of

FIGURE 11.7 A plot of RR Lyrae and Cepheids.

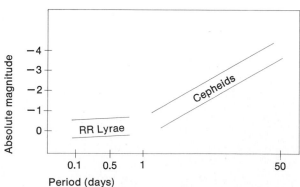

Cepheids in the galaxy, including those of 1-day period; he could find no star of the RR Lyrae type, however, even though such a star should have been of equal magnitude to the 1-day Cepheid.

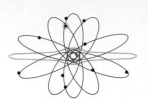

The fact that Baade could not find RR Lyrae stars in the Andromeda galaxy made him question the calibration of absolute magnitudes of the Cepheids. Further study revealed that there are actually two types of Cepheid variables: Type-I Cepheids, found in the disk of spiral galaxies like our own; and Type-II Cepheids, found in globular clusters (along with RR Lyrae variables). Baade made numerous observations using the 5-meter telescope at Mt. Palomar, and he calibrated the 1-day, Type-I Cepheid as having an absolute magnitude of −1.5, which is 1.5 magnitudes brighter than the 1-day Type-II Cepheids Shapley had observed. This difference is equivalent to a fourfold change in luminosity between the Type-I's and the Type-II's. Since the original Cepheid calibration had been used to determine the distance to nearby galaxies, and then the nearby galaxies had been used to estimate distances to other galaxies (and ultimately the size of the universe), a fourfold increase in luminosity estimates produced a doubling of all distance estimates. In other words, when Baade realized that the Type-I Cepheids he had been observing in the disk of the Andromeda galaxy were really four times as bright, he was forced to conclude that they were twice as far away. Remember that the apparent brightness of any light source falls off with the square of the distance.

After the period-luminosity relationship of both Type-I and Type-II Cepheids was determined with greater accuracy, these stars came to play a vital role in estimating distances to galaxies. Let us see how this has been accomplished. Suppose that the astronomer isolates a Type-I Cepheid in the disk of a spiral galaxy and observes its period to be 10 days; its absolute magnitude can be read off Figure 11.8 as −4 (see the dotted lines). Now suppose the same star has an average apparent magnitude of +21. Representing its luminosity as "L," the astronomer may use the following proportion to determine its distance (r):

$$\frac{L(10)}{L(r)} = \left(\frac{r}{10}\right)^2$$

Apparent mag. = +21
Absolute mag. = − 4
Difference =   25

A change of 25 magnitudes is equivalent to a change in luminosity of 10^{10} (see Table 9.1, page 337), therefore,

$$\frac{10^{10}}{1} = \frac{r^2}{10^2}$$
$$r^2 = 10^{12}$$

Thus,

$r = 10^6 = 1$ million parsecs $= 3.26 \times 10^6$ light-years $= 3.26$ million light-years

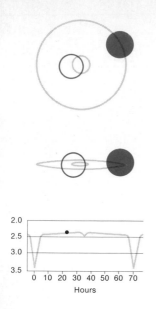

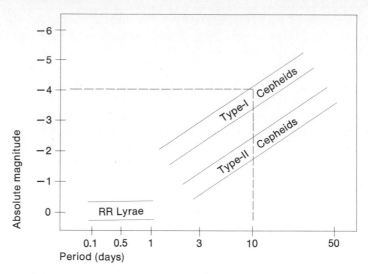

FIGURE 11.8   A modern period-luminosity graph of variables.

LONG-PERIOD VARIABLES

A variable star may be irregular in both its period and its range of magnitude variations. The light curve of Mira Ceti is shown in Figure 11.9. We see that its maximum magnitude is between 3 and 5 and its minimum between 8 and 10. Its period is about 330 days; this does not mean, however, that any single maximum is predictable. What kind of changes in a star can produce a 5-magnitude (hundredfold) change? Mira Ceti is an M-type star with a surface temperature which ranges between 2000° and 2600°K. Such a change in temperature can produce at most a threefold increase in total light output. The second factor that could affect light output is a change in size. Mira Ceti is one of the 8 or so stars whose diameter may be calculated by using an interferometer; its diameter is approximately 300 times that of the sun, and during pulsations it increases to 360 solar diameters, an increase of 20 percent. This expansion produces almost a 50 percent in-

FIGURE 11.9   The light curve of Mira Ceti, an irregular variable.

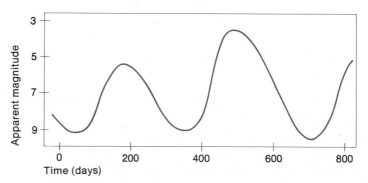

crease in surface area. Temperature and size change, however, do not account for the total variation in light output. The atmosphere perhaps holds additional clues; it has been suggested that atmosphere of the star must also undergo drastic changes.

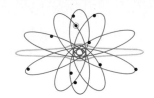

IRREGULAR VARIABLES

Betelgeuse is another example of an *irregular* variable; it shows a variation of one magnitude in a period of 5 to 6 years. Its diameter varies between 500 and 750 solar diameters. The T Tauri stars, still another group of irregular variables, are characterized by their very rapid variations. Stars of this type are usually found in young clusters, often with a certain amount of uncondensed gas (nebulosity) in the star's vicinity. This suggests that the T Tauri stars are probably very young, still contracting, and not yet stabilized as main-sequence stars. T Tauri stars can also be identified by strong hydrogen-alpha emission lines in their spectra, something that is not characteristic of older pulsating stars.

RECURRENT NOVAE

Ancient observers thought that they were seeing the birth of a new star when from time to time a one would suddenly appear where no star had been visible before, and so they called it a *nova*, meaning "new" star (Figure 11.10). Far from being new, the nova is probably a star nearing its old age and is "new" only in the sense that it was dim (or perhaps not detectable at all) before it suddenly brightened without warning. A typical nova may brighten by a factor of 12 magnitudes, corresponding to an increased light output of 60,000 times its former self. We now know that a

FIGURE 11.10   Nova Herculis 1934, showing the large change in brightness between (a) March 10, 1935, and (b) May 6, 1935. (Lick Observatory)

(a)

(b)

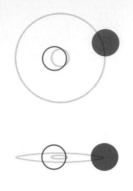

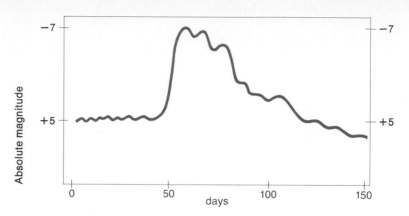

FIGURE 11.11 The light curve of a typical nova.

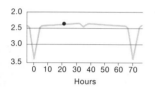

star may repeat such a performance many times. After the occurrence of a nova, astronomers search previous photographs to see what the star looked like before it became a nova. A typical prenova star is a small blue or white dwarf not much larger than the earth but having an absolute magnitude like that of the sun. Figure 11.11 graphically depicts the light curve of a typical nova (note its variability at minimum). The eruption may take place in only a few hours, then require several weeks, months, or years to subside. The nova usually returns to its prior luminosity or only slightly less than that, indicating that the explosion was a superficial one involving only its upper atmosphere. Probably less than one ten-thousandth of its mass is lost. Nova Aquilae 1918, for instance, increased in brigthness 13 magnitudes (from +5 to −8), and a shell of gas was thrown off at a velocity of 1700 km/sec.

By what mechanism can a star brighten by 12 or 13 magnitudes, expel a shell of gas, and then return to its former brightness? It is a very common occurrence for stars to be born in binary pairs, as we saw in Chapter 10. Of the two stars, the more massive one "burns" its nuclear fuels faster than its less massive mate and thus evolves faster. Suppose that the more massive star has evolved to become a hot dwarf star while its mate is still in the red giant stage of its life cycle. As the latter star expands, some portion of its outer layer approaches the dwarf star close enough to experience a stronger gravitational force; as a result, some of the material of the cool giant will be transferred to the dwarf. Such a renewal results in hydrogen-rich gases pouring onto the hot dwarf. At first, an accretion disk forms around the dwarf star, having an appearance somewhat like that of Saturn's rings. While orbiting the dwarf in this disk, material is accelerated to several thousand km/sec. When this material eventually falls onto the surface of the dwarf, it produces temperatures measured in tens of millions of degrees (K) and triggers the gigantic reaction we witness as a nova. This model depends on the presence of a second star (a binary pair) for its validity; the explanation for a supernova, however, is quite different.

SUPERNOVAE

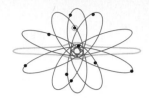

The name *supernova* implies a big nova; the light curve of a supernova may superficially resemble that of a nova, but that is where the similarity ends. The record of supernovae suggests that they are very rare occurrences: only one has been recorded during the past four centuries in the Milky Way galaxy, that occurring in 1604. As you will see in our study of evolving stars (Chapter 13), at a certain point in time the core of a massive star may suddenly collapse, creating an implosion (similar to an explosion, except that the motion is inward). This collapse triggers a release of energy so great that the star may outshine the entire galaxy in which it is located. It would not be unusual for such a star to brighten to 100 million times its normal brightness, a change of 20 magnitudes, then gradually subside to something less than its original magnitude. In the process of the implosion, a shock wave is thought to move outward from the star, driving off a significant portion of the star's mass (several solar masses may be lost) at high velocities (up to 10,000 km/sec) and to produce a visual image like that of the Crab Nebula (Figure 11.12). The star which exploded to form this nebula was recorded by the Chinese in A.D. 1054 as a new star that could be seen in the daytime. Far from being a new star, however, the star that produces a supernova reaction is in a later stage of evolution.

FIGURE 11.12 Supernova in NGC5236 (a) before and (b) after brightening in 1972. (Palomar Observatory Photograph)

(a)   (b)

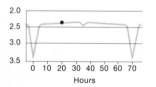

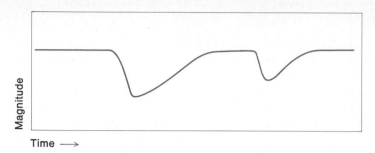

FIGURE 11.13 The light curve of R Coronae Borealis.

R CORONAE BOREALIS STARS

A class of variable stars having a light curve that seems quite different from those that we have studied is the R Coronae Borealis type (Figure 11.13). They do not have a predictable period; they suddenly drop several magnitudes and then slowly return to normal brightness. When at a minimum, some reveal a small shell of gas surrounding the star; the spectrum of this shell is surprisingly similar to that of a planetary nebula. Furthermore, the distribution of these stars is quite similar to that of the planetary nebulae. We cannot be sure of this relationship, but the R Corona Borealis stars form an interesting group and constitute at least a possible predecessor to the planetary nebulae (also to be considered in Chapter 13).

FLARE STARS

The final class of variable stars to be discussed is exemplified by UV Ceti, a flare star. Certain stars may brighten periodically due to flares similar to those that occur on the sun. A flare on the sun is a localized hot spot. A flare is difficult to spot on a star that is very bright, since the proportionate increase in light output is small. By contrast, a flare on an intrinsically faint star, say an M-type dwarf, is much more apparent. Typically, the flaring process creates a very rapid variation in light output, lasting only minutes. If this phenomenon is similar to the localized flare on the sun, it represents a tremendous release of energy in a very short period of time, hence a real explosion—perhaps much more violent than that we observe on the sun.

We have studied a number of special types of stars, some of which surely represent different stages in the life cycle of stars in general. In Chapter 13, we will suggest an evolutionary progression for these various types.

QUESTIONS

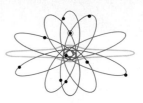

1. How much variation in brightness is shown in Delta Cephei?
2. True or false: Of all the different types of variable stars, the Cepheids can be distinguished by their light curve.
3. What physical change appears to produce the variations of brightness in the Cepheids?
4. What relationship makes the Cepheids a valuable distance indicator?
5. What characteristic of the RR Lyrae stars makes them good distance indicators?
6. The discovery that there are really two types of Cepheids, Type I and Type II, led to a reevaluation of the size of the known universe. Explain this reevaluation.

7. What circumstances are thought to have produced the Crab Nebula?
8. What evidence suggests that a planetary nebula bears little or no relationship to a nova?
9. There is some indication that stars may evolve through several variable stages. Is the T Tauri group thought to be composed of young or old stars? What evidence supports your answer?
10. What distinguished the light curve of the R Corona Borealis stars from that of other variable types?
11. Why is it thought possible for novae to recur over and over, whereas a supernova probably represents a one-time event for a star?
12. RR Lyrae stars are classified as Type-II variables. Explain the meaning of this classification and describe where RR Lyrae stars are usually found.
13. The period-luminosity graph of Figure 11.8 tells us that a 1-day, Type-II Cepheid has an absolute magnitude of -1. Suppose this same star had an apparent magnitude of $+14$, and find its distance.
14. True or false: A flare star can be recognized as showing a variation in its light output only if it is a rather cool star.

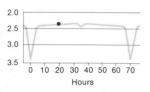

The Interstellar Medium

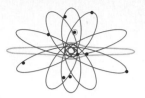

Twelve

Although the space between the stars can be characterized as almost a perfect vacuum (it is a hundred times better than what can be produced in a modern physics laboratory), it still holds a very significant portion of the mass of the Milky Way galaxy and hence of the universe. In our definition of interstellar media, we include atomic gases, molecules, dust, and cosmic rays (sub-atomic particles)—all matter whose individual particles are too small to be seen by the unaided eye. Let us see how each of these categories of material manifests itself.

INTERSTELLAR GASES

The most obvious manifestation of interstellar gases is that of the *emission nebula*. The Orion Nebula, which is visible to the naked eye as a fuzzy patch in the sword of Orion, is an example of such a nebula [see Figure 12.1(b)]. This constellation can be found in the central portion of the winter sky map in Appendix 13 (page 539). A photograph of the Orion Nebula made using the 3-m telescope of Lick Observatory is shown in Figure 12.1(a); it reveals many of the details of its nebular structure and gives an indication of the four very hot stars imbedded in its central region, called the trapezium. The atoms that compose this nebula are excited by the

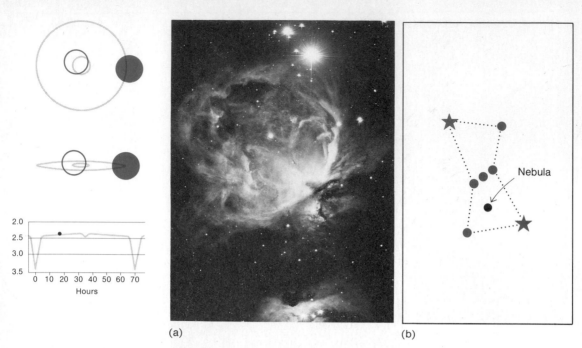

FIGURE 12.1 (a) The Orion Nebula, NGC 1976. (Lick Observatory) (b) The nebula identified in the constellation of Orion.

radiation of these four stars, especially their ultraviolet radiation. Upward electron transitions result from this stimulation, and in fact the atoms in many cases are ionized. We refer to this process as *photoionization*. Upon the recombination of a free electron with the ionized nucleus, downward transitions occur and result in the production of light and/or radio waves, as explained in Chapter 2, page 90. The Orion Nebula is shown in Color Plate 14(b), where something of its composition is revealed. The reddish color originates from the bright red line of the hydrogen spectrum (see Color Plate 1). As shown in Table 12.1, the spectrogram of this nebula, and

TABLE 12.1

THE RELATIVE ABUNDANCE OF OTHER ELEMENTS IN EMISSION NEBULAE[a]	
Helium	1000 atoms
Nitrogen	1 or 2 atoms
Carbon	1 or 2 atoms
Oxygen	2 or 3 atoms
Neon	1 or 2 atoms
Sulfur	1 or 2 atoms

[a]For every 10,000 atoms of hydrogen.

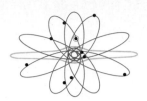

FIGURE 12.2 The Lagoon Nebula in Sagittarius, NGC 6523 (M8). (Lick Observatory)

in fact of most emission nebulae, clearly shows the predominance of hydrogen gas over every other element.

Because the hydrogen atoms are ionized in virtually every emission nebula, such regions are called *H II regions*. Very hot O-and B-type stars can ionize hydrogen (and other) atoms up to a distance of 200 pc. Typically these regions have densities ranging from 1 to 100 atoms per cubic centimeter and high temperatures (10,000°). Figure 12.2 shows yet another emission nebulae called the Lagoon Nebula. Only about 10 percent of interstellar gases may be found in such an ionized state. The remaining gases which are distributed throughout the Galaxy are typically very cool (60°K). At this temperature, hydrogen gas is in a neutral state and such regions are referred to as *H I regions*. Electrons are in their lowest energy level and cannot emit light. Fortunately another atomic process is a work by which the presence of neutral hydrogen may be known.

21-cm RADIO EMISSION

In addition to any other motion that the proton and electron of an hydrogen atom may have, each of these members is spinning on its own axis. Normally the orientation of their spin axes is parallel and pointed in the same direction. On occasion, the spin axis of the electron will flip into the opposite direction, and in the process a small amount of energy will be radiated away in the form of a 21-cm radio signal. Because nearby stars are not required to excite this action, the distribution of neutral hydrogen may

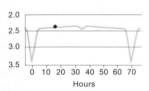

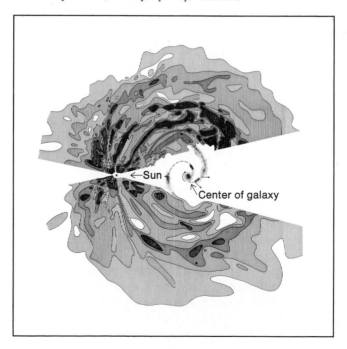

be mapped by its own radiation. To do this, the radio astronomer tunes the radio telescope to receive only the 21-cm signal, as you might tune in a certain radio station. Then, by pointing the telescope in various directions, the astronomer may plot the distribution of neutral hydrogen gas in the Galaxy. The concentration of neutral hydrogen can be judged by the strength of the 21-cm radio signal, and the motions of different regions can be judged by the Doppler effect, as it acts to lengthen or shorten this characteristic wavelength. All of these factors have been used to produce the plot of neutral hydrogen you see in Figure 12.3; the darker regions in the figure represent the highest concentrations.

Gas clouds may reveal themselves in one further way—by the absorption process of upward electron transitions. Because these gases are cool and electrons are in their lowest energy state, any light from a bright star passing through the region would create upward electron transitions from the first energy level. If starlight passes through several clouds on its way to our eyes (telescopes), multiple absorption lines may be evident, each indicating a Doppler effect associated with the motion of the individual clouds. Figure 12.4 shows multiple absorption lines, featuring the K-line of calcium. Because many of the upward electron transitions originate with energy level 1, the absorption lines produced by these transitions are in the ultraviolet portion of the spectrum, and these are obscured by the earth's atmosphere. We have surmounted that problem by orbiting the

FIGURE 12.3 The spiral nature of the Milky Way galaxy as constructed from 21-cm, neutral-hydrogen observations made at Leiden Observatory, Netherlands, and at Radio Physics Laboratory, Sydney, Australia.

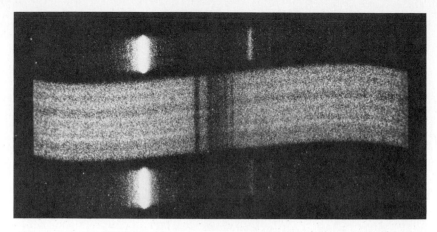

FIGURE 12.4 A multiple-line spectrum indicating several clouds between the source and the observer, each with slightly different line-of-sight velocities. (Kitt Peak National Observatory)

ultraviolet observatories (satellites) such as OAO-Copernicus and the International Ultraviolet Explorer Satellite. These fine tools produce ultraviolet spectrograms (see Chapter 2, page 103). Think of this spectrogram as an extension of the visible spectrum into the ultraviolet. The very fact that hydrogen molecules (H_2) are identified in particular regions of space is an indication that these are rather cold regions, for only under such circumstances can atoms form into molecules. The presence of H_2 molecules is an indication that dust is probably also present in these regions.

INTERSTELLAR DUST

Interstellar dust consists of solid particles that are large compared to the diameter of a hydrogen atom; dust particles are about a thousand times larger. The composition of these dust grains probably includes many elements such as hydrogen, oxygen, carbon, nitrogen, and possibly some silicates. Some grains show a high reflectivity suggesting almost a frosty surface, one in which a mantle of ice covers the grains. Certainly it is cool enough inside of dust clouds (100°K to 20°K) for this to be true. It is not surprising, then, that dust clouds often exhibit themselves in *reflection nebulae*. For example, the Pleiades are enveloped in a nebulosity that might be mistaken for an emission nebula (Figure 12.5); however, the spectrum of the nebulous material is almost the same as that of the stars it surrounds. This indicates that the nebulous material is merely reflecting the light and so must have nature of dust. Gas molecules are not large enough to reflect light nor can they effectively absorb light over the continuous spectrum. Gas clouds absorb a limited number of wavelengths to produce an absorption spectrum, but this does not appreciably dim the light from a source beyond the cloud. Dust clouds, on the other hand,

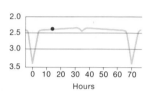

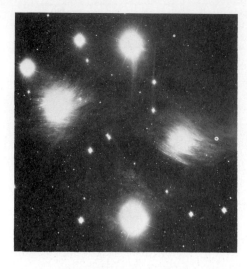

FIGURE 12.5 The nebulosity surrounding the Pleiades is thought to be primarily dust. (Mount Wilson and Las Campanas Observatories, Carnegie Institute of Washington)

appreciably dim such objects. The Horsehead Nebula is an example of such a dust cloud (Figure 12.6).

The North American Nebula is an emission nebula; however, the portion roughly corresponding to the Gulf of Mexico is an absorption nebula, consisting basically of cosmic dust (Figure 12.7). Star counts are appreciably reduced where dust clouds intervene (note the portions cor-

FIGURE 12.6 The Horsehead Nebula in Orion. The dark "horsehead" figure represents a dust cloud that obscures the light from behind. (Palomar Observatory Photograph)

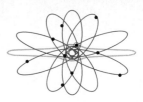

FIGURE 12.7 The North American Nebula in Cygnus. (Lick Observatory)

responding to the Gulf of Mexico and Atlantic Seaboard areas in Figure 12.7). It should also be noted that when we characterize a cloud as a "dust" cloud, it is not exclusively dust; only a small percentage (perhaps 10%) of the mass of such a cloud is actually dust—the remainder is composed of gases that do not glow in the absence of ultraviolet radiation.

The dark central region of the Milky Way, near the constellation of Sagittarius, is known to be a region of cosmic dust that obscures our view of almost all objects behind it (Figure 12.8). This region which is aligned with the center of our Galaxy was called the "Zone of Avoidance" by early observers, for no galaxies were evident in that direction. Modern astronomers are penetrating this zone by using infrared sensitive detectors (for example C.C.D.'s, see page 100). Infrared wavelengths are not so strongly affected by dust; this part of the spectrum has recently revealed new galaxies beyond the center of the Milky Way (see Chapter 16). The presence of obscuring clouds makes the sky look spotty in many regions, and what appears to be a lack of stars in a given area of the sky may signal the presence of such clouds. Concentrations of dust may also appear as small black globules against the background of a glowing nebula (Figure 12.9). In Chapter 13, we will see that these globules are the sites of new star formation. Dust plays a significant role in this process.

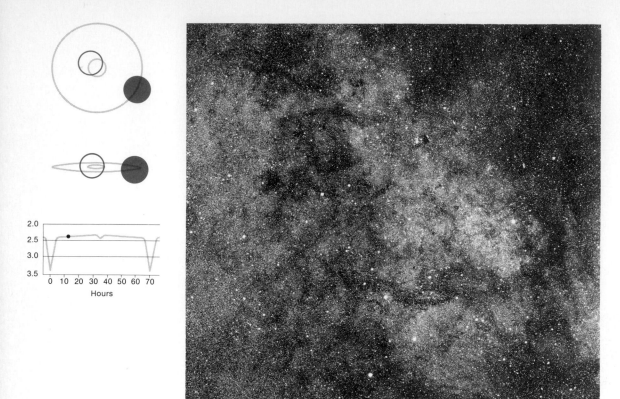

FIGURE 12.8   The Milky Way near Sagittarius. (Lick Observatory)

The starlight that enters a telescope is often found to be polarized. Most astronomers agree that the most likely cause of this polarization is the presence of dust in the path of the light. Interstellar dust grains are believed to have a variety of shapes, some of which may take the form of long strings of molecules. If their form were longer in one dimension (perhaps even needlelike), then their ability to polarize starlight would be understandable. Furthermore, if they existed in a magnetic field, the particles would align themselves and would act like the long narrow crystals that are laid down to form polarizing sunglasses. Astronomers have measured the polarization of starlight in virtually every direction within the Milky Way galaxy and have constructed a map like that shown in Figure 12.10; the length of the lines indicate the degree to which the light is polarized and the direction of the lines indicates the plane of polarization of the electrical component. Although there is some randomness, there appears to be an overall pattern that suggests a general magnetic field within the Galaxy. Furthermore, if a map of the general distribution of interstellar gas were compared to the map of polarization, we would find

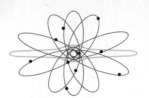

FIGURE 12.9 Dark globules in the Lagoon Nebula (M8). (Lick Observatory)

minor concentrations of that gas in the form of long filaments (and in some cases, loops) that tend to follow the pattern of polarization and, more generally, the pattern of the general magnetic field of the Galaxy (Figures 12.11 and 12.12). Some observers suggest that expanding "shells" are very typical of the interstellar gas and that these produce shock waves through the media. Several shells may have interacted to form the complex filaments we see or a network of intersecting bubbles. Could a star explosion be sufficient to produce such a shell? These ideas, many of which are relatively new and speculative, illustrate the interrelationships among gases, dust, magnetic fields, and the ultimate structure of interstel-

FIGURE 12.10 The polarization of starlight in the Milky Way suggests the overall structure of its magnetic field.

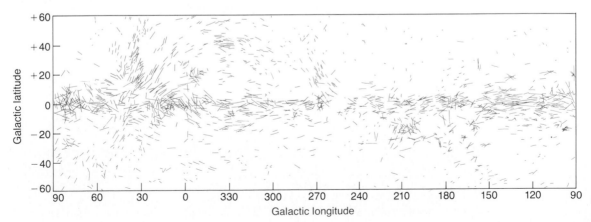

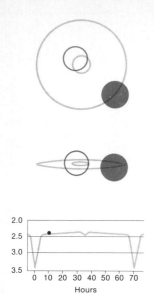

FIGURE 12.11 By its form, the Veil Nebula tends to suggest the presence of a shock wave moving through space. (Lick Observatory)

FIGURE 12.12 Cygnus Loop, a shell of expanding gas thought to have been formed by the explosion of a star. (Palomar Observatory Photograph)

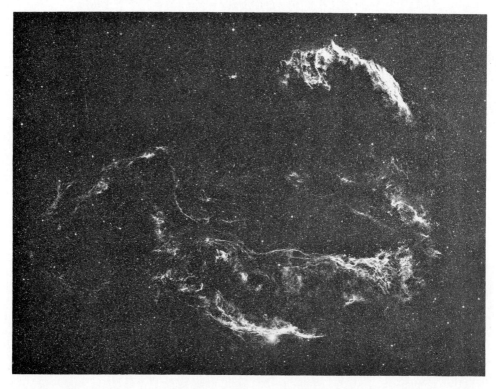

lar media. If dust were not imbedded in the gas, polarization would not occur since gas molecules do not typically polarize light.

In addition to dimming the light of objects beyond these obscuring clouds, cosmic dust also reddens the light that does penetrate. This indicates that the dust absorbs and scatters blue light (short wavelengths) better than red light (long wavelengths), thus leaving the light reddened. The effect of interstellar reddening must be accounted for in the process of determining the spectral classes of stars. The presence of dust in the line of sight causes a star to appear dimmer and redder, and thus more distant than it would otherwise. A correction must therefore be applied, with the amount of correction depending on the amount of reddening. This is determined by comparing the apparent color of a star with its spectral class. For example, if a B-type star (judged by its spectrum) looks like an A-type star (judged by its color), then the observer would compute the amount of reddening and make allowance for this factor in determining its distance. Such striking effects of interstellar dust would seem to suggest that it must constitute a significant part of interstellar material; however, the total mass of dust represents only about one percent of the mass of the interstellar clouds.

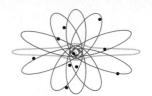

INTERSTELLAR MOLECULES

The first hint that molecules exist in interstellar space came in 1937 with the recognition of certain spectral lines associated with methylidyne radical, an atom of carbon linked to an atom of hydrogen (CH), its ion (CH^+), and cyanogen molecules (CN). The real breakthrough came with the advent of radio astronomy, which ultimately revealed some very complex molecules. One of the reasons radio telescopes are so effective in the study of molecules in space is that radio energy penetrates dust clouds that are opaque to visible light and it is within such dust clouds that the more complex molecules are found.

Molecules have certain inherent abilities to generate radio wavelengths that independent atoms do not have. When two or more atoms bond chemically to form a molecule, the bonds are not static, but rather the atoms vibrate in relation to each other. They also rotate end-over-end or in some other fashion. Any change in this rotation or vibration pattern represents a change of energy state, somewhat like an electron transition or an electron spin-flip. When the molecules change from a higher to a lower state, an electromagnetic disturbance is generated and typically the energy released is in the form of a radio signal. Fortunately, each molecule` has its own characteristic wavelength of radiation. Table 12.2 presents a listing of some interstellar molecules and their characteristic wavelengths. To determine the distribution of methyl alcohol in space, for example, you would tune a radio telescope to 35.9 cm. If you detected an increased radio (noise) signal at that wavelength, you would be detecting methyl alcohol along the line of sight of your telescope since no other molecule emits that wavelength. The abundance of any given element is judged from

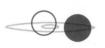

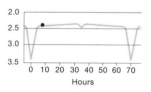

2.0
2.5
3.0
3.5
0 10 20 30 40 50 60 70
Hours

the strength of its spectral "line." In the case of radio emissions, a spectral "line" would look something like that shown in Figure 12.13; this is a profile of the spectral "line" of ammonia in radio emission with its peak at 1.27 cm (rounded to 1.3 cm in Table 12.2).

Certain molecules can be described as masers in interstellar space. Lasers and masers are very similar in their operation. Their names are acronyms for "light amplification by stimulation emission of radiation" and "microwave amplification by stimulation emission of radiation" respectively. Energy levels within molecules can be thought of in a way similar to that which is thought to exist in atoms (see Chapter 2, page 90). In the case of a maser or laser, molecules are "pumped up" (excited) in such a manner that many more atoms exist at a higher energy level than one would expect for a given temperature. This pumping action may result from collisions among the molecules, but it is more likely due to radiation from a cosmic energy source. When a molecule returns as it naturally does

TABLE 12.2

INTERSTELLAR MOLECULES AND THEIR CHARACTERISTIC WAVELENGTHS[a]		
Molecule	Formula	Characteristic Wavelength
Methylidyne	CH	4300 Å
Methylidyne (ionized)	CH+	3958 Å
Cyanogen radical	CN	3875 Å, 2.6 cm
Hydroxyl radical	OH	18.0,6.3,5.0,2.2 cm
Ammonia	NH_3	1.3, 1.2 cm
Water	H_2O	1.35 cm
Formaldehyde	H_2CO	6.6,6.2,2.1,0.2 cm
Carbon Monoxide	CO	2.6 mm
Methyl alcohol	CH_3OH	35.9 cm
Hydrogen Cyanide	HCN	3.4 mm
Cyanoacetylene	HC_3N	3.3 cm
Formic acid	HCOOH	18.3 cm
Silicon monoxide	SiO	2.3, 3.4 mm
Carbon monosulfide	CS	2.0 mm
Formamide	NH_2CHO	6.5 cm
Carbonyl sulfide	OCS	2.5 mm
Methyl Cyanide	CH_3CH	2.7 mm
Isocyanic acid	HNCO	1.36 cm, 3.4 mm
Methylacetylene	CH_3CCH	3.5 mm
Acetaldehyde	CH_3CHO	28.1 cm
Thioformaldehyde	H_2CS	9.5 cm
Methanimine	CH_2NH	5.8 cm
Hydrogen sulfide	H_2S	1.8 mm

[a]Over 50 molecules have now been identified in space and over 40 of those molecules could be termed "organic." There is no reason to believe we have found all of the different molecules that exist in space.

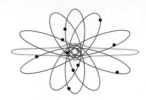

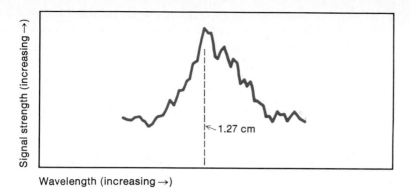

FIGURE 12.13 A radio spectrogram showing one of the emission lines of ammonia. (From Barry E. Turner, Interstellar molecules, *Scientific American* 228 (3), 50–69 (1973))

to a lower energy state, its characteristic wavelength is emitted and thereby the molecule is detected. The maser or laser represents a source of energy to keep the molecule "pumped up" in an excited state. Because the molecules tend to be excited to certain energy states, they typically emit one, or sometimes two, wavelengths instead of several different ones. In the case of the laser, only a single wavelength of light is emitted. The optical behavior of a single wavelength is quite different from that of white light, a mixture of many wavelengths. For instance, a beam of laser light can be sent to the moon with very little spreading, and thus very little loss of its energy.

The hydroxyl molecule (OH) is a particularly good maser when stimulated, and several hundred such sources have been identified. Water (H_2O) masers are also found in many locations. Thus the list of places where we have found water, water ice, or water vapor in the Galaxy and presumably throughout the universe now includes not only the earth, the ice caps of Mars, the rings of Saturn, and the frost of numerous moons—to name a few—but also interstellar space. The molecules listed in Table 12.2 are composed of five predominant elements: hydrogen, oxygen, nitrogen, carbon, and sulfur. These same five elements are the main constituents of the organic molecules of life as we know it. Some of the organic molecules listed in Table 12.2 are components of amino acids, which are components of the proteins essential to life; for example, formic acid (HCOOH) plus methanimine (CH_2NH) yields the amino acid glycine (NH_2CH_2COOH). One might think that the entire universe was in some sense destined for life. We will look at this question more closely in Chapter 18.

GIANT MOLECULAR CLOUDS

Molecules do not have a general distribution throughout the Galaxy but have preferential locations. Typically they are found in *giant molecular clouds*, of which more than 400 are known. The mass of each cloud falls

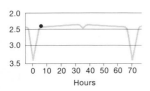

between 100,000 and 300,000 times the mass of the sun, making some of them more massive and more numerous than the globular clusters we studied in Chapter 10. Measuring more than 150 light years across, these giant clouds appear to be distributed somewhat uniformly in the spiral arms of the Milky Way galaxy. How do astronomers detect their presence? While giant molecular clouds are mostly composed of hydrogen molecules (H_2), hydrogen does not radiate the energy by which they are detected. Rather it is a molecule like carbon monoxide that radiates at radio wavelengths. This radiation occurs as a result of collisions between the CO and the H_2 molecules—collisions that create changes in the energy state of the CO molecule. Thus, the CO molecule could be called a "tracer" for the presence of hydrogen.

Within these giant clouds, there exist denser regions (1000 times as dense as the average interstellar material) referred to as globules, with temperatures ranging downward to about 4°K. The presence of dust is thought to be a factor in producing these cooler regions where the slower collision between atoms will increase the probability that molecules will form. These globules, then, should be the natural location for star formation as well.

COSMIC RAYS

A major component of the interstellar media consists of charged particles called *cosmic rays* that have been accelerated to almost the speed of light. These particles include the nuclei of almost every kind of atom, but there is a predominance of hydrogen nuclei (simple protons) and helium nuclei (alpha particles). Also included are electrons, positrons, and neutrinos. The energy of cosmic rays is basically due to their velocity—a small mass traveling at very high velocity possesses high kinetic energy. Imagine an iron nucleus (with 56 protons and neutrons) traveling at 95 percent of the speed of light. Its kinetic energy compares to that of a tennis ball traveling 150 km/hr. If you could catch an iron nucleus with your hand, it would be like catching the tennis ball.

Cosmic rays appear to approach the earth equally from all directions, and as they enter the earth's atmosphere, collisions occur that split various nuclei into smaller (secondary) parts, a process called *spallation*. From the surface of the earth we see primarily these *secondary* particles. The rocket or satellite observatory gives us a better view of the *primary* components as they exist above the earth's atmosphere. Even these primary particles may have experienced some changes as they traveled through interstellar media for millions of years.

As we search for the source of cosmic rays, we naturally look to energetic (violent) processes to explain the tremendous velocities we see in these particles. We know that the sun ejects particles at high speeds especially during a flare, but such energies are too low to explain more than just the low-energy cosmic rays. Furthermore, the flow of cosmic rays is not related to the sun's position. A supernova (an exploding star) has long been suspect, because the energy and particles released are sufficient to explain the energy of the cosmic rays. It is still difficult to explain, however, why we do not sense more cosmic rays from one direction than another. What process could have distributed this radiation? Another possible source lies within the nucleus of the Milky Way. Many very energetic processes are going on there. Charged particles accelerated by any of

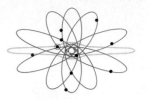

FIGURE 12.14 Cosmic ray abundances as compared to abundances in the solar system.

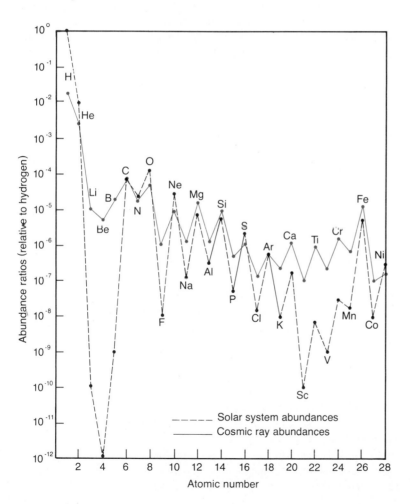

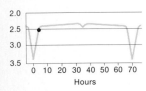

these processes would be influenced by the magnetic field of the Milky Way and most of the cosmic rays would be contained by that magnetic field. Some observers have suggested quasars as a source, since the process whereby they radiate energy equivalent to that of 1000 normal galaxies certainly should be adequate to accelerate particles as well. As a final suggestion, some observers believe that shock waves passing through the interstellar media accelerate those media.

One way in which astronomers are testing their hypotheses is by analyzing the abundance of elements and their isotopes that compose the cosmic rays. The findings can then be compared to the abundance ratios in various sources. Figure 12.14 shows such a comparison between cosmic rays and solar system abundances. Note the similar pattern in most elements except lithium (Li), berylium (Be), and boron (B). Some observers would attribute the higher abundance of these elements in the cosmic rays to the splitting of heavier elements such as carbon (C) and oxygen (O). The solar system abundance ratios are not significantly different from those of nucleosynthesis (the manufacture of heavier elements) in the cores of hot, massive stars. These same massive stars are thought to explode as supernovae at some stage in their life cycle, hence the evidence seems to point toward the supernovae. Only with refinement of abundance and isotope ratios will the question of the source of cosmic rays be settled by this approach.

Albert Einstein stated that our perception of space is influenced by the distribution of matter which produces gravitational fields. We realize that it may also be influenced by the distribution of electrical charges which produce electrical and magnetic fields.

QUESTIONS

1. By what process does an interstellar gas cloud emit its own light?
2. By what process may a nebula produce a radio spectrum?
3. How can the astronomer distinguish a reflection nebula from an emission nebula?
4. Describe the effects that interstellar dust clouds have upon light passing through them.
5. In what sense is the term "cosmic ray" a misnomer?
6. What are possible sources of cosmic rays?
7. What protection does the earth have from continual bombardment by cosmic rays?
8. What did early observers mean by the term "Zone of Avoidance"?
9. How has the modern astronomer penetrated the "Zone of Avoidance" to discover new objects beyond the central bulge of the Milky Way galaxy?

10. How does neutral hydrogen generate a 21-cm radio signal?
11. List the individual elements that form the compounds listed in Table 12.2.
12. Can you explain how molecules found in interstellar space may be distinguished from each other?
13. How is polarization related to the study of dust among the stars?
14. Cosmic "rays" are charged particles like protons and electrons. What distinguishes them from ordinary atomic particles?

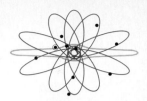

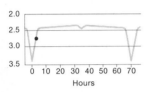

Stellar
Evolution

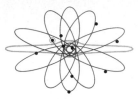

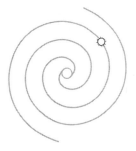

Thirteen

Our study has revealed that stars vary greatly in size, mass, temperature, color, and luminosity. Does this great variety indicate that stars are all different from one another, or does it indicate that stars are merely in different stages of a life cycle? The latter has been shown to be the case; the fundamental reason for the apparent variety in stars is that they undergo change. Like humans, stars go through stages which could be termed pregnancy, birth, youth, adulthood, old age, and death. Only under the rarest circumstances may an astronomer hope to witness a transition from one evolutionary stage to another since the life cycle of an average star extends over billions of years. The astronomer must thus be content to observe stars in different stages and try to arrange them in the proper order.

The problem is somewhat like that which would confront an intelligent being from another planet who had only one day to spend on the earth, and who had no prior knowledge of the life cycle of humans. What approach might this observer use to develop a theory about the life cycle of humans? It could observe humans in all stages of development, but then how could it determine which stage came first? It might judge age by size, or it might judge by other characteristics, such as vigor, metabolism, conditions of the skin, and speech qualities. As a result of these observations,

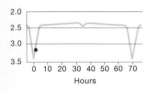

it might deduce the proper life cycle. In a very similar way, the astronomer observes stars in different stages of their life cycle and attempts to place them in the proper order.

To complete the analogy, we must make two assumptions: (1) stars, like humans, have been and are being created continually and hence are of different ages, and (2) stars do change with time, however slowly. The pieces of our puzzle include main-sequence stars, the dwarfs, the giants and supergiants, the pulsating variables, the novae and supernovae, and others. How do we determine which came first? Let us approach this question from a theoretical point of view and decide how a star should behave assuming that it is in essence a ball of gas.

GAS LAWS

Consider a certain number of gas molecules (or atoms) in a container with a plunger-type top. Such molecules are always in motion. The temperature of the gas is a measure of the average velocity of these molecules. The small arrows in Figure 13.1 indicate their direction of motion and the length of the arrows indicates their speed. If the plunger is lowered to reduce the volume but the temperature is held constant [Figure 13.1(b)], the pressure increases. This increase in pressure occurs because the molecules are closer together and hence collision between individual molecules is more frequent. There are also more collisions per square

FIGURE 13.1 Operation of the gas laws: (a) large volume and low temperature produces low pressure; (b) reduced volume and same temperature yields higher pressure; (c) increased temperature and same volume as in (a) yields higher pressure; (d) high temperature and expanding volume results in reduced pressure.

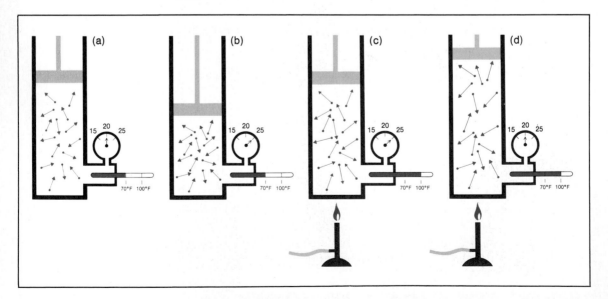

centimeter on the walls of the container and this fact can be read off the pressure gauge which has risen from 20 to 25.

Next, let's compare Figure 13.1(a) with Figure 13.1(c), which shows the gas heated. As the temperature rises, the molecules move faster on the average (as indicated by longer arrows), and this increases the tendency for particles to collide, producing an increase in pressure. Molecules hit the walls of the container with more energy on the average, and this is translated as a reading of 25 on the pressure gauge.

If we started with the conditions of Figure 13.1(c) and raised the plunger as in Figure 13.1(d) until the pressure read 20 again, we would note an increase in volume over that of Figure 13.1(a). When a star is heated, it will tend to expand as long as it behaves like a gas. An expanding gas, however, will typically experience a cooling effect. As a star produces energy at an ever-increasing rate, it heats up which causes it to expand, but expansion produces cooling which tends to slow the rate of energy production and keep it in check.

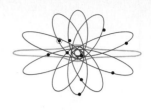

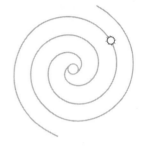

INITIAL ENERGY SOURCE

The very fact that the molecules of a gas cloud are spread out over a given volume and that they attract one another gravitationally represents a form of energy called *potential energy*. If that cloud contracts, it is capable of converting some of this potential energy into *kinetic energy* (energy associated with motion) by accelerating, the molecules and thus heating the gas. As a gas cloud condenses, it not only gets more dense but also increases in temperature owing to this energy conversion. Then, if temperature and pressure are sufficiently high, other sources of energy (thermonuclear) will also become activated.

MODEL BUILDING

The astrophysicist obviously cannot experiment with star formation and evolution in the same way that a chemist or physicist experiments in the laboratory, for he cannot get his hands on a star. Rather, the astrophysicist builds models based on his knowledge of fundamental laws of nature, such as the law of gravity, the laws of motion, and the gas laws. The model of a star also includes a statement of factors such as mass, density, and temperature at various radii within the star. To see how a star might evolve under a given set of circumstances requires mathematical computations entailing billions of separate steps, and this requirement stymied such an approach until the advent of the electronic computer. The modern computer can perform millions of arithmetic steps per second. When programmed to follow the basic laws of nature and instructed as to the initial conditions of mass, density, and temperature, it can trace the evolution of a star and predict its state at any specific time. The theoretical model that results shows a star contracting under the force of gravity, heating, initiating a thermonuclear reaction like the one we saw in the sun, stabilizing,

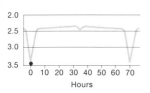

then later growing into a huge but cooler star, and finally taking on one of several forms of death. The question which naturally follows this simulation is, Can we find evidence among the stars that these changes really take place?

FORMATION OF A STAR

Where do we look to find a star in the making? We know that gas and dust, the building blocks for stars, are scattered throughout interstellar space. The density of that material, however, is only 0.1 to 10 atoms/cm^3 and its average temperature is 100°K (−173°C). Under these conditions the atoms would be moving so rapidly (in a random fashion) that there is little likelihood that gravity could arrest their motion and hold them together in a knot. A more likely situation would involve a denser concentration of matter and a cooler temperature. As you can see in Figures 13.2 and 13.3, small dark regions often appear silhouetted against bright nebulae. These globules are dark primarily because they contain not only higher concentrations of gas (10^4 to 10^6 atoms/cm^3) but also dust particles that effectively screen out the ultraviolet light of nearby stars. Temperatures in these dark globules average only 10°K (10° above absolute zero). These regions are also typified by the presence of molecules, as verified by their radio emissions. In fact *giant molecular clouds* exist that may be more significant with respect to star formation than the visible globules themselves.

The slow random motion which is natually associated with low temperatures makes the formation of a star far more likely; however, to say that a cloud of gas and dust will turn itself into a star by its own self-gravity may be a bit too simplistic. When small regions of such a nebula become unstable and start to form a knot of material, that knot experiences pressure forces from within that tend to destroy it. Certain types of external forces, though, may act on the globule to trigger star formation, such as the

FIGURE 13.2 The dark globules in this nebula (IC 2944) in the constellation of Centaurus suggest regions of cool gas and dust, perhaps regions of star formation. (Cerro Tololo Inter-American Observatory)

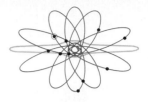

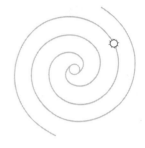

FIGURE 13.3 Dark globules in the Lagoon Nebula (M8). (Lick Observatory)

shock wave produced when certain stars explode as supernovae tend to sweep up material, creating pockets of higher density (see Figures 12.11 and 12.12, page 392). If enough mass of gas is compressed, then its own gravity may be sufficient to overcome internal pressures and it may continue to accrete (pull in) material from the surrounding nebula. Figure 13.4 shows the site of supernova; the figure shows the shell of material thrown off by this stellar explosion. The expanding shell generates a shock wave which produces higher concentrations of material along its borders. New stars are detected by their infrared radiation when they are just warming up.

For years astronomers have recognized young stars that have begun to shine by their own internal energy. Of late, however, they have succeeded in detecting stars earlier in their formation, perhaps up to 1 million years before they will be clearly visible in ordinary light. Assuming that the observers have chosen a dark globule, with molecular radio emissions, their first real clue to a star in the making comes via infrared wavelengths. Infrared radiation penetrates the dust that surrounds a forming star core. It is in fact this dust, heated as a result of the energy given up in the gravitational contraction of the core, that radiates the infrared energy.

As the core continues to heat, say to 500°K, some of the surrounding gases will be ionized, producing an H II atmospheric region. Such a region may be detected by a broad range of radio emissions. The late Bart Bok, who identified far less obvious dark regions (Figure 13.5) for years and associated them with possible star formation, was delighted with the confirmation that is now in progress. With specially sensitized photographic plates, he recorded faint images which penetrate the dust. This particular type of dark globule has been termed a Bok globule in his honor. The

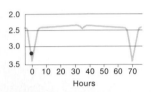

latest approach to the detection of dark globules involves a very sensitive detector called C.C.D. (charge coupled device, discussed on page 100). This is another example of how electronic devices are extending the level of sensitivity beyond that which has been achieved photographically.

As the condensation knot grows and a temperature of about 1500°K is reached, the *protostar* (a star in the making) becomes clearly visible on red-sensitive plates. The photo record of condensations in the Orion Nebula actually shows growth over a period of only a few years (Figure 13.6). This may represent one of the rare examples of a crucial stage in stellar evolution occuring on a time scale which humans can witness. Of course, we are seeing only a very brief segment of a star pregnancy in these photos, for the average time needed for a star (such as the sun) to reach this state is of the order of a million years.

FIGURE 13.4 The Omega Nebula is the result of a supernova explosion. Astronomers suspect that the shock wave produced by such an event may have triggered star formation in the location indicated by an infrared object—a star in the making. (Lick Observatory)

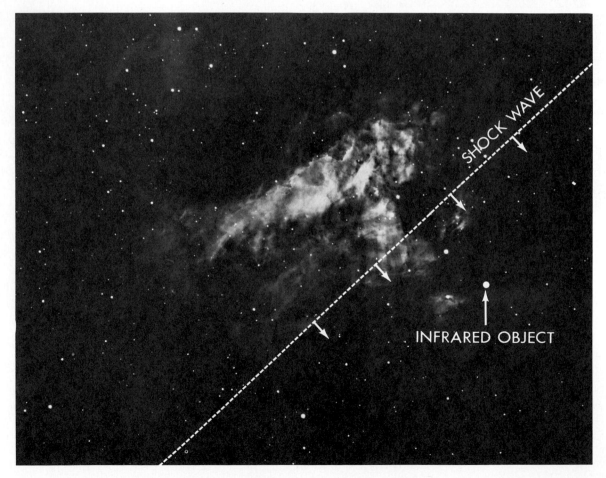

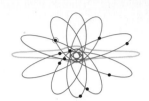

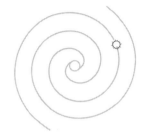

FIGURE 13.5 An enlarged section of the nebula in Monoceros, showing dark regions (see Figure 13.7 for a view of the entire nebula). (Palomar Observatory Photograph.)

FIGURE 13.6 Herbig-Haro objects in Orion, taken (a) in 1947, (b) in 1954, and (c) in 1959, showing growing concentrations. (Lick Observatory)

(a)   (b)   (c)

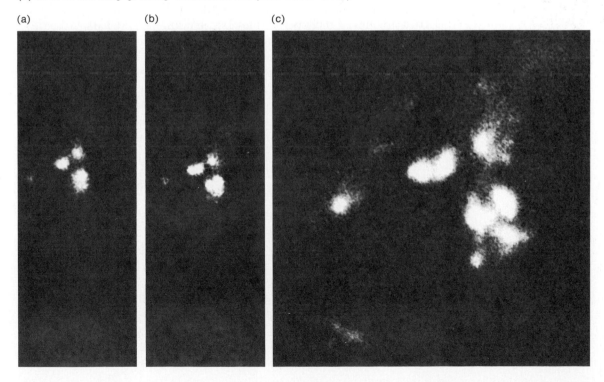

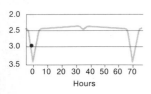

Most models of star formation assume that when enough mass collects to form a knot of material and that knot survives any disruptive influence, the force of gravity will finish the job by simply gathering in more material which will collapse to form the star. There are recent observations which suggest that in the case of more massive (O-and B-type) stars, a rather violent event occurs during their protostar stage. Carbon monoxide molecules typically emit a 2.6 mm radio signal as they make a transition from one rotational state to another. This radio "line" normally would be rather sharp and would indicate the motion of a gas cloud by its Doppler shift. The remarkable new discovery is that the "line" is not sharp, but is broadened by the motion of gases which are flowing generally toward the observer and by the motion of gases which are flowing away from the observer; we are seeing a variety of both red shifts and blue shifts in the spectrum due to the motion of the gases. We are not talking about a mere stellar wind similar to that of the sun (solar wind), for measurements show that more mass may be involved in these stellar flows than in the nucleus that is left behind. The flow consists of some 30 solar masses stretching outward for a full light-year and moving (in some cases) at 60 to 100 km/sec, leaving only some 15 solar masses in the core. There is more (kinetic) energy represented in such a flow than the sun emits in the form of electromagnetic radiation in a million years. In several of the two dozen known cases, the gases are streaming mainly in two opposite directions. Astronomers are hard pressed to give an explanation of the mechanism or source of energy for this streaming. At the moment, these observations do establish that certain potentially very massive stars rid themselves of a great amount of mass before they become full-fledged stars.

With the exception of the unexplained force that causes these stellar flows, the basic force that shrinks the cloud of gas to smaller and smaller dimensions thus forming a star is gravity. Where will this condensation be halted and by what force? The force which tends to counter the force of gravity is pressure. We have noted that pressure gradually increases with a decrease in volume. When a protostar's central pressure has reached a billion earth atmospheres and its central temperature has built up to 10 million degrees Kelvin, the proton-proton cycle begins (see Chapter 8, page 292)—an enormous amount of energy is released, and pressure is increased significantly.

A STAR IS BORN

Although a star can now be detected before it begins to shine, it is not considered truly born until the proton-proton cycle begins; the preceding steps might be classified as its period of pregnancy. The star will continue to contract until the outward force of pressure exactly balances the inward force of gravity at each point within the star, a condition called *hydrostatic equilibrium*. This stabilized state probably does not occur as a simple halting of contraction but rather may be a period in which the star pulsates in size and appears to vary in light output. When astronomers

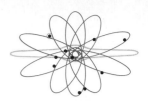

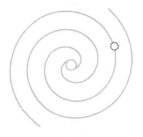

observe certain regions of gas and dust, as in the Monoceros cluster (Figure 13.7), they see a type of variable star which we have already studied, the T Tauri type. These stars show rapid, irregular variations and may be recognized by their typical emission spectra that are thought to originate in the envelope of gas and dust which surrounds these young stars. Most stars, by contrast, have absorption spectra. A plot of the T Tauri stars on the H-R diagram seems to suggest that they may represent an evolutionary state just preceding the stable "main-sequence" stage, for most of them appear just to the right of and above that stable group. In Figure 13.8 the T Tauri stars in the Monoceros cluster (NGC 2264) are shown by the "plus" symbol. Further evidence of the youthfulness of the T Tauri stars is that they are always found in the denser regions of gas and dust.

As a means of visualizing the changes taking place as stars go through their life cycle, astronomers draw evolutionary tracks on an H-R diagram. The tracks presented in Figure 13.9 are due to Margherita Hack. Tracks are shown for stars of differing masses. We have already demonstrated in Chapter 10 (page 358) that more massive stars appear higher on the main

FIGURE 13.7 Nebulosity seen in Monoceros. (Palomar Observatory Photograph)

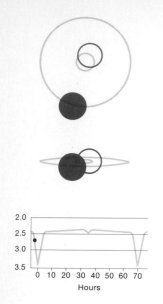

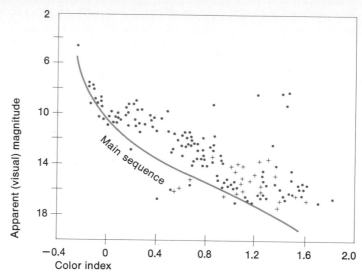

FIGURE 13.8 A plot of the stars in NGC 2264.

sequence. The numbers shown along the main-sequence line indicate the number of solar masses in each star.

As you view these evolutionary tracks in Figure 13.9 and in other figures in this chapter, keep in mind what the changes on the H-R diagram mean. A vertical change upward means a brightening and downward

FIGURE 13.9 The evolutionary track of stars of different masses. (Diagram prepared by Margherita Hack)

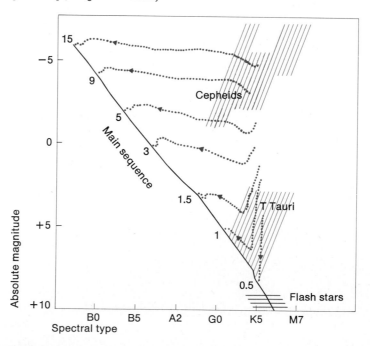

means a dimming. A horizontal change to the left means an increase in surface temperature and to the right means a decrease. Thus an evolutionary track which slopes upward to the left indicates an increase in luminosity (a brightening) accompanied by an increase in surface temperature. This is what we might expect of a star that is heating up but not changing its size appreciably. What must be happening when a track is horizontal and moving to the left? The surface temperature is increasing, yet the luminosity remains unchanged. We conclude that the surface area must be decreasing at the same time.

MAIN-SEQUENCE STAGE

A protostar becomes a main-sequence star when the energy of hydrogen fusion becomes the main source of energy and when a state of hydrostatic equilibrium is reached. The size, surface temperature, and luminosity of the star are basically determined for the major portion of its life cycle. Stars appear to spend at least 95 percent of their visible life in the main-sequence stage. This is verified by the fact that on an H-R plot of any large number of stars selected at random, we find the majority of stars on the main sequence, and there is a direct correlation between the number of stars found in any stage and the length of time stars spend in that stage (Figure 13.10).

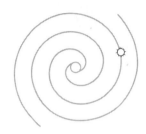

To clarify this idea, let us return to our analogy of the intelligent being from outer space who comes to the earth and tries to decipher the life cycle of humans in one day. It would find less than 1 percent of humankind in cribs, and would therefore conclude that the crib stage is very short. It would find perhaps 3 percent of preschool age, another short period in human life. If youth and adulthood were grouped into one stage, this would account for about 90 percent of the population, indicating that humans spend about 90 percent of their life cycle in the combined state of youth and adulthood. The remaining 6 percent would be the elderly and the dying. The combined youth and adulthood stage would correspond to the main-sequence stage in a star.

The sun typifies a main-sequence star, a stable (nonpulsating) star with a rather constant output of energy derived from the *proton-proton cycle*. Surface activities, such as sunspots and flares, represent only rather superficial variations in output. Not every star is like the sun even in its stable (main-sequence) state. Stars more massive than the sun become much hotter in their core and they experience additional energy-producing cycles. One such cycle is called the *carbon cycle*, or *CNO (carbon, nitrogen, oxygen) cycle*. In this cycle, carbon serves merely as a catalyst and is neither created nor destroyed. In the first step, an atom of carbon fuses with an atom of hydrogen to form an atom of nitrogen-13 and release a gamma ray. The second step represents a decay of nitrogen to carbon-13 by converting a proton to a neutron, releasing a positron and neutrino. Similar processes occur in successive steps. The net effect of this cycle is that four hydrogen atoms are used to make one helium atom, with the same conversion of mass to energy as in the proton-proton cycle. This

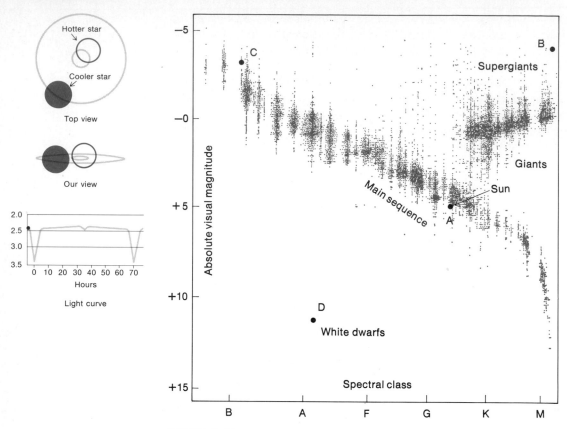

Top view

Our view

Light curve

FIGURE 13.10 The H-R diagram. The points represent approximately 6700 stars of known absolute magnitude. (Yerkes Observatory)

cycle depends on the presence of carbon in a star; nitrogen and oxygen are intermediates in the process.

$$^{12}_{6}C + ^{1}_{1}H \longrightarrow ^{13}_{7}N + \text{gamma ray}$$
$$^{13}_{7}N \longrightarrow ^{13}_{6}C + \text{positron} + \text{neutrino}$$
$$^{13}_{6}C + ^{1}_{1}H \longrightarrow ^{14}_{7}N + \text{gamma ray}$$
$$^{14}_{7}N + ^{1}_{1}H \longrightarrow ^{15}_{8}O + \text{gamma ray}$$
$$^{15}_{8}O \longrightarrow ^{15}_{7}N + \text{positron} + \text{neutrino}$$
$$^{15}_{7}N + ^{1}_{1}H \longrightarrow ^{12}_{6}C + ^{4}_{2}He$$

Throughout the main-sequence stage the star is building a helium core, and this finally leads to rather drastic changes. As the supply of hydrogen available in the core is depleted, the rate of energy production decreases. One might think that the star would be ready to die at this point. Rather than fading away, however, the star undergoes changes that give it new life. With diminished hydrogen burning, the source of internal support is drastically weakened and the helium core contracts under the weight of the overlying layers. The reduction in gravitational (potential) energy which this contraction represents shows up as heat. The increased core temperature in turn excites the surrounding hydrogen shell to even more vigorous burning. The additional energy is expended in the expansion of the outer layer of the star, resulting in an increase in total luminosity which outranks anything the star has achieved before.

Let us look more closely at this process. First, an expansion of gases is generally accompanied by a cooling trend, shown as a movement to the right on the H-R diagram. With such cooling, say from 6000° to 3000°K, the rate at which energy is radiated for each square meter of surface area drops to one sixteenth its previous level, since the luminosity of a star is proportional to the fourth power of temperature and in our example the temperature is one-half $(\frac{1}{2})^4 = \frac{1}{16}$. This drop in luminosity could be compensated for by a fourfold increase in size, since a star's luminosity is directly proportional to the square of any increase in radius. Thus a star like the sun would need only a fourfold increase in radius in order to balance the loss in surface temperature. Observations of some large nearby stars, such as Betelgeuse, show radii ranging up to 750 times that of the sun, so it is not hard to see how the net output of an expanding star could increase despite its loss in surface temperature. If we trace the changes just described on the H-R diagram, they will produce a track which moves to the right as the star cools but upward as its total output of energy increases (see the dashed lines in Figure 13.11, as plotted by Margherita Hack). Because it appears reddish (it is cool on the surface) and because it is large, this kind of star has been dubbed a *red giant*.

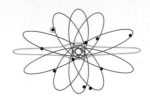

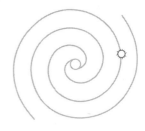

FIGURE 13.11 The evolutionary track of stars of different masses. (Diagram prepared by Margherita Hack)

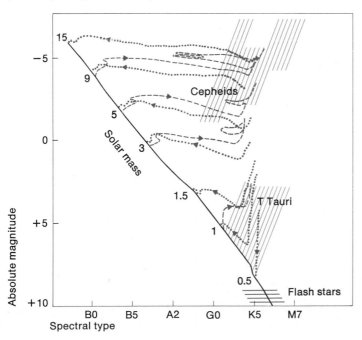

THE RED GIANT

Massive stars do evolve more rapidly because their self-gravity creates much higher central core pressures and temperatures. Their expenditure of fuel proceeds much faster than in stars like the sun, hence more observational evidence exists for these stars. Not very many stars of one solar mass have had time to evolve to the red giant stage, but many stars of three or more solar masses have already reached it. This is indicated by a plot of the stars of an old globular cluster such as M3 (NGC 5272) on the H-R diagram. This cluster is composed of stars of varying masses but all of about the same age; we assume that they are same age because we think the stars in any given cluster have a common origin. We know that the more massive stars plot high on the main sequence, and from Figure 13.12 it is obvious that most of these stars have already evolved away from the main sequence to become red giants. There is a strong possibility that they have followed the tracks outlined in Figure 13.11; in fact, some may have evolved to still later forms, which we have not yet discussed. The stars of one solar mass (or less) evolve so slowly that even in an old cluster like M3 (Figure 13.13) they still remain on the main sequence. A plot of younger clusters in the disk of the galaxy (Figure 13.14) shows the same tendency for the more massive stars to evolve away from the main sequence first. With the exception of M67, these clusters are so young that not even their most massive stars have become red giants.

We must not forget that within the red giant lurks a core that is still increasing in temperature, owing to further contractions. When the core

FIGURE 13.12 A plot of the stars in M3. (Courtesy A. R. Sandage)

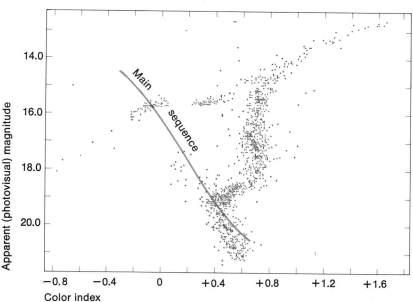

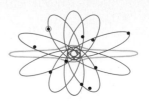

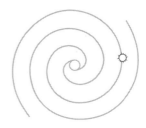

FIGURE 13.13 The globular cluster NGC 5272 (M3). (Lick Observatory)

FIGURE 13.14 The evolutionary tracks of several familiar galactic clusters superimposed on the track of M3.

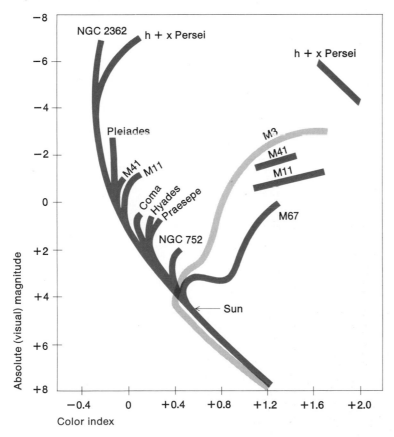

temperature rises to the order of 100 million degrees Kelvin, helium fusion takes place; that is, helium atoms begin to fuse, producing still heavier elements, such as berylium (Be), carbon (C), oxygen (O), and so forth. It should be noted that berylium is quite unstable and should not be expected as a final product, only as an intermediary:

$$^4_2\text{He} + \,^4_2\text{He} \longrightarrow \,^8_4\text{Be} + \text{gamma ray}$$
$$^8_4\text{Be} + \,^4_2\text{He} \longrightarrow \,^{12}_6\text{C} + \text{gamma ray}$$
$$^{12}_6\text{C} + \,^4_2\text{He} \longrightarrow \,^{16}_8\text{O}$$

The star at this point has two sites (layers) of energy production, a hydrogen-burning shell and a helium-burning core (Figure 13.15). This process eventually yields a carbon and oxygen core and may even continue up through an iron core (Figure 13.16).

At this stage, the mass of the star appears to determine the next steps. Let us consider a star of approximately one solar mass. At some point in its contraction and heating, the gaseous core degenerates to a densely packed mass of protons and electrons, which take on the rigidity of a steel ball. If that core were like a gas, it would expand with the increased temperature and start to cool down, controlling the rate of energy production thus acting as a safety valve. Because of its dense nature, however, the core no longer responds like a gas, and without a safety valve the core continues to heat up until the thermonuclear process literally runs away with itself, producing a very explosive reaction called the *helium flash* (see Figure 13.17, point A). We do not literally see such a flash, for most of the energy is absorbed in the expansion of the core to restore its gaseous nature. Hydrogen burning in the shell that surrounds the core appears to be halted by the helium flash, and there is actually a temporary decline in energy output (Figure 13.17, point B). The core once again contracts

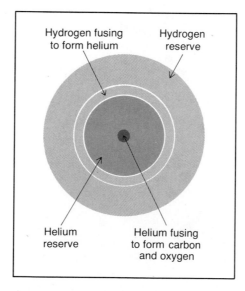

Hydrogen fusing
to form helium

Hydrogen
reserve

Helium
reserve

Helium fusing
to form carbon
and oxygen

FIGURE 13.15 Sites of fusion within the core of an evolving star.

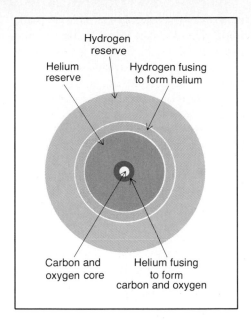

FIGURE 13.16 Sites of fusion within the core of an evolving star in a more advanced stage.

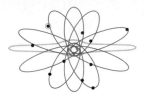

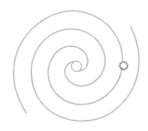

under the influence of gravity and heats up to 100 million degrees Kelvin or more; helium burning then becomes the primary source of stellar energy. In the more massive stars, the degenerate core probably does not develop at this time. Helium burning produces a carbon core and with increased temperatures carbon burning may occur, producing still heavier elements such as silicon.

In the most massive stars this process is thought to continue until an iron core is produced, with each fusion step converting mass to energy. The fusion process stops with iron because iron represents the most stable form in which protons and neutrons can exist. If one expected iron to fuse into yet heavier elements, one would have to supply energy to make that happen.

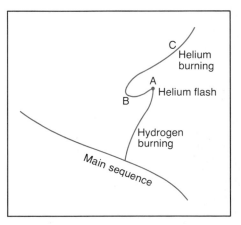

FIGURE 13.17 The evolutionary track of a star as it passes from the hydrogen "burning" stage to the helium "burning" stage.

At this point in the evolution of a very massive star, a most dramatic display takes place. We recall that the size of a star, from the time it first stabilizes, is determined by an interplay between the force of gravity which tends to make the star smaller and the pressure associated with energy production which tends to make it larger. In the iron core, energy production comes to an end; therefore that source of pressure no longer exists and gravity takes over. Within a matter of seconds, the iron core collapses with such force that not even the space within the orbital structure of the atom is preserved—this could be called an implosion. It may be at this moment that "layers" within the iron core fall toward the center at differing rates and as a result an outward moving shock wave is generated. This shock wave is thought to be capable of driving off a large portion of the mass of the star. In the case of a 10 solar mass star, on the order of 85 percent of the mass may be lost. We say that the star goes *supernova*.

THE SUPERNOVA

The first evidence of such an event is a sudden brightening of the star—to ten billion times its previous brightness in just a day or so. If a supernova appears in one of the less luminous galaxies, it may temporarily outshine that entire galaxy. The remains of a supernova recorded in A.D. 1054 by the Chinese is still visible as the Crab Nebula, which continues to expand at the rate of more than 1300 km/sec. From our knowledge of this velocity, together with its apparent increase in size over the years, we have determined its distance to be approximately 6500 light-years (Figure 13.18). We have already noted that a mass several times that of the sun is required to create the conditions which produce a supernova event, so it follows that not all stars will experience this event during their lifetime.

FIGURE 13.18 The Crab Nebula, the remains of a supernova first seen in A.D. 1054. (Palomar Observatory Photograph)

Have you ever wondered how the great variety of atoms were created? We have just explained the process whereby the first 26 elements up through iron are formed by fusing lighter elements. To build yet heavier elements, however, requires energy. The supernova implosion and subsequent explosion is just such a source of energy, for as a result of the shock wave, existing atoms are in very agitated motion and they may collide to form heavier elements. For example, atoms may repeatedly capture neutrons to form heavier isotopes and through beta decay (the ejection of an electron and a neutrino from a neutron, leaving behind a proton), a new, heavier element is formed:

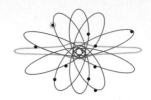

$$^{56}\text{Fe} + \text{neutron} \longrightarrow {}^{57}\text{Fe}$$
$$^{57}\text{Fe} + \text{neutron} \longrightarrow {}^{58}\text{Fe}$$
$$^{58}\text{Fe} + \text{neutron} \longrightarrow {}^{59}\text{Fe}$$
$$^{59}\text{Fe} + \text{beta decay} \longrightarrow {}^{59}\text{Co}$$

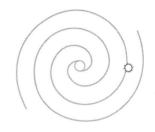

Or a helium nucleus may fuse with an iron atom to form nickel:

$$^{56}\text{Fe} + {}^{4}\text{He} \longrightarrow {}^{60}\text{Ni}$$

Many observers believe that this is the principal way that elements heavier than iron are formed in the universe. At the same time the shock wave of a supernova is the means whereby these elements may be dispersed to mix with the interstellar gases, enriching them for future generations of star formation. Our sun, its system of planets, and our very bodies are a

FIGURE 13.19 Typical light curves for (a) Type I Supernova, (b) Type II Supernova.

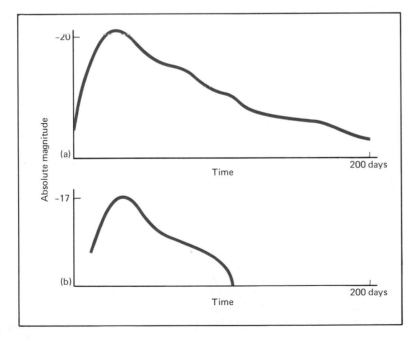

product of star formation from such an enriched cloud of gas and dust. Had our sun been formed during the very early stages of the universe, when only hydrogen and helium were present, life as we know it would not have been possible on the earth. Thus we are suggesting that the atoms which compose your body were created in the cores of very massive stars and in the supernovae events which those stars experienced. Many such events must have occurred before our solar system was formed.

It should be noted that supernovae may be classified into two categories: Type I, those that eject relatively little mass yet which may brighten to a luminosity 10 billion times that of the sun (absolute magnitude of −20); and Type II, those which throw off very significant fractions (1/2 to 3/4) of their mass yet only brighten to −17 magnitude (less than 1 billion times the sun's luminosity). It is the Type II supernova with the larger loss of mass that produces a drastic alteration in the evolution and death of a star. These two types may be easily differentiated by their light curves (Figure 13.19). It is the Type II that we may expect to see most often happening to very massive O- and B-type stars in the spiral arms of our Galaxy.

The way a star dies is largely determined by its mass. This is the subject of the next chapter.

QUESTIONS

1. Describe a likely birthplace of stars and the process that is thought to bring them into being.
2. What evidence suggests that certain stars are relatively young?
3. Not all stars are thought to evolve at the same rate. Which kind appear to evolve most rapidly?
4. Which stage in evolution appears to be the longest and which the shortest for the following stars: variable, T Tauri, red giant, main-sequence? State evidence for your answer.
5. When the evolutionary life cycle of a star is shown on an H-R diagram, a change in location on the diagram does not indicate a change in location in the sky. What does such a change on the diagram indicate?
6. If a star increases in temperature, what change must take place if its internal pressure remains substantially the same?
7. In a gas cloud, gas molecules and dust particles have a certain form of energy which may be converted to heat (kinetic energy) upon collapse of the cloud. Name this form of energy.
8. What is the role of dust in creating a condition in which a star is more likely to form?
9. Name at least two ways in which dust is revealed visually.
10. What role does carbon play in the carbon cycle in stars?
11. Interpret a horizontal movement to the left on an H-R diagram.
12. What evidence do astronomers have that stars spend a large percentage of their time in the main-sequence stage?

13. What is thought to trigger the evolution of a star from the main-sequence stage to the red giant stage?

14. What evidence do we have that the T Tauri variables are associated with very young stars and not with older stars?

15. How is it possible for a star that is getting cooler also to get brighter?

16. When a star runs out of available hydrogen in the core, it should cool down, but instead heats up significantly. Why does this occur?

17. How are the elements heavier than iron manufactured in stars?

18. Why is it not thought possible for the sun to fuse elements as heavy as iron?

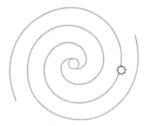

Death
of Stars

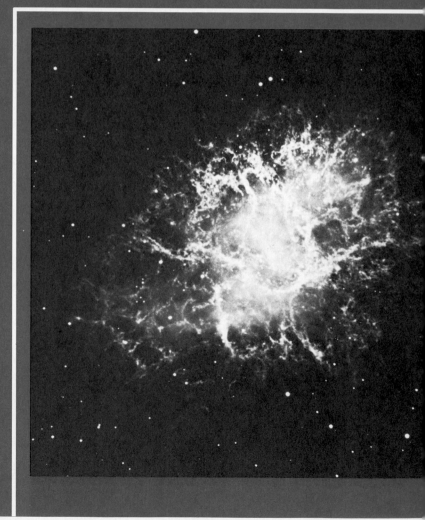

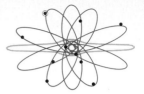

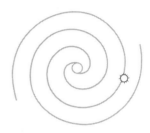

Fourteen

In the last chapter we saw that with the development of a helium core and then a carbon core in a star, higher and higher temperatures are required to ignite nuclear fusion. Eventually the star will simply not have sufficient gravitational force to ignite any further thermonuclear processes; at this point then, nuclear reactions are no longer a source of energy. The star has several sources of energy remaining; its gravitational force is still capable of producing temporary heating through contraction and its rotational energy may still contribute to the production of radiation.

The mass of a star appears to be very significant in determining the way a star dies. Stars of approximately one solar mass are thought to have a rather nonviolent demise. The pressure that derives from the thermonuclear processes holds the star in balance against the force of gravity. But what keeps gravity from crushing this star completely? Nature does provide such a pressure—it is referred to as *electron pressure*. Under the usual gaseous conditions within the star, atoms have plenty of room to maintain their electrons in orbitals, and their average motions are correlated with their temperatures. However, by the time a star shrinks to the size of the earth, atoms are forced so close together that not only are the orbitals of the electrons distorted (essentially destroyed) but the nuclei (protons and neutrons) are forced to assume what amounts to a crystalline-type structure imbedded in a "sea" of electrons. This is now called a *degenerate* gas. The electrons are in rapid motion and this tends to support the structure with no further reduction in size. Furthermore, one of the properties of electrons (stated by the *Pauli Exclusion Principle*) says that no two electrons can exist in the same energy state; thus a limit to further condensation is achieved. The astrophysicist suggests that a star of less than 1.2 solar masses does not have sufficient mass to force its condensation beyond this state. Thus a star of one solar mass would shrink to the size of the earth but would have a density of 2 million g/cm^3 (the sun's current core density is only about 100 g/cm^3).

In the last stages of contraction, the core of the star is heated as gravitational potential energy is given up. That heat is conducted very rapidly to the surface, where the temperature ranges from 50,000°K to 100,000°K. The star is now white hot and relatively small and called a *white dwarf*.

THE WHITE DWARF

The white dwarf state represents a drastic change from the star's former self, for its tenuous atmosphere (outer layers) has been shed. By what mechanism? Many white dwarfs have been observed to be surrounded by large, slowly expanding shells of gas. These objects are called *planetary nebulae* because when first discovered in a small telescope (with rather poor resolution), they resembled the disklike image of a planet. Modern views of two such "planetaries" are shown in Figure 14.1. Measurements reveal that a single planetary nebula would dwarf our entire solar system, being 20,000 to 200,000 A.U. in diameter. Often such a nebula appears somewhat like a ring, however this is an optical illusion because our line of sight near the center passes through less of the shell's thickness; in reality the shell of gas is spherical in form. More than a 1000 planetary nebulae have been catalogued. Their distribution in the Milky Way galaxy resembles a flattened sphere surrounding the nucleus, somewhat similar in distribution to the RR Lyrae stars. This suggests that planetary nebulae and the stars with which they seem to be associated are older objects, well along in their evolutionary development. The planetary nebulae are not the only manifestation of the shedding of atmosphere. Stars seem to shed significant amounts of mass throughout their lifetime and how much they shed not only affects their rate of evolution but ultimately their form of death.

STELLAR WINDS

We have already discovered that the sun relieves itself of at least a million metric tons of mass per second, as a result of the solar wind. Although this sounds like a very high rate of dissipation, it would take a 100 billion years for this to represent a significant loss. When we talk of stellar winds we will refer to loss rates that are several thousand times greater. Although the

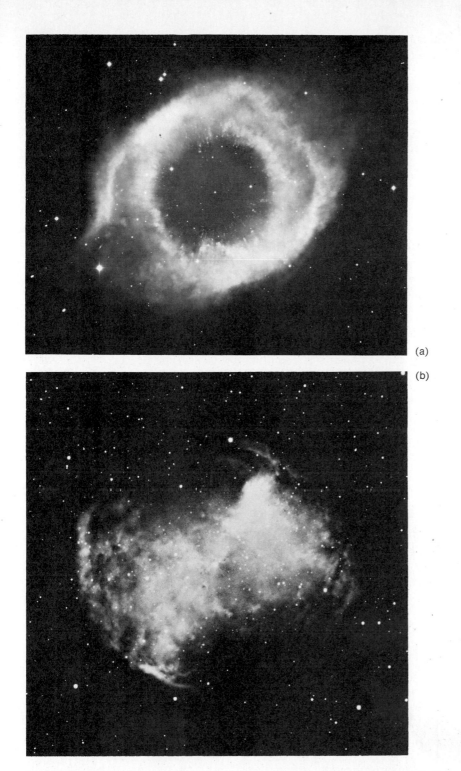

FIGURE 14.1 (a) The planetary nebula in Aquarius, NGC 7293. (b) The Dumbbell Nebula, a planetary nebula in Vulpecula. (Lick Observatory) (Palomar Observatory Photograph)

supernova reaction is most effective in removing mass from a star, stellar winds can remove a significant amount.

A number of red giant stars like Betelgeuse, whose radius is 750 times that of the sun, are known to have huge extended coronas (up to 300 times their own radii). Atoms and dust particles in the outer reaches of such a corona escape the gravitational pull of the star much more easily than in the middle. We now have direct evidence of the outflowing nature of such material in the form of a separate (blue-shifted) spectrum that is superimposed on the spectrum of the star itself. You can see in Figure 14.2 that regardless of the motion of the star, you would see the spectrum of outflowing gases as having a relative blue shift along your line of sight because their motion relative to the star is toward you.

For this outflow to represent any significant loss of mass, we must look for a mechanism beyond (or in addition to) the thermal expansion which seems to account for the solar wind. Infrared studies seem to suggest that dust particles play a significant role since dust particles respond to radiant energy in a wide range of wavelengths, as in the case of the dust tail of a comet (see page 267). If radiation can exert a force on such particles, accelerating them to escape velocity, then the dust particles in turn may transmit some of their kinetic energy to gas molecules and in effect carry off significant amounts of mass in the process.

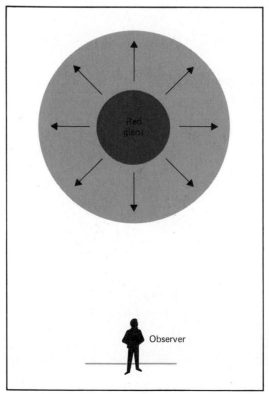

FIGURE 14.2 The expansion of a star will create a blue-shift superimposed on Doppler shift which is due to its motion.

Red giant

Observer

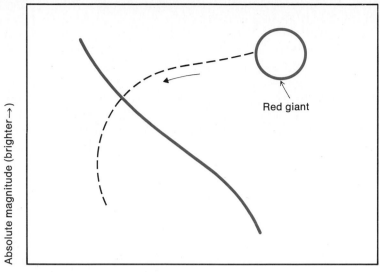

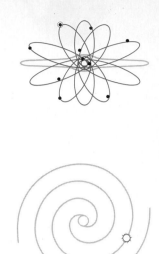

FIGURE 14.3 The evolutionary track of a star moving from the red giant stage to the white dwarf stage.

In the case of very hot stars that radiate mostly in the ultraviolet wavelengths, gas atoms may respond to such radiation and produce velocities of 100 times those seen in red giants. Such velocities in normal-sized stars are essential for escape.

The evolutionary track of a star moving from the red giant stage to the white dwarf stage is shown in Figure 14.3; the star is plotted further to the left on the H-R diagram as the hotter temperature of the core is revealed. As the star shrinks in size, its luminosity decreases and it gradually cools, sending it downward to the right on the H-R diagram, into what will eventually be a yellow-, red-, and black-dwarf death. This final cooling process may require up to a trillion years.

Stars of more than 1.2 solar masses will have quite another form of death.

THE NEUTRON STAR

Let us suppose that a star originally contained four solar masses and that half its mass was expelled in a supernova-type explosion or by stellar winds, leaving a remnant core of two solar masses. Not even the electron pressure we mentioned in connection with the formation of a white dwarf can stop the final collapse of this core. Such a star will likely contract to only 32 km (20 miles) in diameter, and its density will reach a rather staggering figure of 200 trillion g/cm³ (200,000,000,000,000 g/cm³). To conceive of the meaning of such density, let us first visualize a greatly enlarged model of the hydrogen atom. Imagine a single golf ball placed at the

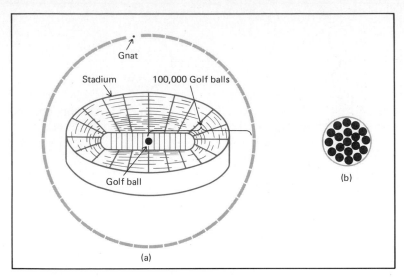

FIGURE 14.4 (a) An enlarged model of a hydrogen atom; (b) a model of a neutron star.

center of a football field (Figure 14.4). This represents the single proton of the hydrogen atom. Now imagine a gnat flying around the outside of the huge stadium which surrounds the playing field. This gnat represents the one electron which is normally a part of the hydrogen atom. If you were sitting in that stadium, you could imagine yourself sitting inside a greatly enlarged hydrogen atom, yet the proportions in terms of size of the particles and the space separating them would be quite accurate. One would certainly conclude that an ordinary atom is mostly empty space and its mass is concentrated in a very small volume of space at the center. To carry the analogy further, two atoms in an interstellar gas cloud would be separated like two stadiums in adjoining towns. In the neutron star, on the other hand, virtually all empty space has been taken up, and the gnats (electrons) have neutralized the charge on the golf balls (protons), producing neutral golf balls (neutrons). Lacking charge, the neutrons do not experience electric repulsion and they pack yet more closely together (they form a tightly packed crystalline structure). Should such a star exist, how would it be likely to manifest itself?

PULSARS

In 1967, while investigating certain sources of radio emission, Jocelyn Bell (a 24-year old graduate student at the University of Cambridge) noticed what she called a "bit of scruff"—a tiny wavering of a radio signal which seemed to recur every time she came back to the same spot in the sky. We know that the earth spins in relation to the stars in a period of 23 hr 56 min, so any source of emission originating outside our own solar system will be detected again in exactly that period, assuming the radio tele-

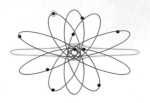

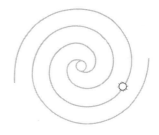

scope is left in a stationary position with respect to the earth. A car driving by her antenna could have caused a similar wavering on the radio record since the firing of automobile spark plugs creates radio noise. The record showed repeated returns of her "scruff" in a period of 23 hr 56 min over and over again, ruling out the possibility of an artificial source.

In modern times, the idea that intelligent beings may exist elsewhere in the universe has become almost an obsession. This possibility entered the minds of the observers with whom Jocelyn Bell first shared her discovery and they dubbed the source as LGMs (Little Green Men). It was not long before many such sources were identified and the conclusion drawn that some natural phenomenon was causing rapidly pulsing radio signals; these signals were named *pulsars*. The first pulsar was called CP 1919, meaning "Cambridge Pulsar located at right ascension 19 hr 19 min." The first nine pulsars to be discovered are listed in Table 14.1. By 1979 over 150 pulsars had been catalogued.

TABLE 14.1

PULSARS[a]				
Pulsar	Right Ascension	Declination	Period (sec)	Distance (parsec)
CP 1919	19hr 19min 37sec	+21 47'	1.337301	126
CP 0950	09hr 50min 29sec	+08 11'	0.253065	30
CP 1133	11hr 33min 36sec	+16 08'	1.187911	50
CP 0834	08hr 34min 22sec	+06 07'	1.273764	128
HP 1506	15hr 06min 50sec	+55 41'	0.739678	200
CP 0328	03hr 28min 52sec	+54 23'	0.714463	270
CP 0808	08hr 08min 50sec	+74 42'	1.292231	60
PSR 1749	17hr 49min 49sec	−28 06'	0.562645	500
PSR 2045	20hr 45min 48sec	−16 28'	1.961663	115

[a] CP indicates Cambridge Pulsar; HP indicates Harvard Pulsar; PSR indicates Pulsar.

A search was made for a visible counterpart to these pulsars, but none was immediately found. Early speculation concerning the signal's origin took many forms. Most theories suggested a pulsating star or a revolving system. Most of the theoretical models devised to account for the very rapid recurrence of the signal were of very small, dense stars or systems. None would account for this next discovery.

In 1969, the team of Coche, Disney, and Taylor at Steward Observatory at the University of Arizona identified a star clearly visible in the Crab Nebula as being a pulsar (Figure 14.5). This was Pulsar NP 0532, with a period of about 0.0333 second—the flashing of its optical counterpart was thus too rapid to be detected by the eye. Special equipment was devised to record its very rapid radio pulses and later, at Lick Observatory at the

FIGURE 14.5 Crab Nebula. (Palomar Observatory Photograph)

University of California, a special rotating aperture was perfected that allowed astronomers to photograph this pulsar and show that it is "on" part of the time and "off" part of the time (Figure 14.6). Pulsar NP 0532 is an 18-mag object, and its flashes are equivalent to 15 magnitudes. Its light curve is detailed in Figure 14.7. No model of pulsating stars was adequate to explain the rapid pulsing of NP 0532. New ideas must be forthcoming.

FIGURE 14.6 Pulsar NO 0532 in the Crab Nebula, photographed using a special rotating disk, showing the pulsar "off" in the left-hand view and "on" in the right-hand view. This object blinks on and off approximately 30 times per second. (Lick Observatory)

 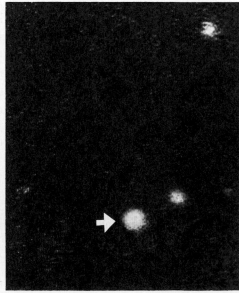

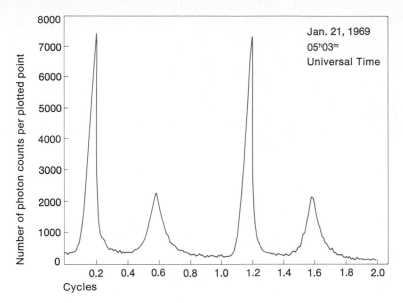

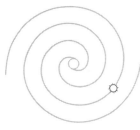

FIGURE 14.7 Light curve of NP 0532. (From R. Lynds, S. P. Maran, and D. Trumbo, Optical identification and observations of the pulsar NP 0532. Astrophysics Journal 155, L 121 (1969). © University of Chicago Press; Kitt Peak Observatory)

Almost every star rotates, even if very slowly. The sun rotates in a period of approximately one month. As a star shrinks in its old age to become a neutron star, its period of rotation is shortened—it spins faster and faster. This happens to a figure skater who starts her spin and then pulls in her arms, only to spin faster. This illustrates a principle of physics called "the conservation of angular momentum." A star which was spinning slowly during its main-sequence or red giant stage thus could be expected to spin very rapidly in its condensed stage. The current model of the pulsar, then, is a rapidly rotating neutron star. According to this model NP 0532 is spinning 30 times per second, its pulses being emitted in step with its rotation. This idea is born out by the very nature of the radiation it emits; the radiation is strongly polarized and continuous in its spectral character. We are no longer seeing the kind of radiation produced by the thermonuclear processes of a normal star but rather by a process characteristic of a cold star, whose remaining energy lies in its rotation and in its possession of a strong magnetic field. Almost all stars possess a magnetic field and as they condense, their fields are strengthened by concentration. The strength of the magnetic field near the surface of a pulsar is thought to be 10^{12} gauss—a trillion times that of the sun.

When electrons move in a magnetic field, they tend to spiral around the magnetic field lines rather than to cut across them (Figure 14.8). This spiraling motion is one of the ways a particle may be accelerated and when a charged particle is accelerated, it emits electromagnetic radiation (recall the discussion of an oscillating charged particle on page 56). This form of radiation is called *synchrotron radiation* because it was first observed in a

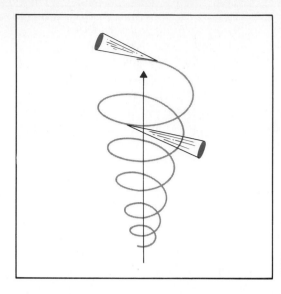

FIGURE 14.8 Electrons tend to spiral along (around) magnetic lines and as a result they emit synchrotron radiation.

particle accelerator called a synchrotron. Both the radio and light pulses from the pulsar in the Crab Nebula are thought to originate from such a process. The strong magnetic field of pulsars is thought to direct radiation into two rather narrow cones of emission (Figure 14.9). Such a star will be sensed only if the observer is actually within one of these cones of radiation. Perhaps only 10 percent or less of all neutron stars that fit this model are aligned in such a way as to be sensed from the earth.

The periods of the pulsars seem to be increasing. This indicates that these objects are rotating more slowly as time passes, giving up some of their rotational energy to the acceleration of electrons, which in turn pro-

FIGURE 14.9 The rotating neutron-star model of a pulsar.

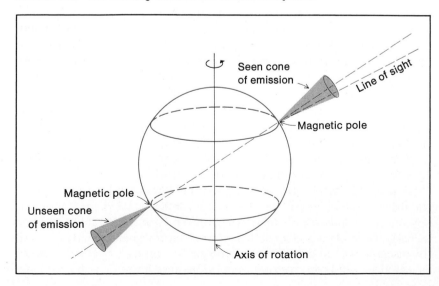

duces radiation. Based on these observations, the duration of the stage of evolution represented by a pulsar seems relatively short, in some cases as little as 1 million years. The primary source of energy for the pulsar is its rotation; thus when that rotation stops or the magnetic field collapses, the object will no longer radiate energy; this would be the last stage in the evolution of such a star. In several cases pulsars have been known to suddenly speed up. This sudden increase in the rate of rotation may indicate that the star has contracted to a smaller volume. Such a contraction would necessitate a rearrangement of the neutrons (somewhat like the collapse of a crystalline structure). This could be termed a "starquake." The increase in the rotational (angular) velocity that would accompany such a reduction of size would act to conserve angular momentum. A change of only 1 centimeter in the radius of a neutron star would be sufficient to explain a measurable change in the period of the pulsar.

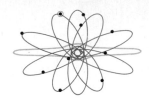

By 1982, more than 300 pulsars had been observed. The most remarkable one was discovered on November 12 of that year by a team of astronomers headed by Donald Backer of Berkeley. In the constellation of Vulpecula, the team found a pulsar with a period of 0.001558 sec, indicating that it rotates 20 times as fast as the Crab pulsar. It is now designated as PSR 1937+214. We know that the surface (and internal) gravity of a neutron star is so great that it seems unlikely such a star could tear itself apart by rapid rotation, yet this one is suspected of being quite close to doing just that. A point on its equator must be traveling at approximately one-eighth the speed of light. Another remarkable characteristic of PSR 1937+214 is the fact that its spin rate is slowing much more than we could predict from our knowledge of other pulsars. The characteristics of this pulsar are not easily explained.

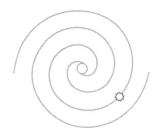

Although pulsars show neither emission nor absorption lines in their spectra, their pulses do show a form of dispersion when observed in a variety of wavelengths. This phenomenon, called *velocity dispersion*, is a valuable tool for studying the interstellar media that causes it. When a pulse of energy is created at the source, all wavelengths are generated simultaneously. In traversing a region of space in which even a small amount of material exists, however, the shorter wavelengths travel more rapidly and arrive ahead of the longer wavelengths. Figure 14.10 illustrates the velocity dispersion due to arrival time delays for a single pulse observed at five different wavelengths. The amount by which these signals are delayed is a direct measure of the number of free electrons (products of ionization) that lie along the line of sight to the pulsar. To the degree that the density of free electrons can be estimated (about 1 electron per 30 cm^3), the velocity dispersion becomes a distance indicator for the pulsars: the more free electrons, the greater the dispersion and the greater the distance. On page 422 we discussed the method whereby the distance to the Crab Nebula and therefore to the Crab Pulsar may be found. This distance has been compared to the velocity dispersion in its pulses, and the relation has been the basis for using velocity dispersion as a distance indicator for pulsars in general.

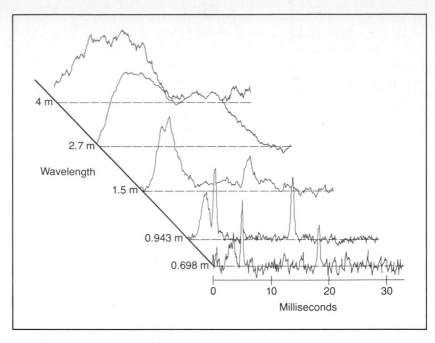

FIGURE 14.10 The velocity dispersion for a single pulse, as observed in five different wavelengths. (Cornell University)

We have been discussing stars in the 1.4 to 3 solar mass range. What about stars that are more massive than 3 solar masses?

THE BLACK HOLE

The life cycle of the more massive stars parallels closely those of neutron stars, but because these more massive stars have even higher levels of gravitational force, temperatures may be developed in their cores of the order of 1 billion degrees Kelvin; elements as massive as iron may be produced before the core collapses. Under the extremely large gravitational forces associated with 10-to 20-solar mass stars, the collapse of such an iron core transcends anything discussed to this point. A significant portion of the star's mass may be thrown off in the accompanying supernova explosion; however, the momentum created in the condensing core cannot be stopped by neutron forces (forces that dictate the stable size of a neutron star), and its inevitable fate is a *black hole*.

Albert Einstein predicted the possibility of such a collapse in his general theory of relativity published in 1915, and Karl Schwarzschild worked out some of the consequences of this theory in 1916. Schwarzschild said that when a star collapses to a certain size (dependent only upon its mass), the radiation from the star could no longer escape, owing to the enormous gravitational forces that would ensue. For example, if a star of three solar masses were to collapse to become a black hole, this discontinuance of radiation would occur when the radius was reduced to 9 km (for a 6-solar mass star, the radius would be 18 km). This does not mean that the collapse would stop at that point, for the collapse is thought to continue into a condition approaching infinite density and infinite tidal forces. The collapsed star is now called a *singularity*.

Let us examine this process in a little more detail. In order to appreciate the effect of increasing surface gravity that accompanies the contraction of a body, imagine first that you lived on the surface of a huge body the size of the sun but with three times the mass. You could shine a bright flashlight in any direction above your horizon and expect its beam to travel essentially in a straight line (Figure 14.11). You might describe its cone of radiation as 180°. Now let the body on which you are standing shrink to a radius of only 10 km. At this point you would find that you could still point your flashlight in an upward direction which is defined by a cone, say of 90°, and expect its beam to travel in essentially a straight line, however something strange appears to happen when you point it along the edge of the cone. The light beam bends into a circle surrounding the body and the photons of light form a sphere of radiation 3.5 km above the surface. If the light beam is directed at yet a lower angle, it will curve under the influence of gravity and be drawn back toward the body. Now let the body shrink to a 9.5-km radius and the cone of radiation will narrow further. When the body shrinks to a 9-km radius, the light cone disappears altogether—no light can leave the body and you are now sitting on a "black hole" (rather you would be spread out like a thin coat of paint over its surface). Furthermore, the body will continue to shrink to the singularity.

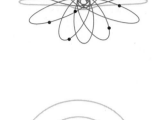
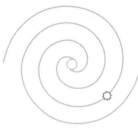

FIGURE 14.11   As a particle passes within the sphere defined by the Schwarzschild radius, on its way toward the singularity, the particle's radiation ceases because it cannot escape the gravitational field of the black hole.

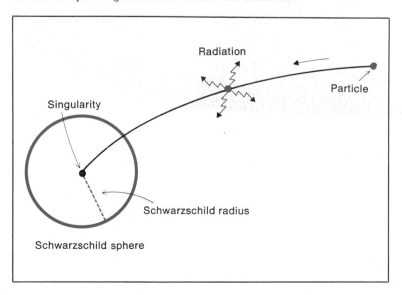

The Schwarzschild radius defines a sphere within which particles and radiation can only fall toward the singularity. In fact, if a particle were approaching a black hole, its radiation would cease as it passed through this sphere, and the event could no longer be witnessed; hence the Schwarzschild sphere is often called the *event horizon* (Figure 14.11). Before reaching the event horizon, a particle would experience rapid acceleration and consequent heating by collision, perhaps to 1 billion degrees Kelvin, so that a portion of its radiation could be expected to fall in the X-ray portion of the spectrum.

These speculations led astronomers to reexamine known X-ray sources and ponder the most likely circumstances under which material would be available to be sucked into black holes. One very natural suggestion is that of a binary star system in which one member is a black hole and the other member is a red giant. The surrounding envelope of the red giant might extend close enough to the black hole so that material from that envelope would be drawn off and fall toward the black hole (Figure 14.12). Instead of flowing directly onto the surface of the black hole, the gases would tend to form an orbiting disk around the ultradense object. Gas particles would orbit at very high velocities and become highly turbulent at certain elevations above the black hole. Frictional heating, which might raise the temperature of the gas to tens of millions of degrees, would result in a condition ripe for X-ray emission. Many astronomers

FIGURE 14.12 Mass exchange in a pair of binary stars. The very strong gravitational field of the black hole attracts material from the outer portions of the expanded red giant.

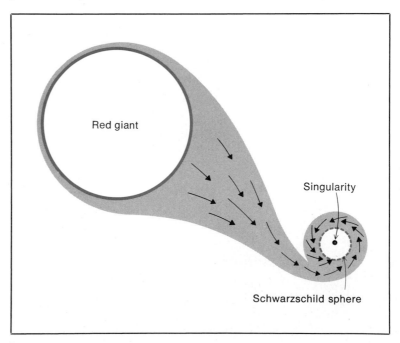

believe that this explains what they are observing in Cygnus X-1, a source of X rays in which the visible star behaves as though it were a member of a binary system where the unseen member has a mass of more than 3 solar masses.

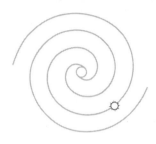

ROTATING BLACK HOLES

From our discussion to this point, it is difficult to imagine a black hole giving up some of its energy to the outside universe. However, a rotating black hole may do so. As a spinning ice skater pulls in her arms, she spins faster and faster, conserving angular momentum. Likewise, a rotating star will spin faster as it collapses just as an ice skater spins faster when the arms are pulled in. As a consequence of this rotation, a deformed region called an *ergosphere* develops outside the black hole (Figure 14.13). A particle that enters the ergosphere might split into two parts, one part entering the black hole and the other ejected with more mass energy than the original particle had. The additional mass energy is derived from the rotational energy of the black hole.

FIGURE 14.13 Rotating black hole. A particle that enters the ergosphere (shaded) may split into two, one part passing through into the Schwarzschild sphere and the other being ejected back outside the ergosphere with more mass-energy than the original particle had.

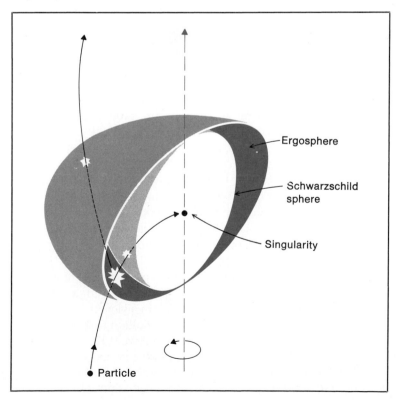

EXPLODING BLACK HOLES

Theorists accepted the idea that nothing can escape the interior of a black hole for quite sometime. It was reasoned that if light cannot escape and if nothing can travel faster than light, then nothing can escape. This reasoning persisted until theorists attempted to fit together several of the most fundamental theories of all time: Einstein's special theory of relativity (1905), Einstein's general theory of relativity (1915), and the quantum theory (1926). The general theory of relativity predicts the collapse of massive stars into what are now essentially confirmed as black holes. This same theory describes gravity in a manner vastly different from Newton's classical approach, but no one had been able to unite the concept of gravity with quantum mechanics. Quantum mechanics basically describes the behavior of particles and energies on the atomic and subatomic scale—the world of the extremely small. It addresses the problem of identifying the location and momentum of subatomic particles given the inevitable disturbance created in the process of trying to observe them. Who would think that a theory concerning the microworld would have any connection with the macroworld of the black hole?

A young British theoretical astrophysicist, Steven W. Hawking, made just such a connection. He reasoned that because the universe is lumpy today, it was probably lumpy when it was formed by the Big Bang event. Hawking proposed that the extreme pressures associated with the birth of the universe could easily have created small black holes with as little as 0.00001 g of mass and a Schwarzschild radius of 10^{-33} cm. A more typical dimension, however, would include a mass of 10^{15} g (a billion tons) and a radius of 10^{-13} cm, the size of a proton. Certainly this represents the size scale to which quantum mechanics is usually applied.

Hawking solved the equations of general relativity for this black hole of 10^{15} g and to his unbelieving amazement his solution predicted the creation and emission of particles and radiation from such an object. In fact, it predicted the same spectrum of particles and radiation as would be seen emanating from a hot object at 120 billion degrees Kelvin. This seemed to violate two long-held theories concerning black holes: that nothing left their core, and that they had a temperature of virtually 0°K. The same equations, however, predicted 0.000001°K for a 3-solar mass black hole and the escape of virtually no particles from such a star. The breakthrough in the explanation of this contradiction came with the recognition that small black holes are different.

If small black holes do exist and if they emit particles and radiation as predicted, then they will tend to waste away (lose mass). Temperature will tend to rise, and eventually this trend will result in a gigantic explosion— the final reaction being equivalent to 10 million, million hydrogen bombs exploding simultaneously. Certainly such an event would seem easily observable, however the distribution of these objects may make their detection more difficult. Their thermal character results in emissions in the range of hard gamma rays with very short wavelengths and at present large orbiting gamma ray detectors are not available. The space shuttle program may remedy this in the near future. Hard gamma rays are also known to produce a reaction in the earth's atmosphere, which may be detected.

The various types of star deaths are summarized in Table 14.2.

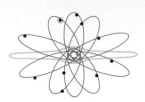

TABLE 14.2

CHARACTERISTICS OF DYING STARS			
	White Dwarf	Neutron Star	Black Hole
Mass at death	up to 1.4 sm	1.4 to 2.5 sm	2.5 sm +
Mass prior to death	up to 1.4 sm	1.4 to 5 sm	5 to 60 sm
Size (average)	size of earth	32 km	8 km +
Density	10^6 g/cm	10^{12} g/cm	infinite

QUESTIONS

1. Describe a neutron star.
2. How does an astronomer sense that he or she is receiving energy from a pulsar as compared to other type stars?
3. What changes have already been observed in the pulsars?
4. Compare the densities of a white dwarf, a neutron star, and a black hole.
5. Explain what is meant by the terms *singularity* and *Schwarzschild radius* in relation to a black hole.
6. Why is it not thought possible for the sun ever to become a black hole?
7. How may a black hole be recognized?
8. What is meant by the *event horizon* of a black hole?

The Milky Way

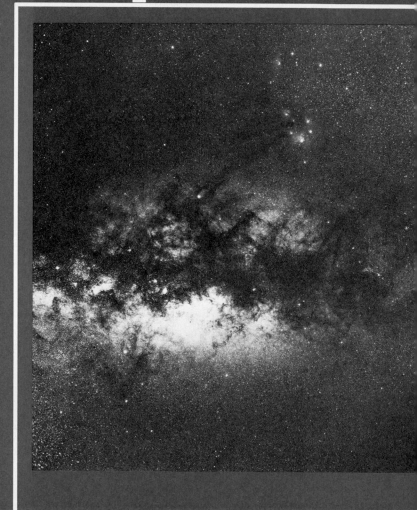

Fifteen

Our perception of the universe took a giant leap forward when the nearby collection of stars, of which the sun is one, was recognized to be only an example of a fundamental unit in the universe, the *galaxy*. By the late eighteenth century, Messier had catalogued more than 100 nebular objects without ever suspecting that some of them were collections of stars completely separated from our own collection, the Milky Way galaxy. By the early nineteenth century, more than 5000 nebulae had been identified, largely as a result of the work of Sir William Herschel and his son, John. The latter published the *General Catalogue of Nebulae* in 1864. By 1908, almost 15,000 nebulae had been listed in a volume called the *New General Catalogue*, together with its supplement, the *Index Catalogues*.

The term *nebula* in those days referred to any object which appeared fuzzy in the telescopes of the time. Sir William Herschel did express his belief that certain objects that he found were collections of stars separated from our own; lacking proof, however, he did not pursue the idea. As late as 1912, Henrietta Leavitt, in her work with the Cepheids in the Magellanic Clouds, did not realize that these "clouds" were galaxies outside our own.

With the installation of the 2.5-m telescope on top of Mt. Wilson in 1917 and the use of photographic methods

FIGURE 15.1 The Andromeda galaxy, a spiral galaxy considered to be very much like the Milky Way galaxy. (Palomar Observatory Photograph)

available by that time, individual stars were being resolved in the nearer galaxies (Figure 15.1), and novae were discovered in them. It was reasoned that if these novae were as bright as some that had been discovered in our own Galaxy and yet appeared very dim to the observer, they must be at very great distances. However, the matter was not settled until 1924, when Edwin Hubble, working at the same Mt. Wilson Observatory in California, discovered Cepheid variables in several nearby galaxies. From the periods of these stars their absolute magnitude was computed, as ex-

FIGURE 15.2 The Milky Way from Sagittarius to Cassiopeia. (Palomar Observatory Photograph)

plained in Chapter 11. They appeared so faint that they were undoubtedly outside our own Milky Way galaxy.

By this time astronomers realized that galaxies exist in a wide variety of shapes. Just a glance at the first few pages of Chapter 16 will show spherical, elliptical, and flattened galaxies, none of which have arms, as well as the spiral galaxies and barred spirals with arms, and the irregulars. It seems reasonable to suspect that our own galaxy looks something like one of these types, but since we are situated within it, determining its shape is not so simple.

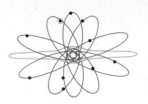

If the Milky Way galaxy were spherical, then we might expect a rather even distribution of stars within it. Let us perform a simple experiment to see if this is the case. Cut out a small cardboard frame with a square opening of 5 cm × 5 cm. Hold the frame at arm's length and count the stars you see within it. Repeat this for different parts of the sky, recording the location and number of stars counted on each occasion. It will become apparent that the distribution of stars is not uniform and that there are higher concentrations of stars in certain parts of the sky. Try the region near Sagittarius (Figures 15.2 and 15.3) or one near Orion, or any area through which runs the fuzzy band of light we refer to as the Milky Way. What are we really looking at when we see this milky band on a clear dark night? A small telescope or even a pair of binoculars will begin to reveal that the band is composed of millions of stars. If we observe the Milky Way over a period of several months (or by staying up all night), we realize that it forms a complete circle in the sky (Figure 15.4). This suggests that we are part of a flattened system of stars.

Where are we in the Milky Way system? The more than 100 known globular clusters will help us answer that question. Recall that globulars contain RR Lyrae-type variables with absolute magnitudes very close to zero. By comparing this absolute magnitude to the apparent magnitude observed for each, astronomers have determined their distances and have plotted their positions in relation to the sun (Figure 15.5). We are obviously not at the center of this distribution of globulars. There is a very high concentration of these objects toward the constellation of Sagittarius. If the sun were assumed to be the center of the galaxy (Figure 15.6), the distribution of globulars would appear very strange in relation to the Galaxy. It seems far more reasonable to assume that the globulars are distributed in some symmetric manner in relation to the galaxy (Figure 15.7), that the

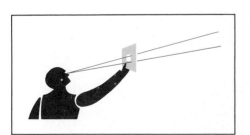

FIGURE 15.3 Counting stars.

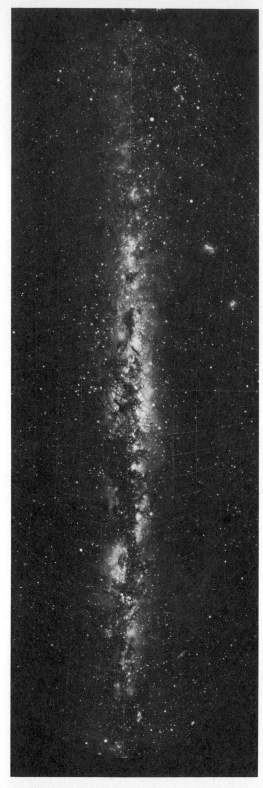

FIGURE 15.4 The Milky Way as shown by Martin and Tatjana Keskula. Sagittarius is in the center, and the two bright objects in the lower right portion are the Magellanic Clouds. (Lund Observatory, Sweden)

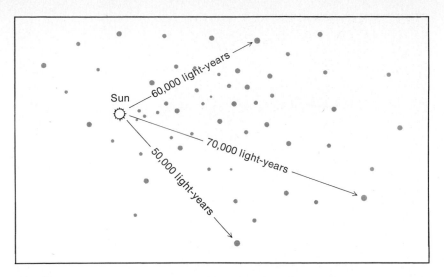

FIGURE 15.5 Distribution of the globulars in relation to the sun.

sun then is located approximately 30,000 light-years from the center of the galaxy, and that the galaxy's overall diameter is approximately 110,000 light-years. The concentration of globulars toward Sagittarius simply points in the direction of the galactic center.

The center, or nucleus, of the Galaxy is identified by the high concentration of stars toward the constellation of Sagittarius. Actually we can not observe the nucleus itself because of intervening clouds of dust that

FIGURE 15.6 If the sun were located at the center of the Milky Way galaxy, the globulars would be thrown off-center in relation to the galaxy.

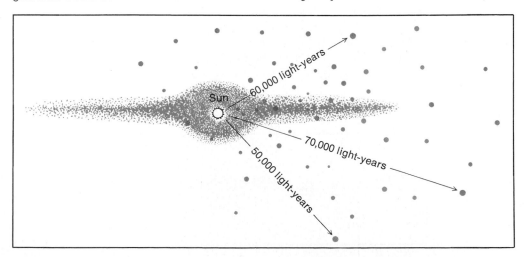

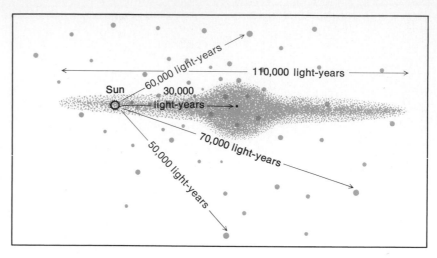

FIGURE 15.7 By assuming that the globulars form a symetric "halo" around the Galaxy, we believe our sun to be offset from the center of the galaxy by 30,000 light-years.

obscure the light. When we look in exactly the opposite direction, toward the bright star Capella (in Auriga, see Winter Sky map), we are looking out through the disk toward the outer edge of the Galaxy (Figure 15.8). The circle of the Milky Way band is marked on the sky maps in Appendix 13.

Next we might ask whether the Milky Way galaxy has arms like the many spiral galaxies we can view, or whether its disk is more uniformly populated with stars. By observing other galaxies of the spiral type, astronomers have noticed that very hot O- and B-type stars populate the arms and that they form a spiral pattern in themselves. In the photograph of the Whirlpool galaxy (Figure 15.9), these bright stars are easily recognized as bright dots. We would expect that ionized (emission) nebulae would accompany these bright young stars, because it is such stars that excite the gases in the first place.

Astronomers have determined the distance to numerous O- and B-type stars within the Milky Way, and a plot of these stars strongly suggests a spiral nature for our Galaxy (Figure 15.10). This particular plot shows only stars which are relatively close to the sun and the pattern for the whole

FIGURE 15.8 The Milky Way from our own point of view (edge-on).

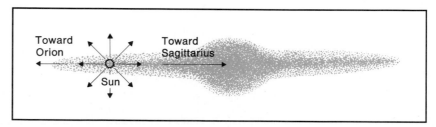

FIGURE 15.9   The Whirlpool galaxy, showing bright O- and B-type stars outlining its spiral arms. (Lick Observatory)

FIGURE 15.10   The optical pattern of the Milky Way galaxy based on a plot of O- and B-type stars in the vicinity of the sun.

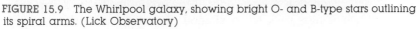

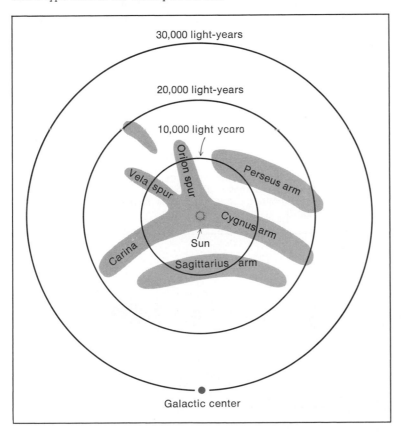

30,000 light-years

20,000 light-years

10,000 light years

Orion spur

Vela spur

Perseus arm

Carina

Cygnus arm

Sun

Sagittarius arm

Galactic center

Galaxy may be extrapolated from this. Plots of emission nebulae also tend to confirm what we are seeing here. Again, the presence of dust in the Milky Way galaxy obscures most of the starlight from the central region and the far side. The Galaxy is much more transparent to radio wavelengths, and in fact, radio signals characteristic of neutral hydrogen may be received from almost the entire Galaxy.

We have seen how hydrogen when excited produces certain characteristic wavelengths in the visible spectrum, such as the hydrogen-alpha line produced by a downward transition of the electron. Neutral hydrogen also produces a characteristic radio "line" which has a wavelength of 21.106 centimeters, simply referred to as the 21-cm signal of hydrogen. The process whereby the hydrogen atom emits or absorbs this particular wavelength is associated with the spin of its nucleus and the spin of its electron. The electron may align its spin axis with that of the proton, in which case it has more energy than when it opposes the spin axis of the proton. Thus, when the electron flips over from an aligned orientation to an opposed orientation, it emits energy corresponding to the 21-cm radio signal. When it flips back to an aligned direction, it absorbs the same wavelength (Figure 15.11).

When an astronomer tunes his radio receiver (telescope) to this 21-cm wavelength, much as we would tune in a local station on an ordinary radio, he can detect hydrogen in the Galaxy. The hydrogen clouds, however, are in motion relative to the sun, as demonstrated by the fact that the typical 21.106-cm signal is not always detected by our receivers at exactly that wavelength; it is Doppler shifted to a shorter wavelength if the cloud is moving toward the sun, or toward a longer wavelength if the cloud is moving away from the sun. Before we can show how productive 21-cm radio observations can be, we must first understand the pattern of revolution of stars within the galaxy itself.

FIGURE 15.11 Electron spin reversal in the hydrogen atom. When the spin axes of the proton and the electron are aligned (a), the atom contains more energy than when they are opposed (b). When the electron flips from state (a) to state (b), it emits a 21-cm radio signal.

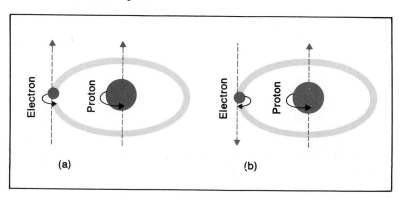

(a)   (b)

ROTATION OF THE MILKY WAY GALAXY

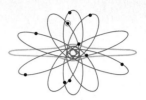

We can begin to describe the rotation of the Galaxy by describing the motion of the sun with respect to some reference system that does not participate in that rotation. The globulars form a natural reference system. They do not participate in the general rotation of the Galaxy but have a motion which carries them in and out of the Galaxy from locations above and below its flattened plane, as shown by the right-hand flip pages beginning on page 345. Using the average motion of the globulars as a reference, we find that the sun moves about the nucleus of the Milky Way galaxy in almost a circular orbit at a speed of between 220 and 250 km/sec, making one complete revolution in 200 million years. But what about the stars closer to the nucleus and those on the outside of the sun's orbit. If most of the mass of the Galaxy were concentrated in the nucleus, one might expect a pattern of motion like that of the solar system; namely stars nearer the nucleus would travel faster and those farther out would travel slower. This is not the case within the Galaxy; rather just the opposite is true. Stars nearer the nucleus, say at a distance of 15,000 light-years, have velocities of the order of 200 km/sec, and stars at a distance of 55,000 light-years, have velocities of the order of 300 km/sec. A plot of velocities versus distances from the nucleus is shown in Figure 15.12. Note that this plot is not a straight line, a fact that seems to suggest that mass is not uniformly distributed throughout the Galaxy. More importantly, the increase in velocities with increased distance from the center indicates that far more mass exists beyond the sun and perhaps beyond the visible galaxy than exists toward the nucleus.

FIGURE 15.12 Rotation curve of the Milky Way galaxy (after Bart Bok). The fact that the graph is not a straight line suggests that the mass is not distributed uniformly in the galaxy.

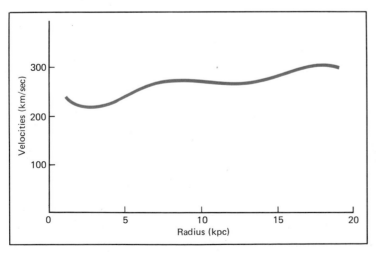

21-cm RADIO OBSERVATIONS SHOW SPIRAL STRUCTURE

Let's suppose, for sake of illustration, that the Milky Way galaxy really does have spiral arms—arms of stars, gas, and dust, with relatively free spaces in between. Further, suppose a radio telescope were directed along a line of sight, as shown in Figure 15.13, so that it crossed five different portions of arms, each containing neutral hydrogen (H I regions). Due to the different velocities of gas clouds and stars in each arm, the radio signal received from the hydrogen gas in each arm would have its own peculiar Doppler shift and therefore each would be received at slightly different wavelengths. A profile of the radio signals from such a system of arms would resemble that shown in Figure 15.14 and indeed this is an actual plot as observed by radio astronomers. Each peak in the plot represents a different arm and the dips represent the relatively free spaces in between the arms. Were the Galaxy simply a flattened system with a rather smooth distribution of hydrogen gas (no arms), then the plot would have appeared as in Figure 15.15. When the Galaxy is scanned in all longitudes, and distances are judged from the velocities indicated, a neutral hydrogen

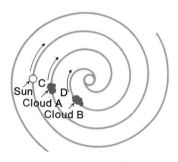

FIGURE 15.13 A model of the Milky Way galaxy in which clouds A and B are assumed to be located in the arms and points C and D represent spaces between the arms.

FIGURE 15.14 The profile of the 21-cm radio signal from a spiral galaxy—the dips correspond to gaps between the atoms of the galaxy.

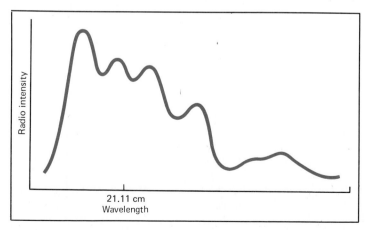

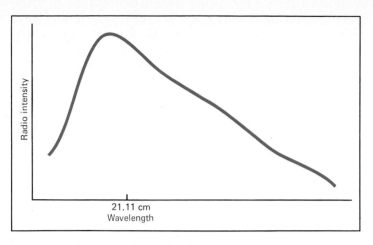

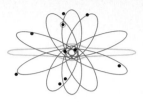

FIGURE 15.15. The profile of the 21-cm radio signal of a flattened galaxy with no arms.

FIGURE 15.16 The structure of the Milky Way galaxy, in the 21-cm radio signal of neutral hydrogen. The darker regions represent higher concentrations of hydrogen. (Leiden Observatory, Netherlands, and Radio Physics Laboratory, Sydney, Australia)

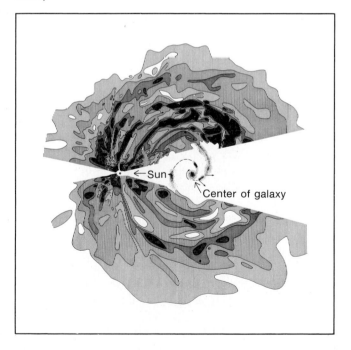

map can be constructed, as shown in Figure 15.16. Here the regions of highest concentration of hydrogen gas are shown as darker and certainly these concentrations suggest a spiral structure.

MASS OF THE GALAXY

While the actual determination of the total mass of the Milky Way galaxy is beyond our immediate scope, we can easily calculate that portion of the mass which is inside the sun's orbit by observing its effect on the motion of the sun. If we treat the central mass of the Galaxy as if it were one body and the sun as another, the problem reduces essentially to a two-body problem like that of the sun and a planet. Using Kepler's laws:

$$m_{gal} + m_{sun} = \frac{(r_{sun})^3}{(p_{sun})^2}$$

where r is measured in astronimical units and p in years. The distance between the galactic center and the sun is approximately 30,000 light-years, which is equivalent to 2×10^9 A.U., and the period of the sun is 2×10^8 years. Since the mass of the sun is negligible in comparison to that of the Galaxy, the term m_{sun} may be dropped:

$$m_{gal} = \frac{(r_{sun})^3}{(p_{sun})^2}$$

Thus the mass of the galaxy inside the sun's orbit would appear to be approximately 200 billion times the mass of the sun. We will be dealing with additional mass in subsequent sections of this chapter. We have already observed that there are far more stars which are less luminous than the sun; thus we can conclude that the sun's mass is above average. We are then led to believe that the number of stars in the Milky Way galaxy exceeds 200 billion.

DISTRIBUTION OF STARS IN THE MILKY WAY GALAXY

We have characterized the Milky Way galaxy as a flattened spiral, its most obvious shape. The observation of certain types of stars within the Galaxy, however, reveals that it really has at least four distinct regions, each with its own shape. As we examine these shapes, let us see if they give us a clue to the possible changes our Galaxy may have undergone before assuming its present shape.

Consider those star types thought to be oldest—stars typical of the globular clusters. The advanced age of such clusters is indicated by the fact that they have already used up their gas and dust in star formation and further, that their more massive stars have evolved off the main sequence. Astronomers refer to stars which inhabit the globulars as Population II objects. These include the RR Lyrae, Type-II Cepheids, and the long-period variables. The globular clusters form an almost spherical halo around the Galaxy, and they may still preserve the earlier shape of the cloud from which the Milky Way galaxy formed (Figure 15.17—see the halo labeled A).

An intermediate system exists between the spherical halo of globulars and the flattened disk where new stars are being formed. The intermediate system is shown in Figure 15.17 as the semiflattened halo labeled B. It consists of semi long-period variables, subgiants, white dwarfs, G- to M-type dwarfs, and planetary nebulae. These types of objects are also believed to be of moderately old age and may represent, by their distribution, the shape of the galactic cloud when it was still in the process of flattening due to rotation.

The third region is the flattened disk system (labeled C in Figure 15.17). It is characterized by very hot, young (O-, B-, and A-type) stars, Type-I Cepheids, supergiants, open clusters, and interstellar gas and dust. Each of these types represents young stars or the material from which new stars are formed. It should be noted that the nucleus of the Galaxy is composed of older Population II objects: RR Lyrae stars, globular clusters, planetary nebulae, and M-type dwarfs; hence it could be classified as yet another region different from the disk.

The fourth region is referred to as the *corona* of the Galaxy, a region that may extend outward for a distance of up to 300,000 light-years from the center of the Galaxy. This region was not at first detected by direct observation, rather it was inferred by the gravitational effect it appears to have on the motions of stars near the outer edge of the visible galaxy. If such a massive corona did not exist, it would be unlikely that these outer

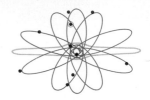

FIGURE 15.17 The distribution of stars in the Milky Way (see text).

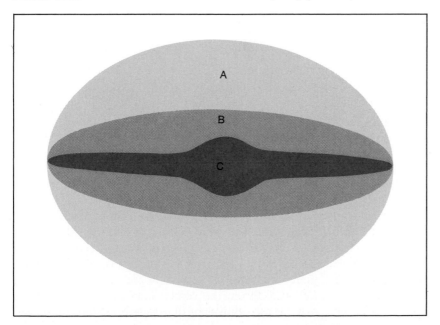

stars would be traveling faster than the inner ones and furthermore, if it did not exist, high velocity stars would be able to escape the Galaxy. Estimates of the mass of the corona run in excess of one trillion solar masses—that is 500 times the figure calculated above.

The presence of a galactic corona is being confirmed directly by means of an ultraviolet sensing satellite called the *International Ultraviolet Explorer (IUE)* which was launched in 1978. The specific component of the corona that is being sensed by this craft is hot gases. The temperature of these gases is of the order of 100,000°K, as indicated by absorptions lines related to carbon atoms that have been ionized three times and silicon atoms that have been ionized five times. The IUE satellite recorded the spectrum of a particular star in the Large Magellanic Cloud, which is itself a galaxy at a distance of about 160,000 light-years. Not only do the absorption lines present in the spectrum of this star reveal the presence of a corona of hot gases around our own Milky Way galaxy, but also a slight Doppler-shifted set of lines reveals a corona of hot gases around the Large Magellanic Cloud. This may indicate that the existence of a corona is typical of galaxies in general. If this is true, it will greatly influence our estimate of the total mass of the universe. Hot gases may not be the only component of the galactic coronas. Cold dust clouds, old dead stars, black holes, low-mass (low-luminosity) stars, and neutrinos are also possible components of the coronas, to mention only a few.

Does this progression of shapes represent an *evolution* of galaxies in the usual sense of the word? Does it mean that elliptical galaxies evolve into spirals or visa versa, as some would suggest? Perhaps not, but it surely suggests that in its very early stages the Milky Way galaxy was spherical, and that owing to rotation it flattened significantly in the first billion years of its existence as a separate entity in space.

THE GALACTIC NUCLEUS

The nucleus of our Galaxy cannot be observed optically. The intervening dust clouds obscure its light so effectively that only about one-trillionth of the potential light gets through. However, we are not observing the nucleus in radio wavelengths, in infrared, and in X rays. The strongest source of both radio and infrared emissions falls in a very restricted region of Sagittarius, with size estimates of 30 to 40 light-years in diameter and mass estimates of 1 to 10 million solar masses. At the very heart of the nucleus may be a super massive black hole which contains a million or more solar masses itself. Material falling onto such an object may cause such violent reactions as to produce X rays and to trigger some of the observed phenomena that follow. For instance, there is an armlike system of H I regions that are moving away from the nucleus at speed of 100 to 200 km/sec. These regions contain gas with a total mass estimated at up to 2 million solar masses. Also, there is a "ring" of molecular clouds at a distance of approximately 600 light-years from the center, expanding at the rate of 150 km/sec. There may be the equivalent of 100 million solar masses in these clouds. There are only two pieces of evidence that demand an explanation in terms of an expulsive force in the core of the Galaxy.

As you continue the study of galaxies in the next chapter, you will see that compact and active nuclei may be very common in galaxies. Particularly note the discussion of BL Lacertae objects.

QUESTIONS

1. The term *nebula* includes what kinds of objects in addition to gas clouds (if used in its most general sense)?
2. What evidence does the casual (naked-eye) observer have that the Milky Way galaxy is not simply a sphere (ball) that surrounds us?
3. What role did the globular clusters play in helping us to find our true location in the Galaxy?
4. How much time would be required for light to travel from a point on the outer edge of our Galaxy to the opposite edge?
5. How do we know that our Galaxy is of a spiral nature? Specifically, how do we know it has "arms"?
6. The sun participates in the rotation of our Galaxy. What is its speed because of this rotation?
7. Indicate whether each of the following would be found in the halo(h) of the Galaxy or in its disk(d): (a) open clusters, (b) globular clusters, (c) Type-II Cepheids, (d) Type-I Cepheids, (e) interstellar gas, (f) interstellar dust, (g) very hot (O- and B-type) stars.
8. True-false: It is more reasonable to assume that the globulars are distributed so that their center coincides with the center of the Galaxy rather than to assume that the sun is in the center.
9. An optical picture of the spiral arms of the Milky Way galaxy has been made using what kind of stars?
10. Why is a radio "picture" of the Galaxy more complete than the optical picture?
11. Neutral hydrogen clouds have differing speeds, depending upon their location in the Galaxy. How does this fact show up on the radio record?
12. What evidence do we have that the Milky Way galaxy was once more like a sphere and then later flattened to become a spiral?
13. What factor probably influenced the Galaxy to become a flattened system?
14. What indirect evidence do we have that the Milky Way galaxy has a very massive (unseen) "corona"?

The Realm
of Galaxies

Sixteen

Suppose that we had the rare privilege of using the Palomar 5-m telescope to take a time-exposure photograph of the sky. Numerous star images would be apparent, and in certain locations more galaxies than individual stars (Figure 16.1). There are estimated to be over 100 million galaxies within the magnitude range of the 5-m telescope, each of which contains millions or billions of stars. More amazing is the fact that these 100 million (plus) galaxies can be classified in a relatively few categories. A degree of uniformity within the universe is suggested by this pattern.

CLASSIFICATION OF GALAXIES

The Milky Way galaxy represents only one of several forms in which galaxies are found. Hubble, in his early work with galaxies during the 1920s, devised an extremely simple classification system for them. While some refinements have been suggested and his system does not include all types of galaxies, it remains in use to this day, essentially unchanged. The three main categories of galaxies include the *elliptical* type, the *spiral* type, and the *irregular* type (Figure 16.2).

The elliptical galaxies are subdivided in groups according to the eccentricity of their shape. A spherical galaxy is designated E0, and galaxies which are more and more

FIGURE 16.1 Cluster of galaxies in the constellation of Hercules. (Palomar Observatory Photograph)

FIGURE 16.2 Types of galaxies: (a) elliptical, NGC 4486; (b) spiral, NGC 2841; (c) irregular, NGC 3034. (Palomar Observatory Photograph)

(a)   (b)   (c)

(a)   (b)

FIGURE 16.3 Elliptical galaxies of (a) the E0-type and (b) the E5-type. (Palomar Observatory Photograph)

eccentric are designated successively E1, E2, . . . , E7. Examples of E0 and E5 ellipticals are shown in Figure 16.3.

The dwarf systems represent a further subdivision of elliptical galaxies; there are at least six examples within 1 million light-years of the Milky Way galaxy. They are intrinsically dim (their luminosity is about 0.001 that of the Milky Way) and sparsely populated, yet they represent one of the most common types of galactic systems in the universe.

The spiral galaxies are flattened systems that by their appearance suggest their actual rotation. Those with a very large, bright nucleus and tightly wound arms are called Sa-type spirals. The Milky Way galaxy is an example of the second classification, the Sb-type, in which the brightness is more evenly distributed between the nucleus and distinct arms. The third classification, type Sc, has a much smaller nucleus and bright, open arms. In fact most of the light of an Sc-type galaxy is concentrated in its arms. These three types are illustrated in Figure 16.4 (a, b, and c). If a spiral galaxy falls between two of these catagories, say between Sa and Sb, then it is designated as Sab.

Within the general spiral class fall the barred spirals, a type which appears to have a barlike distribution of brightness running through its nuclei. The same general classification as for spirals applies here. Those with a high concentration of brightness in the bar and nucleus and tightly wound arms are called SBa, whereas the brightness of the SBb type is more evenly distributed between the arms and the nucleus. The SBc type may have a less distinct nucleus, but it has a bright barlike structure and open arms [Figure 16.4 (d, e, and f)]. Note the SBab type shown in (d) is midway between the SBa and SBb.

The SO-type galaxy shown in Figure 16.5 may at first glance resemble an elliptical type, yet if we could see this galaxy from an edge-on view, it would present a flattened form, like a spiral without arms. That is where the simularity ends, however, for the SO-type galaxy shows none of the gas

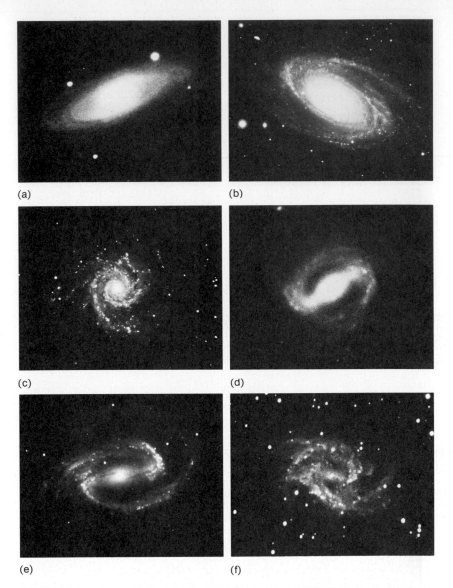

FIGURE 16.4 Spiral galaxies: (a) Sa, NGC 2811; (b) Sb, NGC 3031; (c) Sc, NGC 628. Barred spiral galaxies; (d) SBab, NGC 175; (e) SBb, NGC 1300; (f) SBc, NGC 2525. (Mount Wilson and Las Campanas Observatories, Carnegie Institute of Washington)

or dust characteristic of the spiral. The SO-type galaxy does seem to bridge the gap between the ellipticals and the spirals; however, this should not be interpreted as an indication that it is an intermediate step in the evolution of galaxies from elliptical to spiral or vice versa. In fact, most observers do not think that galaxies evolve from one type to another. The SO-type may represent a galaxy that has passed through another galaxy. If this happened, individual stars would have been so far apart that they would not

FIGURE 16.5 An SO-type galaxy, NGC 1201. (Palomar Observatory Photograph)

have collided nor would their gravitational interaction have been apparent, but any gas and/or dust that they possessed would have interacted and been swept from both galaxies. Since SO galaxies are found in regions of space where other galaxies are relatively close together, this appears to be a plausible explanation for their lack of gas and dust. This absence of gas and dust leaves the SO-type with predominantly type-II (old) stars.

FIGURE 16.6 The Large Magellanic Cloud. (Lick Observatory)

FIGURE 16.7 An Irregular II galaxy, NGC 3034, showing filaments extending 25,000 light-years outward from the nucleus. (Palomar Observatory Photograph)

The irregular-type galaxies are so named because they have no specified symmetry or structure. They fall into one of two subtypes: Irregular 1, illustrated by the Large Magellanic Cloud (Figure 16.6); and Irregular 2, like NGC 3034 (Figure 16.7). Irregular 1 galaxies are characterized by O- and B-type stars and regions of ionized hydrogen. A few examples have a suggestion of spiral arms. Irregular 2 galaxies are characterized by the fact that their stars cannot be resolved in the telescope; hence they must be composed of dim stars; yet they show spectroscopic evidence of gas and dust and they eject material at high speed (as shown by the filaments we see extending outward from the nucleus in Figure 16.7). This particular galaxy could also be classified as an exploding galaxy, of which there are several additional types (see M87 on page 470, and others which follow in this chapter).

In addition to these classifications that are based primarily on the shape and distribution of brightness, there is a further classification that ranks any of the above classifications according to luminosity in the order: I, II, III, IV, and V. For instance, the brightest among the spirals, type a, would carry the designation Sa I, or the brightest among E0s would be E0 I. In either of these cases, it would be quite safe to assume that we are talking about the most massive of galaxies in either category, for mass and luminosity are rather intimately related in the case of the stars which compose these galaxies. It then follows that these could rightfully be called the supergiants within their own categories.

HUBBLE'S CLASSIFICATION SCHEME

Edwin Hubble suggested the orderly classification of galaxies shown in Figure 16.8. Some observers have supposed that this classification scheme

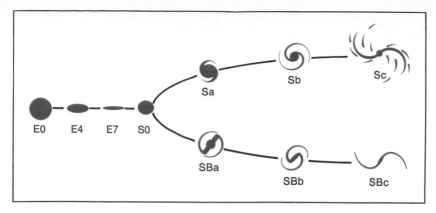

FIGURE 16.8 Hubble's classification of galaxies.

reflects an evolutionary pattern; some suggest that galaxies evolve from right to left in Hubble's diagram, and others suggest the reverse order. It is not clear that galaxies evolve from one type to another at all. As you think about this possibility, consider the physical properties of the ellipticals and the spirals listed in the following table.

TABLE 16.1

A COMPARATIVE ANALYSIS OF GALACTIC TYPES				
	Giant Ellipticals	Dwarf Ellipticals	Irregulars	Spiral
Diameter	50,000 to 200,000 ly	1,000 to 10,000 ly	10,000 to 30,000 ly	50,000 to 200,000 ly
Mass	10^{12} to 10^{14} sm	10^5 to 10^6 sm	10^7 to 10^9 sm	10^9 to 10^{11} sm
Luminosity	$10^{11} \times$ sun	$10^8 \times$ sun	$10^9 \times$ sun	$10^{10} \times$ sun
Dust	Almost none	Almost none	2 to 5%	1 to 2%
Gas (neutral)	Almost none	Almost none	10 to 13%	3 to 4%
Ages of stars	Old	Old	Young	Old (nucleus & halo) Young (disk)

THE LOCAL GROUP

Our own local group of galaxies is only one of many clusterings of galaxies. It consists of at least 28 members that fall within a sphere of approximately 3 million light-years in radius; 10% are spirals, 40% are irregulars, and 50% are ellipticals. Seventeen of these members are shown in Figure 16.9. The Milky Way galaxy is not actually at the center of the group, but it is placed at the center of the diagram to help you visualize distances to the various other members. Table 16.2 lists the same seventeen members to

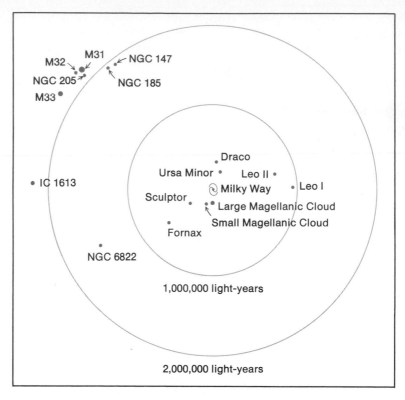

FIGURE 16.9 The local group of galaxies.

TABLE 16.2

THE LOCAL GROUP OF GALAXIES

Galaxy	Type	Absolute Magnitude	Distance (Million Light-years)	Diameter (Light-years)
M31 (Andromeda)	Sb	−21	2.2	100,000
Milky Way	Sb	−20	—	100,000
M33	Sc	−19	2.3	—
Large Magellanic Cloud	Irr	−18	0.16	30,000
Small Magellanic Cloud	Irr	−17	0.18	25,000
NGC 205	E5	−16	2.2	5,000
NGC 221 (M32)	E3	−16	2.2	8,000
NGC 6822	Irr	−15	1.5	9,000
IC 1613	Irr	−15	2.2	16,000
NGC 185	E_2	−15	1.9	8,000
NGC 147	E6	−15	1.9	10,000
Fornax System	E3 (Dwarf)	−13	0.6	22,000
Sculptor System	E3 (Dwarf)	−12	0.27	7,000
Leo I Systems	E4 (Dwarf)	−10	0.9	5,000
Draco System	E2 (Dwarf)	−10	0.33	4,500
Leo II System	E0 (Dwarf)	−10	0.75	5,200
Ursa Minor System	E4 (Dwarf)	−9	0.22	3,000

show the great range in mass, diameter, and luminosity represented by this small group. Within the local group we see almost every type of galaxy represented except the SO type.

CLUSTERING OF GALAXIES

Stars are usually found in galaxies and not wandering around in space separate from the fundamental grouping we call a galaxy. Likewise, it appears that galaxies are not found individually but in clusters of galaxies; examples include the Hercules cluster shown in Figure 16.1 and the Coma cluster shown in Figure 16.10. In fact, clusters of clusters, or super-clusters, are thought exist as well. If this is true, then the Universe can be thought of as a cluster of super-clusters. Our local group, together with the Virgo cluster and the Coma cluster (and perhaps more clusters) may consitute one of approximately 50 super-clusters that have been tentatively identified. The center of our super-cluster appears to be located near the Virgo cluster and we are moving at approximately 500 km/sec about that center. (How many motions does the earth participate in? Add this one to your list.)

The Coma cluster exemplifies what has been termed a *regular cluster* because it possesses a characteristic spherical shape with a concentration of galaxies toward the center, and its bright galaxies include mostly

FIGURE 16.10 The Coma Cluster of galaxies, showing a predominance of ellipticals. (Palomar Observatory)

elliptical and SO types. Regular clusters of galaxies are thought to possess a very large total mass, on the order of 10^{15} solar masses. The Hercules cluster, shown in Figure 16.1, represents an *irregular cluster* with less symmetry, no distinct central concentration, and mostly spirals and irregular-type galaxies. Because irregular clusters typically contain fewer galaxies, their total mass is approximately 10^{13} solar masses.

These total mass estimates are based on stars, dust, and gas that can be detected in the individual galaxies that make up a given cluster. These figures, however, are not large enough to explain the gravitational attraction required to hold such a group of galaxies in a cluster. We speak of the galaxies of a cluster being gravitationally bound to each other, but in order for this to be the case, more mass (which is yet undetectable) must exist in those galaxies. Could it be that many of these galaxies have "coronas" like that recently discovered for the Milky Way galaxy? If so, then our previous assessment of their mass must be multiplied by a factor of five or more, and that is the factor needed to explain why they have not dispersed in the billions of years of the universe's existence (15 to 20 billion years).

By this time, we no longer have the idea of a perfectly smooth distribution of matter throughout the universe, but rather we are getting the picture of a lumpy distribution. We wonder if there is anything between those lumps we call galaxies or clusters of galaxies.

INTERGALACTIC MEDIA

Gases and dust are probably what composes intergalactic media. How shall we search for this media? We recall that neutral hydrogen can be found by its 21-cm radio signal. This, however, depends on collisions between atoms to reverse the electron spin, and the density of this gas in intergalactic space is probably so low that collisions do not occur very often. Neutral hydrogen can absorb energy in the ultraviolet by an upward electron transition from energy level 1 to energy level 2, producing a spectral line at 1216 Å. Neither of these indications have been observed. Dust present in intergalactic space should produce its characteristic reddening, but no such effect has been observed. The only positive indication comes from X-ray astronomy. Very hot gases that may exist as coronae around the visible portion of galaxies, where temperatures are on the order of 100 million degrees Kelvin, are capable of emitting X rays; several "hot spots" have been found associated with "rich" clusters of galaxies. Certainly the closest approximation to empty space is in the spaces between the clusters of galaxies.

MEASURING THE DISTANCE TO GALAXIES

Distances within our local group are relatively easy to determine. Recalling Henrietta Leavitt's work with Cepheid variables (see page 376), and the calibration of the period-luminosity relationship by Shapley and Baade, we understand that the distance to any galaxy may be determined if both the period and the apparent magnitude of a single Cepheid in that galaxy can be determined. This method, then, is limited only by the astronomer's ability to resolve an individual Cepheid and this is possible out to about 10 million light-years.

The very brightest O- and B-type stars in other spirals like our own are assumed to have approximately the same absolute magnitude as the brightest O- and B-type stars in our own Galaxy; thus they too can be used as distance indicators. It is possible to measure the apparent magnitude of such stars up to a distance of at least 75 million light-years. When their apparent magnitude is compared to their known absolute magnitude, distance may be computed using:

$$\frac{L(10)}{L(r)} = \frac{r^2}{(10)^2}$$

where L is luminosity and r is distance. This same approach can be used with supernovae which are seen brightening in another galaxy. The astronomer assesses the type supernova, measures its apparent magnitude, assumes its absolute magnitude to be similar to the same kind of supernova when it occurs in our galaxy, and computes it distance using the above formula. For example, if a supernova appeared in a distant galaxy·and an apparent magnitude of $+10$ and an absolute magnitude of -15 was typical of that type, this means that it would show an increase in brightness of 25 magnitudes if brought to 10 parsecs, which is equivalent to an increase in luminosity of 10^{10}. Substituting and solving for r, we get:

$$\frac{10^{10}}{1} = \frac{r^2}{10^2}$$
$$r^2 = 10^{12}$$
$$r = 10^6$$
$$r = 1 \text{ million parsecs}$$
$$r = 3.26 \text{ million light-years}$$

This method is limited to the observation of supernovae and being able to measure their apparent magnitude. Because a supernova is so bright, this method will stretch our ability to measure the distance to galaxies much farther than the O- and B-type stars. However, even this method falls far short of measuring the size of the universe. What other methods exist for determining distances?

Consider a regular cluster containing many elliptical galaxies. It seems reasonable to assume that the brightest ellipticals in one cluster of galaxies have about the same luminosity as the brightest ellipticals in another cluster. Some degree of uniformity in the universe is assumed once again. If we know the distance to a nearby cluster using one of the previous methods, we can compute the distance to the more distant galaxy by comparing the apparent magnitudes of their brightest galaxies—the dim-

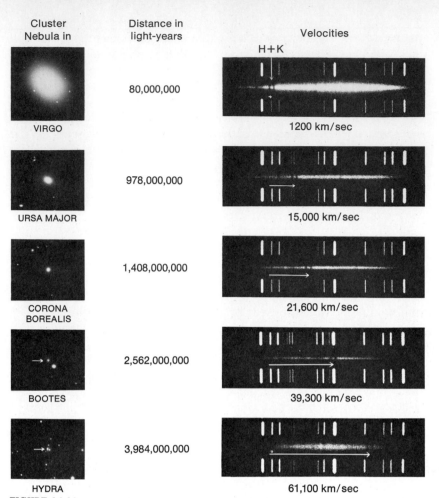

Cluster Nebula in	Distance in light-years	Velocities
VIRGO	80,000,000	1200 km/sec
URSA MAJOR	978,000,000	15,000 km/sec
CORONA BOREALIS	1,408,000,000	21,600 km/sec
BOOTES	2,562,000,000	39,300 km/sec
HYDRA	3,984,000,000	61,100 km/sec

FIGURE 16.11 Successively more distant galaxies, whose distances are estimated by their apparent magnitude and thus are subject to reevaluation. The spectrogram of each galaxy is shown on the right with the Doppler shift due to recessional velocity indicated by the arrows. Arrows indicate shift for calcium lines H and K. (Mount Wilson and Las Campanas Observatories, Carnegie Institute of Washington)

FIGURE 16.12 The relationship between velocity and distance.

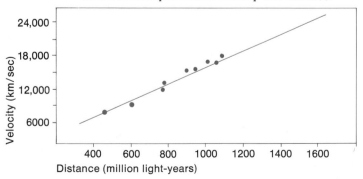

mer the galaxy, the farther away it is since its light falls off with the square of the distance. If you look at the five photographs in Figure 16.11, you will see this point illustrated as well as another. The apparent size of the galaxies appear more and more reduced due to distance. Again, assuming that we have picked galaxies of similar size (usually the brightest are also the biggest in the elliptical class), their apparent diameters correlate directly with their distance—the smaller they look, the farther away they are. Figure 16.12 contains a great amount of information useful for yet another method of distance determination.

HUBBLE'S RED-SHIFT LAW

In 1929 Edwin Hubble made a startling discovery. As he recorded the spectra of many galaxies, he noticed a relationship between the amount by which the spectral lines were red-shifted and the apparent magnitude of the galaxies. In general, the dimmer galaxies had larger Doppler shifts toward the red end of the spectrum. The photographs and spectrograms in Figure 16.11 show this relationship. The red shift of each galaxy is shown by the horizontal arrow on the spectrograms; it is very short in the first, but lengthens with each successive spectrogram.

In an attempt to refine this relationship, Hubble made two assumptions. He assumed that the red shift of each galaxy is due to its motion away from the observer and that the larger red shifts represent proportionately greater recessional velocities. He also assumed that the dimmer galaxies are farther away. Converting red shift to velocity of recession and estimating the distance to nearby clusters of galaxies, he plotted several points on a graph similar to Figure 16.12. A relationship is evident from the diagonal distribution of these points; however, he made an error in assessing the distance to the nearby clusters of galaxies, and as a consequence he obtained an incorrect value for the expansion rate, namely 165 km/sec for each million light-years. The value turned out to be about 10 times too large based upon a modern determination of the distances. Figure 16.12 shows this modern determination. As you can see, it indicates a rate of recession of 15 km/sec for each million light-years. This is called *Hubble's constant*—a misnomer, because its value is subject to change with time. If the rate of expansion of the universe is slowing down, then the value of Hubble's constant will be smaller in the future. Furthermore, its present value is subject to an accurate determination of the distance to certain clusters of galaxies. For instance, if the distance to the Virgo cluster is not 80 million light-years as shown in Figure 16.12, then Hubble's constant is not 15 km/sec for each million light-years. Hubble's relationship can be expressed simply as:

$$H=\frac{v}{r}$$

where v is the velocity of recession and r is the distance to the cluster of galaxies. Hubble's red-shift law expressed the fact that the distance to any given cluster of galaxies appears to be directly proportional to its velocity of recession (based upon its red shift). While there is good confirmation for this relationship in the case of normal clusters of galaxies, controversy exists as to whether the red-shift law applies to quasars—objects whose red shifts indicate recessional velocities near the speed of light if Hubble's red-shift law applies.

RADIO GALAXIES

In 1931 engineer Karl Jansky of the Bell Telephone Laboratories accidentally discovered radio signal coming from outer space. Did these signals merely represent a general background of radio noise, or were they associated with separate, distinct sources? This question was answered by Walter Baade in 1951 when, using the 5-m telescope at Mt. Palomar, he identified optically Cygnus A, a galaxy some 700 million light-years distant as the source of radio energy. The nature of this source was not realized until Baade's identification. This opened a new era for astronomy, the era of radio observations.

Cygnus A is a highly unusual galaxy, for its radiation in radio wavelengths is about one million times greater than that of a normal galaxy. Yet to be discovered is the source of this prodigious outpouring of energy; perhaps a gravitational collapse or an enormous explosion is taking place. The emissions seem to arise from two regions about 50,000 light-years to either side of Cygnus A (Figure 16.13).

Radio sources are classified as extended or compact according to their apparent diameter. Extended sources have an angular diameter of over 1 second of arc, and they probably range in diameter from 10 to 100 parsecs. Compact radio sources have an angular diameter of less than 1 second of arc and actual diameters of less than 10 parsecs (Figure 16.14).

Thousands of discrete radio sources have been found, many of

FIGURE 16.13 (a) Cygnus A, a radio galaxy. (b) The radio map of Cygnus A shows that most of the radio energy is emitted from two regions widely separated from the optical image in the center of this view. (Palomar Observatory Photograph)

(a)   (b)

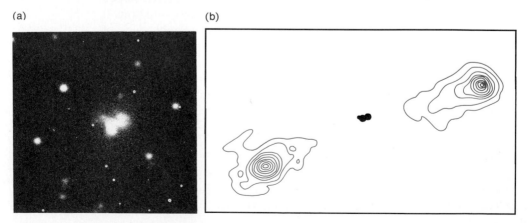

which have been identified with visible galaxies. A *normal radio galaxy* emits about one millionth as much radio energy as it does optical energy. A *peculiar radio galaxy* emits about 100 times as much radio energy as the normal radio galaxy; M87 (Figure 16.15) is an example of an elliptical galaxy that emits approximately 100 times as much radio energy as the normal emitter NGC 1068. The radio emission seems to come from two regions at equal distances on either side of the galaxy. A short-exposure photograph shows a jetlike appendage, the light from which is highly polarized. This indicates a synchrotron emission, a result of high-energy electrons moving in a magnetic field. The bright knots of the jet suggest several violent explosive events. The mass of M87 is over 100 times that of the Milky Way galaxy, and it is now known to have a central nucleus of about 5 billion solar masses, which is so dense that it may be likened to a galactic "black hole." The infall of material onto that central core may be responsible for its tremendous outpouring of energy.

SEYFERT GALAXIES

In 1943 Carl K. Seyfert of Mt. Wilson Observatory described a class of galaxies that superficially resembled a normal spiral (Figure 16.16). The characteristics that seemed to separate these particular spirals included a very small, bright nucleus that showed bright emission lines within its spectrum; this was unusual, since the spectra of most galaxies show only broad absorption lines.

The radiation of several Seyfert galaxies has been observed to vary

FIGURE 16.14 Radio map of the sky showing intense sources near the galactic equator. (Ohio State University Radio Telescope)

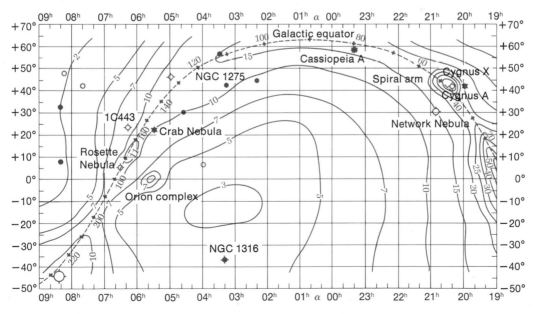

FIGURE 16.15 An elliptical galaxy in Virgo, NGC 4486 (M87), with a jetlike appendage. (Lick Observatory)

FIGURE 16.16 A Seyfert galaxy. NGC 4151. (Palomar Observatory Photograph)

greatly in a period of only a few months. In particular the X-ray radiation from NGC 6814 has been observed to vary in a period of less than ten minutes. Several other Seyferts are strong radio emitters, again showing distinct variations in output. These properties, together with their general starlike appearance, suggest that they are *very* compact sources. The energy distribution of Seyfert galaxies over a wide spectrum of wavelengths is shown in Figure 16.17. Notice how similar this output curve is to that of a quasar.

QUASARS

With the methods of radio astronomy firmly established, the search for new radio objects continued. Early in the 1960s, observers discovered radio sources that appeared to be more starlike in dimension than galactic. These sources were termed *quasi-stellar radio sources* (contracted to *quasars*). One such object was called 3C273 (Number 273 in the *Third Cam-*

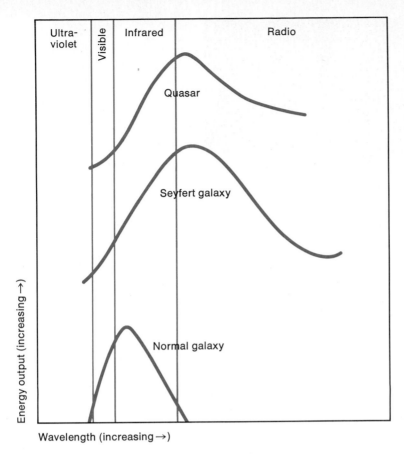

FIGURE 16.17 The energy output of a quasar as compared to a Seyfert galaxy and a normal galaxy.

bridge Catalogue of Radio Sources). Attempts were made to find the optical image of 3C273; however, the ability of a radio telescope to pinpoint a source is rather limited. Because the radio wavelengths are long compared to that of light, the resolution of even the larger radio telescopes is rather poor (see Chapter 2). A unique approach to locating 3C273 precisely was used by Cyril Hazard or Parkes Radio Observatory in Australia. On a certain day when the moon was to pass directly in front of the radio source blocking the radio signal, Hazard observed the exact time of the eclipse and reappearance of the radio source. Using his knowledge of the exact position of the moon at all times, he then pinpointed the radio source. As soon as this information reached Maarten Schmidt at Palomar, he was able to locate the optical image of 3C273 on earlier photographs. It was starlike in size but fuzzy, and it had a jet extending out a distance almost equal to its diameter (Figure 16.18).

Schmidt was puzzled by the spectrum of this object that showed an unfamiliar placement of spectral lines. He finally realized that he was see-

FIGURE 16.18 Quasar 3C273. (Palomar Observatory Photograph)

ing lines of hydrogen that should appear in the blue portion of the spectrum but which had been red-shifted so far that they appeared in the red portion.

If the red shift of the spectral lines can be explained entirely as a phenomenon related to the speed with which an object is receding from the observer, then Schmidt had found an object receding at a rate of 45,000 km/hr. Furthermore, if this object was to fit Hubble's general pattern of red shift among the distant galaxies, then it had to be placed at a distance of 3 billion light-years.

Schmidt's analysis of the red shift was soon confirmed by his colleague Beverly Oke, who showed that the hydrogen-alpha line, normally a red line, had been shifted into the infrared portion by an amount equal to Schmidt's estimate. But this was only the beginning, for today over 1500 quasi-stellar objects are known, including many which are radio-quiet, having been recognized by their unique emission spectra. In the spectra of quasar 0Q172 such a large red shift has been found as to indicate a recessional velocity greater than 90 percent of the speed of light. If this object is to fit the Hubble red-shift pattern, then it must be placed at a distance of 18 billion light-years.

These observations set the stage for a controversy that was to last many years. On the one hand, some astronomers believed that they were seeing the most distant objects ever seen by man—objects so far distant that they were looking backward in time to what was perhaps almost the beginning of the present universe (Figure 16.19). If this were the case, then they were also looking at the brightest objects ever witnessed. They were hard pressed to explain a source of energy so great that it could be recorded 10 to 18 billion light-years away. No ordinary process in nearby galaxies could match the outpouring of energy necessary to be detectable at such distances, an energy output that must be thousands of times that of a normal galaxy. Other astronomers say that these objects are closer and therefore need not be so bright.

Dr. Halton Arp of Hale Observatories may be characterized as the leader of this latter school of thought. He has observed that quasars are not distributed—either as to position in the sky or as to red shift—in the

(a)

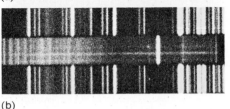

(b)

FIGURE 16.19 (a) Source of radio noise, 3C295, in the constellation of Bootes. (b) Its very large red shift of spectral lines may indicate that it is one of the more distant objects to the known universe. (Palomar Observatory Photograph)

uniform way that one would expect if they were actually the more distant objects of an expanding universe. Dr. Arp finds alignments and groupings of quasars and other compact objects in apparent proximity to large elliptical and spiral galaxies. We say "apparent" because this is where the controversy lies. Various objects may appear close together in the sky yet be separated by vastly different distances from the observer. How can we tell if they are actually close to one another? Dr. Arp looks for bridges of material that physically connect the various objects and he looks for gravitational interaction between the members. He feels that it is too much of a coincidence for several objects to be so nearly aligned (Figure 16.20), especially when this alignment is so closely correlated with the radio-emission lobes of the central galaxy (Figure 16.21). Often the quasars or compact galaxies appear on both sides, as though they were ejected from the primary galaxy. Could this explain how NGC 7331 (an Sb spiral) could have a red shift indicating a recessional velocity of only 800 km/sec while the red shifts of the compact objects which appear in the same region indicate velocities between 6000 and 8000 km/sec? What of the five galaxies in Stephan's Quintet (see Figure 16.22) which fall within 0.5° of NGC 7331? Four of these objects have velocities in the range of 5700 to 6700 km/sec, whereas the fifth has an indicated velocity of 800 km/sec. If NGC 7331 and the compact galaxies which appear around it are actually close to each other, then we must look for a different explanation of their different red shifts—an explanation that does not depend simply upon an expanding universe in which larger red shifts correlate with greater distances. While his critics suggest that he is simply seeing what he wants to see, Dr.

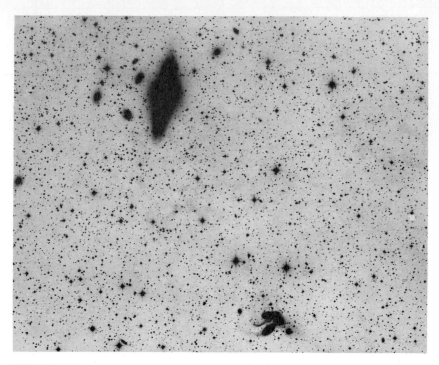

FIGURE 16.20 A positive print showing a clustering of Arp objects in the region around NGC 7331 (a large spiral galaxy) including Stephan's Quintet in the lower right. (H. Arp, Palomar Observatory Photograph)

Arp would retort that there are far too many examples of the type cited to pass the matter off that lightly.

Let us return now to the possibility that quasars are very small distant objects. Some have suggested that they range from 1/25 to 1/100 the size of a normal galaxy, yet radiate up to 1 million times more energy than the average galaxy. Quasars have been seen to vary in energy output in less than ten minutes. Let's take as an example a quasar that shows variation in one week. We must conclude that this object is relatively small in order to show this variability. Theoretically an object which shows variations in as little as a week cannot be much larger than 1 light-week in radius, or such variations would be smoothed out due to the fact that light from the nearest surface of the object reaches us ahead of the light from middle sections. If an object were even a light-month in radius, at least one month would be required to receive signals from all the various distances within it and over that interval of time a variation of 1 week's duration would be wiped out. In order to appreciate how very small a quasar may be, one light-week is equivalent to approximately 15 times the diameter of the solar system. How could such small objects emit the tremendous amount of energy indicated, if they are in fact situated 10 to 18 billion light-years away?

Thermonuclear reactions that take place in the stars of normal galaxies simply do not release sufficient energy to be seen at these vast cosmic

FIGURE 16.21   A radio map of the region surrounding the Sb Spiral galaxy NGC 7331, with Stephan's Quintet shown in the lower right. The shape of the radio emission lobes tends to suggest a connection between NGC 7331 and the other galaxies shown in the same region.

FIGURE 16.22   The five galaxies in Stephan's Quintet lie within 0.5° of NGC 7331 (see Fig. 14.20). (Lick Observatory)

distances. We must look for more energetic reactions. One source of energy that could be released in a relatively short period is gravitation. If the core of a supermassive galaxy collapsed, with the nuclear particles falling together, huge amounts of energy would be released in a very short time. The result of such a collapse could be the ejection of surrounding materials, such as the ejection of jets from several quasars. If electrons ejected by such an event move in a magnetic field, they generate synchrotron radiation. Such radiation is emitted from quasars and is capable of releasing large amounts of energy. It has also been observed that quasars emit prodigious amounts of energy in X-ray wavelengths, particularly quasars which are also good emitters of radio.

The characteristics of quasars remind us of the bright core of a very massive galaxy, one which has a black hole at its center consisting of several million solar masses, with material being pulled toward its surface. This model seems consistent with both the amount of energy needed and with the variability that has been observed. We usually think of black holes as the latter stage of certain stars, yet as we view quasars that are up to 18 billion light-years distant, we should remember that we are looking backward in time 18 billion years—we are seeing these objects as they existed that long ago. If a quasar is really the core of a supermassive galaxy with a black hole at its center, then it must have evolved to that state in a rather short time as compared to the Milky Way. Perhaps such an object was formed very early in the history of our expanding universe, at a time when the material of the universe was more densely packed, and furthermore it may have ceased radiating before the sun was formed.

THE DOUBLE QUASARS

When one thinks of the general distribution of galaxies and quasars throughout the sphere of the sky, it would seem highly unlikely that two quasars with identical spectra and identical red shifts would appear as pairs or triplets right next to each other, and yet several such cases have been found since 1979. It is now quite apparent that another galaxy may just happen to line up between a single quasar and ourselves, for example, and its gravitational effect on light from the quasar, may be causing the quasar to appear in several positions from our point of view. The idea of such a *gravitational lens* is diagrammed in Figure 16.23. If you trace an imaginary ray of light from the quasar, you can see that by the time it has been bent it strikes our plates at an angle other than its true direction. This is similar to the effect of our own atmosphere that makes stars near the horizon appear higher in the sky; of course in the case of the atmosphere, refraction, not gravitation, effects the light. If the intervening galaxy were perfectly lined up between the quasar and ourselves, the totality of all rays would produce a ring of light on our plates, but the probability of this perfect alignment is rather low. We do see double and triple images however as a result of the gravitational lens effect.

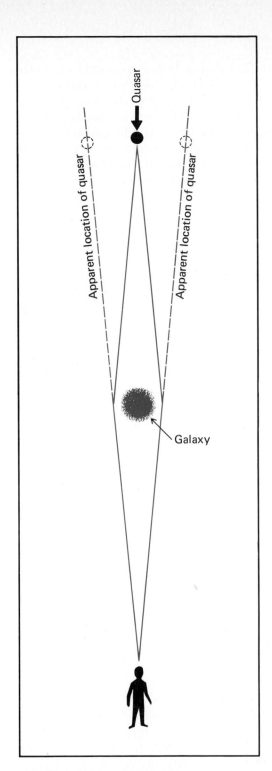

FIGURE 16.23 Gravitational lens.

BL LACERTAE OBJECTS

The BL Lacertae Objects were identified in 1929 as a variable star in the constellation of Lacertae. It became the model for a whole collection of fuzzy variable "stars" called BL Lac for short. In 1969 the discovery of the radio emission of these objects was the first clue that they had been incorrectly classified, for only galaxies and quasars have radio emissions of this type. Obviously, if a red shift could be found, it would settle the matter of whether BL Lacs belong to the Milky Way or not. No sharp spectral features were apparent, however, and the question was not settled at that time. With improved techniques, weak emission and absorption lines are being found, and their red shifts indicate that the distances to the BL Lac Objects must rank with those of the quasars. This only compounds the questions about the quasars raised in the last section, for if located at the great distances indicated, then these objects are even brighter than the quasars and more variable. One might class them among the most violent quasars. Thus BL Lac was first identified as a lowly variable star in the Milky Way and then became one of the most actively studied objects in the entire universe.

A recent observation of BL Lacertae itself has revealed that this object is located at the center of a normal galaxy, and because of the very faint nature of that surrounding galaxy, the great "cosmological" distance of BL Lacertae is confirmed: it really is at the great distance indicated by its red shift. How can any object produce enough energy to be seen at such a great distance? The variability both in light output and in polarization indicates a very small nucleus of the order of a light-day. It appears that the emanations of the nucleus simply overpower those of the surrounding material, hence astronomers may be "seeing" closer to the heart of a violent quasar as they study the BL Lacs.

A MODERN CLASSIFICATION OF GALAXIES

Hubble's classification of "normal" galaxies will continue to be useful; however, another scheme seems necessary to account for the very active (violent) natures of many of the objects being observed today. It would appear that the BL Lacertae Objects are the most violent of all objects in the universe, and that their most intense activity is concentrated in a relatively small nucleus. Quasars follow as a close second, both those with radio emissions and those without. The Seyferts would constitute the third classification, followed by the "normals."

Any classification scheme seems to raise the question of a possible evolution among galaxies. Could a BL Lac Object be a quasar evolving into an elliptical or vice versa? There appears to be very little evidence to support any unified theory of evolution, however perhaps our view of the universe may give a clue. When we look at very distant objects such as BL Lacs and quasars, we are seeing them as they existed in the past, in the early stages of the universe. Their density was much greater then, and this density may have influenced the nature of galaxies as they formed. The very active small cores we see in these objects may have been a part of their early nature. As we look to nearer and nearer objects, we are seeing these objects in a later stage of development, because light doesn't take so long to reach us. We see bright nucleii in certain middle-distance galaxies, such as Seyferts, but not so active as the quasars. We also see at near and middle distances the "normal" ellipticals with no apparent core emphasis. Some observers have suggested that a Seyfert is merely a nearby quasar— perhaps a quasar that has evolved with time to become a Seyfert. Likewise, a comparison can be made between the elliptical and the BL Lac, which has an elliptical around its violent core.

While this discussion does not solve the question of evolution, it does show the interrelationship of many different factors and the complexity of the astronomer's effort to read the past history of the universe.

COSMOLOGY

Cosmology is the study of the universe as a whole—its large-scale structure, its organization, and its history. In a sense, some of the early models of the universe discussed in Chapter 1 represent cosmologies; however, they were very limited views. What was the chain of events by which the view of the cosmos was expanded and an interest in its history developed.

In 1915 Albert Einstein gave science one of its most profound theories: *the general theory of relativity*. He realized that in order to describe an event, both the three dimensions of space and the fourth dimension of time are necessary—that we live in a world of space-time. To illustrate this point consider a full description of a trip that takes us from one location to another in three-dimensional space (like traveling up a mountain) and also consumes a certain time interval (the fourth dimension of our full description). In his general theory of relativity, Einstein presented a description of gravity by stating that a mass (such as a star) curves space-time in its vicinity and furthermore that it is this curvature of space-time that "directs" the motion of any object passing nearby. All of these ideas are expressed in a set of equations beyond the level of this text, which have served as a foundation for numerous cosmological models.

In 1922 Alexander Friedmann found a solution to Einstein's equations for a certain time in the past history of the universe. His solution predicted a state of very high density and very high temperature. One might liken the high density to that of the singularity of the black hole (see Chapter 14, page 433). Some would visualize this as the "primordial atom." It is thought that this superdense, superhot "atom" exploded, producing the "primordial fireball"—the Big Bang, the name given to the theory today. Theorists have tried to picture the nature of those first moments after the explosion occurred and in the following paragraph we will relate some of their thoughts about the *first second* of time.

The temperature is thought to have been about 100 billion degrees

Kelvin at the moment of the Big Bang, and all energy was in the form of radiation. Within a very small fraction of a second, however, the expanding universe had cooled sufficiently for some of that energy to be converted to matter. There may have been hundreds of different kinds of particles created, but two categories, called quarks and leptons, seem to be crucial to the next stage. Within another small fraction of the first second, these particles were formed into protons, neutrons, and electrons.

The theory further suggests that when the universe had cooled to 3 billion degrees Kelvin, atomic nuclei began to form. The hydrogen nuclei occurred in two isotopic forms: H-1, hydrogen containing one proton; and H-2, deuterium (heavy hydrogen) containing one proton and one neutron. The amount of deuterium formed is inversely proportional to the density of the matter at the time. In a very dense state, little deuterium is left because most of it reacts with other nuclei to form helium. In a less dense situation, more of the deuterium will be left intact. Thus the deuterium abundance as observed today is considered an important clue to the early state of the universe. While deuterium is very unstable at high temperatures, its formation was the steppingstone to the production of helium-3 and helium-4. The universe at this point could be characterized by the presence of ionized matter (protons, deuterium nuclei, helium nuclei, free electrons), neutrinos, and electromagnetic radiation (photons) interacting intensely with each other. Within 35 minutes of the Big Bang, the temperature had cooled to 300 million degrees Kelvin and all nucleosynthesis had stopped; no more helium or heavier nuclei would be formed at this time. Only after about 100,000 years did continued expansion result in cooling to about 4000 degrees Kelvin, at which temperature free electrons could combine with the nuclei to form neutral atoms of hydrogen (73 percent), deuterium (trace), and helium (27 percent). With continued cooling due to expansion, clusters of galaxies formed, and the motion of these clusters still demonstrates the expansion of the universe today. The Big Bang theory assumes that galaxies formed in the first 100 million years.

In 1924 Edwin Hubble first demonstrated the existence of galaxies outside our own Milky Way galaxy when he recognized that the Andromeda Nebula was not a part of our galaxy. Only within the last half century have we known that the universe is composed of clusters of galaxies, the Milky Way being only one member of such a cluster. In 1929 Hubble discovered the red-shift law, a relationship which plays an essential role in testing cosmological models. Hubble was joined by astronomers such as M. L. Humason and N. U. Mayall in his study of the red shifts of distant clusters of galaxies, and by 1936 two additional discoveries had been made, both of which were essential to the development of cosmological models:

1. These observers noticed that the brightest elliptical galaxies of one cluster had virtually the same absolute magnitude as the brightest ellipticals of any other cluster. This was significant because these bright ellipticals could be used as an accurate distance indicator—the dimmer they appeared, the more distant they were. With knowledge of both their absolute magnitude

and their apparent magnitude, their distance could be found using the relation discussed previously:

$$\frac{L(10)}{L(r)} = \frac{r^2}{10^2}$$

But certain corrections must be applied, as we shall see, and the process is not as simple as it may appear.

2. In all the astronomer's searching, clusters of galaxies appeared to be randomly distributed with regard to distance appeared to be randomly distributed with regard to distance and with regard to direction. Clusters of galaxies do not favor one part of the universe over another part (on the big scale). This was the first observational evidence that the universe appears to be *homogeneous* (similar in density throughout) and *isotropic* (the same in all directions). This was a very significant observation, for it suggested that we do not live in some unusual part of the universe, but rather that what we see from our location is typical of the entire universe. This observation formed the basis of an assumption which is at the heart of most modern cosmological models. Called the *cosmological principle*, it is stated:

The universe, on a large scale, appears the same from any location.

This suggests that the different parts of the universe may have had a common origin and lends hope that its history can be discovered.

By about 1950 this cosmological principle had been expanded by a group at the University of Cambridge, headed by Hermann Bondi and Thomas Gold, and later by Fred Hoyle. They suggested that not only does the universe appear the same from any location, but it also appears the same at any time (past, present, or future). This concept, referred to as the

FIGURE 16.24 The Big Bang theory.

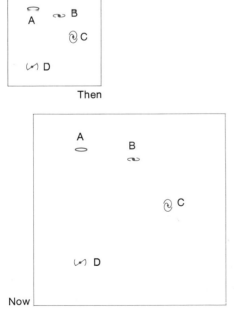

Then

Now

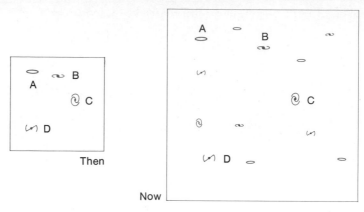

FIGURE 16.25 The steady state theory.

perfect cosmological principle, formed the basis for the "steady state" model. This model does not deny the expansion of the universe, but rather it predicts its acceleration. In order to keep the distribution (density) of the universe constant, it suggests that as galaxies move apart, new matter comes into being. Because it is assumed that the universe has always looked the same, there need be no creation event equivalent to the primordial fireball. The Big Bang theory and the steady state theory are illustrated in Figures 16.24 and 16.25. As we shall see, there are several lines of evidence against the steady state theory and in favor of the Big Bang theory. One of the most convincing is the discovery of a form of radiation which may be a consequence of the primordial fireball.

COSMIC BACKGROUND RADIATION

If the universe began as a gigantic fireball, characterized by a temperature of the order of a billion degrees Kelvin, and if for a period of time a state of thermal equilibrium prevailed, its energy output would have a definite pattern, resembling that of the sun or stars in general. Because of the expansion of the universe, that radiation today should appear to have cooled and should still exhibit itself as a background radiation equivalent to that of a very cold blackbody, one of 3°K (−270°C). If this radiation could be found, then it would be direct evidence of the fireball (Figure 16.26).

About 20 years after it was initially predicted, blackbody radiation was found quite accidentally by two scientists of the Bell Telephone Laboratory, Arno H. Penzias and Robert W. Wilson. In 1965, they were testing very sensitive radio communication equipment, which seemed plagued by a form of radio noise. When theoretician Robert H. Dicke heard of the discovery, he devised twin receiving antennae, one of which was attached to a reference source that was known to radiate at 4.2°K, and the other of which was pointed skyward. When the sky radiation was compared to the reference source, it became evident that the radio noise that Penzias and Wilson had discovered was equivalent to the radiation of a black body at

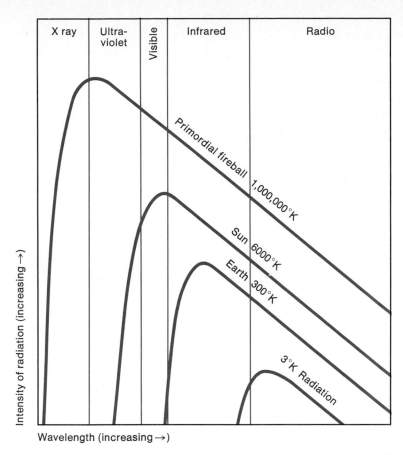

FIGURE 16.26 The black-body radiation curves for the primordial fireball and for 3 K radiation, the latter falling almost entirely in the radio portion of the spectrum.

3°K. Furthermore, it can be received from any direction. In the years that have followed, measurements have been taken in a variety of wavelengths, and this background radiation appears to fit the predicted curve quite closely. Sufficient evidence is now available to strongly suggest the existence of the fireball. The steady state theory provides no convincing explanation of the 3°K blackbody radiation.

This cosmic background radiation forms one of the best frames of reference by which the earth's motion in space may be specified. Because the radio antenna records an increase in temperature when pointed toward the constellation of Leo, observers compute for the earth a velocity of approximately 400 km/sec in that direction.

THE PRIMORDIAL ATOM: INFINITE OR FINITE?

To think of the matter of the entire universe being confined in the remote past to a very small volume—to a singularity—does not appear consistent with our observation of cosmic background radiation. If this radiation

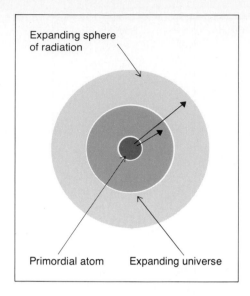

FIGURE 16.27 Radiation from a very small primordial atom.

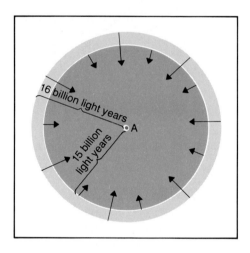

FIGURE 16.28 Radiation from an infinite primordial atom.

originated from only a limited region of space, it would have traveled outward at the speed of light and would not be detectable today (Figure 16.27).

Consider an infinite primordial atom which everywhere experienced the very high temperature of the Big Bang fireball, each point radiating energy in every direction. Subsequently a receiver located at point A would receive the signal which had originated at points just far enough away to be arriving now. For instance, we live about 15 billion years after the Big Bang; therefore we should be receiving a signal which was generated from places 15 billion light-years away—that is, from all points on an sphere of radius 15 billion light years (Figure 16.28).

If we could observe this background radiation a billion years hence, we would then sense it coming from points 16 billion light-years away. The radiation would appear cooler because it had been red-shifted more than we see it today. This model presents an infinite universe both before and after the Big Bang and does away with any need to visualize its point of origin.

CURVATURE OF SPACE-TIME

Several variations of the two models of cosmology exist. Whereas the steady state model implies that the rate of expansion is speeding up (accelerating), variations of the Big Bang model imply a slowing down (deceler-

FIGURE 16.29 The flight of a ball (shotput) thrown upward may be described in various ways depending upon the force applied: (a) oscillating (bouncing)—if only a small initial force is applied; (b) parabolic—continually rising but also slowing down; (c) hyperbolic—continually rising at a greater rate than in (b); (d) accelerating—continually rising and speeding up, due to an added propulsive force (here applied by a rocket).

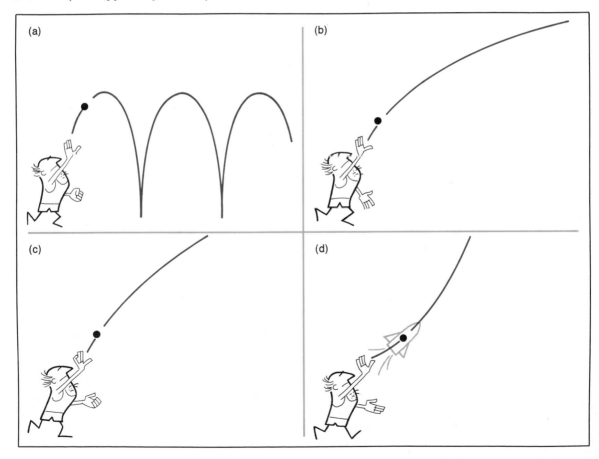

ation) of the rate of expansion. Different rates of deceleration are possible, however. In order to visualize these various possibilities, let us suppose that a ball is tossed upward with varying forces, imparting different initial velocities in each case. Our normal experience with such a toss suggests that the force of gravity will decelerate (slow down) and eventually stop the upward flight of the ball. The ball will then accelerate as it returns to earth, only to bounce upward again and again [Figure 16.29 (a)]. We might characterize its flight as an *oscillation*. On the other hand, if it were possible to give the ball a higher initial velocity, it might just overcome the force of gravity and leave the earth forever [Figure 16.29 (b)]. It would experience deceleration, but its flight would be arrested only at some great distance from the earth (at infinity). The physicist would describe the flight of the ball under such conditions as *parabolic*. If still a greater force were used, the flight of the ball would appear as in Figure 16.29 (c), continually decelerating, yet climbing higher without limit. Such a flight is termed *hyperbolic*. A fourth possibility involves the continual application of a force, say by means of a rocket attached to the ball [Figure 16.29 (d)]. The force of the rocket engine produces an acceleration, and the ball is carried upward at increasing speeds.

These four cases are analogous to various ways in which the universe might be expanding. The explosive force of the Big Bang event may have sent the material of the universe outward at such a rate as to allow gravity to bring it to a halt and collapse upon itself; this is the *oscillating model*. A plot of the scaling factor of the universe, as time goes by, would appear as in curve (a) in Figure 16.30. Notice the use of the term "scaling factor" in place of the word "radius," for the universe may not have a radius if it is infinite. The term "scaling factor" is a measure of the distances which are opening up between clusters of galaxies in an expanding universe, and when compared to time this gives a measure of the rate of expansion. Had the outward force, at the time of the Big Bang, been just sufficient to overcome the force of gravity, the universe would continue to expand but would be slowing in its rate of expansion as shown in curve (b). The shape of such a curve is said to be *parabolic*. Curve (c) illustrates the expansion of the universe if still greater initial velocity had been imparted. This, too, represents a decelerating universe but is said to be *hyperbolic*.

The fourth model, that of an *accelerating expansion*, is illustrated by curve (d) in Figure 16.30. This represents a necessary condition of the steady state model; we have no indication that it is correct, however. The best determination of the Hubble "constant" at present is represented by the straight (solid) line in Figure 16.30; points on each curve having the same slope as that reference line have been made to coincide (see the point labeled "now"). As you can been made to coincide (see the point labeled "now"). As you can see, each model predicts that Hubble's "constant" is not really constant but rather is changing. If it is changing as in curve (a) in Figure 16.30, then the universe will oscillate; projecting the curve backward into the past, we see that it has only been 11 billion years since the last Big Bang. If the change in the rate of expansion is more accurately described by curve (b), then we must look back in time about 13 billion years for the Big Bang. Curve (c) indicates an even earlier beginning—about 15 billion years ago. If Hubble's "constant" were really constant, its straight-line graph would indicate an age for the universe of 20 billion years. Curve (d) the accelerating universe, requires a still earlier beginning.

How do these models relate to Einstein's concept of the curvature of space-time? To describe the evolving universe mathematically, he introduced a constant k into his equation. Depending upon the value of this constant, he termed the curvature of the universe either positive (if k is positive), zero (if k is zero), or negative (if k is negative). Positive curvature of space-time correlates with the oscillating model, zero curvature with the parabolic model, and negative curvature with the hyperbolic and the accelerating models.

Because it is impossible for us to visualize four-dimensional space in our three-dimensional world, we might try to visualize each of these curvatures of space-time by considering models of one less dimension. Our representation of a four-dimensional surface of zero curvature in three-

FIGURE 16.30   The curves associated with the four models of the universe are shown with their present state made to coincide at the "now" point: (a) oscillating universe; (b) parabolic universe; (c) hyperbolic universe; and (d) accelerating universe. The straight reference line represents the best estimate of the Hubble "constant" at the present time. As is clear from this drawing, the "start-up" time of the universe is dependent upon the model that best describes its evolution.

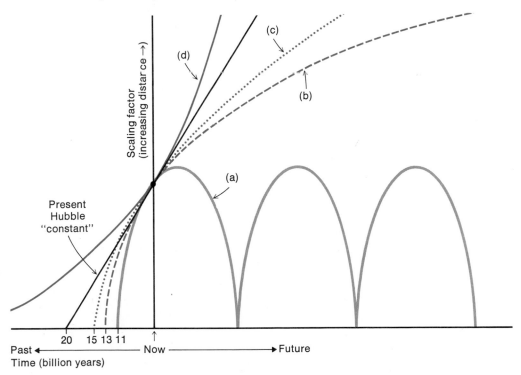

space, would be a plane (flat surface). If dots (each of which represents a cluster of galaxies) are distributed uniformly over such a surface, one may predict the ratio of the number of dots in successive circles of radius 1, 2, 3, and so on. The area of a circle is proportional to the square if its radius ($A = \pi r^2$), therefore the area of successively larger circles will have the ratio of $1:4:9:$ and so on (the squares of the corresponding radii). Likewise we would expect the ratio of counts of dots (clusters of galaxies) in successive circles to be the same. On the other hand, if the dots are distributed uniformly over a surface of positive curvature, like that of a sphere, then successively larger radii produce less than the expected ratio, say $1:3.6:8:$ and so on. This is graphically demonstrated by the gaps which develop in the flattening process shown in Figure 16.31(b), resembling a flattened orange peel. Negative curvature produces just the opposite effect; circles of successively larger radii produce more than the expected number of

FIGURE 16.31 The curvature of space may be visualized as: (a) flat, wherein we would find a uniform distribution of galaxies; (b) positive, wherein galaxies would appear to thin out with greater distances; or (c) negative, wherein the distribution of galaxies would appear more dense at greater distances.

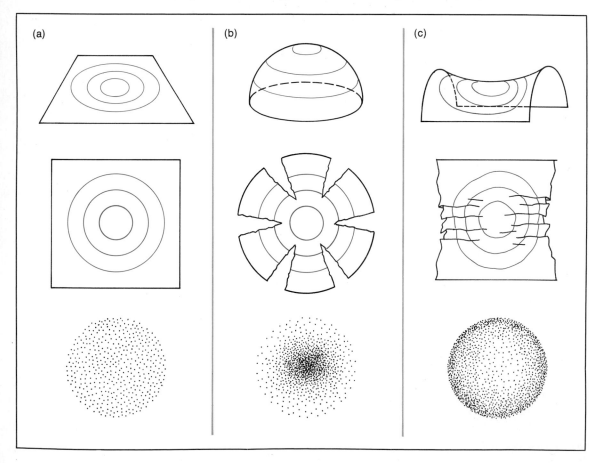

dots, for when a surface of negative curvature (hyperbolic or saddle-shaped) is flattened, wrinkles occur because of excess material. The ratio of dots may approach 1:4.5:10: and so on. When this whole idea is projected back into four-dimensional space-time, one would expect the density of clusters of galaxies to be uniform (to have a predictable ratio in terms of increasing distances) in the case of zero curvature, to thin out with increased distance in the case of positive curvature, or to increase with increased distance in the case of negative curvature (Figure 16.31, bottom row). If this analysis is correct, the astronomer should be able to tell which kind of a universe we live in.

TESTS FOR COSMOLOGICAL MODELS

The effect of space-time curvature does not become apparent merely by our counting nearby galaxies or clusters of galaxies. The astronomer is therefore concerned with clusters of galaxies that are beyond the reach of optical telescopes—that is, with radio galaxies. Recently a number of astronomers on the staff of Ohio State University completed a survey of over 8000 radio sources. They assumed that the radio sources that are located billions of light years away are just as bright as those which are nearer. Based upon that assumption, their plot showed a thinning out with increased distance, somewhat resembling that of Figure 16.31(b), indicating a positive curvature. When very distant galaxies are viewed, however, we are also looking backward to the time when the light left those sources (billions of years ago), and perhaps galaxies were intrinsically brighter (or dimmer) in the early stages of the universe. The very fact that such an evolutionary factor must be considered illustrates the uncertainty with which the astronomer must regard this survey of radio sources. Perhaps we may confirm or deny the results of the Ohio survey by taking still another approach, one typified by the work of Allan Sandage and William A. Baum.

Let us recall how Hubble first recognized the relationship of red shift to distance. He plotted the red shift of galaxies against their apparent magnitudes (see Figure 16.12, page 466). Extending this procedure to the more distant clusters of galaxies, astronomers have produced the plot shown in Figure 16.32. The way in which the four models of the universe—oscillating, parabolic, hyperbolic, and accelerating—have been interpreted by cosmologists is shown superimposed upon this plot. You can see that the plot of nearby galaxies (lower left-hand corner) is not sufficient to differentiate among the models. As more and more distant galaxies are being added, however, there appears to be a fairly close correlation with the oscillating or positive curvature model. Is this evidence conclusive? Perhaps not, for there is a fair amount of scattering in the points, and furthermore the location of the points on the graph is subject to correction with the advancement of techniques for measuring apparent magnitudes and for correcting these measurements for various effects. There is little chance, however, that the plot will change sufficiently to

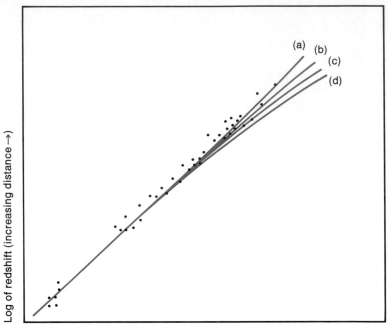

FIGURE 16.32 Hubble diagram—a plot of distant galaxies shows a fair amount of scattering among the four theoretical models of the universe: (a) oscillating; (b) parabolic; (c) hyperbolic; or (d) accelerating (steady state). Although there is some tendency at present to favor the oscillating model, more data are necessary for a conclusive decision.

correspond to the accelerating universe. This observation has virtually sounded the death knell for the steady-state theory.

We may summarize the primary task of the cosmologist as that of finding two numbers: the Hubble "constant" and the actual rate at which it is changing. The search is challenging the very limits of the art today. However, still one further test may be applied in determining the curvature of space-time.

MASS-ENERGY DENSITY OF THE UNIVERSE

Einstein predicted in his theory of relativity that the curvature of space-time is proportional to the mass-energy density of the universe. It is possible to compute the density associated with each model, and these figures are tabulated in Table 16.3.

At present, the best estimate of the density of the universe is 3×10^{-31} g/cm^3, about one-thirtieth of the mass-energy necessary to create an oscillating condition. This estimate is based upon the mass of all luminous material (stars, luminous nebulae, and the like), together with an estimate of the nonluminous gas and dust in the galaxies and of all the cosmic rays.

TABLE 16.3

MASS-ENERGY DENSITY ASSOCIATE WITH VARIOUS MODELS OF THE UNIVERSE	
Cosmological Model	Mass-energy Density
Oscillating	More than 10^{-29} g/cm^3
Parabolic	Exactly 10^{-29} g/cm^3
Hyperbolic	Less than 10^{-29} g/cm^3

Recent X-ray studies suggest a background of X-ray radiation which may be originating from nonluminous sources. Infrared astronomers are identifying large numbers of cool red dwarfs, unknown a few years ago and we must also throw in the mass of all the black holes we think exist. Another very significant addition to our understanding of the mass-energy of the universe may come from the huge (massive) invisible corona that is now thought to surround our Milky Way galaxy. If all galaxies (or just the spirals) have massive coronas, they could multiply the known mass by a significant factor. The galaxies may be even more massive than previously thought if they have super massive (black hole-type) centers. Astronomers are on the threshold of new techniques and new discoveries which may provide clues to enough mass-energy to close the universe—that is, to provide enough self-gravity to eventually stop the expansion and create an oscillating condition. For this to be true, the universe must contain approximately thirty times as much mass as stated at the beginning of this section.

DEUTERIUM ABUNDANCE

On page 480 we discussed the formation of deuterium during the Big Bang event. Because the percentage of material that remained in the form of deuterium (heavy hydrogen) after the event depended upon the density of that early universe and because its present density is also related to that early density, the percentage of deuterium present in the universe today constitutes a test for cosmological models. The deuterium abundance found indicates that the early density was too low to yield a closed universe; hence by this test the universe is "open"—that is, it will expand forever.

The case for the open universe appears strong, although most astronomers would agree that the tests are not conclusive the search thus goes on—we cannot say for sure which model represents the universe. For that matter, who is to say that any of the proposed models is correct?

THE UNIVERSE IN INFRARED

In late 1983, there was unleashed a flood of new observations by the IRAS infrared satellite—from rings of dust and new objects in our solar system to galaxies at great distances. IRAS sees things that are too cool to shine in visible light. Objects that are totally cold still emit electromagnetic radiation in the infrared portion of the spectrum. The sensitivity of the detectors that are an essential part of the IRAS satellite is greatly enhanced by cooling them to almost absolute zero ($2.5°K$) using superfluid helium. With only a limited supply of liquid helium, the satellite was doomed to failure in approximately 12 months time; but in that time astronomers were able to survey virtually the entire sky several times over. Of the more than 200,000 infrared sources that were found, some correspond to familiar objects and others have no visible counterpart.

Sites of star formation probably account for many of the images that were recorded. Astronomers believe that dust plays an essential role in the formation of stars and that it is the dust that is revealed by infrared detectors. Furthermore, as material contracts to form a star, much of its potential gravitational energy is converted to heat. Therefore a warm cocoon around a forming star would radiate in the infrared.

Three rings that are primarily composed of dust have been found in the solar system: one that lies between Mercury and the sun; one that lies between the orbits of Mars and Jupiter; and one that exists at a distance far beyond Pluto, approximately 9 billion miles from the sun. But such disks of dust are not limited to our solar system. Both the bright star Vega and the super giant star Betelgeuse are among the approximately fifty stars that appear to have disks of material surrounding them—suggesting that solar systems may exist around many of these. In fact, several solid bodies have been identified as being gravitationally bound to Vega. If we believe that our galaxy is not unusual, then this discovery points to the fact that solar systems may be very common in the entire universe. At least it seems that disks of material around stars are quite common.

A glance at the white light photograph shown in Figure 15.2 reveals that a large amount of the central portion of the Milky Way galaxy is obscured by dust lanes that have a dark appearance. Infrared wavelengths penetrate such regions of dust and reveal objects within the core of the galaxy and galaxies beyond. For instance, huge molecular clouds have been found in the direction of Sagittarius, together with features resembling cirrus clouds in the region that represents the flattened portion of our galaxy. In a smaller survey that only covered 1 percent of the sky, more than eighty galaxies were found. Many of these were barely visible on the most sensitive photographic plates, and we must conclude that most of the energy is radiated in the infrared. A preliminary thought suggests that star formation in these galaxies may account for the infrared excess.

Five new comets have been discovered, and the dusty envelope of others has been identified. In addition, an object called 1983 TB has an orbital motion that suggests that it was once a comet and that it may be the primary body associated with the Geminid meteor shower. This object, at perihelion, passes closer to the sun than any other known object and periodically crosses the earth's orbit as well, classifying it among the Apollo objects.

It should be noted that only about 1 percent of the total information from the IRAS satellite has been processed at this time. As each new astronomical satellite is launched, it seems to perform beyond our greatest expectation and open up new vistas of understanding about the universe.

QUESTIONS

1. What distinguishes an SO-type galaxy from the highly flattened elliptical galaxy or the Sa-type spiral galaxy?
2. Astronomers estimate that the Milky Way galaxy contains at least 100 billion stars. How many galaxies do they estimate exist in the universe?
3. Sketch the following types of galaxies: (a) Sa; (b) Sc; (c) SBb; (d) E3.
4. What general type of galaxy is most prevalent in our local group?
5. In order to use a Cepheid variable to find the distance to a galaxy, what must the observer be able to determine in relation to that Cepheid?
6. Under what conditions is it safe to assume that because a galaxy looks dimmer, it must be farther away?
7. What relationship did Edwin Hubble find between the distances to galaxies and the red shift in their spectra?
8. Can we apply the red-shift law that Hubble developed to every distant object in the universe without questioning its validity for that object?
9. Normal galaxies emit much less radio energy than light energy, but certain types of galaxies have a very high output of radio energy. Name several of these types.
10. Is it possible for a radio galaxy to emit as much as a million times as much radio energy as a normal galaxy?
11. What does the contracted word *quasar* stand for in full?
12. What characteristics do all quasars have in common?
13. What evidence exists for the idea that the quasars are the most distant and therefore the brightest objects ever sensed by astronomers? What evidence is there that quasars may be relatively nearby objects?
14. If the quasars are very distant objects, then what may be their source of energy?
15. In what sense is time related to distance in all astronomical observations?
16. Is there more evidence that elliptical galaxies evolve into spirals or that spirals evolve into ellipticals? Or is it more likely that galaxies do not evolve except in their early stage of formation? Substantiate your answer.
17. What factor probably determines the ultimate shape of a galaxy?
18. If our universe is expanding, as indicated by the fact that all distant objects have red shifts, are we necessarily at the center of that universe? Would you still see the same explanation from any point of view in the universe? See if you can prove your answer by drawings.
19. What would have to be happening in the universe if the steady-state model were to be satisfied?
20. Within which theories of cosmology does the density of the universe change?
21. What recent findings tend to substantiate the Big Bang theory?
22. What kind of matter-energy has recently been discovered that may add to our knowledge of the total mass-energy of the universe?

Relativity

Seventeen

Our description of the universe is continually changing—it is often turned upside down. The description that prevailed during the eighteenth and nineteenth centuries was due largely to the work of Sir Isaac Newton. The physical laws that he set forth in the late seventeenth century were considered valid for over 200 years; they still serve to describe our everyday experiences. When Newton asserted, "I hold time and space to be absolute," he meant that two (or more) observers would produce the same value when measuring the time and/or space interval between two events, even if the observers were moving with respect to each other. He thought that the time and the distance between two events could be treated separately, each being independent of the relative motion of the observers. Let us examine why these assumptions are incorrect.

RELATIVE MOTION

There are very simple forms of relativity we live with each day. Suppose a special train car has been equipped for bowling (see facing page) and suppose that car is passing a station, traveling at 80 km/hr in an eastward direction. A woman bowls a ball at 60 km/hr in an eastward direction. Two observers, one standing on the station platform and the other traveling in the moving train car, each describe the velocity of the ball. The observer in the car says that the ball is traveling at 60 km/hr in an eastward direction, whereas the observer on the platform says that the ball is traveling 140 km/hr in an eastward direction (the algebraic sum of 80 and 60 km/hr). Why are their descriptions different from one another? The reason is that the two observers are in motion relative to each other. We will see that Einstein's special theory of relativity has a lot to do with the fact that space-time intervals are quite different for observers who are moving relative to one another. Let us consider further the example of the bowling alley in the train car. How would each observer have described the velocity of the bowling ball if the woman had bowled the ball in the opposite direction—toward the west? The observer in the moving car would now say that the ball was traveling at 60 km/hr in a westward direction but the observer on the platform would have described the ball as moving eastward at 20 km/hr. They would have described the ball as moving in opposite directions. Again the platform observer's value represents simply the algebraic sum of the two velocities, in this later case 80 km/hr minus 60 km/hr equals 20 km/hr in an eastward direction ($80 - 60 = 20$ km/hr). One description is just as good as the other, for there is no fixed frame of reference by which motion is judged. We get used to treating the surface of the earth as fixed, but that is only for convenience. Imagine still another observer suspended in space on one side of the solar system. That observer would see a point on the earth's equator rotating at almost 1500 km/hr and the earth revolving around the sun at approximately 100,000 km/hr and that observer's description of the velocity of the bowling ball would be the algebraic sum of the four velocities mentioned in this illus-

FIGURE 17.1 Both the observer in the rocket and the observer on the ground measure the same speed of light.

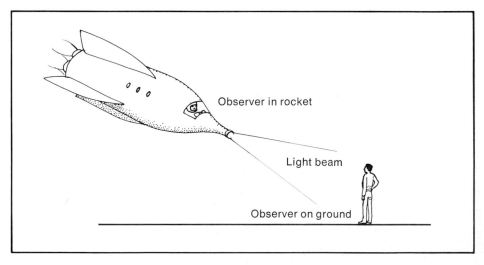

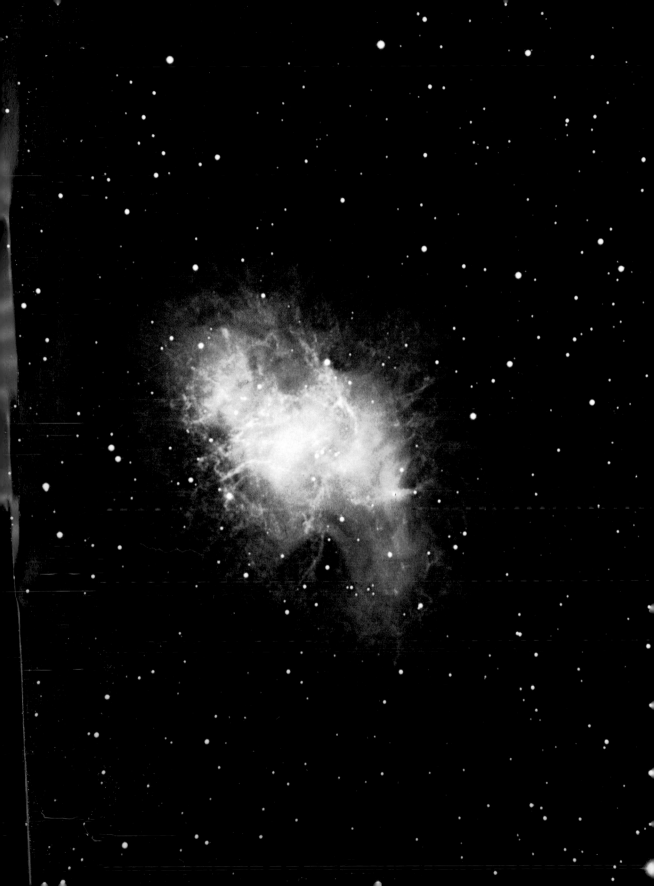

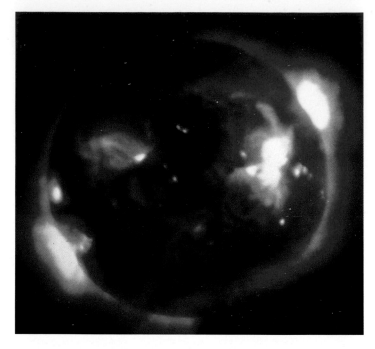

PLATE 8 (a) An X-ray photograph of the sun taken aboard Skylab with the S-054 X-ray Spectrographic Telescope on May 28, 1973, revealing the corona with temperatures ranging upward to 1 million degrees Kelvin. The photo shows the whole range of coronal features in a broad spectral range. The active regions, bright points, interconnecting loops, filament cavities, coronal holes, and other features seen in the photograph are produced by the interaction of the sun's magnetic field and the ionized gas of the corona. (American Science and Engineering, Inc./NASA)

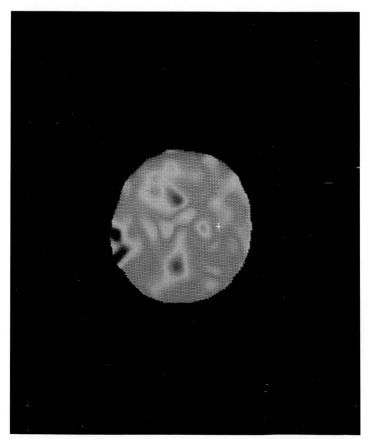

PLATE 8 (b) A speckle photograph, showing details on the surface of the red giant, Betelgeuse. (Kitt Peak National Observatory)

<PLATE 9 The Crab Nebula, the remains of a supernova first seen in A.D. 1054. The gases in this object are still expanding very rapidly. (Copyright 1959 California Institute of Technology and Carnegie Institute of Washington)

PLATE 10 The Pleiades, an open cluster in the constellation of Taurus. The nebulosity that surrounds these young stars is thought to be primarily dust. (Copyright 1961 California Institute of Technology and Carnegie Institute of Washington)

(a)

PLATE 11 (a) Large Magellanic Cloud and (b) Small Magellanic Cloud—galaxies neighboring our own Milky Way but which are visible only from southern locations. (Cerro Tololo Inter-American Observatory)

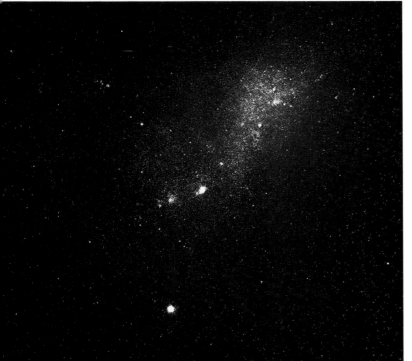

(b)

PLATE 12 Eta Carinae (NGC 3372), an emission nebula seen in the southern sky, as photographed in 1974 by the 1.5-m Schmidt camera. This emission nebula resulted from the explosion of a supernova recorded in 1843. Embedded in this nebula are numerous hot blue stars; the ultraviolet radiation of these stars excites the atoms of the nebula, causing it to emit light. (Cerro Tololo Inter-American Observatory) ➤

(a)

PLATE 14 (a) Gaseous emission nebula and an associated galactic cluster in Serpens (M16; NGC 6611). Compact dust clouds, seen as dark globules, may be stars forming. Strong turbulence has been detected in the dark lanes. A 4-m telescope photo taken in 1974. (Kitt Peak National Observatory) (b) The Great Nebula (M42; NGC 1976), a gaseous emission nebula in Orion. Visible to the naked eye in Orion's sword, M42 has been known since the beginnings of recorded astronomy. Composed of mainly hydrogen gases, the Orion Nebula has a critical density about that required for star formation, and some star formation is believed to be occurring here. A 4-m telescope photo taken in 1974. (Kitt Peak National Observatory)

(b)

PLATE 15 The Great Galaxy in Andromeda (M31; NGC 224), with satellite galaxies NGC 205 and 221. (Copyright 1959 California Institute of Technology and Carnegie Institute of Washington) >

tration. We could think of almost endless vantage points from which our description would vary.

We might begin to think that the description of a moving object is always a simple algebraic sum of separate velocities, but this is not so when we deal with the speed at which all forms of electromagnetic radiation including light, travel. Let us suppose that the train car is a rocket and the bowling ball is a light beam coming from the front of a rocket ship (Figure 17.1). Light travels at 300,000 km/sec. Let us suppose that the rocket ship is traveling toward earth at the rate of 100,000 km/sec (such a speed is not possible under modern technology). According to our previous experience, we would think that the observer on earth would measure the velocity of light in that beam as 400,000 km/sec, the algebraic sum of the two velocities. However, whenever the velocity of light is measured by the observer on the ground it is found to be 300,000 km/sec, the same value as that which would be obtained by an observer riding in the rocket ship. The velocity of the vehicle that carries the source of light has absolutely nothing to do with the velocity of that light. A beam of light shining from the rear of the rocket ship would also have the same speed (in the opposite direction).

To see how this phenomenon was discovered, consider the late nineteenth-century experiment of Michelson and Morley. Observers of the eighteenth and nineteenth centuries, upon realizing that light is a wave phenomenon, were convinced that electromagnetic disturbances needed a medium through which their wave-like nature could be propagated. They called the medium the *ether* and thought of it as an absolute frame of reference through which the earth and other celestial bodies moved. If this were true, then an *ether wind* should be "blowing" past the earth due to the earth's motion due to rotation and revolution. In 1881, Michelson tried to measure this ether wind. He repeated his experiment with Morley in 1887.

MICHELSON-MORLEY EXPERIMENT

To illustrate the idea behind the experiment, consider a boat that makes different speeds relative to the land, depending upon its direction of travel on a river (Figure 17.2). Suppose the river flows steadily at 3 km/hr and that the boat can make 5 km/hr in still water. When going upstream, the boat will make only 2 km/hr ($5 - 3 = 2$ km/hr) relative to the land. Going downstream, it will make 8 km/hr ($5 + 3 = 8$ km/hr) relative to the land. Going across the stream, the boat has to head slightly upstream to compensate for the drift, and its velocity is computed by the Pythagorean relationship, $v^2 = (5)^2 - (3)^2 = 25 - 9 = 16$, $v = 4$ km/hr. These velocities are borne out by actual experiments, and the relationships that are expressed thereby seem to hold true for velocities encountered in such a situation.

Note this interesting fact: If the boat made a trip of 8 km upstream and then returned, it would take a total of 5 hours:

$$\text{time upstream} = \frac{8\,\text{km}}{2\,\text{km/hr}} = 4\,\text{hr}$$

$$\text{time downstream} = \frac{8\,\text{km}}{8\,\text{km/hr}} = 1\,\text{hr}$$

$$\text{round-trip time} = 5\,\text{hr}$$

However, if the same distance were covered across the river and back, it would require only 4 hours:

$$\text{time across} = \frac{8\,\text{km}}{4\,\text{km/hr}} = 2\,\text{hr}$$

$$\text{time back} = \frac{8\,\text{km}}{4\,\text{km/hr}} = 2\,\text{hr}$$

$$\text{round-trip time} = 4\,\text{hr}$$

The reason an extra hour is taken on the up and back round-trip as compared to the across and back round-trip is that more than half the time was spent at the slower speed going upstream.

Michelson and Morley devised an instrument whereby they treated the rate at which the earth "flows" through the ether as the rate of the river, and the speed of light as the speed of the boat in still water. In Figure 17.3, you can see that they caused the light to reflect back and forth and in certain cases to be partially transmitted through half-silvered mirrors. If the ether really does exist and the earth moves through it, then the upstream-downstream round trip should take longer than the across-stream round trip. As the two beams are brought back to the eye, they interfere

FIGURE 17.2 When a boat moves upstream, its speed is slowed by the current; when it moves downstream, its speed is increased by the current, relative to the ground. In crossing the river, the boat must head slightly upstream in order to travel directly across the river.

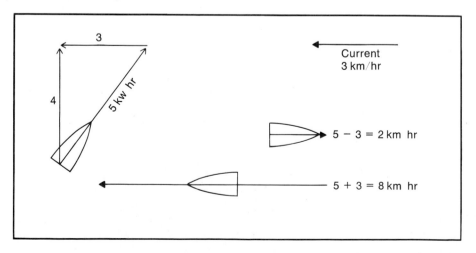

with one another, producing alternating light and dark bands. If the device is rotated 90°, interchanging the roles of the two beams, then one would expect a change in the interference pattern, because the time required for each beam to travel the set distance is changed. The result of the Michelson and Morley experiment, to everyone's amazement, was that the interference pattern remained unchanged, even though their device was capable of showing the expected change. The experiment showed that the velocity of light is constant—it is not affected by the relative motion of the source, and that the ether wind may not exist. The results of this experiment produced a revolution in thought, and ultimately confirmed Einstein's thoughts on relativity.

THE LORENTZ CONTRACTION

Some thinkers remained bent on preserving the idea of the ether. Typical of this school were George F. Fitzgerald of Ireland and Hendrik A. Lorentz of Holland, who, working independently, suggested that any device moving through the ether was shortened in the direction of its motion by just the right amount necessary to allow light to travel up and back in the same time as across and back. The *Lorentz contraction* is expressed mathematically as follows:

$$l = l_0 \sqrt{1 - \frac{v^2}{c^2}}$$

where l_0 is the length "at rest" relative to the earth, v is the velocity of the earth through the ether, and c is the speed of light. Lorentz and Fitzgerald held such a change in length to be a physical contraction caused by the

FIGURE 17.3 The Michelson-Morley experiment.

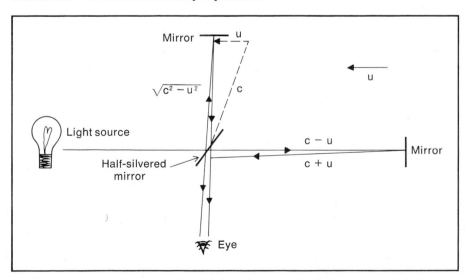

pressure of passage through the ether. A new concept expressed by Albert Einstein, however, viewed the change as occurring only in the perceived length.

ALBERT EINSTEIN

Einstein's name is synonymous with the term *relativity* and rightly so, for he was truly interested in separating aspects of observation that are relative to the observer from those that are independent of the observer and therefore "absolute." Einstein (Figure 17.4) felt no commitment to preserving the idea of the ether; rather he assumed that the ether wind did not exist. He postulated that the speed of light in a vacuum is constant, regardless of the motion of the source or the observer. Thus it would be impossible to

FIGURE 17.4 Albert Einstein.

detect the motion of the earth in its orbit by means of an experiment using light. This postulate would explain the negative results of the Michelson-Morley experiment, although Einstein was not even aware that the experiment had been performed. Einstein's thinking was independent of any given experiment. Rather it stemmed from a more fundamental assumption: that no experiment can be performed to demonstrate the motion of an object if that object is moving uniformly (at a constant speed in a straight line).

Suppose, for example, you were riding in a train car that had no windows and that rode so smoothly that you felt no bumping or lurching. How would you know that you were moving with respect to the ground? You might try bowling a ball down the aisle to strike a set of pins at the end of the car; however, nothing about the car's motion would be revealed by such an experiment—the ball would seem to move in precisely the same way as if the car were standing still, and the pins would be struck with the same force. One might conclude that the laws of nature inside a moving car are just the same as those experienced in a stationary car. This was Einstein's second postulate: The laws of nature are the same for observer A and for observer B, even though they are moving uniformly in relation to one another. If this assumption is to hold true and the speed of light is to remain constant, then certain other quantities must differ for observers who are moving relative to one another.

A THOUGHT EXPERIMENT

Consider the following thought experiment, even as Einstein must have done. Because light is the carrier of information, if you traveled away from a clock tower at the speed of light, no new information could reach you about that clock. The hands of the clock would seem frozen in place. Time would have stopped for you, as measured by that clock. If this is true, then an observer traveling at a speed a little less than the speed of light relative to the clock tower will see the hands of that clock move more slowly than does the person who stands at its base. If the observer who is moving rapidly away from the clock tower is wearing a wristwatch, he or she will see that watch running at its usual pace, whereas the person at the base of the tower will see that wristwatch running slowly. Which observer is correct? Both are correct, for there is no absolute time by which their observations can be compared. Rather, the rate at which time is measured in any given frame of reference is relative to the motion of the observer with respect to that frame. If an observer moves with the clock, the time it keeps can be labeled its "proper time." If, however, the observer moves in uniform motion, with velocity v, with respect to the clock, its rate will be slowed—its minutes and hours lengthened (dilated)—as follows:

$$t = \frac{t_0}{\sqrt{1 - \dfrac{v^2}{c^2}}}$$

where t_0 is proper time and c is the speed of light. Time dilation has been observed directly in the case of radioactive subatomic particles. When moving at high velocities relative to the observer, they decay more slowly than does a similar particle that is at rest relative to the observer. Since the radioactive decay rate of a given particle is as regular as the most accurate clock, we say that its clock runs slow from our point of view owing to its high relative velocity.

Based on his fundamental postulates, and working independently of Lorentz and Fitzgerald, Einstein also derived a relationship between the "proper" length of a rod and the length which would be perceived by an observer who was moving uniformly at a velocity v with respect to that rod. His results were in agreement with those of Lorentz:

$$l = l_0 \sqrt{1 - \frac{v^2}{c^2}}$$

where l is the perceived length, l_0 is the "proper" length, and c is the speed of light. Einstein said simply that two observers, moving uniformly relative to each other, will perceive the passage of time and distance between two points differently, hence time and distance are relative whereas the speed of light and the laws of nature are not relative. How much shorter is the length of a given rod, as perceived by an observer who is moving past it at half the speed of light, as compared to the length of the rod as perceived by an observer moving with the rod?

$$l = l_0 \sqrt{1 - \frac{v^2}{c^2}}$$

$$\text{Let } v = \frac{1}{2} c$$

$$l = l_0 \sqrt{1 - \frac{(0.5c)^2}{c^2}}$$

$$l = l_0 \sqrt{1 - 0.25}$$

$$l = l_0 \sqrt{0.75} = 0.866 \, l_0$$

The rod is perceived to be approximately 13 percent shorter for the moving observer (0.866 is 13 percent is less than 1.000). If you could move down a street at half the speed of light, all the buildings would measure skinny and tall, for in the direction of your travel the widths of the buildings diminish by 13 percent while the heights would appear normal to you. The contraction effect applies only in the direction of relative motion.

One further quantity must be considered, that of mass. It has been effectively demonstrated that the mass of an object increases with higher and higher velocities relative to the observer. Electrons have been accelerated to speeds of 99.999 percent of the speed of light. At such speeds objects become harder and harder to accelerate, indicating that their mass is significantly greater than when they were "at rest" relative to the observer. At velocity (v) with respect to the observer, mass (m) is given by:

$$m = \frac{m_0}{\sqrt{1 - \dfrac{v^2}{c^2}}}$$

Where m_0 is the "rest mass" of the object and c is the speed of light. Mass is thus also a relative quantity. What is the source of this increase in mass? To continuously apply a force, energy must be continuously expended. As the object gains velocity relative to the observer, it also gains kinetic energy, and hence its rest mass plus its energy due to motion make up its total mass-energy. Einstein's statement of the equivalency between energy and mass, namely $E = mc^2$, was derived directly from the relationship between (m) and (m_0) stated above. (The mathematics of this derivation is beyond the level of this text.)

It would be very helpful at this point to try to begin to think of mass and energy as simply two different manifestations of the same thing. This equivalency is illustrated by the sun's conversion of a portion of its mass to energy every second (see Chapter 8). At the time of the Big Bang, radiant energy converted to mass (particles) when the universe had cooled sufficiently. In our everyday experience of winding a watch, we add mass-energy every time we wind the watch and it loses mass-energy as it runs down.

The concepts we have presented thus far are expressions of Einstein's 1905 theory of special relativity, "special" in that restrictions were placed on its application. It describes only those situations in which two observers are traveling at constant velocities with respect to each other, and it does not cover the highly probable situation in which one observer is accelerating with respect to the other. Remembering that accelerations can take several forms—speeding up, slowing down, or changing direction we need only think of the influence of gravity to envision situations where Einstein's special theory would not apply.

In 1916 Albert Einstein made one of the most significant contributions ever made to science. His general theory of relativity is a fundamental attempt to explain gravity by means of equations which describe how space-time is affected by presence of a massive object (actually any object) and how objects will move in the vicinity of such an object. To present the equations or to attempt to solve them would be beyond the level of this text; in fact, their solution has challenged the finest minds of the world. Let us be content to review some of the predictions that are derived from certain solutions and see how they have been confirmed.

The general theory predicts that light will appear to be bent as it passes a massive object—or more accurately, space-time will be warped near a massive object in such a way that a light ray that follows the most efficient path will appear to us to be bent. Its most efficient path is not a straight line as in unwarped space-time. This predicted effect was witnessed in 1919 when stars near the sun were photographed during a total solar eclipse. When the apparent positions of these stars were compared to their positions photographed months earlier, small deviations in position were observed (Figure 17.5). The light from stars that appeared near the edge of the sun had been bent by an angle that was within 15 percent of Einstein's predicted value (1.75 seconds of arc).

A by-product of the Viking mission to Mars was a confirmation of this same phenomenon. When Mars passed behind the sun (at conjunction), as its radio communication signal passed near the sun, it appeared to be bent. A very careful observation of the degree to which the radio signal was bent, utilizing Viking landers, Viking orbiters, and two radio antennas on earth, confirmed Einstein's prediction to an accuracy of 1 percent. Furthermore, the idea of the gravitational lens, presented on page 476, illustrates the bending of light by the warping of space-time.

Moving beyond Kepler's description of the elliptical orbits of the planets, Einstein's equations predict a slight deviation from a perfectly closed orbit. They predicted that the perihelion (closest point to the sun) of each planet's orbit would gradually occur in new places. This effect, referred to as the precession of the perihelion, is illustrated in Figure 17.6 for Mercury. Mercury's perihelion point moves through an angle which is 43 seconds of arc per century more than had been predicted by Newton. The observed precession is in close agreement with Einstein's prediction and thus appears to strengthen his theory.

Also coming out of the general theory is the prediction that massive stars will collapse in their old age and become so dense as to produce a surface gravity that will prevent any form of radiation from leaving the star; photons of electromagnetic radiation respond to the warping of space-time by the presence of massive objects. When radiation can no longer escape such an object, then it becomes a black hole. Astronomers are quite certain that such objects exist. An example is Cygnus X-1 (see Chapter 14 to review the idea of black holes).

Lest we conclude that the general theory is proven beyond any shadow of doubt (something that never happens in science), we should note that there may be alternate explanations for any or all of these experimental results. In fact, rival theories concerning gravitation do exist, and their predictions are quite similar to Einstein's. Experimental evidence is not yet sufficiently accurate to allow us to test one against the other and

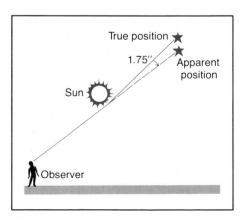

FIGURE 17.5 The gravitational effect on light.

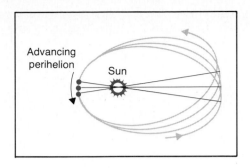

FIGURE 17.6 The advancing peri-helion of Mercury.

decide among them. We may, however, have a great deal of confidence in both the special and general theories.

The one expressed goal of Einstein's career was to develop a unified field theory in such a manner as to describe gravity and electromagnetic radiation with a single set of equations. This goal was not realized in Einstein's lifetime. This goal has motivated many humans and will continue to do so until the problem is solved.

QUESTIONS

1. If a train travels eastward at 80 km/hr and a passenger runs down the aisle toward the rear of the train (westward) at 6 km/hr, how will someone on the ground describe the passenger's motion as he passes the observer.

2. How will someone seated in the train describe the runner's motion in Question 1?

3. Why are the answers to Questions 1 and 2 different?

4. If modern rockets (and/or space probes) can achieve speeds no greater than 10 km/sec, what fraction of the speed of light does this represent?

5. When the Michelson-Morley apparatus was rotated 90°, the roles of the beams were interchanged. Explain what this sentence means.

6. What changes did Lorentz predict in moving objects?

7. Einstein said that it is impossible to determine the motion of an object, say in a smooth riding railroad car with no windows, by performing an experiment in that car. What would happen if one threw a ball straight upward in a moving car? Wouldn't it fall toward the rear of the car? If not, why not?

8. Einstein's correction for the observed (measured) length of a rod, moving with respect to the observer, appears to be the same (mathematically) as that of Lorentz. What was different about their interpretation of the meaning of that correction?

9. Determine the measured length of a rod which has a "proper" length of one meter, if the observer is moving at three-fourths the speed of light (0.75c) with respect to the rod. (Answer: $\dfrac{\sqrt{7}}{4}$ m)

10. How does the equation $E = mc^2$ relate to the sun's source of energy?

11. What limitation is placed on the special theory of relativity?

12. List several specific predictions of Einstein's general theory that have apparently been verified.

Extraterrestrial (ET) Life

Eighteen

It would never have occured to ancient observers to ask, "Are we alone in the universe?" Their universe consisted only of a limited, flat earth with a canopy of stars overhead. Not until humans realized that the sun is a star, and a rather ordinary star at that, did they begin to see the possibility of other planetary systems and therefore extraterrestrial life. From the time Sciaperelli first thought he saw "canali" on Mars and Percival Lowell turned these "canali" into canals dug by the little Martians in order to irrigate their crops, the human imagination has been fired with the thought that we might not be alone. As recounted in Chapter 5, our particular search for life on Mars was met with only negative results. If we had found life on Mars, certainly it would have qualified for the title "extraterrestrial" life, for the word itself means other than on earth. We will use the abbreviation ET for this word and add an L (ETL) for Extraterrestrial Life, or an I (ETI) for Extraterrestrial Intelligence. Since we are already involved in a Search for Extraterrestrial Intelligence, this fact is conveyed by the letters SETI. The first SETI project was called Ozma and it began in 1960 with an attempt to detect radio signals coming from two nearby stars, Tau Ceti and Epislon Eridani. The fact that we have not found signs of ETL or ETI on the moon, on Mars, in the clouds of Jupiter, or on Saturn's Titan really has not dampened the spirit of searchers, it has simply pushed them to explore farther into space—beyond our solar system.

Our searching could be carried out most thoroughly if we could actually travel to a star, land on one or more of its planets that are remotely hospitable, and search for life there. You will probably not be surprised by the statement that this approach would be difficult, if not impossible, to carry out under our present understanding of the distances involved and of the limitations on speeds that may be attained by a spacecraft that could transport us. If we were able to travel at one-tenth the speed of light, it would still take us almost 90 years to make one round trip to the nearest star, Alpha Centauri. If the star we wished to visit was 100 light-years distant, then we would have to spend 2000 years journeying there and back—many generations would have to be raised on the way. After reading the last chapter on relativity, you can see that when our technology is sufficiently advanced to travel at 99 percent of the speed of light, time for us in the spacecraft will pass more slowly and a round trip for us will be reduced to a little less than 40 years. By the time we return, all those left behind will have aged slightly more than 200 years and all our friends will be dead. The greatest speed yet attained by any of our spacecraft is just over 100,000 mi/hr (28 mi/sec), and this represents only 15 thousandths of 1 percent of the speed of light. As discouraging as this sounds, we should not give up hope, for we must merely find a new source of energy that will continue to accelerate the spacecraft throughout the first half of its journey to the distant star. Ideas for such a propulsion system include hydrogen fusion, anti-matter being combined with matter, and an ion exchange system. In the meantime, however, we will be looking to a mode of communication that does travel at the speed of light, namely electromagnetic radiation.

Radio signals travel through empty space at the speed of light, and we already have the technology to put information onto those signals. But at what wavelength shall we listen and broadcast? The full range of wavelengths which are classified as radio wavelengths is very broad and it would be less than desirable to try to cover this entire range. A portion of the band is plagued with a form of background radiaton (radio noise) which is generated by electrons being accelerated by the magnetic fields of the Galaxy—we wouldn't want to compete with these rather long radio waves. Fortunately, there is a limited range of wavelengths that might be a natural choice and which lie in a rather quiet portion of the spectrum, namely the range from about 18 cm to 21 cm (actually a little bit broader). Why should these wavelengths be special? You can guess the answer to the 21 cm figure; it is the signal produced by the most prevalent element in the entire universe, neutral hydrogen. Any ETI would undoubtedly know that and might also choose that wavelength because of its significance. This was the wavelength chosen for the SETI project called Ozma. The wavelength of 18 cm is characteristic of the radiation of the hydroxyl (OH) radical found in interstellar space. When you place an OH-radical near an H-atom, you are very likely to form H-O-H or H_2O (a water molecule). While this line of reasoning doesn't really prove anything, the range of wavelengths between 18 cm and 21 cm has been dubbed the "water hole" of the universe, the place where ETs get together to exchange tales of their exploits. Because we have found water to be so prevalent in many extraterrestrial situations, we think that it may be prevalent throughout the universe and if this is true, the ETs may see the same line of reasoning and choose to broadcast and listen on a variety of wavelengths in this band. We do have the capability of beaming strong radio signals in any desired direction and listening on a variety of wavelengths simultaneously. But now we ask the big question.

IS IT WORTH THE EFFORT?
WHAT ARE THE CHANCES THAT ETI EXISTS?

To answer these questions and several allied ones, let us go back to a conference held by interested scientists in 1961 at the National Radio Astronomy Observatory in Greenbank, West Virginia. They answered the question of how many communicating ETI "societies" exist in the Milky Way galaxy by the following equation:

$N = R f_p\, n_e\, f_l f_i f_c\, L$ where:
N = number of communicating societies in the galaxy
R = rate at which stars are being formed in the galaxy
f_p = fraction of stars that have planets
n_e = number of planets/star suitable for life
f_l = fraction of planets suitable for life that actually possess life
f_i = fraction of planets that possess life that also possess intelligent life
f_c = fraction of planets that possess intelligent life that also have the desire to communicate
L = time during which civilization will be in a communicative stage

Just to see how this equation works, let's pick some values out of the hat and plug them in.

$R = 10$ stars/year (200 billion stars formed in 20 billion years)
$f_p = 0.1$ (only 1 star out of 10 has planets)
$n_e = 2$ (a typical star has 2 planets which are suitable for life)
$f_l = 0.1$ (of those suitable for life, only 1 out of 10 will possess life)
$f_i = .0.5$ (of those that sustain life, one half will possess intelligent life)
$f_c = 0.9$ (of those that sustain intelligent life, 9 out of 10 will want to communicate)
$L = 10,000$ years (the ability and desire to communicate will last for 10,000 years)
$N = 10\ (0.1)\ (2)\ (0.1)\ (0.5)\ (0.9)\ (10,000) = 900$ societies

Because we have been rather pessimistic in several choices, this figure of 900 would be considered in general to be rather pessimistic. The range generally found in the literature is from 1 (we are alone) to 100,000,000 communicating ET societies in the Milky Way galaxy. Now it is our purpose to show how the value for each factor in this equation may be estimated. You should try your own values and produce a value for N.

The value for R rests on fairly substantial knowledge, the total number of stars in the galaxy and the length of time for them to be formed (namely the age of the universe).

The value for f_p expresses the probability that stars other than our sun will also have planets. To what extent are planets a natural consequence of star formation? Taking our own solar system as a model, we believe that the planets formed from the same condensation of material that made the sun. As a huge volume of gas and dust condensed, its rotation component was increased as a function of conserving angular momentum, a portion of the matter flattened into a disk surrounding the protosun and the planets formed from this disk of material. It seems that this could happen to almost any star; however, we are aware that very massive (hot) stars may blow such a disk away before planets are formed in that disk. Furthermore we think that binary stars may perturb any planets to the point of ejecting them from the system, and if these two possibilities are ruled out, our probability is reduced to single stars with less than about 2 solar masses, hence the value of f_p should range between 0.5 and 0.001 perhaps. Astronomers are actively engaged in searching for such solar systems, especially those which might surround nearby sunlike stars. The long focal-length refractor telescope is particularly well suited for this work, not because they expect to actually detect planets visually, but rather they expect to detect small perturbations (a wavelike path in the sky) in the star as a result of the gravitational effect of one or more planets which orbit around the star. The positive results which are now being seen for three nearby stars. Barnard's star, Lalande 21185, and 61 Cygni, seem to give us hope for a higher value in the case of this factor (f_p). The dedication of certain large refracting telescopes to this effort is highly commendable.

The value for n_e represents a guess as to the number of planets that will sustain life in a given solar system. Again, using our own system as a model, what is the value of n_e? As far as we know, it is one. The earth is the only planet in the solar system capable of sustaining life. What other planet(s) of our system come closest to such a condition? Mercury and Venus are far too hot and Jupiter and those beyond are far too cold, so only Mars remains a possibility. The temperature on Mars is survivable to man, but how about its atmosphere? Its own mass is not sufficient to hang onto a significant amount of atmosphere, even though it is cooler than the earth and that should help. Furthermore, what atmosphere exists is not of the right composition for us to breath (it is CO_2). Factors that restrict life zones in a planetary system thus include distance from the sun (and therefore its temperature), mass of the planet (its ability to hold onto an atmosphere and this may affect temperature through a "greenhouse" effect), and a suitable composition of atmosphere. To be suitable for life, a planet must possess the necessary elements (both in its atmosphere and in its soil) to make life possible. This suggests that the planet must be created out of a cloud of material that has been enriched by elements heavier than helium, say by the expulsion of these elements in a supernova explosion. We see a natural tendency for carbon atoms to bond with hydrogen, oxygen, nitrogen and sulfur to form amino acids, for these are found where life has not existed, formed simply by chemical reaction. These are the building blocks for life, but they do not constitute life. One of the most fundmental differences between a nonliving molecule and a living organism is the presence of a coded message in each cell of that organism, whereby each cell can reproduce itself (Figure 18.1).

Notice that we have been more restrictive than is necessary just because we are assuming ETs to be similar to ourselves. If we do not make this assumption we must revise many of our considerations.

The value assigned to f_l expresses one's estimate of the probability

FIGURE 18.1 The DNA double helix molecule.

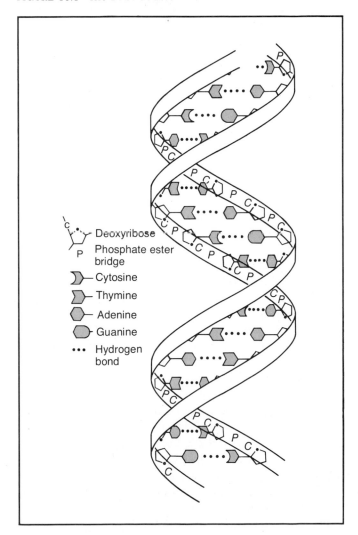

Deoxyribose
Phosphate ester bridge
Cytosine
Thymine
Adenine
Guanine
Hydrogen bond

that life exists wherever it is suitable for life to exist. Scientists do not have direct evidence of how nonliving molecules became living organisms; hence the value you assign to this term will ultimately be determined by your own philosophy of the origin of life. It may range from 1 (meaning that life will exist wherever it is suitable for life to exist) to 0.001 or less (meaning that life is a very unlikely happening).

The term f_i specifies the probability that if life exists, intelligence also exists. Intelligence goes beyond the instinct which appears to direct life in most animals (there are exceptions such as the chimpanzee and the dolphin). Intelligence implies an ability to think about different alternatives and to choose the one considered best; thus intelligence provides a much higher level of adaptability. The range of values appropriate for this factor is just about as wide as in the case of the previous term and may again depend upon your own philosophy as to the source of intelligence.

The term f_c reflects our thinking about the connection between intelligent life and curiosity. It measures the likelihood that any ETI would want to communicate with suspected aliens, if they exist. Would they be asking, "Are we alone?", just as humans appear interested in doing now? I gave this a rather high value because I think their intelligence is fed by curiosity. What value would you assign?

The term L measures the length of time a communicating society will survive and keep on wanting to communicate. To measure L for ourselves, we wouldn't even start counting time before about 50 years ago, for until that time we did not possess the technology whereby we might both send and receive radio signals and thus communicate with intelligent ETs. Now that we have both powerful transmitters and sensitive multi-channel receivers, the question boils down to how long we can survive and also how long we can sustain our interest in searching. As an example, we have chosen 10,000 years, but again you may estimate this number from your own perspective.

Before leaving the factor L, let's pursue this question in a slightly different manner. Because the nearest ETI may easily be 1000 light years-distant, two-way communication is virtually out of the question. For instance, if we sent a message to such an ETI via radio (traveling at the speed of light), it would take a thousand years to get there and the answer a thousand years to get back; that's not much of a dialogue. To even assume that they would receive our message assumes that they are intelligent beings, technologically developed, and want to communicate (at least they are listening). It also assumes that their technology will exist when our signal gets there. If their technology "peaked" ten thousand years ago, will it survive another thousand years? Or alternatively, if they have not developed a technology yet, will it happen while our signal is on its way? This discussion should point out the fact that it would not be most advantageous if every ETI technology peaked the the same time, rather, like everything in the Galaxy (or in the Universe) as an ongoing process. It should also be clear at this point that the longer a communicating ET society can last (the larger the value of L), the greater the chances are for contact and the less critical the time of the "peaking" of that technology.

With most of the factors in this equation being so uncertain, you can appreciate the wide range of values that are possible for N. For what value of N (assuming you think that N is greater than 1) would you personally agree to an sizeable expenditure of public funds to support a search for ETI—say an annual expenditure equal to that spent on chewing gum?

WHAT LANGUAGE WOULD WE USE?

The answer is obviously not English or French or Russian, although these and more were included on the Voyager record. There is one language that is almost "universal" among the peoples of the earth and therefore might rank high among ETIs. It is the language of mathematics. In a very real sense, mathematics is a language—it is the language with which we communicate with computers—the language in which information is stored, manipulated, and spewed back out again by those computers. Realizing that computers are nonintelligent entities, it should be easier to develop a common language with an ETI, for both they and ourselves are motivated to accomplish this task.

Yet another line of reasoning gives us hope for an understandable language. It is the fact that specialists have been able to "break" almost any code that has been written and every language of the world is translatable.

WHAT SHALL WE SAY TO ETIs?

In a sense we have been "talking" to them for years in perhaps an unintentional manner, namely our radio and TV signals have been leaking into space for approximately 50 years. This means that a sphere with a 50 light-year radius surrounds the earth that is full of radio signals and TV pictures, and it is growing at the rate of one light-year each year. In another 50 years, we will have covered every ETI society within 100 light-years of the earth. They can pick up "Caruso" or "Uncle Miltie" or check out "Days of Our Lives." I'm not sure that I would want this sampling to represent our total existence, would you?

We have also sent two artifacts of ourselves on spacecraft that have now left (or soon will leave) the solar system. If an ETI should happen to recover Pioneer 10 or 11, at some distant point in time and space, it would find a plaque depicting our bodily forms and a pictograph that will tell where we are in the universe. On both the Voyager spacecraft, we included a record (resembling the modern video record) on which are recorded greetings from 60 countries (in their own languages), photographs of the earth and ourselves, and 90 minutes of the world's greatest music.

Assuming that we are probably dealing with an one-way message and cannot carry on an ordinary dialogue, what message would you devise for potential ETI consumption? What message would you like to receive from them?

HOW SHALL IT BE DONE?

If money were no object, I'm sure we would pursue a plan something like the Cyclops plan proposed a number of years ago. Figure 18.2 will give you an idea of its magnitude: Several dozens of large radio antennae arranged so as to concentrate huge quantities of energy in a rather narrow beam, but costing at least 10 billions of dollars. A group of Soviet astronomers have proposed that a huge radio antenna be placed into solar orbit capable of mapping the entire cosmos in three dimensions for the purpose of detecting the signals of ETI. More reasonable would be a commitment of a certain percentage of existing telescope time for such a search. We now have the technlogical skill to receive many channels of information simultaneously and, with the aid of properly programmed computers, to recognize signals from intelligent beings. Therefore, it is only a slight reassessment of priorities that is needed to begin the search in earnest. If such a commitment is made, perhaps we will find that we are not alone in the universe.

FIGURE 18.2 The Cyclops proposal for a radio telescope array. (NASA-Ames)

QUESTIONS

1. Why does it seem reasonable to assume that many other solar systems may exist in the Milky Way galaxy?

2. If the probability of stars having planets is one in a million, what do you think may be the probability of life existing elsewhere in the universe? Substantiate your view.

3. What is function of the DNA molecule in living cells?

4. Suppose intelligent beings existed on a planet 1000 light-years away and they possessed a radio technology compatible with ours. How would time delays affect our effort to carry on a dialogue? How would you solve the problem?

5. Why would the radio band between 18 cm and 21 cm be a natural one for communicating with ETIs?

6. Devise a code or language which one might use to communicate with ETIs.

Glossary

aberration of starlight The apparent displacement in the location of a star due to the orbital motion of the earth.

absolute magnitude The apparent magnitude of a star if viewed from a distance of ten parsecs.

absolute zero The temperature at which the linear motions of all molecules stop. A temperature of $0°K$ (Kelvin), equivalent to $-273°C$ (Celsius) and $-460°F$ (Fahrenheit).

absorption spectrum Dark lines on the background of a continuous spectrum.

acceleration The change in velocity. It may be an increase or decrease in velocity, or a change in direction.

achromatic lens A lens system composed of two or more elements that are designed to correct for chromatic aberration.

active sun The sun during times of unusually large numbers of sunspots, flares, and other events.

albedo The percentage of light that a planet or moon reflects.

alpha particle A positively charged particle that consists of two protons and two neutrons; hence, a helium nucleus.

altitude The angle at which an object appears above the horizon as measured along its vertical circle.

angle of incidence The angle between the incoming ray and the normal (perpendicular) to the reflecting or refracting surface.

angstrom (Å) A unit of length equal to 10^{-10} m, used to measure very small wavelengths.

angular diameter The angle that the diameter of an object makes as measured at the observer's eye.

angular distance The angle between two objects as viewed on the celestial sphere.

annular eclipse An eclipse of the sun that occurs when the apparent diameter of the moon is not as great as the apparent diameter of the sun, thus leaving a ring of sunlight showing around the moon.

antimatter Particles that appear to possess properties opposite to those of matter.

aphelion A point in the orbit of a planet at which it is farthest from the sun. (*Helios* is a Greek word for sun.)

apogee A point in the orbit of an earth satellite at which it is farthest from the earth. (*Ge*, *gee*, and *geo* are combining forms—suffixes or prefixes—meaning earth.)

apparent magnitude A measure of the brightness of a star or other celestial object as seen from earth.

apparent solar day The interval between two successive transits of the sun's center across the observer's meridian.

Arctic Circle The parallel of latitude $66.5°$ N. Within this circle the sun is not seen on the day of the winter solstice.

ascending node The point on the orbit of a body at which it crosses the celestial equator from south to north.

association A very loose cluster of stars that are thought to have a common origin.

asteroid A small body in orbit around the sun; a minor planet or planetoid.

astrology A study of the supposed influence of the positions of the sun, moon, planets, and stars upon human affairs.

astrometric binary A binary-star system in which only one component is visible but in which the presence of the second component is deduced from the perturbations (disturbances) that it produces upon the orbital motion of the first.

astrometry The branch of astronomy that is primarily concerned with the accurate measurement of positions and motions of stars.

astronomical unit (A.U.) The average distance between the earth and the sun. By international agreement, 1 A.U. = 92,870,000 miles, or approximately 149,790,000 km.

astronomy The science whereby celestial objects are described according to their location, motion, size, composition, and appearance.

astrophysics The branch of astronomy that applies the methods and tools of physics to the study of celestial objects.

atmospheric refraction The bending of light rays from celestial objects due to refraction by the earth's atmosphere. This phenomenon is most noticeable near the observer's horizon.

atom The smallest particle of an element that retains the properties which characterize that element.

atomic mass unit One-twelfth the mass of an atom of the most common form of carbon (^{12}C); approximately the mass of the hydrogen atom.

atomic number The number of protons in the nucleus of a given atom.

atomic weight The average mass of an atom of a given element as measured in atomic mass units.

aurora The light display that is produced by ionized atoms, usually in the polar regions; the northern and southern lights—*aurora borealis* and *aurora australis*.

autumn equinox The point on the celestial equator at which the sun crosses from north to south.

azimuth The angle measured eastward along the horizon from the north point to the vertical circle which passes through a given object.

Balmer lines The series of spectral lines, either bright or dark, that are produced by electron transitions up from energy level 2 or down to energy level 2, in the hydrogen atom. These lines lie in the visible portion of the spectrum.

barred spiral A galaxy characterized as having a "bar" (armlike extensions) through its nucleus. Spiral arms extend from the ends of the bar.

barycenter A point around which two objects that lie in each other's gravitational field seem to orbit. It constitutes a center of mass of the system.

beta particle A negatively charged particle; an electron.

"big bang" theory A theory concerning the evolution of the universe, which states that the expansion of the universe is the result of a primeval explosion.

binary star A double-star system in which the components revolve around a common point situated between them, their barycenter.

black body Theoretically, a body that is a perfect radiator, for instance, one that absorbs and reemits all radiation which falls on it.

black dwarf Thought to be the final stage in the evolution of some stars, a state in which all energy of the star has been exhausted and in which it no longer radiates.

black hole A star which has collapsed under the influence of gravity to such an extent that its surface gravity prevents further radiation of energy.

blink microscope An instrument in which two different photographs may be viewed alternately. If the two photographs represent the same region of the sky but were taken at differing times, then stars that have moved or changed in their apparent brightness may be easily recognized in the instrument. Stars that have moved will seem to jump back and forth. The image of stars that have varied in brightness will appear to change in size on the photograph.

Bode's law (More correctly called the *Bode-Titius relationship*.) A sequence of numbers that approximate the distances from the sun to the planets, measured in astronomical units.

Bohr atom A model of the atom, devised by Niels Bohr, that depicts the electrons in orbit around the nucleus.

bolide A very bright meteor or fireball, sometimes accompanied by a sound.

bolometric magnitude A measure of the total radiation of a star as received above the earth's atmosphere, measured in the full spectrum of electromagnetic radiation.

bright-line spectrum An array of colorful lines against a dark background, produced by an excited, low-pressure gas.

brightness A measure of the actual luminosity of an object.

calorie A unit of heat energy, the amount needed to raise the temperature of 1 g (1 cm^3) of water 1°C.

candlepower A unit of light intensity.

carbon cycle The series of nuclear reactions, involving carbon, that transforms hydrogen into helium.

cardinal points The four main points of the compass: north, east, south, and west.

Cassegrain reflector A telescope that utilizes a convex secondary mirror to bring the light rays to a focus near the primary lens. In order that the image formed at this point may be viewed, a hole is made in the primary mirror and an eyepiece inserted there.

Cassini's division A wide gap in the ring system of Saturn, between the outer and middle ring.

celestial equator The projection of the earth's equator onto the sky; hence a great circle that is 90° from each pole in the sky.

celestial mechanics A branch of astronomy that deals primarily with the motions and mutual gravitation of objects in space.

celestial meridian The great circle through the celestial poles and the observer's zenith.

celestial navigation The art of finding one's way from one point to another on the earth's surface by means of observation of the position of the sun, moon, planets, and stars.

celestial poles Extensions of the earth's axis of rotation; points on the celestial sphere about which the sky appears to rotate daily.

celestial sphere An apparent sphere of very large radius, centered on the observer, upon which the location of the stars may be specified.

center of gravity The point through which gravity seems to act no matter how the object is turned. The center of mass of the object.

centripetal force A force, directed toward the center of curvature, that diverts a body from a straight path into a curved one.

Cepheid variables A class of pulsating stars that vary in light output.

Ceres The first asteroid (minor planet) to be discovered; also the largest known.

chromatic aberration A defect in a simple lens whereby light of different colors is brought to different focal points.

chromosphere That portion of the sun's atmosphere which lies directly above the photosphere.

circumpolar stars Those stars near the celestial poles which always appear above the observer's horizon.

cluster of galaxies A grouping of galaxies composed of hundreds or thousands of member galaxies.

cluster of stars A grouping of stars that is held together by mutual gravitation; they thus possess a common motion.

cluster variables A large class of pulsating variable stars that have a period of less than one day and are usually found in globular star clusters.

color index (C.I.) The difference between the photographic and the visual magnitudes of a star: C.I. = $m_p - m_v$.

coma A defect in a telescope by which the rays of light that enter the telescope at an angle to its axis are not brought to the same focal point but form *comma*-like (,) images.

coma (of a comet) The fuzzy, gaseous component of the comet's head.

comet A swarm of bodies composed of frozen gases with solid particles at their centers. Comets usually have very elongated orbits around the sun.

compound A substance composed of two or more elements.

concave lens or mirror A lens or reflecting surface curved somewhat like the inside of a sphere.

conduction The transfer of energy (usually heat energy) by the direct passing of energy from atom to atom.

configuration A particle form or arrangement of the sun, moon, planets, and/or stars.

conic section A curve formed by cutting a circular cone with a flat surface. Such a cut may produce a circle, ellipse, parabola, or hyperbola.

conjunction A lining up of celestial objects so that they appear to have the same right ascension.

constellation A group of stars that by its shape suggests an object, person, or animal. Today, a constellation includes a definite region of the sky around such a configuration.

continuous spectrum The uninterrupted band of color produced by a heated solid, liquid, or gas under pressure.

convection The transfer of energy (heat) by the motion of the medium that "carries" the energy.

convex lens A lens in which one or both surfaces are curved like the outside of a sphere and thus are thicker at the center than near the outside.

Copernican system A system of planets that revolve around the sun (heliocentric).

core The central portion of a planet or of any celestial body.

corona The outer atmosphere of the sun, seen only when the central disk of the sun is covered.

corona (of the galaxy) The halo or sphere of objects that surround the central nucleus of the Milky Way galaxy.

coronagraph An instrument for photographing the outer atmosphere of the sun by artificially covering the disk

of the sun's image at the focal plane of the telescope.

cosmic rays High-energy particles, largely protons, that strike the earth.

cosmogony The study of the origin and evolution of the material in the universe.

cosmological principle The assumption that, in general, the universe would appear the same from any location within itself.

cosmology The study of the organization and structure of the universe and its evolution.

cosmos The entire universe, seen as an orderly, self-inclusive system.

Coudé focus A system by which an arrangement of secondary mirrors directs the light gathered by the primary mirror along the polar axis of the telescope to the Coudé focal point. This point remains fixed as the telescope moves to follow a star.

crater A depression in the surface of the earth or moon.

crepe ring The innermost visible ring of Saturn; a dark ring.

crescent That phase of the moon which shows it less than half full; also applicable to Venus, or any planet when viewed from space.

crust The outer layer of the earth, moon, or other such body.

dark nebula A gas or dust cloud that obscures the light of stars and galaxies behind it.

declination The smallest angle between a given object and the equator.

deferent The larger orbit in the Ptolemaic system, along which the center of the epicycle of a planet supposedly moves.

deflection of starlight The bending of light as it passes close to a massive object; a bending due to gravity.

density The mass of an object divided by its volume and usually measured in grams per cubic centimeter.

descending node The point along the orbit of an object at which it passes from north to south of the celestial equator or of some other reference plane.

deuterium (Also called "heavy hydrogen.") A form of hydrogen in which the nucleus of each atom contains one proton and one neutron.

diffraction The absorption and reemission of light as it passes an object or passes through a small opening.

diffraction grating A system of finely ruled lines that diffract the light and by interference produce a spectrum.

diffraction pattern A series of bright and dark lines produced by the interference of light.

diffuse nebula A bright or dark nebula of irregular shape; not of the planetary form.

diffusion (of light) The scattering of light from an irregular surface.

direct motion The typical motion of

a planet—west to east—as seen against the background of stars.

disk (of our Galaxy) The flattened portion of the Milky Way galaxy; hence, the spiral arms.

disk (of a planet) The round shape of a planet as seen in a telescope; of measureable size.

dispersion The separation of white light into its component colors by means of refraction or diffraction.

diurnal circle The apparent path of an object in the sky during one day, due to the earth's rotation.

diurnal motion The daily apparent motion of all objects, due to the earth's rotation.

Doppler shift The change in observed wavelength of sound, radio, or light, due to the motion of the source, the observer, or both.

double star A star system that is composed of two stars, each influenced by the other's gravitational field. Their dual nature is revealed by telescopic or spectroscopic observation.

dwarf (star) A star of less than average mass and luminosity; a small star.

dyne A unit of force in the metric system—the force necessary to give a 1-g mass an acceleration of 1 cm/sec/sec.

earthquake A release of stresses that have been stored within the earth's crust, producing a movement of the crust.

earthshine (earthlight) That light which is reflected by the earth's atmosphere and which illuminates the dark portion of the moon.

eccentric A point within a circle but off-center.

eccentricity A measure of the degree to which an ellipse is elongated. It may be found by dividing the distance between the foci by the length of the major axis.

eclipse The partial or total darkening of an object that passes within the shadow of another object.

eclipse path The path on the earth's surface swept out by the shadow of the moon during a total eclipse of the sun.

eclipse season A period of time during which eclipses of the moon and sun may take place, approximately one month in duration and occurring approximately 6 months apart.

eclipsing binary (stars) A binary-star system in which the plane of revolution of the two stars is seen almost edge-on. The light of each star is periodically diminished by the passage of the other star in front of it.

ecliptic The plane of the earth's orbit projected onto the sky. The apparent path of the sun on the celestial sphere.

electromagnetic radiation A disturbance that is transmitted from its source to the observer by means of changing electrical and magnetic fields. Includes radio, infrared, visible light, ultraviolet, X rays, and gamma rays.

electromagnetic spectrum The full array of electromagnetic disturbances (radio, infrared, visible light, ultraviolet, X rays, gamma rays).

electron A subatomic particle that carries a negative charge and is thought to move about the nucleus of the atom.

element Any one of more than 100 fundamental substances that can not be broken down into simpler forms by chemical processes.

elements of an orbit Those particular quantities that describe the size, shape, and orientation of the orbit of a body in space. Such quantities are used to determine the location of the body at any given time.

ellipse A closed curve that may be obtained by passing a plane completely through a circular cone; a curve that describes the orbits of bodies in space.

elliptical galaxy A galaxy whose visible shape is that of an ellipse.

elongation The apparent angle between the sun and the specified object, measured at the observer's eye. (The *elongation* of Mercury is 14° when it appears in the sky 14° from the sun.)

emission line A bright spectral line produced by a downward transition of an electron.

emission nebula A bright cloud of gas that has been excited and is producing its own light.

emission spectrum A system of bright lines produced by an excited, low-pressure gas.

energy The capacity to do work.

energy level (in the atom) The possible energies that an electron may possess if excited by some external source such as light or an electrical current.

ephemeris A table that states the position of celestial bodies at various times.

epicycle In the Ptolemaic system, a small circle, the center of which moves along the deferent. The planet moves on the epicycle, thus the motion of the planet might be described as a circle moving on a circle.

equation of time The difference between the apparent solar time minus the average (mean) solar time; the amount of time by which the noonday sun appears on the observer's meridian, either ahead or behind the average noon time.

equator A great circle on the earth lying halfway between the poles of the earth.

equatorial mount A telescope mounting designed so that one axis is parallel to the earth's axis and hence only one motion is necessary to drive the telescope to compensate for the earth's rotation.

equinox Either of the two intersection points of the celestial equator and the ecliptic.

erg A unit of energy in the metric system; the amount of energy expended when a force of 1 dyne moves an object through a distance of 1 cm.

escape velocity The velocity at which a body overcomes the pull of gravity of another body and moves off into space.

evolution (of the cosmos) Progressive changes in the universe or in objects found in it.

excitation The process whereby the electrons of an atom are given greater energy than they normally possess.

extragalactic Beyond our own Milky Way galaxy.

eyepiece A lens system that magnifies the image formed by the objective lens at the prime focus.

filar micrometer A device attached to a telescope, at the eyepiece, to measure small angles between stars.

fireball A spectacular meteor usually visible for several seconds.

fission The breaking up of heavier atomic nuclei into two or more lighter atomic nuclei.

flare A sudden, temporary brightening of a given region on the sun's surface. It represents a tremendous outpouring of energy.

flare star A star that increases suddenly and unexpectedly in brightness.

flash spectrum A bright-line spectrum produced by the lower atmospheric "layers" of the sun and seen only for an instant before and after the total phase of a solar eclipse.

flocculi Bright regions usually just above sunspots and visible only in a spectroheliogram. (Also called *plages*.)

fluorescence Occurs when light of one wavelength is absorbed and then reemitted at another wavelength. Ultraviolet light may excite certain atoms to produce visible light.

focal length The distance from the center of a lens or mirror to the prime focus of the telescope.

focal ratio (f-number) The focal length of a lens or mirror divided by its diameter.

focus The point at which converging rays of light meet.

forbidden lines Spectral lines that are not usually produced under laboratory conditions but may be produced under certain conditions of space, for instance, very low pressure.

force The influence which can change the speed and/or direction of an object.

Foucault pendulum A device employing a swinging pendulum, used to prove the rotation of the earth.

Fraunhofer lines Absorption (dark) lines in the spectrum of the sun (or star).

frequency The number of waves that pass a given point per second.

fringes The alternate bright and dark regions produced by the interference of light.

full moon The phase of the moon that occurs when it is directly opposite the sun in the sky, revealing its fully lighted face.

fusion The process whereby heavier elements are created by means of nuclear reactions within lighter ones.

galactic center The point around which a given galaxy rotates. In the Milky Way galaxy, this point is located toward the constellation of Sagittarius.

galactic cluster An open cluster of stars situated in the spiral arms of the Galaxy.

galactic equator A great circle on the celestial sphere that indicates the plane of the Galaxy.

galactic poles Points on the celestial sphere 90° from the galactic equator.

galaxy (Capitalized when our own Galaxy, the Milky Way, is meant.) A fundamental collection of stellar material, usually containing millions to hundreds of billions of stars.

gamma rays Electromagnetic disturbances of wavelength shorter than X rays and carrying the most energy of all such disturbances.

gauss A unit of magnetic flux density; a measure of the strength of a magnetic field at a given point.

gegenschein (counterglow) A region directly opposite the sun that appears to glow, perhaps due to the reflection of sunlight from interplanetary particles.

geomagnetic field The earth's magnetic field.

geomagnetic poles The points on the earth's surface that seem to possess the same magnetic properties as the poles of a bar magnet.

giant (star) A star of very large radius.

gibbous The phase of the moon or a planet during which it appears more than half full but less than full.

globular cluster A large globe-shaped system of stars, usually found in the halo that surrounds the nucleus of the Galaxy.

globule A small, dense, dark nebula; possibly a cloud of gas that is in the process of forming a star.

granulation (of the photosphere) The mottled appearance of the visible "surface" of the sun, due to rising columns of hot gases.

gravitation The mutual force of attraction that masses exert on each other.

gravitational constant (G) A number that allows us to express the force of gravitational attraction in terms of the masses of the objects involved and the distance separating these objects.

When the mass is measured in grams and the distance in centimeters, then the force may be expressed in dynes; specifically $G = 6.668 \times 10^{-8}$ dyne-cm²/gm².

gravitational energy The energy that may be released by the partial or total collapse of a system.

gravity (of the earth) The force of attraction that the earth exerts on a given object. The weight of an object.

great circle The largest circle that can be drawn on a sphere; a circle that divides a sphere into two equal parts.

greatest elongation The largest angle of separation between the sun and either Mercury or Venus.

Greenwich meridian A portion of a great circle on the earth that passes through the poles and through a given point at the Royal Greenwich Observatory in England. This meridian is also referred to as the "prime meridian," from which the longitude of any point on the earth is measured.

Gregorian calendar The modern calendar, which was introduced by Pope Gregory XIII in the sixteenth century.

H I (hydrogen-one) region A region in space in which neutral hydrogen gas is found.

H II (hydrogen-two) region A region in space in which ionized hydrogen gas is found.

half-life The time required for one-half of the atoms in a sample of a radioactive element to change spontaneously into another element.

halo (of the Galaxy) System of globular clusters, stars, and gas that surrounds the nucleus of the Galaxy.

harmonic law Kepler's third law of planetary motion: *The cubes of the semimajor axes of the planetary orbits compare in the same way as the squares of the periods of the planets.*

harvest moon The full moon nearest the time of the autumn equinox.

head (of a comet) That portion of the comet, exclusive of any tail, which includes the nucleus and coma.

heavy elements Elements that have many protons and neutrons in their nuclei. (The term sometimes refers to any element other than hydrogen or helium.)

heliacal rising The simultaneous rising of any object with the sun.

heliocentric system A sun-centered system.

Hertzsprung-Russell (H-R) diagram A plot of a group of stars according to their absolute magnitude and spectral class (temperature).

horizon (celestial) A great circle 90° from the observer's zenith.

horizontal system A system by which the location of an object is specified in terms of its angle above the horizon (altitude) and the angle it makes relative to due north (azimuth).

Azimuth is measured in an easterly direction from due north along the horizon.

hour angle The angle between the zero hour circle and any given hour circle.

hour circle A great circle on the celestial sphere that runs through the celestial poles, hence is perpendicular to the celestial equator.

Hubble constant A number that expresses the apparent relationship between the distance to the galaxy and its speed of recession.

hydrostatic equilibrium A condition in which the inward gravitational attraction exactly balances the outward pressure force at every point within a star.

hyperbola A curve that may be produced by cutting a circular cone with a plane parallel to the axis of the cone. It may represent the path of certain comets.

hypothesis An idea that is assumed to be true and upon which a given theory is built.

ideal gas A gas that obeys certain laws that describe the relationships between pressure, volume, and temperature.

image A representation of an object that can be seen by the eye, usually because light rays are brought to a focus.

inclination (of an orbit) The angle between the plane of the orbit of a given body and another specified reference plane, such as that to the celestial equator or the galactic equator.

Index Catalogue (IC) A listing of star clusters, nebulae, and galaxies that supplements the New General Catalogue (NGC).

index of refraction A number found by dividing the speed of light in a vacuum by its speed in a given transparent substance. This number would then be the index of refraction of the transparent substance.

inertia The property of matter which resists any change in its velocity: an object at rest tends to remain at rest; an object in motion tends to remain in motion in a straight line.

inferior conjunction The configuration of Mercury or Venus when that planet is directly between the earth and the sun.

inferior planet A planet whose orbit lies between that of the earth and the sun: thus, Mercury and Venus.

infrared An electromagnetic radiation of wavelength just longer than red light.

intensity The brightness of a source of light.

interference The reinforcing or canceling of waves, as in light or water waves.

interferometer An optical device by

which the diameters of the largest nearby stars may be determined.

International Date Line A line opposite the prime (Greenwich) meridian, at approximately 180° longitude. When a traveller crosses this line, the date is changed by one day.

interstellar dust Microscopic dust-like grains that exist in the space between stars.

interstellar gas Very diffuse gas that exists in the space between stars.

interstellar lines Dark spectral lines that are due to absorption by interstellar gases. These lines are seen against the spectrum of the star itself.

interstellar matter Gas and dust in the space between stars.

ion An atom that has gained or lost one or more electrons; hence, a charged atom.

ionization The process whereby an atom gains or loses electrons.

ionosphere A region of the earth's upper atmosphere in which many atoms are ionized.

irregular galaxy A galaxy of irregular shape, neither spiral nor elliptical.

irregular variable A star whose variations in brightness do not occur at regular intervals.

island universe The name once given to galaxies.

isotope One of several forms that atoms of a given element may take, each varying in the number of neutrons contained in the atomic nucleus.

Jolly balance An apparatus used to measure the mass of the earth. (May also refer to an instrument used to determine the density of a given object.)

Jovian planet A Jupiter-like planet; hence, one of relatively low density—Jupiter, Saturn, Uranus, and Neptune.

Julian calendar A calendar based on the apparent motions of the sun, introduced by Julius Caesar.

Julian day The number assigned to a given day (in the Julian calendar) based on a starting time of January 1, 4713 B.C.

Jupiter The fifth planet from the sun and the largest in the solar system.

Kepler's laws Three basic statements that describe the motions of the planets.

kinetic energy The energy that a body possesses due to its motion.

Kirchhoff's laws Three statements that describe the formation of the continuous, the emission, and the absorption spectra.

Kirkwood's gaps Voids in the spacing of particles in the rings of Saturn or in the asteroid belt.

latitude An angular measure of a point on the surface of the earth that indicates its distance north or south of the equator.

law of areas Kepler's second law of planetary motion: A line joining any given planet with the sun sweeps out equal areas in equal times.

law of the red shift For distant galaxies, their recessional velocities are proportional to their distance. Their recessional velocities are measured by their red shift, hence the red shift becomes an indicator of distance.

law of reflection The angle of reflection is equal to the angle of incidence.

law of refraction A ray of light is bent toward the normal when passing from one medium to another in which it travels more slowly.

law of universal gravitation Any two objects in space experience a mutual and equal force of attraction. This force is proportional to the product of their masses and inversely proportional to the square of the distance separating the objects.

leap year A year containing 366 days; occurs once every four years.

librations (of the moon) Motions that over a period of time allow the observer to see more than one-half the moon's surface.

light An electromagnetic disturbance that is visible to the eye.

light curve A graph that indicates the variations in light output of a variable star or an eclipsing binary system as they occur over a period of time.

light-gathering power (of a telescope) A measure of the amount of light a telescope can collect.

light-year The distance light travels in one year—approximately 9.5 trillion kilometers (6 trillion miles).

limb The visible edge of a planet, the moon, or the sun.

limb darkening The phenomenon of decreased brightness of the sun near its edge.

limiting magnitude The dimmest magnitude observable in a given telescope under specified conditions.

line of apsides Major axis of an ellipse.

line broadening Any phenomenon that results in the broadening of the spectral lines. (Pressure, for example, may cause line broadening.)

line of nodes The line connecting two nodes of an orbit; hence, a line that lies both in the plane of the object's orbit and also in the plane of another's orbit.

local apparent noon The time at which the center of the sun is on the observer's meridian.

Local Group The cluster of galaxies to which the Milky Way galaxy belongs.

local mean noon The time at which the average sun is on the observer's meridian.

local mean time The hour angle of the average sun.

local meridian An imaginary line running north and south through the observer's zenith (in the sky) or through the observer's position (on earth).

longitude The angle between the prime meridian and a meridian through the point on the surface of the earth. This angle is measured along the equator.

long-period variable A variable star whose period of variation exceeds 100 days.

luminosity A measure of actual brightness; the sun may be used as a reference star and other stars compared to it.

lunar eclipse An eclipse of the moon, occurring when the moon passes within the earth's shadow.

lunar month A period of time ($29\frac{1}{2}$ days) based upon the cycle of phases of the moon, i.e. full moon to full moon.

Lyman series Those lines in the spectrum which occur in the ultraviolet portion, created by upward or downward electron transitions that originate or terminate with the first (lowest) energy level of the hydrogen atom.

magnetic field A region of space where magnetic forces may be detected.

magnetic pole One of two points on a magnet (or in a magnetic field) where the greatest density of lines of force exist.

magnifying power The number of times that the apparent size of an object is increased by viewing it through a telescope, as compared to naked-eye observation.

magnitude A number that is used to indicate the brightness of an object, either apparent or absolute, as specified.

main sequence A line on the H–R diagram that represents the majority of stars.

major axis (of an ellipse) A line drawn through the foci of an ellipse; also, the longest line that may be drawn joining two points of an ellipse. It may refer specifically to the length of such a line.

major planet One of the four largest planets (Jupiter, Saturn, Uranus, and Neptune).

mantle (of the earth) The layer of the earth that lies between the crust and the core.

maria Latin for "seas" (singular: *mare*); name of certain regions, once thought to be sealike, on the moon or Mars.

Mars The fourth planet from the sun.

maser Acronym for **m**icrowave **a**mplification by **s**timulated **e**mission, a device utilizing natural oscillations of atoms or molecules for amplifying electromagnetic waves in the microwave region of the spectrum.

mass A measure of the amount of atomic material in an object.

mass-luminosity relation The observed relationship that, in general, the more massive stars are, the brighter they are.

mean solar day The period of time between successive passages of the mean (average) sun across the observer's meridian; the average length of an apparent solar day.

mean sun An imaginary sun that moves at a constant speed along the celestial equator, making one circuit of the sky in one year.

Mercury The planet nearest the sun.

meridian A great circle on the earth that passes through a given point and both poles; a great circle on the celestial sphere that passes through the observer's zenith and both celestial poles.

Messier Catalogue (M) A listing of nebulae, star clusters, and galaxies compiled by Charles Messier in 1787.

meteor The bright streak of light that occurs when a particle from outside the earth's atmosphere enters that atmosphere and is heated due to friction.

meteor shower Numerous meteors that seem to radiate from a given point in the sky. Showers usually occur when the earth passes through collections of material left in the path of a comet.

meteorite That portion of a meteoroid which survives its flight through the earth's atmosphere and strikes the ground.

meteoroid A meteor particle while it is still in space, without relation to the phenomenon it causes once it has entered the earth's atmosphere.

micrometeorite A very small meteoroid which does not create friction sufficient to burn up in the atmosphere, but which falls to the earth.

micrometer (μm) A unit of length equal to 10^{-6} m.

Milky Way A diffuse band of light composed of millions of stars and nebulae. It encircles the sky and represents the flattened disklike portion of the Milky Way galaxy (our Galaxy).

minor axis (of an ellipse) The smallest diameter of an ellipse.

minor planet An asteroid, ranging in size from several hundred kilometers to less than 1 km in diameter.

Mohorovičić discontinuity The boundary between the crust and the mantle of the earth, named for the Yugoslav geologist Andrija Mohorovičic.

molecule A combination of two or more atoms; the smallest particle of a substance that still retains all the properties of that substance.

momentum Mass multiplied by the velocity of an object; a measure of the state of motion of an object.

monochromatic Limited to one color or wavelength.

moving cluster (of stars) A group of stars that are moving in the same direction and at the same speed.

n-body problem A problem that involves the gravitational effects of several bodies on each other.

nadir A point opposite the observer's zenith.

nautical mile The average length of an arc on the earth's surface that makes an angle of one minute (1/60 degree) at the center of the earth.

navigation The art of finding one's position on the earth by means of celestial observations.

neap tides The less extreme tides that occur when the moon is near first or third quarter phase.

nebula A cloud of gas or dust in space; the term may sometimes refer to galaxies or star clusters (because of their fuzzy appearance).

nebular hypothesis The theory that the solar system was formed from a nebula.

Neptune The eighth planet from the sun.

neutrino A neutral particle of little or no mass that carries away energy from a nuclear reaction.

neutron A particle, within the nucleus of an atom, that has no charge but has mass approximately equal to that of a proton.

New General Catalogue (NGC) A listing of nebulae, star clusters, and galaxies that succeeded the one compiled by Charles Messier.

new moon That phase of the moon which occurs when it is between the earth and the sun.

Newtonian reflector A reflecting telescope that, by use of a flat diagonal mirror, brings the rays of light from a distant object to a focal point near the side of the tube.

Newton's laws The three statements set forth by Sir Isaac Newton regarding the motions of objects.

node The intersection points of the plane of the moon's or planet's orbit with another plane, such as that of the earth's orbit or the celestial equator.

north celestial pole The point on the celestial sphere that is determined by the extension of the earth's axis in a northerly direction.

north point A point on the observer's celestial horizon directly under the north celestial pole.

nova A star that suddenly brightens and then fades again, hence is seen as a "new" star.

nuclear Refers to the nucleus of the atom.

nuclear fission A process of nuclear change that results in lighter elements being formed from heavier ones.

nuclear fusion A process of nuclear change that results in heavier elements being formed from lighter ones.

nucleon A constituent of the nucleus of an atom: a neutron or proton.

nucleus (of an atom) The central portion of the atom, containing almost the entire mass of the atom.

nucleus (of a comet) The collection of solid (frozen) particles that compose the head of the comet.

nucleus (of a galaxy) The central, more dense portion of the galaxy.

nutation The small variations in the movements of the earth's poles. The principal movement is precession.

objective The main lens or mirror in a telescope, used to bring rays of light to a focal point.

oblate spheroid A sphere, such as the earth, that is flattened by rotation.

oblateness A measure of the degree to which a sphere is flattened; a number obtained by dividing the difference between the major and minor axes by the major axis itself.

obliquity of the ecliptic The angle between the ecliptic plane and that of the celestial equator.

obscuration The absorption of starlight by interstellar dust.

occular An eyepiece.

occultation The passing of one object behind a larger one, say, the passing of a moon behind Jupiter.

open cluster A loosely formed cluster of stars, usually found in the disk of the galaxy.

opposition The configuration of a planet when it is directly opposite the sun as seen from earth. The planet is seen 180° from the sun in the sky.

optical binary Two stars that merely appear close together because they happen to line up from the observer's point of view. They may actually be separated by a great distance and therefore they do not influence each other gravitationally.

optics A branch of physics that deals with light and its properties.

orbit A closed path along which a body moves as it revolves around a point in space.

orbital plane The plane in which a given body moves in its orbit; an imaginary flat surface that is determined by the motion of the body.

outer planet One of the planets beyond the asteroid belt (Jupiter, Saturn, Uranus, Neptune, and Pluto).

ozone layer A "layer" of the earth's atmosphere, composed of a special form of oxygen (O_3), which filters much of the ultraviolet radiation of the sun.

parabola A curve that is made by cutting a circular cone with a plane parallel to one of its elements (edges).

paraboloid A concave surface, the cross section of which is a parabola. The shape used for the primary mirror of most reflecting telescopes.

parallax The apparent shift in position of an object due to the motion of the observer.

parallax (stellar) The apparent shift of a star against the background of more distant stars due to the motion of the earth around the sun; the angle subtended (cut off) by the radius of the earth's orbit (1 A.U.) at the distance of the star.

parsec The distance to a star that exhibits 1 second of heliocentric parallax (1 parsec = 3.26 light-years).

partial eclipse An eclipse of the moon or sun in which the object is not completely obscured.

penumbra That portion of the shadow of an object in which the light for an extended source is not completely obscured.

penumbral eclipse An eclipse of the moon in which the moon merely passes through the penumbra of the earth's shadow.

perfect radiator A black body—a body that absorbs all radiation falling upon it and reemits all the radiation.

periastron That point in the orbit of a member of a binary star system at which it is nearest its companion star.

perigee The point in the orbit of an earth satellite at which it is nearest the earth (-*gee*).

perihelion The point in the orbit of an object that revolves around the sun (*helios*) at which it is nearest the sun.

period The interval of time necessary to complete one rotation, one revolution, or one cycle.

period-luminosity relation The relationship between the period and the absolute magnitude of certain variable stars.

periodic comet A comet whose orbit is elliptical and hence one that returns to perihelion at regular intervals. A comet whose return is predictable.

perturbation Any gravitational disturbance that causes a body to deviate from its primary orbital path. Such disturbances may be caused by the presence of a third object.

phases (of the moon or planet) Changes in the portion of the illuminated "face" of the moon or planet that is visible from the earth.

photoelectric effect The emission of electrons from the surface of a substance caused by light striking it.

photoelectric magnitude A measure of the brightness of an object as indicated by a photomultiplier.

photographic magnitude A measure of the brightness of an object as indicated on a blue-sensitive photographic plate.

photometry The science of measuring the apparent brightness of celestial objects.

photomultiplier A light-sensitive cell in which the electric current generated by light is amplified so that it can be more easily and accurately measured.

photon A unit of electromagnetic energy; a certain quantity of light energy.

photosphere The apparent (visible) "surface" of the sun; the layer of the sun from which light seems to radiate.

photovisual magnitude A measure of the apparent brightness of an object, using film that is sensitive to the same region of the spectrum as the human eye.

plage A bright region just above the sun's surface as seen in a spectroheliogram.

Planck's constant A number that relates the energy "carried" by a photon of light to its wavelength.

planet One of the nine main bodies that revolve around the sun and reflect its light; any similar body revolving around a star in another (possible) solar system.

planetarium A projection device that is capable of creating an artificial sky on a domed ceiling and showing the motions of celestial objects greatly speeded up in time.

planetary nebula A spherical shell of gas that surrounds a very hot star and is expanding relatively slowly.

planetoid A minor planet or asteroid.

Pluto The ninth planet from the sun.

polar axis (of a telescope) That axis, set parallel to the earth's axis, about which the telescope turns to compensate for the earth's rotation.

polarization A filtering process in which only those light rays whose disturbances lie in a given plane are allowed to pass.

Population I and II (stars) Two classes of stars that appear to be quite different in evolutionary state and in location within the Galaxy: type-I stars are found primarily in the spiral arms of the Galaxy, whereas type-II stars are found elsewhere in it.

positron A particle that has approximately the same mass as the electron but that carries a positive charge.

potential energy That type of energy which an object possesses because of its position; the capacity of an object to do work by reason of its position.

pound A unit of force (not mass) in the English system.

precession The slow gyration of the earth's axis, which sweeps out a circle in the sky over a 26,000-year period. This motion of the earth causes a continuous change of the polar positions in the sky.

precession of the equinoxes The slow westward shift of the equinoxes along the ecliptic due to the precession of the earth.

prime focus The point at which the objective of a telescope brings the light rays to a focus without the use of any secondary mirrors or lens.

prime meridian The meridian that runs through the Royal Observatory of Greenwich, England. The longitudes of points on the earth are measured from this meridian.

primeval atom A single mass composed of all the matter of the universe.

primeval fireball The expanding ball of matter that resulted from the explosion of the primeval atom as depicted in the "big bang" theory of the origin of the universe.

Principia Newton's great work, in which he described the motions of objects under the influence of gravity: *Philosophiae Naturalis Principia Mathematica*.

prism A triangular shape (of glass or other transparent material) that is utilized to disperse light into its spectrum.

prominence A protrusion from the limb of the sun which appears as a flame or loop, best seen in the light of hydrogen.

proper motion The rate at which a star's position in the sky changes, measured in seconds of arc per year.

proton One of the basic subatomic particles that compose the nucleus, carrying a positive charge.

proton-proton cycle An atomic reaction that occurs in the core of a star whereby four hydrogen nuclei combine to form a helium nucleus. This reaction is the source of energy of the star (or sun).

protostar The mass of material that is in the process of forming a star.

pulsar An object that emits brief pulses of radio energy and that has also been observed optically by use of special photographic equipment; possibly a very dense (neutron) star that is spinning very rapidly.

pulsating variable A variable star that changes size at regular intervals, its variation in light output being directly related to its variation in size.

quadrature The configuration of a planet or moon as seen 90° from the sun.

quantum mechanics The study of the structure of atoms and how they interact with one another.

quarter moon A half-full moon as seen when it appears 90° from the sun, one-quarter or three-quarters of the way around its orbit.

quasar Contraction of the term *quasi-stellar radio source;* a starlike object that has a very large red shift in its spectrum, hence is presumed to be very distant—perhaps the most distant object yet known—and probably a galaxy that is emitting much more energy than is normal.

quiet sun The sun at a time of very low activity.

R Corona Borealis variables The class of variable stars that exhibit irregular and sudden decreases in brightness.

RR Lyrae variables The class of variable stars that have periods of less than one day.

radar telescope A radio telescope that is also capable of sending a radio signal into space and then listening for its echo (reflection).

radial velocity That part of the velocity of an object which is measured along the observer's line of sight.

radiant (of a meteor shower) A point in the sky from which a number of meteors seem to originate during a meteor shower.

radiation The process whereby energy is transferred from one point to another through empty space.

radiation pressure The small force that electromagnetic radiation exerts on matter which it intercepts.

radio astronomy That branch of astronomy which is primarily concerned with receiving and analyzing the radio energy received from celestial objects.

radio telescopes A large parabolic reflector that collects the radio energy from one region of the sky and concentrates (focuses) that energy at a focal point. This energy is then amplified and recorded by electronic equipment.

radioactive element An element whose nucleus spontaneously disintegrates to produce a lighter element. Energy is also released in this process.

rays (lunar) A system of bright streaks that seem to radiate from certain craters on the moon.

reaction force The equal but opposite force that accompanies every force.

real image An image, formed at the focus of a telescope, that can be photographed; an image formed by light rays that converge after passing through a lens or after reflecting from a mirror.

red giant A very large, cool star.

red shift The shifting of spectral lines toward the red end of the spectrum due to the relative motion of the source away from the observer.

reddening (interstellar) The reddening of starlight as the result of the scattering of blue light when the light of a star passes through clouds of gas and dust in space.

reflecting telescope A telescope in which the primary objective is a concave mirror; a telescope that depends on the principle of reflection for its operation.

reflection The process whereby the direction of travel of light rays is changed by an optical surface.

reflection nebula A cloud of interstellar dust that is visible because it reflects starlight.

refracting telescope A telescope that depends on the principle of refraction for its operation, hence a telescope that has a lens or lens system as its objective.

refraction The bending of light as it passes from one transparent medium to another of different density.

relativity A theory formulated by Albert Einstein that deals with the measurement of various events as observed by two different observers, themselves in motion.

resolution The ability of a telescope to separate objects that appear close together; its ability to show detail.

retrograde motion (of a planet) The apparent westward motion of a planet as seen against the background of stars.

reversing layer (of the sun) A thin layer of solar atmosphere, just above the photosphere, that produces the dark-line spectrum of the sun.

revolution The motion of a body around a given point in space (for example, the earth's *revolution* around the sun).

right ascension The smallest angle between the zero hour circle and a given celestial object.

rill (or rille) A crevasse in the surface of the moon.

rotation The spinning of a body on its own axis.

saros An 18-year cycle during which the circumstances that produce similar eclipses recur.

satellite Any body that revolves around a larger body (the moon, for instance, is a satellite of the earth).

Saturn The sixth planet from the sun.

scale (of an image) The size of an extended image compared to its apparent size in the sky measured in centimeters (or inches) per degree.

Schmidt camera A telescope that utilizes both a spherical mirror and a weak refracting lens to produce a camera that is capable of photographing a wide field of stars.

science The branch of knowledge which seeks systematically to describe phenomena of nature.

scientific method An approach in which the researcher first observes certain pertinent phenomena, then formulates a theory that seems to be consistent with those observations, and finally tests his theory by determining whether it will accurately predict future events.

Sculptor-type system A very small elliptical galaxy similar to the galaxy in Sculptor.

secondary mirror A mirror that reflects the light gathered by the primary mirror; a mirror second in line.

"seeing" conditions Those conditions within the earth's atmosphere which affect the quality of image formed in the telescope.

seismic waves Vibrations that travel through the interior of the earth due to earthquakes.

seismograph An instrument that records the time, type, and strength of seismic (earthquake) waves.

seismology The science that deals with the origin and transmission of seismic waves in the earth.

semimajor axis One-half the major axis of an ellipse. It represents the average distance from the sun to a given planet or comet.

separation The angular distance between two stars in a visual binary system.

shell star A star that is surrounded by a sphere or shell of gas.

shower (of meteors) Numerous meteors that seem to radiate from a given point in the sky. Such showers usually occur when the earth passes through collections of meteoric material left in the path of a comet.

sidereal day The length of a day as measured by the successive passages of any given star across the observer's meridian. A day as measured by the stars.

sidereal month The length of time required for the moon's revolution around the earth as measured by the stars.

sidereal revolution The period of revolution of one body around another with respect to the stars.

sidereal time Star time; the hour angle of the vernal equinox; the right ascension of the observer's meridian at the given time.

sidereal year The time required for the earth's revolution around the sun with respect to the stars.

siderite An iron-nickel meteorite.

singularity The central core of a black hole, characterized by infinite density and infinite tidal forces.

small circle Any circle on the surface of a sphere that is smaller than a great circle.

solar activity Prominences, sunspots, plages, flares, etc.; activities that occur on or above the photosphere of the sun.

solar apex The direction in which the sun is moving with respect to the average motion of the nearest stars.

solar constant The amount of solar radiation received at the distance of the earth, measured in ergs per square centimeters per second.

solar day The average time required for two successive passages of the sun across the observer's meridian.

solar eclipse An eclipse of the sun.

solar flare A sudden outburst of energy from the sun, causing a brightening of a given region, usually near a sunspot.

solar parallax The angle subtended (cut off) by the equatorial radius of the earth as seen from a distance of 1 A.U.

solar system the system of all objects that revolve around the sun: the planets, moons, comets, meteoroids, etc.

solar time Time as based on the sun; the hour angle of the sun plus 12 hr.

solar wind The outflow of particles from the sun.

solstices Either of the two points on the ecliptic where the sun reaches its maximum declination north or south of the equator; the longest and the shortest days of the year.

south celestial pole A point on the celestial sphere determined by extending the earth's axis southward until it intersects that sphere.

south point The point of intersection of the observer's meridian with his southern horizon.

space motion The velocity of a star with respect to the sun.

specific gravity The density of a given body or substance compared to that of water; numerically equal to the density of the body measured in grams per cubic centimeter.

spectral class The classification of a star with respect to characteristics of its spectrum.

spectrogram The photograph of a spectrum.

spectrograph The instrument used to photograph a spectrum.

spectroheliogram The photograph of the sun, taken in the light of a single spectral line of an element such as hydrogen.

spectroheliograph The instrument used to photograph the sun in the light of a single spectral line of an element such as hydrogen or calcium.

spectroscope An instrument in which an observer may view the spectrum of an element, a star, or the sun.

spectroscopic binary star A star system in which the true binary nature of the system is revealed by the periodic shifting of spectral lines. Such a system can not be separated optically.

spectroscopic parallax A method whereby the distance to a star is determined by observing its spectral characteristics, converting this to absolute magnitude by use of an H–R diagram, and comparing that to the star's apparent magnitude.

spectrum The rainbow of colors produced when light is dispersed by refraction or diffraction.

spectrum analysis The determination of such characteristics of a light source as velocity, temperature, and pressure by studying the spectrum of the source.

spectrum binary A system of stars whose true binary nature is revealed by the presence of spectral lines associated with two stars of different temperatures.

speed The rate at which the distance to an object changes without regard to its direction of travel.

spherical aberration A defect in a lens or mirror that is due to its incorrect shape. Light rays that pass near the center of the lens (or mirror) are brought to a different focus as compared to those which pass near the outer edge.

spicule A jet of hot material rising in the atmosphere of the sun.

spiral arms (of a galaxy) The curved, armlike structures that surround the nucleus of certain galaxies.

spiral galaxy A flattened galaxy composed of a central nucleus and a system of arms which spiral out from that nucleus.

sporadic meteor A meteor that does not appear to be associated with a known shower of meteors.

spring tides The most extreme tides produced when the moon, sun, and earth are aligned, that is, when the moon is new or full.

standard time The time used within a given time zone, computed as the average solar time for that zone.

star A spherical mass of gas that radiates various forms of energy owing to nuclear reactions within its core.

star cloud A region of the sky in which the stars are so close together that they appear as a luminous cloud.

star cluster A grouping of stars that is held together by mutual gravitation; the stars thus possess a common motion through space.

star map A map showing the positions and magnitudes of stars, designed to be held over the observer's head.

steady state theory The theory that the universe has always been as it is today, that matter is being continually created to replace matter which is converted into energy. The density of the universe would thus remain at the same level.

Stefan's law The assertion that the total amount of energy radiated from a body in a given time depends upon the absolute temperature of the body, raised to the fourth power.

stellar evolution The life cycle of a star. Stars change in size, pressure, luminosity, and structure.

stellar parallax The angle subtended (cut off) by the radius of the earth's orbit (1 A.U.) at the distance of the star.

stratosphere The layer of the earth's atmosphere between the troposphere and the ionosphere.

subdwarf star A star that, owing to its smaller size, is less luminous than a main-sequence star of the same spectral class.

subgiant star A star of luminosity between that of a normal giant and a main-sequence star of the same spectral class.

sublimate A process whereby a solid turns directly into a gaseous state without passing through a liquid state, for instance, Dry Ice.

summer solstice The point on the ecliptic where the sun appears farthest north of the equator; the longest day of the year.

sun The star about which the earth revolves.

sunspots Regions of the sun that appear dark because they are temporarily cooler than the surrounding region.

sunspot cycle The period over which both the number and location (latitude) of the sunspots vary (approximately 11 years).

supergiant A large star of very high luminosity.

superior conjunction A configuration that occurs when a planet appears to line up with the sun on its (the sun's) far side. This term is used only in reference to Mercury or Venus.

superior planet A planet whose orbit is beyond that of the earth.

supernova An exploding star that temporarily increases in luminosity, perhaps a million times brighter than it was before the eruption.

synchrotron radiation A type of radiation that results from charged particles being accelerated by a magnetic field.

synodic month The time required for the moon to complete its cycle of phases (29.5 days).

synodic period The time required for a planet (or moon) to move from a given configuration back to that same configuration again, as seen from earth; for example, the period between successive oppositions of a superior planet.

syzygy A lining up of any three celestial objects, as in a conjunction or opposition.

T Tauri stars A class of variable stars that show very rapid and irregular pulsations.

tail (of a comet) The gases that are forced away from the head of the comet by the solar wind.

tangential velocity That part of a star's space velocity which is perpendicular to its radial velocity; a measure of the speed with which a star crosses the observer's line of sight.

tektites Glasslike objects that have traveled in the earth's atmosphere. These objects are thought to have been formed by meteorite impact on either the earth or the moon.

telescope An optical instrument that makes possible the observation and photographing of objects too dim and/or too distant to be seen with the naked eye; hence, a light-gathering device.

telluric (spectral) lines Spectral lines that are produced by elements in the earth's atmosphere.

temperature A measure of the average speed with which the molecules of a substance (or atoms of a gas) are moving.

terminator The line between the sunlit and dark portion of the moon (or planet).

terrestrial planet An earthlike planet, similar to the earth in density (Mercury, Venus, Mars).

theory A set of ideas that are consistent with observed phenomena.

thermocouple A device for measuring the intensity of infrared radiation.

thermodynamics A branch of physical science that deals with the way in which heat moves from one body to another.

thermonuclear reactions Nuclear changes that result from high-temperature and high-pressure conditions.

tide The deformation of land and/or water masses by the differential gravitational attraction of another body. The moon and sun create tides on the earth; the earth and sun create tides on the moon.

time zone A zone on the earth's surface, approximately 15° wide, within which the hour used is uniform.

ton (English) A unit of force (weight) equivalent to 2000 pounds.

ton (metric) A unit of force (weight) equivalent to 1 million grams.

total eclipse (1) An eclipse of the sun during which the disk of the moon completely covers the photosphere of the sun. (2) An eclipse of the moon during which the moon lies completely within the umbra of the earth's shadow.

trail (of a meteor) The temporary luminous streak produced by the passage of a meteoroid through the earth's atmosphere.

transit (1) The passage of a body across the face of a larger body. (2) The passage of a body across a given meridian. (Mercury transits the sun. The sun transits the prime meridian.)

transverse wave A wave in which particles are disturbed in a direction perpendicular to the direction in which the wave is traveling.

triangulation A process whereby an inaccessible side of a triangle may be determined from the measurement of accessible sides and angles.

Trojan minor planet An asteroid that orbits the sun in approximately the same orbit as Jupiter but is located 60° ahead of or behind the planet as viewed from the sun.

Tropic of Cancer The parallel of latitude that lies 23.5° north of the equator, the limit of the sun's apparent travel in a northerly direction.

Tropic of Capricorn The parallel of latitude that lies 23.5° south of the equator, the limit of the sun's apparent travel in a southerly direction.

tropical year The time required for the earth to make one revolution around the sun as measured by the vernal equinox (approximately 365.25 days).

troposphere The layer of the earth's atmosphere just above the earth and extending to an elevation of about 15 km.

twinkle The apparent changes in the brightness and color of a star due to the motion of the earth's atmosphere.

ultraviolet radiation That part of the electromagnetic spectrum with wavelengths just shorter than visible light, approximately in the range of 100 to 4000 Å.

umbra (1) The completely dark central portion of a shadow. (2) The darkest portion of a sunspot.

Universal Time Average Greenwich time.

universe All of space that is occupied by matter and/or radiation.

Uranus The seventh planet from the sun.

Van Allen belts Regions that surround the earth consisting of high-energy charged particles whose motions are directed by the earth's magnetic field.

variable star A star that exhibits changes in luminosity and/or color; a pulsating star.

vector A quantity that has both magnitude and direction.

velocity A vector quantity that denotes both the speed and direction of motion.

velocity of escape The velocity at which a body overcomes the pull of gravity of another body and moves off into space.

Venus The second planet from the sun.

vernal equinox The point on the celestial equator at which the sun crosses on its way northward; an intersection of the celestial equator and the ecliptic; the position of the sun on March 21.

vertical circle A great circle that passes through the zenith of the observer and is perpendicular to the horizon.

visual binary A binary system in which the two components are visible as separate stars in a telescope.

volume A measure of the amount of space occupied by an object.

Vulcan An imaginary planet once thought to orbit between Mercury and the sun.

walled plain (of the moon) A very large crater on the moon.

wandering of the poles A shifting of the body of the earth in relation to its axis of rotation.

watt A unit of power, equivalent to 10 million ergs of energy used up in 1 sec.

wavelength The distance from any point on a wave to the next similar point on the succeeding wave, as crest to crest.

weight A measure of the force exerted on one object by another due to gravity; specifically, the gravitational force exerted by the earth on a given mass.

west point A point on the celestial horizon that is located 270° from the north point, measured in a clockwise direction.

white dwarf An old star that has collapsed due to its exhausted fuel supply and yet has a large portion of its original mass; hence, a very dense, hot star.

Widmanstätten figures A definite pattern of crystal formations often seen in the interior of a polished meteorite.

Wien's law A statement relating T, the temperature of a body (in degrees Kelvin), to λ, the wavelength of its maximum radiation: $\lambda = 3000 \ \mu m / T$.

winter solstice A point on the ecliptic at which the sun reaches its maximum distance south of the celestial equator; the shortest day of the year.

Wolf-Rayet stars A class of very hot stars that eject shells of gas at high velocity.

X rays Electromagnetic radiation of short wavelength between that of ultraviolet rays and of gamma rays.

year The time required for one revolution of the earth around the sun.

Zeeman effect The splitting or broadening of spectral lines, which indicates the presence and strength of magnetic fields at the source.

zenith A point on the celestial sphere directly over the head of the observer.

zodiac A band on the celestial sphere that is centered on the ecliptic and contains the twelve constellations usually associated with astrology.

zodiacal light A faint light seen along the ecliptic, possibly due to sunlight being scattered by interplanetary dust.

zone of avoidance A region toward the center of the Milky Way galaxy where few—if any—other galaxies are seen, since they are obscured by clouds of gas and dust.

Appendixes

Appendix 1 Temperature Conversion Charts

FAHRENHEIT	CELSIUS	KELVIN	EXAMPLES
27 million °F	Approximately 15 million °C	15 million °K	Core of the sun
10,337°F	5727°C	6000°K	Surface temperature of the sun
700°F	371°C	644°K	Probable maximum temperature of Venus
212°F	100°C	373°K	Boiling point of water
100°F	38°C	311°K	Normal body temperature
68°F	20°C	293°K	Normal room temperature
32°F	0°C	273°K	Freezing point of water
0°F	−18°C	255°K	
−100°F	−73°C	200°K	Minimum temperature on the earth's surface
−198°F	−128°C	145°K	Polar cap of Mars
−230°F	−146°C	127°K	Average temperature of Saturn
−297°F	−183°C	90°K	Dark side of Mercury
−459°F	−273°C	0°K	Absolute zero; all translational molecular motion stops

Appendix 2 International System of Units (SI) with English Equivalents

INTERNATIONAL SYSTEM		ENGLISH SYSTEM

Length (units most commonly used):

1 micrometer (μm)	= 0.000001 m	
1 millimeter (mm)	= 0.001 m	= 0.03937 in.
1 centimeter (cm)	= 0.01 m	= 0.3937 in.
1 meter (m)	= 1.00 m	= 39.37 in.
1 kilometer (km)	= 1000 m	= 0.6214 miles
1 megameter (Mm)	= 1,000,000 m	
	1.6093 km	= 1 mile
	2.5400 cm	= 1 in.

Note: SI units are based on powers of ten (see Appendix 3), and the prefix indicates the power to be taken. The following list of prefixes may be used with any unit [for instance, *kilo*gram (kg), a unit of mass; or *nano*second (nsec), a unit of time]:

pico (p)	= 10^{-12}	centi (c)	= 10^{-2}	kilo (k)	= 10^3
nano (n)	= 10^{-9}	deci (d)	= 10^{-1}	mega (m)	= 10^6
micro (μ)	= 10^{-6}	deka (da)	= 10^1	giga (g)	= 10^9
milli (m)	= 10^{-3}	hecto (h)	= 10^2	tera (t)	= 10^{12}

INTERNATIONAL SYSTEM		ENGLISH SYSTEM

Mass:

1 milligram (mg)	= 0.001 g	
1 gram (g)	= 1.000 g	= 0.0022046 lb*
1 kilogram (kg)	= 1000 g	= 2.2046 lb*
	453.6 g	= 1 lb = 16 oz*
	28.3495 g	= 1 oz*

Time:

1 nanosecond (nsec)	= 0.000000001 sec
1 microsecond (μsec)	= 0.000001 sec
1 millisecond (msec)	= 0.001 sec
1 second (sec)	= 1.0 sec

* The weight equivalent at sea level on earth.

Appendix 3 Powers-of-ten Notation

In writing very large or very small numbers, it is convenient to use the following system of notation:

$$10^1 = 10$$
$$10^2 = 10 \times 10 = 100$$
$$10^3 = 10 \times 10 \times 10 = 1000$$
$$10^4 = 10 \times 10 \times 10 \times 10 = 10{,}000$$

Following this pattern:
$$10^{12} = 1{,}000{,}000{,}000{,}000$$

A light-year is approximately equivalent to 6,000,000,000,000 miles, which could be written $6 \times 1{,}000{,}000{,}000{,}000$ miles, or 6×10^{12} miles—a much simpler notation.

In a very similar way:
$$10^{-1} = 0.1 = 1/10$$
$$10^{-2} = 0.01 = 1/100$$
$$10^{-3} = 0.001 = 1/1000$$

Following this pattern:
$$10^{-7} = 0.0000001$$

The wavelength of blue light is approximately 0.0000005 m, but this number is equal to 5×0.0000001, therefore it may be written 5×10^{-7} m.

Summary:

If given 7×10^9, move the decimal nine places to the right, which produces 7,000,000,000.

If given 7×10^{-9}, move the decimal nine places to the left, which produces 0.000000007.

Appendix 4 Constants with Useful Approximations[a]

Pi $(\pi) = 3.14159$
$\cong 22/7$
Velocity of light, $c = 2.99793 \times 10^{10}$ cm/sec
$\cong 300{,}000$ km/sec
$\cong 186{,}000$ mi/sec
Constant of gravitation,
$G = 6.67 \times 10^{-8}$ dyne-cm^2/g^2
Angstrom unit, $\text{Å} = 10^{-10}$ m
Astronomical unit, A.U. $= 1.49598 \times 10^{11}$ m
$\cong 150{,}000{,}000$ km
$\cong 93{,}000{,}000$ mi

Parsec $\simeq 206{,}265$ A.U.
$= 3.262$ light-years
Light-year $= 9.4605 \times 10^{15}$ m
$\cong 9.5 \times 10^{12}$ km
$\cong 6 \times 10^{12}$ mi
Mass of the sun, $m_s = 1.991 \times 10^{33}$ g
Mass of the earth, $m_e = 5.98 \times 10^{27}$ g
Mass of the proton, $m_p = 1.672 \times 10^{-24}$ g
Mass of the neutron $= 1.674 \times 10^{-24}$ g
Mass of the electron $= 9.108 \times 10^{-28}$ g
Charge of the electron
$= 1.60 \times 10^{-19}$ coulomb
Avogadro's number
$= 6.025 \times 10^{23}$ molecules/mole
Planck's constant $= 6.625 \times 10^{-34}$ joule-sec
Acceleration due to gravity $= 980$ cm/sec^2
$= 32$ ft/sec^2

[a] Round-number approximations are indicated by $\cong$.

Appendix 5 Orbital Data of the Planets

PLANET	SYMBOL	SEMIMAJOR AXIS (A.U.)	SIDEREAL PERIOD	SYNODIC PERIOD	ECCENTRICITY OF ORBIT	INCLINATION OF ORBIT	AVERAGE ORBITAL SPEED (km/sec)
Mercury	☿	0.387	87.96 days	116 days	0.2056	7.0°	47.8
Venus	♀	0.723	224.7 days	584 days	0.0068	3.4°	35.0
Earth	⊕	1.000	365.26 days	—	0.0167	0.0°	29.8
Mars	♂	1.524	687.0 days or 1.88 years	780 days	0.0934	1.8°	24.2
(Ceres[a])	①	2.77	4.60 years	467 days	0.0765	10.6°	17.9
Jupiter	♃	5.20	11.86 years	399 days	0.0484	1.3°	13.1
Saturn	♄	9.54	29.46 years	378 days	0.0557	2.5°	9.7
Uranus	♂ or ♅	19.18	84.01 years	370 days	0.0472	0.8°	6.8
Neptune	♆	30.06	164.79 years	367.5 days	0.0086	1.8°	5.4
Pluto	♇	39.44	248 years	366.5 days	0.2502	17.2°	4.7

[a] An asteroid.

Appendix 6 Physical and Rotational Data for the Planets

PLANET	DIAMETER km	DIAMETER E[a] = 1	MASS (E[a] = 1)	DENSITY (WATER = 1)	PERIOD OF ROTATION	INCLINATION OF EQUATOR TO ECLIPTIC[b]	ALBEDO	SURFACE GRAVITY (E[a] = 1)	VELOCITY OF ESCAPE (km/sec)
Mercury	4,880	0.38	0.05	5.2	58^d 15^h	<28°	0.13	0.39	4.3
Venus	12,100	0.95	0.82	5.3	243^d 4^h	177°	0.76	0.90	10.3
Earth	12,750	1.00	1.00	5.5	23^h 56^m	23°27′	0.39	1.00	11.2
Mars	6,800	0.53	0.11	3.8	24^h 37^m	23°59′	0.18	0.38	5.1
Jupiter	142,800	11.23	317.9	1.3	9^h 50^m	3°	0.50	2.58	59.5
Saturn	120,660	9.41	95.2	0.7	10^h 39^m	26°45′	0.61	1.11	35.6
Uranus	51,800	3.98	14.6	1.3	12 to 24^h	98°	0.61	1.07	21.4
Neptune	49,500	3.88	17.2	1.7	18 to 22^h	29°	0.62	1.40	23.6
Pluto	3,000	0.23	0.002	1.1	6^d 9^h 17^m	?	0.50	?	?

[a] E is the earth.
[b] An inclination greater than 90° indicates retrograde rotation.

Appendix 7 Satellites of Planets

PLANET	SATELLITE	DISCOVERER	MEAN DISTANCE FROM PLANET (km)	SIDEREAL PERIOD (DAYS)[a]	INCLINATION OF ORBIT TO PLANET'S EQUATOR	DIAMETER OF SATELLITE (km)[b]	APPROXIMATE MAGNITUDE AT OPPOSITION
Earth	Moon		384,405	27.322	23.5	3476	−12.5
Mars	Phobos	A. Hall (1877)	9,380	0.319	1	27	11.5
	Deimos	A. Hall (1877)	23,500	1.262	2	12	12.0
Jupiter	XVI (J-3)	Voyager (1979)	127,949	0.295	0	40	
	XIV (J-1)	Voyager (1979)	129,300	0.298	0	35	
	V Amalthea	Bernard (1892)	180,500	0.498	0	270	13.0
	XV (J-2)	Voyager (1979)	257,256	0.674	1	80	
	I Io	Galileo (1610)	421,800	1.769	0	3630	5.5
	II Europa	Galileo (1610)	671,400	3.551	0	3130	5.7
	III Ganymede	Galileo (1610)	1,070,000	7.155	0	5270	5.0
	IV Callisto	Galileo (1610)	1,884,000	16.689	0	4820	6.3
	XIII Leda	Kowal (1974)	11,100,000	239	26.7	10	20
	VI Himalia	Perrine (1904)	11,470,000	250.57	27.6	170	14.0
	X Lysithia	Nicholson (1938)	11,710,000	260	29.0	20	19.0
	VII Elara	Perrine (1905)	11,740,000	260.10	24.8	80	17.5
	XII Ananke	Nicholson (1951)	21,200,000	617.0r	33.0	20	18.5
	XI Carme	Nicholson (1938)	22,600,000	692.5r	16.0	24	19.0
	VIII Pasiphae	Melotte (1908)	23,500,000	735.0r	35.0	20	17.5
	IX Sinope	Nicholson (1914)	23,700,000	758.0r	27.0	22	19.0
Saturn	1980S28	Voyager (1980)	137,700	0.602	0	40	
	1980S27	Voyager (1980)	139,400	0.613	0	140	
	1980S26	Voyager (1980)	141,700	0.629	0	100	
	1980S3	Voyager (1980)	151,420	0.694	0	120	
	1980S1	Voyager (1980)	151,470	0.695	0	200	
	Mimas	Herschel (1789)	185,500	0.942	0	400	12.0
	Enceladus	Herschel (1789)	238,000	1.370	0	500	12.0
	Tethys	Cassini (1684)	294,700	1.888	0	1060	10.5
	1980S13	Voyager (1980)	294,700	1.888	0	34	
	1980S25	Voyager (1980)	294,700	1.888	0	32	
	Dione	Cassini (1684)	377,400	2.737	0	1120	11.0
	1980S6	Voyager (1980)	378,100	2.737	0	36	
	Rhea	Cassini (1672)	527,100	4.518	0	1500	10.0
	Titan	Huygens (1655)	1,222,900	15.945	0	5150	8.3
	Hyperion	Bond (1848)	1,481,000	21.277	0	400	13.0
	Iapetus	Cassini (1671)	3,560,900	79.331	14.7	1460	11.0
	Phoebe	Pickering (1898)	12,960,000	550.45r	30	220	14.0
			Note: Four more small moons have been tentatively identified from Voyager imagery.				
Uranus	Miranda	Kuiper (1948)	130,000	1.41r	0	400	19.0
	Ariel	Lassell (1851)	191,000	2.52r	0	1200	15.0
	Umbriel	Lassell (1851)	266,000	4.14r	0	1000	16.0
	Titania	Herschel (1787)	436,000	8.71r	0	1800	14.0
	Oberon	Herschel (1787)	583,400	13.46r	0	1600	14.0
Neptune	Triton	Lassell (1846)	355,500	5.88r	20	4000	14.0
	Nereid	Kuiper (1949)	5,567,000	359.88	27.7	500	19.0
Pluto	Charon	Christy (1978)	20,000	6.39		1000	20.0

[a] The notation "r," when shown with the sidereal period of the satellite, indicates that the satellite orbits the planet in retrograde motion.
[b] Satellites smaller than 500 km are probably not spherical in shape; therefore numbers indicate only the largest dimension.

Appendix 8 The Twenty Brightest Stars

STAR	RIGHT ASCENSION (1950) (h) (m)	DECLINATION (1950) (°) (')	DISTANCE (PARSECS) (pc)	PROPER MOTION ('')	SPECTRA OF COMPONENTS[a,c] A	B	C	VISUAL MAGNITUDES OF COMPONENTS[b,c] A	B	C	ABSOLUTE VISUAL MAGNITUDES OF COMPONENTS[c] A	B	C
Sirius	6 42.9	−16 39	2.7	1.32	A1V	wd	—	−1.47	+7.1	—	+1.4	+10.5	—
Canopus	6 22.8	−52 40	30	0.03	F0Ib	—	—	−0.72	—	—	−3.1	—	—
α Centauri	14 36.2	−60 38	1.3	3.68	G2V	K5V	M5V	−0.01	+1.5	+10.7	+4.4	+5.8	+15
Arcturus	14 13.4	+19 27	11	2.28	K2III	—	—	−0.06	—	—	−0.3	—	—
Vega	18 35.2	+38 44	8.0	0.34	A0V	—	—	+0.04	—	—	+0.5	—	—
Capella	5 13.0	+45 57	14	0.44	G0II	M1V	M5V	+0.09	+10.2	+13.7	−0.7	+9.5	+13
Rigel	5 12.1	−8 15	250	0.00	B8Ia	B9	—	+0.10	+6.6	—	−6.8	−0.4	—
Procyon	7 36.7	+5 21	3.5	1.25	F5IV-V	wd	—	+0.38	+10.7	—	+2.7	+13.1	—
Betelgeuse	5 52.5	+7 24	200	0.03	M2Iab	—	—	+0.41v	—	—	−5.5	—	—
Achernar	1 35.9	−57 29	20	0.10	B5V	—	—	+0.47	—	—	−1.6	—	—
β Centauri	14 00.3	−60 08	90	0.04	B1III	—	—	+0.63	—	—	−4.1	—	—
Altair	19 48.3	+8 44	5.1	0.66	A7IV,V	—	—	+0.77	—	—	+2.2	—	—
α Crucis	12 23.8	−62 49	120	0.04	B1IV	B3	—	+1.39	+1.9	—	−4.0	−3.5	—
Aldebaran	4 33.0	+16 25	16	0.20	K5III	M2V	—	+0.86v	+13	—	−0.2	+12	—
Spica	13 22.6	−10 54	70	0.05	B1V	—	—	+0.91	—	—	−3.6	—	—
Antares	16 26.3	−26 19	120	0.03	M1Ib	B4V	—	+0.92v	+5.1	—	−4.5	−0.3	—
Pollux	7 42.3	+28 09	12	0.62	K0III	—	—	+1.16	—	—	+0.8	—	—
Fomalhaut	22 54.9	−29 53	7.0	0.37	A3V	K4V	—	+1.19	+6.5	—	+2.0	+7.3	—
Deneb	20 39.7	+45 06	430	0.00	A2Ia	—	—	+1.26	—	—	−6.9	—	—
β Crucis	12 44.8	−59 24	150	0.05	B0.5IV	—	—	+1.28	—	—	−4.6	—	—

[a] The Roman numerals after the spectral classifications have the following meanings: Ia or Ib, *supergiant;* II or III, *giant;* IV, *subgiant;* V, *main-sequence star* (see pages 229–230). The notation ''wd'' indicates *white dwarf.*

[b] The notation ''v'' following the magnitude indicates a *variable star.*

[c] When entries are shown in both A and B columns, the star is known to be a binary system. When an entry is also shown in the C column, the system is known to have three components.

Appendix 9 The Messier Catalogue of Nebulae and Star Clusters

M	NGC	RIGHT ASCENSION (1950) (h) (m)		DECLI-NATION (1950) (°) (')		APPARENT VISUAL MAGNITUDE	DESCRIPTION
1	1952	5	31.5	+21	59	8.4	Crab Nebula in Taurus; remains of supernova
2	7089	21	30.9	−1	02	6.4	Globular cluster in Aquarius
3	5272	13	39.8	+28	38	6.3	Globular cluster in Canes Venatici
4	6121	16	20.6	−26	24	6.5	Globular cluster in Scorpius
5	5904	15	16.0	+2	16	6.1	Globular cluster in Serpens
6	6405	17	36.8	−32	10	5.3	Open cluster in Scorpius
7	6475	17	50.7	−34	48	4.1	Open cluster in Scorpius
8	6523	18	00.1	−24	23	6.0	Lagoon Nebula in Sagittarius
9	6333	17	16.3	−18	28	7.3	Globular cluster in Ophiuchus
10	6254	16	54.5	−4	02	6.7	Globular cluster in Ophiuchus
11	6705	18	48.4	−6	20	6.3	Open cluster in Scutum
12	6218	16	44.7	−1	52	6.6	Globular cluster in Ophiuchus
13	6205	16	39.9	+36	33	5.9	Globular cluster in Hercules
14	6402	17	35.0	−3	13	7.7	Globular cluster in Ophiuchus
15	7078	21	27.5	+11	57	6.4	Globular cluster in Pegasus
16	6611	18	16.1	−13	48	6.4	Open cluster with nebulosity in Serpens
17	6618	18	17.9	−16	12	7.0	Swan or Omega Nebula in Sagittarius
18	6613	18	17.0	−17	09	7.5	Open cluster in Sagittarius
19	6273	16	59.5	−26	11	6.6	Globular cluster in Ophiuchus
20	6514	17	59.4	−23	02	9.0	Trifid Nebula in Sagittarius
21	6531	18	01.6	−22	30	6.5	Open cluster in Sagittarius
22	6656	18	33.4	−23	57	5.6	Globular cluster in Sagittarius
23	6494	17	54.0	−19	00	6.9	Open cluster in Sagittarius
24	6603	18	15.5	−18	27	11.4	Open cluster in Sagittarius
25	(4725)[a]	18	28.7	−19	17	6.5	Open cluster in Sagittarius
26	6694	18	42.5	−9	27	9.3	Open cluster in Scutum
27	6853	19	57.5	+22	35	7.6	Dumbbell Planetary Nebula in Vulpecula
28	6626	18	21.4	−24	53	7.6	Globular cluster in Sagittarius
29	6913	20	22.2	+38	21	7.1	Open cluster in Cygnus
30	7099	21	37.5	−23	24	8.4	Globular cluster in Capricornus
31	224	0	40.0	+41	00	4.8	Andromeda galaxy
32	221	0	40.0	+40	36	8.7	Elliptical galaxy; companion to M31
33	598	1	31.0	+30	24	6.7	Spiral galaxy in Triangulum
34	1039	2	38.8	+42	35	5.5	Open cluster in Perseus
35	2168	6	05.7	+24	21	5.3	Open cluster in Gemini
36	1960	5	33.0	+34	04	6.3	Open cluster in Auriga
37	2099	5	49.1	+32	33	6.2	Open cluster in Auriga
38	1912	5	25.3	+35	47	7.4	Open cluster in Auriga
39	7092	21	30.4	+48	13	5.2	Open cluster in Cygnus
40	—	12	20	+58	20	—	Close double star in Ursa Major
41	2287	6	44.9	−20	41	4.6	Loose open cluster in Canis Major
42	1976	5	32.9	−5	25	4.0	Orion Nebula

[a] Index Catalogue (IC) number.

M	NGC	RIGHT ASCENSION (1950) (h) (m)	DECLI-NATION (1950) (°) (′)	APPARENT VISUAL MAGNITUDE	DESCRIPTION
43	1982	5 33.1	−5 19	9.0	Northeast portion of Orion Nebula
44	2632	8 37	+20 10	3.7	Praesepe; open cluster in Cancer
45	—	3 44.5	+23 57	1.6	The Pleiades; open cluster in Taurus
46	2437	7 39.5	−14 42	6.0	Open cluster in Puppis
47	2422	7 34.3	−14 22	5.2	Loose group of stars in Puppis
48	2458	8 11	−5 38	5.5	Open cluster in Hydra
49	4472	12 27.3	+8 16	8.5	Elliptical galaxy in Virgo
50	2323	7 00.6	−8 16	6.3	Loose open cluster in Monoceros
51	5194	13 27.8	+47 27	8.4	Whirlpool spiral galaxy in Canes Venatici
52	7654	23 22.0	+61 20	7.3	Loose open cluster in Cassiopeia
53	5024	13 10.5	+18 26	7.8	Globular cluster in Coma Berenices
54	6715	18 51.9	−30 32	7.3	Globular cluster in Sagittarius
55	6809	19 36.8	−31 03	7.6	Globular cluster in Sagittarius
56	6779	19 14.6	+30 05	8.2	Globular cluster in Lyra
57	6720	18 51.7	+32 58	9.0	Ring Nebula; planetary nebula in Lyra
58	4579	12 35.2	+12 05	8.2	Barred spiral galaxy in Virgo
59	4621	12 39.5	+11 56	9.3	Elliptical spiral galaxy in Virgo
60	4649	12 41.1	+11 50	9.0	Elliptical galaxy in Virgo
61	4303	12 19.3	+4 45	9.6	Spiral galaxy in Virgo
62	6266	16 58.0	−30 02	6.6	Globular cluster in Ophiuchus
63	5055	13 13.5	+42 17	10.1	Spiral galaxy in Canes Venatici
64	4826	12 54.2	+21 57	6.6	Spiral galaxy in Coma Berenices
65	3623	11 16.3	+13 22	9.4	Spiral galaxy in Leo
66	3627	11 17.6	+13 10	9.0	Spiral galaxy in Leo; companion to M65
67	2682	8 48.4	+12 00	6.1	Open cluster in Cancer
68	4590	12 36.8	−26 29	8.2	Globular cluster in Hydra
69	6637	18 28.1	−32 24	8.9	Globular cluster in Sagittarius
70	6681	18 40.0	−32 20	9.6	Globular cluster in Sagittarius
71	6838	19 51.5	+18 39	9.0	Globular cluster in Sagitta
72	6981	20 50.7	−12 45	9.8	Globular cluster in Aquarius
73	6994	20 56.2	−12 50	9.0	Open cluster in Aquarius
74	628	1 34.0	+15 32	10.2	Spiral galaxy in Pisces
75	6864	20 03.1	−22 04	8.0	Globular cluster in Sagittarius
76	650	1 38.8	+51 19	11.4	Planetary nebula in Perseus
77	1068	2 40.1	−0 12	8.9	Spiral galaxy in Cetus
78	2068	5 44.2	+0 02	8.3	Small reflection nebula in Orion
79	1904	5 22.1	−24 34	7.5	Globular cluster in Lepus
80	6093	16 14.0	−22 52	7.5	Globular cluster in Scorpius
81	3031	9 51.7	+69 18	7.9	Spiral galaxy in Ursa Major
82	3034	9 51.9	+69 56	8.4	Irregular galaxy in Ursa Major
83	5236	13 34.2	−29 37	10.1	Spiral galaxy in Hydra
84	4374	12 22.6	+13 10	9.4	S0 type galaxy in Virgo
85	4382	12 22.8	+18 28	9.3	S0 type galaxy in Coma Berenices
86	4406	12 23.6	+13 13	9.2	Elliptical galaxy in Virgo

M	NGC	RIGHT ASCENSION (1950) (h) (m)	DECLI-NATION (1950) (°) (′)	APPARENT VISUAL MAGNITUDE	DESCRIPTION
87	4486	12 28.2	+12 40	8.7	Elliptical galaxy in Virgo
88	4501	12 29.4	+14 42	10.2	Spiral galaxy in Coma Berenices
89	4552	12 33.1	+12 50	9.5	Elliptical galaxy in Virgo
90	4569	12 34.3	+13 26	9.6	Spiral galaxy in Virgo
91[b]	4571(?)	— —	— —	—	
92	6341	17 15.6	+43 12	6.4	Globular cluster in Hercules
93	2447	7 42.4	−23 45	6.0	Open cluster in Puppis
94	4736	12 48.6	+41 24	8.3	Spiral galaxy in Canes Venatici
95	3351	10 41.3	+11 58	9.8	Barred spiral galaxy in Leo
96	3368	10 44.1	+12 05	9.3	Spiral galaxy in Leo
97	3587	11 12.0	+55 17	12.0	Owl Nebula; planetary nebula in Ursa Major
98	4192	12 11.2	+15 11	10.2	Spiral galaxy in Coma Berenices
99	4254	12 16.3	+14 42	9.9	Spiral galaxy in Coma Berenices
100	4321	12 20.4	+16 06	10.6	Spiral galaxy in Coma Berenices
101	5457	14 01.4	+54 36	9.6	Spiral galaxy in Ursa Major
102[b]	5866(?)	— —	— —	—	
103	581	1 29.9	+60 26	7.4	Open cluster in Cassiopeia
104	4594	12 37.4	−11 21	8.3	Spiral galaxy in Virgo
105	3379	10 45.2	+13 01	9.7	Elliptical galaxy in Leo
106	4258	12 16.5	+47 35	8.4	Spiral galaxy in Canes Venatici
107	6171	16 29.7	−12 57	9.2	Globular cluster in Ophiuchus

[b] Items of doubtful identification.

Appendix 10 The Greek Alphabet

A	α	alpha	H	η	eta	N	ν	nu	T	τ	tau
B	β	beta	Θ	θ	theta	Ξ	ξ	xi	Υ	υ	upsilon
Γ	γ	gamma	I	ι	iota	O	o	omicron	Φ	φ	phi
Δ	δ	delta	K	κ	kappa	Π	π	pi	X	χ	chi
E	ε	epsilon	Λ	λ	lambda	P	ρ	rho	Ψ	ψ	psi
Z	ζ	zeta	M	μ	mu	Σ	σ	sigma	Ω	ω	omega

Appendix 11 The Constellations

CONSTELLATION NAME[a]	DESCRIPTION	POSITION IN SKY		CONSTELLATION NAME[a]	DESCRIPTION	POSITION IN SKY	
		R.A.[b]	DEC.[b]			R.A.[b]	DEC.[b]
Andromeda	Princess of Ethiopia	1ʰ	+40°	Leo Minor	The Lion Cub	10ʰ	+35°
Antlia	The Air Pump	10ʰ	−35°	Lepus	The Hare	6ʰ	−20°
Apus	The Bird of Paradise	16ʰ	−75°	Libra	The Beam Balance	15ʰ	−15°
Aquarius	The Water Bearer	23ʰ	−15°	Lupus	The Wolf	15ʰ	−45°
Aquila	The Eagle	20ʰ	+5°	Lynx	The Lynx	8ʰ	+45°
Ara	The Altar	17ʰ	−55°	Lyra	The Lyre	19ʰ	+40°
Aries	The Ram	3ʰ	+20°	Mensa	The Table Mountain	5ʰ	−80°
Auriga	The Charioteer	6ʰ	+40°	Microscopium	The Microscope	21ʰ	−35°
Boötes	The Bear Driver	15ʰ	+30°	Monoceros	The Unicorn	7ʰ	−5°
Caelum	The Sculptor's Chisel	5ʰ	−40°	Musca	The Fly	12ʰ	−70°
Camelopardus	The Giraffe	6ʰ	−70°	Norma	The Carpenter's Square	16ʰ	−50°
Cancer	The Crab	9ʰ	+20°				
Canes Venatici	The Hunting Dogs	13ʰ	+40°	Octans	The Octant	22ʰ	−85°
Canis Major	The Greater Dog	7ʰ	−20°	Ophiuchus	The Serpent Holder	17ʰ	0°
Canis Minor	The Lesser Dog	8ʰ	+5°	Orion	The Great Hunter	5ʰ	+5°
Capricornus	The Sea Goat	21ʰ	−20°	Pavo	The Peacock	20ʰ	−65°
Carina	The Keel (of Argo Navis)	9ʰ	−60°	Pegasus	The Winged Horse	22ʰ	+20°
				Perseus	The Hero	3ʰ	+45°
Cassiopeia	Queen of Ethiopia	1ʰ	+60°	Phoenix	The Phoenix	1ʰ	−50°
Centaurus	The Centaur	13ʰ	−50°	Pictor	The Painter's Easel	6ʰ	−55°
Cepheus	King of Ethiopia	22ʰ	+70°	Pisces	The Fishes	1ʰ	+15°
Cetus	The Sea Monster	2ʰ	−10°	Piscis Austrinus	The Southern Fish	22ʰ	−30°
Chamaeleon	The Chameleon	11ʰ	−80°	Puppis	The Stern (of Argo Navis)	8ʰ	−40°
Circinus	The Compasses	15ʰ	−60°				
Columba	The Dove (of Noah)	6ʰ	−35°	Pyxis	The Compass Box (of Argo)	9ʰ	−30°
Coma Berenices	Berenice's Hair	13ʰ	+20°				
Corona Austrina	The Southern Crown	19ʰ	−40°	Reticulum	The Net	4ʰ	−60°
Corona Borealis	The Northern Crown	16ʰ	+30°	Sagitta	The Arrow	10ʰ	+10°
Corvus	The Crow (or Raven)	12ʰ	−20°	Sagittarius	The Archer	19ʰ	−25°
Crater	The Cup	11ʰ	−15°	Scorpius	The Scorpion	17ʰ	−40°
Crux	The Southern Cross	12ʰ	−60°	Sculptor	The Sculptor's Workshop	0ʰ	−30°
Cygnus	The Swan	21ʰ	+40°				
Delphinus	The Dolphin	21ʰ	+10°	Scutum (Sobieski)	The Shield (of John Sobieski[c])	19ʰ	−10°
Dorado	The Swordfish	5ʰ	−65°				
Draco	The Dragon	17ʰ	+65°	Serpens	The Serpent	17ʰ	0°
Equuleus	The Foal	21ʰ	+10°	Sextans	The Sextant	10ʰ	0°
Eridanus	The River	3ʰ	−20°	Taurus	The Bull	4ʰ	+15°
Fornax	The Laboratory Furnace	3ʰ	−30°	Telescopium	The Telescope	19ʰ	−50°
				Triangulum	The Triangle	2ʰ	+30°
				Triangulum Australe	The Southern Triangle	16ʰ	−65°
Gemini	The Twins	7ʰ	+20°	Tucana	The Toucan	0ʰ	−65°
Grus	The Crane	22ʰ	−45°	Ursa Major	The Greater Bear	11ʰ	+50°
Hercules	Hercules	17ʰ	+30°	Ursa Minor	The Lesser Bear	15ʰ	+70°
Horologium	The Clock	3ʰ	−60°	Vela	The Sail (of Argo Navis)	9ʰ	−50°
Hydra	The Water Serpent	10ʰ	−20°				
Hydrus	The Water Snake	2ʰ	−75°	Virgo	The Maiden	13ʰ	0°
Indus	The American Indian	21ʰ	−55°	Volans	The Flying Fish	8ʰ	−70°
Lacerta	The Lizard	22ʰ	+45°	Vulpecula	The Fox	20ʰ	+25°
Leo	The Lion	11ʰ	+15°				

[a] Constellations with declinations between −50° and −90° are difficult or impossible to see from the United States.
[b] R.A., right ascension; Dec., declination.
[c] King John III of Poland (1624–1697).

Appendix 12 The Periodic Table

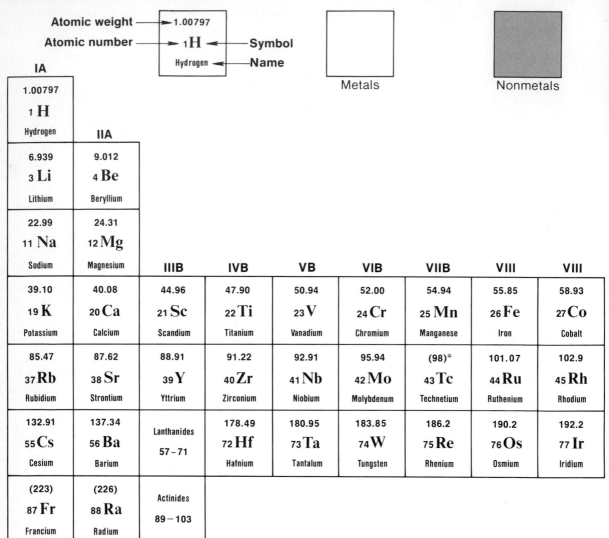

Atomic weight → 1.00797

Atomic number → ₁H ← **Symbol**

Hydrogen ← **Name**

☐ Metals

▨ Nonmetals

IA

1.00797 ₁ H Hydrogen	**IIA**							
6.939 ₃ Li Lithium	9.012 ₄ Be Beryllium							
22.99 ₁₁ Na Sodium	24.31 ₁₂ Mg Magnesium	**IIIB**	**IVB**	**VB**	**VIB**	**VIIB**	**VIII**	**VIII**
39.10 ₁₉ K Potassium	40.08 ₂₀ Ca Calcium	44.96 ₂₁ Sc Scandium	47.90 ₂₂ Ti Titanium	50.94 ₂₃ V Vanadium	52.00 ₂₄ Cr Chromium	54.94 ₂₅ Mn Manganese	55.85 ₂₆ Fe Iron	58.93 ₂₇ Co Cobalt
85.47 ₃₇ Rb Rubidium	87.62 ₃₈ Sr Strontium	88.91 ₃₉ Y Yttrium	91.22 ₄₀ Zr Zirconium	92.91 ₄₁ Nb Niobium	95.94 ₄₂ Mo Molybdenum	(98)* ₄₃ Tc Technetium	101.07 ₄₄ Ru Ruthenium	102.9 ₄₅ Rh Rhodium
132.91 ₅₅ Cs Cesium	137.34 ₅₆ Ba Barium	Lanthanides 57–71	178.49 ₇₂ Hf Hafnium	180.95 ₇₃ Ta Tantalum	183.85 ₇₄ W Tungsten	186.2 ₇₅ Re Rhenium	190.2 ₇₆ Os Osmium	192.2 ₇₇ Ir Iridium
(223) ₈₇ Fr Francium	(226) ₈₈ Ra Radium	Actinides 89–103						

Lanthanides →

138.91 ₅₇ La Lanthanum	140.12 ₅₈ Ce Cerium	140.91 ₅₉ Pr Praseodymium	144.24 ₆₀ Nd Neodymium	(145) ₆₁ Pm Promethium	150.4 ₆₂ Sm Samarium	151.96 ₆₃ Eu Europium
(227) ₈₉ Ac Actinium	232.04 ₉₀ Th Thorium	(231) ₉₁ Pa Protactinium	238.03 ₉₂ U Uranium	(237) ₉₃ Np Neptunium	(242) ₉₄ Pu Plutonium	(243) ₉₅ Am Americium

Actinides →

*Atomic weights in parentheses indicate radioactive isotopes of longest half-life.

VIII	IB	IIB	IIIA	IVA	VA	VIA	VIIA	O
							1.00797 $_1$H Hydrogen	4.0026 $_2$He Helium
			10.811 $_5$B Boron	12.01 $_6$C Carbon	14.01 $_7$N Nitrogen	15.999 $_8$O Oxygen	18.998 $_9$F Fluorine	20.182 $_{10}$Ne Neon
			26.98 $_{13}$Al Aluminum	28.09 $_{14}$Si Silicon	30.97 $_{15}$P Phosphorus	32.06 $_{16}$S Sulfur	35.45 $_{17}$Cl Chlorine	39.95 $_{18}$Ar Argon
58.71 $_{28}$Ni Nickel	63.54 $_{29}$Cu Copper	65.37 $_{30}$Zn Zinc	69.72 $_{31}$Ga Gallium	72.59 $_{32}$Ge Germanium	74.92 $_{33}$As Arsenic	78.96 $_{34}$Se Selenium	79.91 $_{35}$Br Bromine	83.80 $_{36}$Kr Krypton
106.4 $_{46}$Pd Palladium	107.9 $_{47}$Ag Silver	112.4 $_{48}$Cd Cadmium	114.82 $_{49}$In Indium	118.69 $_{50}$Sn Tin	121.75 $_{51}$Sb Antimony	127.6 $_{52}$Te Tellurium	126.90 $_{53}$I Iodine	131.30 $_{54}$Xe Xenon
195.09 $_{78}$Pt Platinum	196.97 $_{79}$Au Gold	200.59 $_{80}$Hg Mercury	204.37 $_{81}$Tl Thallium	207.19 $_{82}$Pb Lead	208.98 $_{83}$Bi Bismuth	(210) $_{84}$Po Polonium	(210) $_{85}$At Astatine	(222) $_{86}$Rn Radon

157.25 $_{64}$Gd Gadolinium	158.9 $_{65}$Tb Terbium	162.5 $_{66}$Dy Dysprosium	164.9 $_{67}$Ho Holmium	167.3 $_{68}$Er Erbium	168.9 $_{69}$Tm Thulium	173.0 $_{70}$Yb Ytterbium	175.0 $_{71}$Lu Lutetium
(247) $_{96}$Cm Curium	(247) $_{97}$Bk Berkelium	(251) $_{98}$Cf Californium	(254) $_{99}$Es Einsteinium	(253) $_{100}$Fm Fermium	(256) $_{101}$Md Mendelevium	(253) $_{102}$No Nobelium	(257) $_{103}$Lr Lawrencium

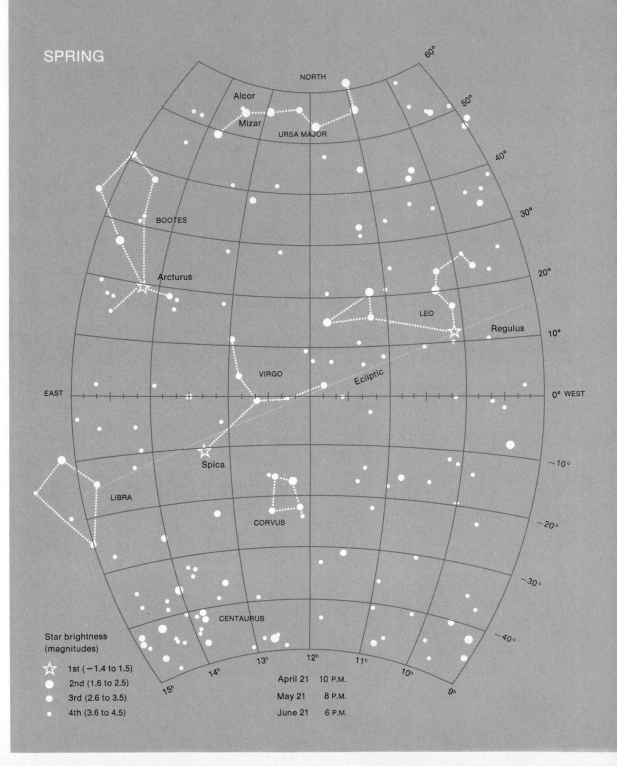

SPRING

Star brightness
(magnitudes)

☆ 1st (−1.4 to 1.5)

● 2nd (1.6 to 2.5)

● 3rd (2.6 to 3.5)

· 4th (3.6 to 4.5)

April 21   10 P.M.

May 21   8 P.M.

June 21   6 P.M.

These star maps show the brighter stars and the prominent constellations as they appear on the dates and at the times indicated. To use these maps, face the south and hold the book overhead with top of the map toward the north and the right-hand edge toward the west. The brightest stars are indicated by the star symbol (☆) and the names are indicated. (Star maps were designed by the author.)

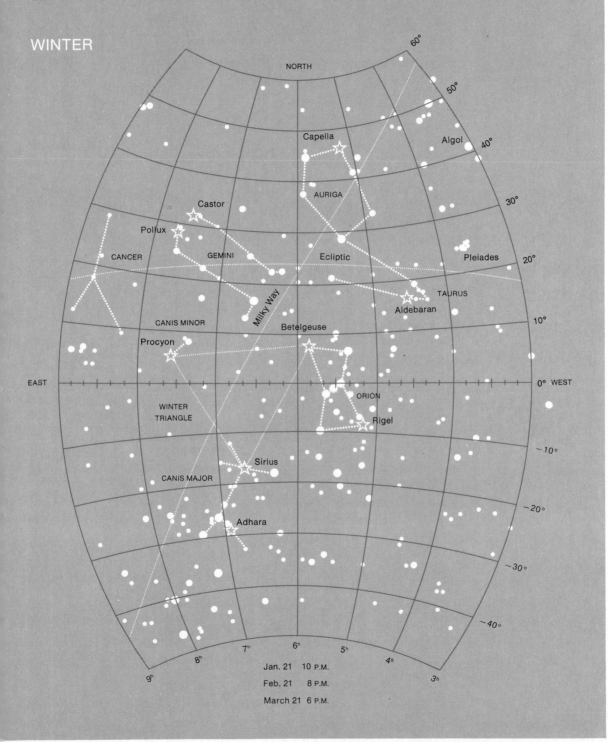

WINTER

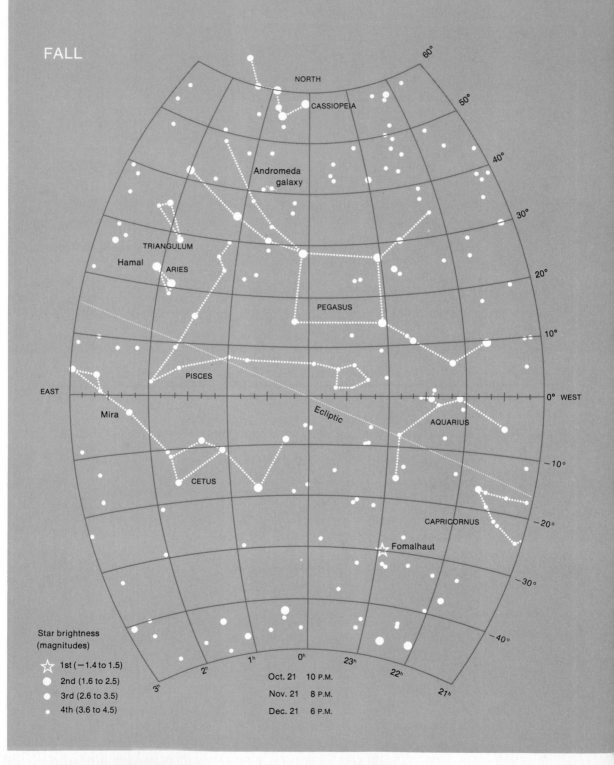

FALL

NORTH

CASSIOPEIA

Andromeda
galaxy

TRIANGULUM

Hamal

ARIES

PEGASUS

PISCES

EAST

Mira

Ecliptic

AQUARIUS

0° WEST

CETUS

CAPRICORNUS

Fomalhaut

Star brightness
(magnitudes)

☆ 1st (−1.4 to 1.5)

● 2nd (1.6 to 2.5)

● 3rd (2.6 to 3.5)

· 4th (3.6 to 4.5)

3ʰ 2ʰ 1ʰ 0ʰ 23ʰ 22ʰ 21ʰ

60° 50° 40° 30° 20° 10° 0° −10° −20° −30° −40°

Oct. 21   10 P.M.

Nov. 21   8 P.M.

Dec. 21   6 P.M.

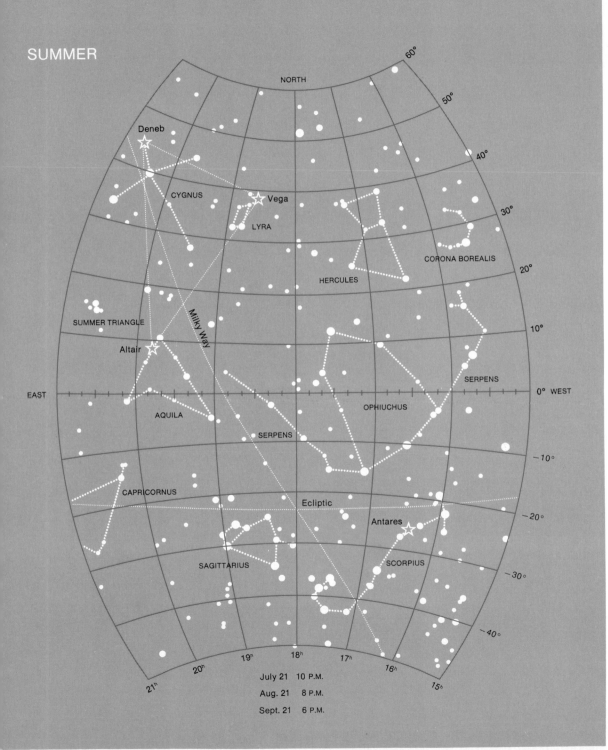

SUMMER

July 21 10 P.M.
Aug. 21 8 P.M.
Sept. 21 6 P.M.

INDEX

A

Aberration:
  chromatic, 73
  spherical, 73
Absolute magnitude, 334
Absorption spectrum, 88
Abundance of elements in cosmic
  rays, 396
Accretion, 160
  planetary origin, 182
Action-reaction law, 41
Adams, John, discoverer of Neptune,
  245
Adonis, an asteroid, 259
Age of Aquarius, 142
Age of earth, 121
Albedo, of a planet, 161, 198
Alcor, member of a multiple system,
  353
Alexander the Great, 18
Alexandria, 18, 111
Allende meteorite, 274
Almagest, 22
Alpha (α) Centauri, nearest star, 19,
  55
Alpha particle, a helium nucleus,
  396
Alphonsine tables, 23
Altitude-Azimuth system, 141
Amino acids in meteorites, 273
Ammonia, in Jupiter's atmosphere,
  225
Analemma, 135
Anaxagoras, Greek philosopher, 16
Anaximander, Greek philosopher, 16
Anaximenes, Greek philosopher, 16
Andromeda galaxy, 48, 440, Plate 15
Angle of incidence, 61
Angle of reflection, 61
Angle of refraction, 63
Angstrom unit (Å), 59
Angular momentum, conservation of,
  253
Annular eclipse, 174
Anorthosite, moonrock, 157
Antarctic Circle, 133
Aperture of lens or mirror, 83
Aphelion of a planet, 189

Apollo, another name for Mercury,
  189
Apollo asteroids, 259
Apparent brightness, 331
Apparent magnitude, 332
Aquinas, St. Thomas, 22
Arabs, astronomers of the Middle
  Ages, 22
Arctic Circle, 133
Arecibo 304.8-m radio telescope, 96
Aristarchus, heliocentric idea, 18
Aristotle Greek philospher, 18
Arp, Halton, 472
Associations, 365
Asteroids, 184
  belt, 223
  discovery, 258
  listing, 259
  origin, 261
Astrology, 10
Astrometric binaries, 353
Astronomical unit, 287
Atmosphere of earth, 122
Atom:
  energy levels, 90
  model, 89, 292
Atomic mass unit (a.m.u.), 195
Aurora australis (southern lights),
  127
Aurora borealis (northern lights),
  127
Australites, tektites, 282
Azimuth, 141

B

Baade, Walter, 374, 465
Babylonia, early observations, 6
Background radiation (3°K), 482
Barnard's Star, 325
Barred spiral galaxies, 457
Barringer Crater, 278
Barycenter, 147, 356
Basalt, moonrock, 157
Baum, William A., 489
Betelgeuse, an irregular variable,
  375

Big Bang theory, 479
Big Dipper, 325
  flip pages, 280
Binary star system, 350
  astrometric, 353
  contact, 355
  eclipsing, 354
  flip pages, 182
  mass, 356
  optical, 352
  origin, 358
  spectroscopic, 352
  visual, 352
Binary system:
  earth–moon, 147
  flip pages, 182
Binoculars, 86
Birth of a star, 406
Blackbody radiation, 304
Black hole, 432
  exploding, 436
  rotating, 435
Blink microscope, 247
BL Lacertae Object, 478
Bode-Titius relationship, 184, 257
Bohr, Niels, 89
Bolide meteor, 273
Bolometric magnitude, 336
Bondi, Hermann, 481
Brady, Joseph, 250
Brahe, Tycho, 27
Breccia, moonrock, 157
Bright-line spectrum, 88
Brightness of stars, 331, 333
Bruno, Giordano, 23
Bursters, X-ray, 363
Butterfly pattern of sunspots, 309

C

Calcium, ionized, 305
Calendar, 143
Callisto, moon of Jupiter, 232, Color
  Plate 5(d)
Carbonaceous asteroids, 261
Carbon (CNO) cycle, 410
Carbon-14 dating, 310
Carpenter, Roland, 194
Cassegrain focus of a telescope, 75

Cassini's division in rings of Sat-
  urn, 232
Catalogue of celestial objects:
  General Catalogue of Nebulae,
    439
  Index Catalogue of Nebulae, 438
  Messier Catalogue of Nebulae,
    270, 439, Appendix 9
  New General of Nebulae, 439
  Radio Sources (Cambridge, Eng-
    land), 470
Celestial:
  equator, 138
  poles, 138
  sphere, 138
Cepheid variables, 370
  distance indicators, 372
  Type I and II, 375
Ceres, first known asteroid, 257
Charge-coupled device (C.C.D.), 84,
  100
Charon, moon of Pluto, 249
China, early observations, 5
Chiron, an asteroid, 251
Chromatic aberration, 73
Chromosphere of the sun, 298
Clusters, 359
  of galaxies, 463
  globular, 360
  motion, 328
  origin, 360
  types, 360
Color index of stars, 336
Comets:
  chemical composition, 265
  discovery, 269
  flip pages, 235
  Halley's, 261
  how to report a discovery, 270
  Jovian group, 261
  Kohoutek, 260
  life expectancy, 267
  Oort's theory, 263
  origin, 263
  properties, 264
  record, 269
  spin, 268
  West, 268, Color Plate 7
Configurations of the planets, 187
Conic sections, 75
Conservation of angular momentum,
  253
Constellations, 8
Contact binaries, 355

Continuous spectrum, 89
Convective envelope of the sun, 296
Copernican system, 25
  flip pages, 115
Copernicus, Nicolaus, 23
Coriolis effect, 130
Coronagraph, a view of the sun, 299
Coronal "holes" of the sun, 300
Corona of the Galaxy, 451
Corona of the sun, 299
Cosmic background radiation (3°K),
    482
Cosmic dust, 394
Cosmic rays, 394
  abundance, 395
Cosmological models, tests for, 489
Cosmological principle, 481
Cosmology, 479
Coudé focus of a telescope, 75
Crab Nebula, 377, 416, Color Plate 9
Craters, meteorite, 278
Crust of earth, 119
Curvature of space-time, 485
Cyclops proposal, 514
Cygnus A, a source of radio noise,
    468
Cygnus X-1, a source of X-rays, 435

D

Dark Ages, 22
Dark-line spectrum, 89, Color
    Plate 1
Day:
  sidereal, 133
  solar, 133
Death of a star, 420
Declination of a star, 138
Deferent, 20
Degenerate gas, 422
Deimos, moon of Mars, 219
Delta (δ) Cephei, a variable star, 367
Democritus, a Greek philosopher,
    18
Density of common materials, 117
De Revolutionibus Orbium Coeles-
    tium (Copernicus), 24
Deuterium, 292, 480
  abundance, 491
Diamond ring effect in a solar
    eclipse, 172
Dicke, Robert H., 482

Differential gravitational forces, 167
Differentiation of moon, 149
Diffraction of light, 65
Dispersion, 64
Distances:
  to galaxies, 465
  to planets, 32
  to stars, 323
  to sun, 288
Diurnal motion of earth, 129
DNA, code of life, 511
Doppler effect, 91
Dual nature of light, 68
Dwarf stars, 422
Dyce, R. B., 190

E

Earth:
  age, 121
  atmosphere, 122
  blue sky, 124
  crust, 119
  density, 116
  diameter (Eratosthenes), 111
  ionosphere, 123
  layers, 118
  magnetic field, 125
  mass, 114
  motions, 144
  plates of crust, 119
  revolution, 130
  rotation, 127
  seasons, 132
  stratosphere, 123
  tilt, 132
  troposphere, 122
  Van Allen Belts, 127
  viewed by "Martians," 203
  volume, 114
Earth–moon system:
  flip pages, 182
Earthquakes, 119
Eccentricity, 31
Eclipse:
  annular (ring), 174
  lunar, 175
  observing (safely), 174
  paths, 172
  seasons, 176
  solar, 170
  warning, 174

Eclipsing binaries, 354
  flip pages, 410
Ecliptic, 132
Egypt, early observations, 10
Einstein, Albert, 107, 432, 479, 500
  thought experiment, 501
Electromagnetic spectrum, 51, 92
Electron, 89, 397
  spin, 446
Ellipse, 30
Elliptical galaxies, 457
Elongation of planets, 187
Emission nebulae, 381
Emission spectrum, 89
Enceladus, a moon of Saturn, 239
Encounter theory (planets), 182
Energy for star formation, 401
Epicycle, 20
Equation of time, 135
Equator, 135
Equinox, 12
  vernal, 139
Eratosthenes, of Alexandria, 111
Ergosphere of rotating black hole,
  435
Eros, an asteroid, 257
Eta Carinae, remains of supernova,
  Color Plate 12
Ether, 464
Europa, moon of Jupiter, 231
Evening star (Venus), 194
Evolution of stars, 398
Exploding black holes, 436
Extended source, 81
Extraterrestrial life, 506

F

"Falling off with the square of the
  distance," 46
Fireball meteor, 271
Flare star, 378
Flares on the sun, 314
Flip pages:
  binary system, 182
  comet in motion, 235
  Copernican system, 115
  eclipsing binaries, 410
  globular clusters, 345
  proper motion of stars, 280
  Ptolemaic system, 25
  retrograde motion, 114

rotation of the Milky Way galaxy,
  345
Focal length of a lens or mirror, 70
Focal ratio (f-stop) of a telescope,
  83
Focus:
  Cassegrain, 75
  Coudé, 75
  of an ellipse, 31
  Newtonian, 75
  prime, 75
Foucault pendulum, 127
Fraunhofer lines in the spectrum, 87
Frequency of waves, 58
Friedmann, Alexander, 446
Fusion reaction, in the sun, 291

G

Galaxies, 47
  classification, 455, 478
  cluster, 463
  distance, 465
  elliptical, 455
  irregular, 455
  local group, 461
  radio, 468
  Seyfert, 469
  spectrograms, 469
  spirals, 455
Galaxy (the Milky Way), 438
  arms, 444
  corona, 451
  disk, 442, 450
  evolution, 452
  halo, 361
  mass, 450
  nucleus, 443, 452
  Population II objects, 451
  rotation, 447
  velocity of the sun, 447
Galilean satellites of Jupiter, 229
Galileo, 35
  telescope, 36
  phases of Venus, 194
Gamma-ray astronomy, 106
Ganymede, moon of Jupiter, 232
Gas laws, 400
General Catalogue of Nebulae, 439
General theory of relativity, 503

Geocentric parallax, 323
Geographic north pole, 125
Geographos, an asteroid, 259
Glare (reflected light), 60
Globular clusters, 360, 441
    density, 362
    flip pages, 345
Globules in nebulae, 406
Gold, Thomas, 481
Goldstein, Richard, 194
Goldstone 64-m radio telescope, 94
Gram, unit of mass, 113
Granulation of the photosphere, 294
Gravitational lens, 476
Gravitational wave astronomy, 107
Gravity, 42, 115
Great Red Spot of Jupiter, 228
Greeks, early concepts of the uni-
        verse, 16
Greenhouse effect, 188
Greenwich, England, 136
Gregorian calendar, 143

H

Hack, Margherita, 407, 411
Half-life of uranium, 122
Halley's Comet, 261
Halo of the Milky Way galaxy, globu-
        lar clusters, 441
Hawking, Steven, 436
Hazard, Cyril, 471
HEAO (High Energy Astronomical
        Observatory), 105
Helin, Eleanor, asteroid searcher,
        258
Heliocentric model of solar system,
        19
Heliocentric parallax, 323
Helios, Greek sun god, 16
Helium:
    burning, 293
    flash, 414
Herbig-Haro objects, 407
Hermes, an asteroid, 257
Herschel, Sir William, 351
    discovery of Uranus, 240
    early thoughts, 439
Hertzsprung-Russell (H-R) diagram,
        339, 410
Hipparchus, 19
    catalogue of stars, 19

classification of star brightness
        (magnitudes), 331
Homogeneous universe, 481
Horsehead Nebula, 386
Hour circles, 139
"Houses," signs of the zodiac, 131
Hoyle, Fred, 481
H-R (Hertzsprung-Russell) diagram,
        339, 410
Hubble, Edwin, 440, 460, 467, 480
Hubble's classification of galaxies,
        460
Hubble's red shift law, 467
Humason, M. L., 480
Hven, Brahe's island observatory, 27
Hydrogen:
    burning, 293
    H II regions, 383
    neutral, 383
    nuclei, 394
Hydrogen-alpha line in the spec-
        trum, 89
Hydrostatic equilibrium, 406
Hyperbolic universe, 486

I

Icarus, an asteroid, 257
Image:
    brightness, 83
    magnification, 85
    resolution, 84
    size, 81
Incident ray, 61
Index Catalogue (IC) of Nebulae,
        438
Index of refraction, 63
Indocrinites, tektites, 281
Inertia, 39
Inferior planets, 187
Infrared astronomy, 100, 102
Infrared sky, 347
Inquisition, 23
Interference of light, 66
Interferometry, radio, 96
Intergalactic media, 465
International System (SI) of Units,
        56, Appendix 2
Interstellar dust, 385
Interstellar gas, 381
Interstellar molecules, 391
Io, moon of Jupiter, 230

Ion, a charged atom, 123
Ionosphere of earth, 123
IRAS, infrared satellite, 102
Irons, meteorite, 276
Irregular galaxies, 460
Irregular variables, 375
Isotropic universe, 481

J

Jansky, Karl, 416
Jovian group of comets, 261
Jovian planets, 185, 222
Jupiter:
    atmosphere, 225
    belts and zones, 227
    internal structure, 227
    magnetic field, 225
    moons, 229
    photo, Color Plate 4
    properties, 224
    rotation, 228
    surplus energy, 225

K

Kepler, Johannes, 30
    laws, 32
Kilogram, 56
Kilometer, 56
Kinetic energy, 401
Kohoutek, a comet, 263
Kowal, Charles, 250

L

Large Magellanic Cloud, 459, Color
    Plate 11(a)
Laser beam to the moon, 148
Latitude on earth, 135
Leavitt, Henrietta, 370, 439
Leonids, meteor shower, 275
Leverrier, Joseph, discoverer of Nep-
    tune, 245
Life cycle of a star, 399
Light:
    diffraction, 65
    dispersion, 64
    dual nature, 68
    interference, 66
    polarization, 59
    reflection, 61
    refraction, 62
    wave nature, 56
Light curves:
    of binaries, 354
    of variables, 370
Light ratios, 333
Light year, 48, 55
Limb of the sun, 297
Limitations in a telescope, 72
Local group of galaxies, 461
Local meridian, 5
Local standard of rest, 328
Lodestone (magnetite), 125
Longitude on earth, 135
Lorentz, Hendrik A., 499
Lorentz contraction, 499
Lowell, Percival, 245

M

Magellanic clouds, nearby galaxies,
    370
Magnetic field:
    of earth, 125
    of moon, 161
    of sun, 306, 311
Magnification of a telescope, 85
Magnitude:
    absolute, 334, 373
    apparent, 19, 373
    bolometric, 336
    photographic, 336
    visual, 336
Main-sequence stars, 338, 411
Maria of the moon, 151
Mariner space probes:
    to Mars, 205
    to Mercury, 191
    to Venus, 196
Mars:
    atmosphere, 213
    "canals," 204
    Mariner probes, 205
    Mariner Valley, 207
    moons, 219
    motion, 30
    Olympus Mons, 206

photo, Color Plate 3
polar region, 217
relief map, 208
search for life, 216
surface features, 207
Mascons of the moon, 151
Mass, 113
of binaries, 356
of Milky Way galaxy, 450
unit of, 56
Mass-energy density of the universe,
490
Mass-energy relationship, 293
Mass-luminosity relationship, 356
Maunder Minimum, sunspot cycle,
310
Maximum elongation of planets, 187
Mayall, N. U., 480
Mean solar day, 133
"Mean sun," 135
Mercury:
atmosphere, 192
Mariner probe, 191
motions, 189
radar observations, 190
surface features, 193
synodic period, 190
temperature, 192
Mesopotamia, early observations, 6
Mesosphere, 123
Messier, Charles, catalogue of nebulae, 270, 439, Appendix 9
Meteorites:
age, 278
craters, 278
micrometeorites, 283
physical properties, 276
types, 276
Meteoroids:
orbit, 275
origin, 275
Meteors:
fireball (bolide), 271
observing, 272
showers, 273
Meter (m), unit of length, 56
Methane, in Jupiter's atmosphere,
225
Metonic cycle, an eclipse cycle, 177
Metric system, 56, Appendix 2
Michelson, Albert A., 55
rotating mirror experiment, 55
Michelson-Morley experiment, 497
Micrometeorites, 283

Mid-Atlantic ridge of earth, 120
Milky Way galaxy, 438, 442
Minor planets (see Asteroids)
Mira Ceti, a variable star type, 374
Mizar A and B, a multiple star system, 353
Model building, 401
Molecules, interstellar, 391
Monoceros cluster, 407
Month, lunar, 7
Moon:
age, 160
crust, 150
farside, 157
first quarter, 152
highlands, 151
librations, 166
magnetic field, 161
maria, 151
mascons, 151
origin, 162
phases, 4, 164
revolution (sidereal), 163
revolution (synodic), 163
rilles, 157
rocks, 157
surface features, 151
synchronous rotation, 165
temperature, 161
terminator, 154
third quarter, 154
tidal action, 166
Moonquakes, 150
Moons:
Jupiter, 229, Color Plate 5
Mars, 219
Neptune, 245
Pluto, 249
Saturn, 237
Uranus, 244
Motion of stars, 325
Mount Palomar Observatory, 77
Mount Wilson Observatory, 439
Moving clusters, 328
Multiple-Mirror telescope, 78
Multiple star system, 350

N

Nebulae, 47
Neptune:
discovery, 243

moons, 243
properties, 243
Neutrinos from the sun, 293
Neutron star, 425
   model, 430
New General Catalogue of Nebulae,
   439
Newton, Sir Isaac, 39, 112
   laws of motion, 39
Newtonian focus of a telescope, 75
Node points (of moon), 176
Normal (perpendicular) to a surface,
   61
Normal radio galaxy, 469
North American Nebula, 386
North Magnetic Pole, 125
Northern lights (Aurora borealis),
   127
North star (Polaris), 129
Novae, 375
Nucleosynthesis, 480
Nut, Egyptian sky goddess, 16

O

Objective lens, 72
Oke, Beverly, 472
Olber, Heinrich, 255
Olympus Mons (Nix Olympica) on
   Mars, 206
Oort, J. H., comet cloud theory, 188
Opposition of planets, 188
Optical binary systems, 352
Orbiting Astronomical Observatory
   (OAO-C), 305
Origin of:
   binaries, 358
   clusters, 360
   comets, 265
   moon, 162
   planets, 182, 250
Orion Nebula, 360, Color Plate
   14(b)
Oscillating universe, 486
Oxygen, ionized, 305
Ozone layer of earth, 123

P

Pallas, an asteroid, 256
Palomar 200-inch telescope, 77
Parabolic reflector, 75
Parabolic universe, 486
Parallax:
   geocentric, 323
   heliocentric, 323
   spectroscopic, 343
   stellar, 19, 130
Parallels of latitude, 136
Parsec, a unit of distance, 324
Paths of eclipses, 172
Pauli Exclusion Principle, 422
Peculiar radio galaxies, 469
Peculiar velocity of stars, 328
Penumbra:
   of an eclipse shadow, 169
   of a sunspot, 308
Penzias, Arno H., 482
Perfect Cosmological Principle, 482
Perihelion of a planet, 189
Period-luminosity relationship, 371
Pettengil, G. H., 190
Phases:
   of the moon, 163
   of Venus, 37
Philolaus, Greek philosopher, 18
Phobos, moon of Mars, 219
Photoelectric effect, 68
Photoionization, 382
Photometry, 331
Photomultiplier, 332
Photon of light, 69
Photosphere of the sun, 297
Pioneer Venus probe, 199
Pisa, Leaning Tower of, 38
Plages of the sun, 312
Planetary nebulae, 422
Planet-X, 240
Planets:
   aphelion, 189
   configurations, 187
   elongation, 187
   escape velocity, 187
   evolution, 252
   inferior, 187
   Jovian, 185
   motions, 20

opposition, 188
origin, 182, 252
perihelion, 189
Planet-X, 250
properties, 185
relative size, 186
rotation, 183
spacing, 183
synodic period, 189
terrestrial, 185
Plates of the earth's crust, 119
Plato, the Academy of, 18
Pleiades cluster, 360, 388, Color
      Plate 10
Pluto:
   discovery, 247
   eccentricity, 247
   inclination of orbit, 247
   moon (Charon), 249
   properties, 248
   rotation, 249
Point source, 81
Polaris (North Star), 129
Polarization:
   of light, 59
   of starlight, 388
Population I and II stars, 451
Positron, 394
Potential energy, 401
Power of a telescope, 85
Precession:
   of the earth, 142
   of the equinoxes, 142
Primary cosmic particles, 394
Prime focus of a telescope, 75
Prime meridian, 136
Primordial atom (fireball), 479, 483
Prominences of the sun, 313
Proper length, 502
Proper motion of stars, 325
   flip pages, 280
Proper time, 501
Proton, spin, 446
Proton-proton cycle, 292, 411
Protoplanet theory, 182, 252
Protostar, 347, 404
Protosun, 252
Proxima Centauri, nearest star, 324
Ptolemaic system, 21
   disproved, 37

flip pages, 280
Ptolemy, Claudius, 21
Pulsar, 348, 426
Pythagoras of Samos, 16

Q

Quasars (quasi-stellar radio sources),
   95, 470

R

Ra, Egyptian sun god, 16
Radar:
   astronomy, 99
   observations of Mercury, 190
   observations of Saturn's rings, 234
   observations of Venus, 196
Radiant of meteor showers, 274
Radiation, forms of, 51
Radiation curves of stars, 346
Radiation zone of the sun, 296
Radioactive elements, 121
Radio astronomy, 93
Radio interferometry, 96
Radio galaxies, 468
Random-capture theory, planets, 182
R Corona Borealis variables, 378
Recurrent novae, 375
Red giant stars, 412
Redshift, Doppler effect, 92, 436
Reflecting telescope, 73
Reflection of light, 61
Reflection nebulae, 385
Refraction of light, 62
Refractor telescope, 71
Relative motion, 495
Relativity:
   general theory, 479, 503
   special theory, 503
Resolution in a telescope, 84, 352
Retrograde motion:
   flip pages, 114
   of planets, 20
Revolution:
   of earth, 130

of moon, 162
of planets, Appendix 5
Rigel, magnitude and distance, 335
Right ascension of a star, 138
Rilles of the moon, 157
Rings:
  of Jupiter, 233
  of Saturn, 233, 241
  of Uranus, 244
Roche limit, in rings of Saturn, 235
Roemer, Ole, navigator-observer, 52
Rotating black holes, 435
Rotation:
  of earth, 127
  of the Galaxy, 447, flip pages,
    345
  of planets, Appendix 6
  of sun, 311
Royal Greenwich Observatory, 136
RR Lyrae variables, 372, 422

S

Sagittarius, points to center of the
    Galaxy, 444
Sandage, Allan, 489
Saros, an eclipse cycle, 177
Satellites (moons) of planets, Appen-
    dix 7
Saturn:
  moons, 235–241
  photo, Color Plate 6
  Pioneer and Voyager probes, 235
  radar observation, 234
  revolution, of rings, 234
  rings, 233, 235
Scattering of light, 124
Schiaparelli, Giovanni (Mars), 204
Schmidt, camera, 80
Schmidt-Cassegrain telescope, 81
Schmidt, Maarten, 471
Schwarzschild, Karl, 432
Schwarzschild radius and sphere,
    433
Sea of Tranquility, moon, 158
Seasons of earth, 132
Second (sec), unit of time,
    Appendix 2
Secondary cosmic particles, 394
Seeing, 85
Semimajor axis of an ellipse, 31

Seyfert galaxy, 469
Shapley, Harlow, 371, 465
Siberian event, 280
Sidereal:
  day, 133
  month, 163
  period of a planet, 189
Signs of the zodiac, 131
Silicon, ionized, 306
Singularity of a black hole, 433
Sirius A and B, a binary system, 354
Size of stars, 344
Sky (blue) of earth, 124
Sky maps, four seasons, Appendix
    13
Small Magellanic Cloud, 370
Solar activity, a summary, 318
Solar constant, 301
Solar day, 133
Solar eclipse, 170, Color Plate 8(a)
Solar spectrum, 88, 298, 301
Solar wind, 300
Solstice, summer and winter, 12
SO-type galaxies, 459
Space Telescope (ST), 104
Space-time, 503
Space-time curvature, 485
Space velocity of stars, 327
Spallation, by cosmic rays, 394
Spectral class of stars, 338
Spectrograms, Color Plate 1
Spectrograph, 87
Spectroheliograph, 304
Spectroscopic binary system, 352
Spectroscopic parallax, 343
Spectrum:
  absorption (dark-line), 89
  continuous, 88
  emission (bright-line), 88
  solar, 301
Speed of light, 55
Spherical aberration, 73
Spicules of the sun, 313
Spin of the electron, 446
Spiral galaxies, 455
Starquake, 431
Stars:
  absolute magnitude, 334
  associations, 365
  binary systems, 351
  birth, 406
  black holes, 432
  brightness, 331

B-type, 444
Cepheid variables, 368
clusters, 359
color, 337
death, 420
evolution, 398
evolutionary tracks, 407
flare type, 378
general, 320
great variety, 346
local standard of rest, 328
magnitude, 332
main sequence, 340, 409
mass, 356
motion, 325
multiples, 350
neutron, 425
O-type, 444
peculiar velocity, 328
Population I and II, 451
pressure classes, 342
proper motion, 325
pulsars, 348, 426
radial velocity, 91, 327
radiation curves, 346
RR Lyrae, 372
size, 344
space velocity, 327
stellar winds, 422
supergiants, 340
temperature, 336
variable, 366
white dwarf, 422
Star trails, 129
"Steady State" theory, 482
Stellar parallax, 19, 130
Stellar winds, 422
Stonehenge, England, 14
Stone meteorites (aerolites or chondrites), 276
Stony-iron meteorites (siderolites), 276
Stratosphere of earth, 123
Sublimation of polar caps (Mars), 214
Sun:
  angular momentum, 253
  chromosphere, 298
  convective envelope, 296
  core, 295
  corona, 299
  coronagraph, 299
  density, 291
  diameter, 289
  eclipse, 170
  elements, 303
  flares, 314
  fusion, 291
  granulation, 296
  layers, 295
  limb, 297
  magnetic field, 306
  mass, 291
  motion, 328
  photosphere, 297
  physical characteristics, 289
  plages, 312
  prominences, 313
  proton-proton cycle, 292
  radiation zone, 296
  rotation, 311
  solar constant, 301
  solar wind, 127, 299
  source of energy, 293
  spectroheliogram, 304
  spectrum, 298
  spicules, 313
  sunspots, 308
  velocity in the Galaxy, 447
  X-ray view, Color Plate 8(b)
  Zeeman effect, 307
Sunspot cycle, 308
Supergiant stars, 340
Superior planets, 187
Supernovae, 387, 416
Syene, Egypt, 111
Synchronous rotation of the moon, 165
Synchrotron radiation, 429, 476
Synodic period of a planet, 189
System International (SI system), 56, Appendix 2

T

Tangential velocity, 325
Technology, 512
Tektites, 281
  age, 283
Telescope:
  Arecibo, Puerto Rico, 96
  console, 101
  designs, 70

focal length, 70
f-ratio (*f*-stop), 83
Galileo's, 36
Goldstone (radio), 94, 194
Kitt Peak 4-m, 76
Kuiper Airbourne, 101
limitations, 72
magnification, 85
multiple-mirror, 78
Palomar 5-m, 77
quality, 81
radio, 94
reflector, 73
refractor, 71
resolution, 84
Schmidt camera, 80
Schmidt-Cassegrain, 81
Space Telescope, 104
USSR 6-m, 78
Telluric lines, spectrum, 301
Temperature scales, Appendix 1
Temperature of stars, 336
Terminator of moon, 154
Terrestrial planets, 185
Tethys, moon of Saturn, 239
Thales of Miletus, Greek philosopher, 16
Third Cambridge Catalogue of Radio Sources, 471
Three-degree (3°) Kelvin background radiation, 482
Tides:
    on earth, 166
    spring and neap, 168
Tilt of the earth, 132
Time, 133
Time zones, 137
Titan, moon of Saturn, 237
Tombaugh, Clyde, discoverer of Pluto, 247
Triangulation, 322
Triton, moon of Neptune, 245
Trojan asteroids, 260
Tropical year, 143
Tropic of Cancer and Tropic of Capricorn, 133
Troposphere of the earth, 122
T Tauri variables, 375, 407
Turbulence of atmosphere, 85
Twenty-one centimeter (21-cm) radio radiation, 446, 508
Twilight, 125

U

UHURU, X-ray satellite, 105
Ultraviolet astronomy, 103
Umbra:
    of an eclipse shadow, 169
    of a sunspot, 308
Universe:
    early concepts, 16
    homogeneous, 481
    hyperbolic, 486
    isotropic, 481
    mass-energy density, 490
    oscillating model, 486
    parabolic model, 486
Universal time, 137
Uranium-238, 122
Uranus:
    properties, 242
    rings, 244
    tilt, 243

V

Van Allen belts of earth, 127
Variable stars:
    Cepheids, 368
    irregular, 375
    R Corona Borealis, 376
    RR Lyrae, 372
    T Tauri, 375
Velocity:
    dispersion, 431
    of escape, 86
    of light, 55
Venus:
    atmosphere, 200
    clouds, 198
    Mariner probe, 196
    phases, 37, 194
    Pioneer probes, 199
    radar observations, 196
    USSR probes, 197, 203
Vernal equinox, 138
Very Large Array (VLA) radio telescope, 98
Viking 1 and 2, to Mars, 212
Virtual image, 62
Visual binary system, 352
Volume, 112
Vredefort, meteorite crater, 280
Vulcan, an imagined planet, 250

W

Water hole, 18 to 21 cm radio band, 509
Water in space, 393
Wavelength, 58
Wave nature of light, 56
White dwarf stars, 341, 422
Widmannstätten lines in meteorite, 276
Wien's Law, 347
Wilson, Robert W., 482
Wind, stellar, 422

X

X-ray astronomy, 105, 348
X-ray bursters, 363
X-ray radiation, 434

Y

Year, length, 7, 132

Z

Zeeman effect on the sun, 307
Zenith of an observer, 5, 138
Zero hour circle, 139
Zodiac, signs, 131
Zone of avoidance, 387